POPULATION VIABILITY ANALYSIS

POPULATION VIABILITY ANALYSIS

Steven R. Beissinger and Dale R. McCullough, Editors

THE UNIVERSITY OF CHICAGO PRESS   CHICAGO & LONDON

The University of Chicago Press, Chicago 60637
The University of Chicago Press, Ltd., London
© 2002 by The University of Chicago
All rights reserved. Published 2002
Printed in the United States of America

11 10 09 08 07 06 05 04 03 02   1 2 3 4 5

ISBN: 0-226-04177-8 (cloth)
ISBN: 0-226-04178-6 (paper)

Library of Congress Cataloging-in-Publication Data

Population viability analysis / Steven R. Beissinger and Dale R. McCullough,
    editors.
        p. cm.
    Includes bibliographical references and index.
    ISBN 0-226-04177-8 (cloth : alk. paper) — ISBN 0-226-04178-6
    (pbk. : alk. paper)
        1. Population viability analysis.   I. Beissinger, Steven R.
    II. McCullough, Dale R., 1933–

    QH352.5 .P66 2002
    577.8'8—dc21

                                                        2001052271

CONTENTS

FOREWORD: RAISING THE BAR

New technologies often channel cultures in directions that are ultimately detrimental to nature because they deepen and broaden the human footprint. This is the expected outcome, given that the purpose of technology is to leverage power and increase efficiency and productivity. Many would agree, for example, that nature, not to mention society, would be better off without the automobile.

Until recently, the major technological drivers of social and environmental change were industrial, machine-based, or "heavy" technologies. Now, informational, or "light," technologies such as microprocessors, computer software, and biotechnology are changing all facets of society and nature more profoundly than physical machines alone ever could. Population viability analysis (PVA) is a light, informational technology and, like other inventions, may have unforeseen consequences as described below.

This book is a great leap forward in describing the technical dimensions of PVA. The reader will discover the latest advances, including methods for dealing with uncertainty in the demographic variables and in the results (e.g., improvements in sensitivity analysis, incorporation of Bayesian statistical methods, and the much-needed integration of PVA and adaptive management). The skillfully rendered chapters also serve as signposts for more comprehensive and realistic PVAs that explicitly incorporate spatial factors, that integrate genetics and demography more systematically (the VORTEX computer model has pioneered this), and that begin to treat catastrophic events more realistically. There is room for much improvement in all areas—better integration of genetics and demography in models, for example—and one goal of the book is to point out deficiencies. The book achieves this goal admirably.

As PVA becomes more sophisticated and complex, it raises the technical bar for practitioners. We applaud higher standards, but complexity isn't always beneficial. In fact, a good case could be made that PVA is sufficiently robust now and is unlikely to get much better at predicting the future or identifying critical variables of viability. Notwithstanding the inevitability of the ratcheting of complexity, we should also be mindful that elegance and simplicity are traditionally esteemed in science.

What about PVA in the larger scientific and cultural contexts? While PVA has increased our ability to predict the fates of small or declining populations, its popularity may also have delayed the development of equally fertile approaches to conservation, such as methods for analyzing multispecies interactions that affect ecosystem diversity and resilience. For example, PVA is easily subverted to finding minimal criteria for short-term viability, as in recovery goals for endangered species. This abuse has distracted population ecologists from the more important problem of ecologically effective numbers and distributions. This is no reason to desist from PVA; it is merely a warning about conceptual entrenchment. Conservation biology best serves conservation when it advances along a broad front.

How is PVA faring financially and politically? At least with regard to politics, PVA is certainly visible, and its salience in controversial endangered-species issues has drawn the attention of friends and foes alike. The careers of PVA practitioners owe much to controversies over vulnerable species.

Another dimension of culture is ethics. Although the technical trajectory of PVA is impelled by science, at its core PVA is (or should be) a tool for the salvation of nature. The drivers of PVA's development have been laws and treaties that recognize the right to persistence of species; examples are the Convention on International Trade in Endangered Species of Wild Fauna and Flora (CITES) and the U.S. Endangered Species Act (ESA). Unlike any other contemporary law, the ESA expanded the circle of moral considerations beyond humanity, bestowing the right to persistence even on plants, insects, and mollusks, although few observers anticipated its social reach and political significance.

We tacitly assume, therefore, that PVA will do more good than harm, primarily because it incorporates the best science of the day to level the playing field of land-use policy. This should help to address the current imbalance between the vital needs of all species and the material ambitions of human beings. We must hope that the increasing technical sophistication and political relevance of PVA will benefit the campaign to protect nature. Sadly, however, unintended consequences bedevil all major inventions.

Originally, PVA was viewed as an unbiased, analytical tool that would cut through the fog of uncertainty surrounding viability. But endangered-species controversies touch many social nerves, including those that affect private property rights, industrial polluters, resource extractors, agribusiness, and suburban developers. Vested interests are rarely passive when profits are threatened and will attempt to alter PVAs to suit their needs. Additionally, there is a tendency in all of us to want to avoid conflict and to be seen as "reasonable," "team members," and "players."

The danger is a kind of moral drift—the gradual relaxation of standards and a lowering of the ethical bar that causes, over time, a tilt toward less conservative guidelines and conclusions—or a blunting of the lance of PVA. Thus, we should be aware of certain principles that apply to any kind of science and that could affect someone's profits.

- Your results may be used to advance political agendas.
- Your results may be gratuitously misinterpreted.
- Your results may cause harm to species and ecosystems.
- Your results may be contaminated by self-interest (yours and others'), in part because political objectives and personal agendas will affect your methods, analyses, interpretations, and conclusions.

There is no absolute immunity against moral drift. Some protection is afforded by attention to the social context and the biases intrinsic to all economically driven interests, particularly if we adhere to the "do no harm" principle. Beyond such generic prescriptions, we should be as mindful about maintaining precautionary guidelines as about the technical nuances of stochasticity, elasticity, and inbreeding depression. For example, it is arguable that (1) no continental population of large mammals has ever persisted long enough to evolve into a distinct species with a census size less than tens of thousands; (2) PVAs that fail to incorporate catastrophe and genetics favor development; and (3) a 5% risk of extinction of a species, if real, is unacceptable.

For good or ill, PVA is socially embedded, and its practitioners are vested with ethical responsibilities. As physicians to the "others," we are called upon to raise the bar morally as well as technically. Conservation biology is a challenge for the heart as well as the mind.

I wish to acknowledge Richard Frankham, L. Scott Mills, and the editors for their comments on an earlier draft.

Michael E. Soulé

EDITORS' PREFACE

This book is about a set of methods for evaluating the risk of extinction and potential recovery strategies for wild populations of plants and animals. Extinction is a natural process and an important part of the history of life on Earth. There have been five periods of very high rates of extinction of species in the past 500 million years, due to asteroids crashing into Earth, volcanic eruptions, climate change, sea level shifts, and continental drift. These events have changed the biota of our planet in irreversible ways, diverting the composition of communities from previous courses and directing evolution along new pathways. After such events Earth was truly a different biological realm, one not predictable from its earlier composition.

Once again Earth may be facing an extinction event, but this time because of our own behavior. As human populations grow and make insatiable demands for materials, food, and wealth, natural landscapes have been transformed and life histories of plants and animals interrupted. The by-products of this transformation foul land, water, and air, and change the climate in ominous ways. Some wild species are favored by this transformation—there have always been opportunists that flourish from disturbance—but most suffer. For example, about two-thirds of the crayfish and mussel species in the United States, one out of every nine of the world's 9,500 species of birds, and two-thirds of the world's primate species are considered to be threatened or endangered. They have a high chance of becoming extinct in the next 50 to 100 years due primarily to habitat destruction, overexploitation, introduction of exotic species, and pollution. Shifting from a rich diversity of species with complex interconnections to a biota characterized by a few species in great abundance and many in dire straits has consequences for both ecosystems and people about which we can only speculate.

Ecologists have long recognized this problem. Periodically, the occasional, lonely voice could be heard crying in the wilderness: Aldo Leopold, Rachel Carson, Paul Ehrlich, and others were heard briefly by society throughout the 1940s, 1950s, and 1960s, but were quickly forgotten in the rush to prosperity. With growing prosperity, leisure time, and discretionary income, however, Americans began to hear the message, and the early 1970s marked a turning point of concern for the environment in the United States, with Earth Day and the passage of the Endangered Species Act and similar measures. With the enactment of the Convention on Biological Diversity in 1992 in Rio de Janeiro, Brazil, protection of biological diversity became a part of nearly every national agenda and began to challenge unlimited extraction and exploitation as a concern of society.

A key scientific development that helped in small ways to catalyze this change occurred in the 1980s with the emergence of the field of conservation biology. Despite resentment of some members of older, well-established resource-management professional societies, conservation biology was something new. The approach of the traditional natural resource societies was akin to that of public health. Their goal was the prevention of disease through good practices. The approach of conservation biology was more like modern medicine. The goal was to treat sick patients and rehabilitate them to health. This was an important theme of the late Graeme Caughley's influential essay on the state of species conservation, in which he contrasted the traditional "declining population" with the newer "small population" paradigm. Both are needed in a natural world with a mix of healthy and sick patients.

Making management decisions about threatened and endangered species demanded a tool, like a crystal ball, that would allow us to see or predict the future in order to know what actions need to be taken to ensure that species would persist. Management often floundered when expert judgment was the sole basis of such assessments. The variables were too complex and the interactions too unpredictable. In the 1980s the intersection of statistical methods and computing power led inevitably to population viability analysis (PVA). PVA comprises a set of analytical and modeling approaches for assessing the risk of extinction. Life history, demography, and genetics of a species are integrated with environmental variability to project the future course of populations.

PVA quickly became established in the 1990s as an important way to analyze the probability of extinction. Several computer software programs appeared that made it easy to run complex stochastic simulations, with or without good data. PVAs were being used to predict the probability of extinction for highly endangered and threatened species 50, 100, and even 1,000 years into the future. PVA workshops for hundreds of species were held around the world by the Captive Breeding Specialist Group. There was even hope that PVA could answer important theoretical and practical questions, such as how much habitat needs to be protected to prevent extinction and in what configuration.

The attention given to PVA exposed its weaknesses as well as its strengths. Being probabilistic means the potential outcomes of a PVA could be many, whereas the history of a species has only one course. Multiple replications could specify the array of possible population trajectories, but which one would the actual population take? How far into the future should we project, when our models are based on current demographic rates, and errors multiply with each time step? What assumptions about future conditions of habitat, climate, and other physical factors dare we make? Do we really have an accurate estimate for the mean and variance of important demographic rates? How should genetics be incorporated into demographic models? How should uncertainty in model inputs be treated? How should PVA models be tested? When should PVAs be conducted? What makes a good PVA? And finally, what constitutes a PVA?

PVA is one of conservation biology's most important tools, and periodically it is important to ask how well we are using that tool. Abraham Maslow is credited with saying that, if your only tool is a hammer, every problem starts to look like a nail. Are we trying to do too many jobs with one tool? PVA has its critics and defenders, but where is the science of PVA, given recent developments and debates? We have about 20 years of experience using PVA, which is only a blink in time compared to the duration that we want to conserve species, but it is long enough for a major reassessment. That we have begun a new millennium and are on a path leading to a different kind of world adds to the timeliness of a reevaluation.

This situation prompted us to organize an international symposium to address the status and future of PVA as a tool for species conservation. The symposium, sponsored by the Western Section of The Wildlife Society and the University of California, Berkeley, was held on March 15–17, 1999, at the Town and Country Resort Hotel in San Diego, California. We invited 25 world leaders in the field of PVA and species conservation, solicited 100 additional contributions through a general call for papers, and attracted 350 participants to the symposium. As usual, many of the important exchanges occurred in the hallways and over food and drink.

The symposium, by all accounts, was highly successful, and this book is the result. It is not a transaction of the meeting, as not all presentations, even some that were invited, are included and others were solicited. All papers were subjected to peer review comparable to those of leading journals in the field. As editors, we have purposefully tried to shape the contributions to achieve a coherent whole in the book.

The reader will quickly discover that there is not a grand synthesis and meeting of minds about PVA. PVA is and will remain for some time a work in progress, and various prognostications are put forward about its future. We have not shied from this divergence of opinion and, in fact, redesigned the last part of the book to feature various views about the value and uses of PVA. But for now, PVA is—so to speak—the only game in town, and as such we must apply it as wisely as we can until it matures or is replaced by a better method. History, and the specific courses taken by species and populations, will be the final arbiter.

When we organized the symposium, we thought we were the only ones planning such an assessment of PVA. We since learned of the collection of papers being edited by Per Sjögren-Gulve and Torbjörn Ebenhard published in 2000 as volume 48 of *Ecological Bulletins*. Given the evolving state of PVA, we welcome this diversity to inform new and improved applications of the technique. We all serve in the cause of conservation of biodiversity.

Many people contributed to the production of this book and the success of the symposium. Margaret Jaeger provided indispensable help in every phase of this project, from assisting with symposium details and communicating with authors and reviewers to copyediting and reading page proofs. William Hull, Barbara Kermeen, and Bill Stanley of the Western Section of The Wildlife Society helped organize, publicize, and make the symposium a success. Contributions to the book were ably reviewed by Fred Allendorf, Paul Beier, Mark Boyce, Peter Brussard, Mark Burgman, Tim Clark, Daniel Doak, Barny Dunning, John Emlen, Alan Franklin, Wayne Getz, Daniel Goodman, Susan Haig, Ilkka Hanski, Susan Harrison, Philip Hedrick, Ray Hilborn, Douglas Johnson, Susan Kalisz, Bruce Kendall, Robert Lacy, Russell Lande, William Lidicker, Donald Ludwig, David Maehr, Eric Menges, Scott Mills, James Nichols, Barry Noon, Len Nunney, Hugh Possingham, Katherine Ralls, Martin Raphael, Michael Reed, George Roderick, Hal Salwasser, Fred Samson, Brett Sandercock, Charles Schwartz, Per Sjögren-Gulve, Barbara Taylor, Jeff Walters, Robin Waples, Robert Wayne, Gary White, and Anne York. Financial support for the symposium and the production of this book was received from the Western Section of The Wildlife Society and from the A. Starker Leopold Chair in Wildlife Ecology at the University of California at Berkeley, held by McCullough.

PART 1

OVERVIEW OF
POPULATION
VIABILITY
ANALYSIS

This part introduces the reader to the history and essential ideas that make up the endeavor we call PVA. It is intended to overview the theory and concepts that provide a foundation for understanding extinction processes and to discuss how PVA models are used and what controversies have resulted.

The history and development of PVA is reviewed by Steven Beissinger. He traces how PVA, propelled by recently enacted legislation requiring protection of viable populations, grew rapidly through the 1980s with the advent of personal computers and user-friendly software, and how PVA approaches have changed from models emphasizing genetics over demography to ones that considered demography but not genetics. This chapter presents a road map to the remainder of the book, and overviews the current controversies in the construction, parameterization, use, and interpretation of PVA models.

The process of extinction has both stochastic and deterministic components. Russell Lande thoroughly reviews the role of stochasticity in extinction. He covers important components of demographic and environmental stochasticity, and what is known about stochastic genetic processes, including genetic drift, effective population size, inbreeding, and mutational meltdown. Lande emphasizes the need to understand how stochastic forces interact with deterministic forces to erode a population. The role of deterministic versus stochastic forces was the topic of Graeme Caughley's often-cited essay that is revisited by Mark Boyce. The small-population and declining-population paradigms offer different insights into the process of extinction but are related. Considering one without the other results in an incomplete picture of what is threatening a population and what to do about it. Fred Allendorf and Nils Ryman revisit the role of genetics in PVA, advocating a stronger effort to incorporate genetics into nearly all kinds of PVA models. They emphasize that ignoring inbreeding and genetic variation in assessing the long-term viability of populations will lead to underestimation of the size of viable populations.

The metapopulation concept revolutionized the way conservation

3

biologists think about space and population viability. Two chapters examine the role of spatial population subdivision in the assessment of viability for animals and plants. Ilkka Hanski explores the role of metapopulation processes and models for animals in highly fragmented landscapes. He shows how the incidence function approach to estimate population turnover and extinction has been successfully applied to butterfly populations and presents a refinement of the Levins model, using analytical equations to evaluate the qualitative consequences of changes in landscape structure. Susan Harrison and Chris Ray evaluate whether metapopulation processes are important to plant population dynamics. They conclude that relatively few examples of plant metapopulations exist and that care must be taken in applying the concept. They illustrate the use of logistic regression as an alternative to the incidence function for evaluating metapopulation dynamics.

An examination of the role of PVA in conservation policy concludes this section. Mark Shaffer and colleagues review how PVA has been used in the courts and in policy making. They conclude that PVA has not yet reached its potential for influencing decision making. Furthermore, some of the weaknesses of PVA, especially the dearth of good demographic data for most species, have been recognized for several decades. They call for a well-funded effort by government agencies and nongovernmental organizations to study a set of representative species in a variety of habitats to rectify this problem.

1

Population Viability Analysis: Past, Present, Future

Steven R. Beissinger

Population viability analysis (PVA) has become a cornerstone of conservation science. It is both a process and a tool that has been used to create theory, to analyze data and project population trends, and to make policy decisions. Viability is a pervasive concept that has contributed almost as much to theory as it has to application. PVA has provided a framework for understanding how stochastic events and processes affect the chances of extinction (Lande, chap. 2 in this volume). The role of stochasticity in extinction has been called the *small-population paradigm* (Caughley 1994). Few scientists challenge the notion that genetic, demographic, and environmental stochasticity and catastrophes can result in extinction, although the relative role of each process is often debated (Lande 1988, 1993; Allendorf and Ryman, chap. 4 in this volume). Yet it is the way in which we apply viability concepts to evaluating risks and recovery strategies for populations, species, and ecosystems in trouble that has sometimes engendered controversy over both scientific theories and public policy (Mann and Plummer 1999; Shaffer et al., chap. 7 in this volume).

Perhaps these disagreements stem partly from differences in the way that viability processes interact with the processes of recovering endangered species, which Graeme Caughley (1994) so elegantly defined as the *declining-population paradigm*. The declining-population paradigm consists of determining if there is a population trend, identifying what element of the organism's demography is depressed, determining the environmental or intrinsic factors that cause the depressed rate, and then examining what can be done to correct these limiting factors. In this paradigm, stochasticity is rarely identified as a specific cause of poor reproduction or survival, perhaps with the exception of some genetic factors like inbreeding. The declining-population paradigm and the small-population paradigm offer complementary ways of analyzing the same problem, and they may lead to similar or contradictory conclusions (Boyce, chap. 3 in this volume).

This chapter reviews the past, present, and future of PVA from the perspective of a conservation biologist who works at the interface of academia and management and advocates good science rather than a particular conservation philosophy. In doing so, I try to bridge the worlds of model builders and legacy builders. I begin by examining the history and growth of PVA over the past 20 years and review how it has evolved to date. I next consider the criticisms of PVA and how it has been applied in conservation. I conclude by speculating about how PVA might look in the future.

HISTORY AND GROWTH OF PVA

While the history of PVA is short (Soulé 1987a), the nature of PVA has changed greatly over time. At the beginning of the 1960s, prior to the passage of the first U.S. Endangered Species Act in 1966, the most threatened and publicly visible endangered species in the United States were the California condor (*Gymnogyps californianus*), whooping crane (*Grus americana*), tule elk (*Cervus elaphus nannodes*), and black-footed ferret (*Mustela nigripes*). It is notable that PVA did not contribute to the recovery of any of these species, except after the fact for the condor and ferret, although deterministic demographic models that estimated the rate of annual population growth had been used to make management recommendations for the whooping crane (Miller and Botkin 1974) and for the California condor (Mertz 1971).

PVA was spawned by a controversy in the late 1970s over the Yellowstone National Park population of grizzly bears (*Ursus arctos*). The controversy centered on the recommendations made by wildlife researchers John and Frank Craighead and their colleagues regarding the consequences of closing the park dumps to the bear population. Craighead et al. (1973) used a computer model to show that mortality from the park's approach to problem bears was driving the population to extinction, which led to a review by a National Academy of Sciences committee (Cowan et al. 1974). As a member of the committee, Dale McCullough developed an alternative model that led to differing predictions about the grizzly bear's fate and which was subsequently published (McCullough 1978, 1986). Both models were deterministic models, although McCullough (1986) and Avrin (1976) investigated stochastic recruitment. A model for the grizzly bear that included both demographic and environmental stochasticity was later developed by Mark Shaffer in 1978 as part of his Ph.D. dissertation at Duke University, and to him goes the credit for the first true PVA.

Shaffer's approach (1981) was a new direction in the use of models for conservation, because he developed a stochastic population simulation that incorporated chance events—specifically demographic and environmental stochasticity—and produced extinction probabilities. His model also estimated a minimum viable population size (MVP) by varying the initial number of individuals to find the smallest population size with a 95% chance of remaining extant after the simulated 100-year period (Shaffer and Samson 1985).

The MVP concept was seductive and had arrived at the right moment. The National Forest Management Act of 1976 mandated that harvesting policies and management actions of the U.S. Forest Service (USFS) must sustain viable populations of terrestrial vertebrates. Hal Salwasser and Fred Samson of the USFS were looking for ways to estimate viable population size (Samson 1983; Salwasser et al. 1984; Samson et al. 1985), and they encouraged the agency to incorporate PVA approaches in forest planning. At the same time, viability concepts greatly appealed to decision makers in the U.S. Fish and Wildlife Service who had to implement the Endangered Species Act. They needed a method to quantify risk of extinction to evaluate if a species should be listed as threatened or endangered, to indicate if a federal project adversely affected a species during section 7 consultations, and to determine when a species had recovered and could be proposed for delisting.

Shaffer's MVP paper was followed by the publication of Frankel and Soulé's *Conservation and Evolution* in 1981. Here for the first time genetic approaches to evaluate the viability of populations were detailed in one place, building on the work of Denniston (1978). The short-term and long-term effects of inbreeding and genetic drift were placed in a framework that clarified their erosive impacts on genetic diversity. Here is where the 50/500 rule for effective population sizes, which had recently been proposed by Franklin (1980) and Soulé (1980), was stated in a manner that became etched into the conservation consciousness, for better or worse, as one of the few rules of thumb that conservation biologists have produced to judge viability.

Soon papers that estimated MVP size began to appear (Schonewald-Cox et al. 1983; Lacava and Hughes 1984; Lehmkuhl 1984; Shaffer and Samson 1985; Menges 1986; Reed et al. 1988). Most of these works employed genetic approaches to determine MVPs by adapting and parameterizing demographic equations for estimating effective population size that had been part of classical population genetics theory. I suspect that demographic applications of PVA lagged behind genetic applications because of the difficulty and time involved in developing Monte Carlo computer applications that were needed to produce MVPs based on stochastic population simulations. Let's not forget that in the mid-1980s the personal computer was just making its appearance on the commercial market, and programming in FORTRAN for mainframe computers using decks of cards could hardly be called user-friendly.

In 1986 the work of Michael Gilpin and Michael Soulé put PVA on firmer footing. Gilpin and Soulé (1986) broadened the definition of PVA when they used it as a heuristic concept to examine the many forces that interact in vortices that can affect the viability of a population, including genetic factors. Although they designated the term "Population Vulnerability Analysis" for this approach, the term never became widely used. However, the idea of PVA as a process of risk analysis emerged, where hazards are identified, risks are considered, and a model is developed in the final step.

The Gilpin and Soulé paper made its debut in 1985 at the Ann Arbor, Michigan, meeting where the Society for Conservation Biology was established. Prior to that meeting, there was a workshop that gathered a few scientists together to work on the MVP concept. The result was a series of papers published in *Viable Populations for Conservation* (edited by Soulé [1987b]) that established PVA as both a process and a technique. Concurrently, Gilpin and Soulé were working on the first metapopulation PVA model for a threatened water snake (Soulé 1989).

PVA applications grew rapidly in the late 1980s and early 1990s, largely due to two factors. First, zoological parks quickly employed the power of genetic viability concepts to help them manage small populations in captivity to minimize the loss of genetic diversity. Robert Lacy, Jon Ballou, Kathy Ralls, Ulysses Seal, and others developed workshops and computer software programs to assist zoos in managing captive populations. The application of the small-population paradigm has been extremely successful in this regard. Second, demographic PVAs proliferated rapidly with the development of a variety of computer software packages, such as VORTEX, the RAMAS series, and ALEX, that made Monte Carlo population simulations relatively easy to perform (Lindenmayer et al. 1995). These software packages put PVA on the desktop computer of every interested conservation biologist and empowered users who were computer literate but who were not computer programmers. The advent of these computer programs, as much as any other single factor, promoted the rapid proliferation of PVA as a tool for evaluating conservation problems.

As the power of personal computers has grown over the past decade, approaches for modeling populations have changed. Ease of computation led to exploring a greater number of scenarios in simulation models and to more sophisticated sensitivity analyses (Mills and Lindberg, chap. 16 in this volume). Also, in the past 20 years PVA models have become more complex as we began to incorporate spatial processes. Models grew from single populations to metapopulation models that treated demography at the level of a homogeneous patch, then matured to grid models that created spatially explicit addresses for groups of individuals that allowed them to take on demographic characteristics based on the surrounding grid squares, and finally ripened to the current trend of individual-based models that track each individual as it lives, dies, and moves among landscape elements (Beissinger and Westphal 1998).

The increase in model complexity was driven by the need to consider a variety of processes and scales that affect both organisms and management decisions, and by the growth of GIS and remote sensing technologies. The "rediscovery" of Levins's metapopulation concept (1969, 1970) and its application to conservation biology spawned a new spate of theory, more field studies, and sophisticated modeling efforts. Furthermore, Pulliam's work (1988) on sources and sinks reminded us that demography could vary across a landscape. Hanski (chap. 5 in this volume) and Harrison and Ray (chap. 6 in this volume) summarize advances in this field and the applicability of metapopulation models to animal and plant populations, respectively.

With new technologies, conservation biologists now had the tools to make relatively detailed habitat maps. All that was required to build models that were spatially explicit was the ability to construct complex computer programs. The increased complexity of spatially explicit and individual-based models was seductive because they could be applied to specific conservation situations, such as the loss of particular patches to habitat destruction or the role of dispersal corridors (Dunning et al. 1995). However, increasing the complexity of models also greatly inflated the number of model parameters that had to be estimated (Beissinger and Westphal 1998; Ralls et al., chap. 25 in this volume). It is an unusual endangered species for which we have enough data to estimate most parameters in these models. Such species are either inherently easy to study or their welfare is threatened by the extraction of very valuable resources, which results in large financial investments in field research. Even in these cases, data that are critical to model outcomes—such as mortality rates in relation to movement—may not exist.

CURRENT ISSUES IN THE USE OF PVA

Despite the prevalent position of PVA in conservation science, there has been growing concern over how PVA models are used for making conservation decisions (Boyce 1992; Ralls and Taylor 1997; Beissinger and Westphal 1998; Groom and Pascual 1998; Reed et al. 1998). Criticisms have been raised over the manner that inputs, assumptions, and structure of demographic PVA models affect their ability to predict the fate of populations with accuracy and precision. Four dominant causes of errors result in uncertainty in the outcomes predicted by PVA models:

(1) poor data, (2) difficulties in parameter estimation, (3) weak ability to validate models, and (4) effects of alternative model structures.

Rarely have detailed field studies with adequate sample sizes been used for developing mean estimates of vital rates, and causes and timing of mortality are seldom known. In animals, survival is often the most difficult vital rate to measure precisely because it must be distinguished from the probability of resighting (Nichols 1992). Although sophisticated mark-recapture statistical techniques have been developed to yield accurate estimators of survival and test for differences among individuals of different ages or stages (Lebreton et al. 1992; White et al., chap. 9 in this volume), they require a minimum of three years of study to estimate probability of resighting and survivorship for a single year, and more years to develop precise estimators when the probability of resighting is low. Unfortunately, lack of good survival data can complicate estimation of population change and extinction rates, because elasticity and sensitivity analyses suggest that population change in long-lived organisms is usually most affected by changes in adult survivorship (Boyce 1992; Silvertown et al. 1993; Pfister 1998; Sæther and Bakke 2000). Estimating survival of adults is much less of a problem in plant studies, since they are nonvagile, but survival of seeds and mating success are correspondingly difficult to measure (Doak et al., chap. 15 in this volume). In addition, survival is difficult to estimate for perennials that experience dormancy. As a way to circumvent this problem, mark-recapture models were recently adapted to estimate dormancy and survival rates for rare perennials (Shefferson et al. 2001).

It is equally challenging to develop robust variance estimators for stochastic models (White et al., chap. 9 in this volume). We still require formal definitions of demographic and environmental stochasticity for age- and stage-structured models, which are well tested and confirmed, to incorporate into PVA software (Sæther and Engen, chap. 10 in this volume). Furthermore, accurate estimates for variance of vital rates may require demographic measurements made over many years to sample the range of environmental variation. Rare events, such as 1-in-50- or 100-year droughts, floods, fires, or storms, likely have large effects on variance estimates and population viability (Ludwig 1996, 1999). Long-term studies have shown that estimated variance in population size does not begin to asymptote until after 8 to 20 years, if at all (Pimm 1991). Use of data from short-term studies will usually underestimate the variance in vital rates. However, the effect of short study periods may be offset because estimates of variance derived from field studies include sampling error, which results in an overestimate of variance (Beissinger and Westphal 1998; White et al., chap. 9 in this volume).

Dispersal rates are an important component of metapopulation and spatially explicit models, but our understanding of dispersal is very poor for most species (Hanski, chap. 5 in this volume; Harrison and Ray, chap. 6 in this volume). Consequently, dispersal rules are often coarse caricatures of biological reality due to the difficulty of empirically determining dispersal distances, age of dispersers, and mortality during dispersal. Errors are most exacerbated for species with low success in finding suitable habitat, precisely the situation for many endangered species in fragmented landscapes (Kareiva et al. 1997).

Once a PVA model is constructed and parameterized, it is unwise to place great confidence in its outputs until the model has been tested to determine its accuracy and ability to make predictions (Oreskes et al. 1994; Bart 1995). PVA models are rarely tested against independently gathered field data, so validating their primary prediction, the probability of extinction, is very difficult (Belovsky et al. 1999, chap. 13 in this volume). True validation would require tracking scores of replicate populations that experienced similar conditions and comparing predicted to observed frequencies of extinction, something that is far better suited to the laboratory than to the field. Comparing the average population projection to a time series of population size (e.g., Brook et al. 2000) does not verify value of environmental stochasticity or the magnitude and frequency of catastrophes used in the model, which are a primary cause of differences among replicate model runs and the likelihood of extinction. Usually the best that can be done is to test model assumptions or secondary predictions other than time to extinction, such as estimates for means and variances of vital rates, the distribution of individuals on landscapes, or movement rules. Attempting to confirm secondary predictions could lead to real improvements in the predictive ability of PVA models.

Uncertainty is just about the only certainty in PVA. How we deal with uncertainty in making decisions with PVA models is the subject of much ongoing work. Incorporating Bayesian statistical approaches into PVA is an alternative to frequentist statistics that offers a new way to incorporate uncertainty into model building (Wade, chap. 11 in this volume). Taylor et al. (chap. 12 in this volume) examine the performance of this approach for listing species under ESA and determining what level of threat they should be assigned. Alternatively, one can construct a prediction interval for comparison with model results (Sæther and Engen, chap. 10 in this volume). Still another approach to uncertainty is to take a decision-analysis perspective to examine costs and benefits and then rank management options (Possingham et al., chap. 22 in this volume).

In the absence of data, sensitivity analyses are often recommended to determine which parameters most affect model outcomes and require more study or better parameter estimation. Sensitivity analyses, such as elasticity, have also been used to rank options for conservation actions (Mills and Lindberg, chap. 16 in this volume). However, such analyses do not indicate what factors are causing populations to decline, and elasticity estimates are not value free, but depend on the vital rates used (Caswell 1996, 2001). If a demographic rate is depressed due to the effects of a limiting factor, its matrix element(s) will have a smaller elasticity value and would wrongly suggest that stages associated with this element are less important for management than other elements (Beissinger and Westphal 1998). A little-appreciated insight from sensitivity analyses is that recovery will require more time in populations that are declining due to limiting factors operating on elements or variables with low sensitivity than on those with high sensitivity. From elasticity or other kinds of sensitivity analyses, one can identify the potential management strategy that leads to the fastest population recovery, but one cannot conclude which factors limit population growth.

An important conceptual issue is the need to incorporate genetics in our understanding of population viability and the application of PVA. While genetics was central in many early applications of PVA, the pendulum swung during the last decade to demographic applications, and presently genetics is not well integrated into PVA. Most would agree that this was a good trend, but the momentum in thinking is shifting again, and it is time to reexamine the proper blend of genetics and demography (Allendorf and Ryman, chap. 4 in this volume). Examining the interaction of genetics and demography is problematic, in part because of the difficulty of translating genetic currencies of heterozygosity and diversity into a probability of extinction. We still have few direct studies that link genetic diversity to individual survival and fitness. Even rarer are demonstrations of the role of genetics on the viability of populations. Just considering the negative effects of inbreeding, it can be difficult to say how genetics should be incorporated into demographic models; for example, should it affect juvenile mortality, adult mortality, or litter or clutch size? It can be hard to know the magnitude of the impacts of inbreeding, or what demographic characteristic it will affect.

There are, however, a number of ways that genetics can be better incorporated into PVA. Hedrick (chap. 17 in this volume) describes how molecular genetics has been able to answer a variety of questions that are notoriously difficult for demographic studies to tackle. Estimates of effective population size (N_e) have been one of the main genetic approaches used to assess viability. Yet there are still improvements that can be made to estimating N_e, and this concept can be better integrated into PVA (Waples, chap. 8 in this volume). Another useful genetic approach, pedigree analysis, has often been used with captive animals but has rarely been employed with wild populations, where it could make important contributions (Haig and Ballou, chap. 18 in this volume). Do we need to develop a genetic PVA that would estimate genetic goals to complement management directives from demographic PVAs?

THE FUTURE OF PVA

Given the issues facing PVA, what will be its future? Single-species conservation approaches like PVA are sometimes pitted against multispecies conservation approaches currently in vogue such as identification and preservation of biodiversity hot spots, landscape conservation planning, and ecosystem conservation. Do we really need PVA in the age of GAP analysis to identify where reserves should be placed, campaigns to establish border-to-border wilderness, implementation of ecosystem management approaches such as President Clinton's Northwest Forest Plan, and complex reserve selection algorithms?

Yes! PVA appears unlikely to go extinct in the near future because of the kinds of challenges that conservation biologists and decision makers need to address. The problems of managing small or declining populations of threatened species inside and outside of reserves that motivate us to construct formal models and conduct PVAs will not only continue into the foreseeable future, but are likely to grow in prominence as the human population continues to expand and its enterprises threaten more species.

There are, however, disparate views of the directions that PVA might evolve. It seems natural to incorporate some PVA approaches as part of a risk-management framework, and decision makers need and want the kind of information that PVA can provide (Goodman, chap. 21 in this volume). Optimists suggest expanding PVA from static to dynamic models that incorporate human factors like land use, economics, and politics that affect the future of the earth's biological diversity (Lacy and Miller, chap. 23 in this volume). With increased computing and remote sensing capabilities, it is easy to envision models that will become more complex and be limited primarily by ability of field biologists to provide critical data (Beissinger et al. 2002). Pessimists, however, argue that PVA will decline in importance because PVA models cannot and will never be able to accurately predict extinction rates, and past attempts to implement truly adaptive management have a mixed record at best (Ludwig and Walters, chap. 24 in this volume).

If PVA models are not ready to make forecasts of the future trajectory of a population 25 years down the road, let alone the 50, 100, or 1,000 years that models have been used to project, what should their role be in making management decisions? Some uses of PVA interpret model output in an absolute fashion by assuming that the models are predicting the fate of a population (Beissinger and Westphal 1998). Other uses of PVA compare relative differences among scenarios to rank model alternatives. It makes sense to treat applications of PVA that interpret model output in an absolute and relative fashion differently (Ralls et al., chap. 25 in this volume).

Currently, PVA lacks standards or guidelines for use. What is a PVA? What is a viable population? There is no established definition of what passes for a respectable PVA, or a consensus of what constitutes a viable population in terms of time horizons or levels of acceptable risk of extinction. What type of PVA model is appropriate, given the many types of models that can be used? What makes a good PVA? When should we use PVA instead of alternative approaches for gaining understanding, estimating risk, and making decisions? Ralls et al. (chap. 25 in this volume) discuss these issues and offer initial suggestions for establishing criteria to judge the merits of a PVA and guidelines for its use.

However PVA is used in the future, the added scrutiny that PVA has recently received from scientists has already raised the bar for what constitutes a good PVA. It will become harder to use guesses or poorly documented inputs in PVA models instead of rigorously developed estimates for vital rates and other model parameters. Models that have not thoroughly explored the sensitivity of model outputs to changes of each parameter and explored how changes in model structure affect model outcomes are unlikely to be well received. Likewise, models that do not incorporate or analyze the role of uncertainty on model predictions are certain to fail.

In conclusion, the future of PVA depends very much upon the way that conservation biologists develop and apply PVA models, and how they sell model results to policymakers and managers. PVA has had few successes in the policy process, but the need remains as great as ever (Shaffer et al., chap. 7 in this volume). Laws mandate that decisions must be made on the fate of species. How PVA reinvents itself and evolves in the future is up to us.

LITERATURE CITED

Avrin, D. E. 1976. A numerical model of the Yellowstone grizzly bear population and its management implications. M.S. thesis, University of Michigan, Ann Arbor, Michigan.

Bart, J. 1995. Acceptance criteria for using individual-based models to make management decisions. *Ecological Applications* 5:411–420.

Beissinger, S. R., J. R. Walters, D. G. Catanzaro, K. G. Smith, J. B. Dunning Jr., S. M. Haig, B. R. Noon, and B. M. Stith. 2002. Modeling approaches in avian conservation and the role of field biologists. *Current Ornithology* 17.

Beissinger, S. R., and M. I. Westphal. 1998. On the use of demographic models of population viability in endangered species management. *Journal of Wildlife Management* 62:821–841.

Belovsky, G. E., C. Mellison, C. Larson, and P. A. Van Zandt. 1999. Experimental studies of extinction dynamics. *Science* 286:1175–1177.

Boyce, M. S. 1992. Population viability analysis. *Annual Review of Ecology and Systematics* 23:481–506.

Brook, B. W., J. J. O'Grady, A. P. Chapman, M. A. Burgman, H. R. Akçakaya, and R. Frankham. 2000. Predictive accuracy of population viability analysis in conservation biology. *Nature* 404:385–387.

Caswell, H. 1996. Second derivatives of population growth rate: calculation and applications. *Ecology* 77:870–879.

———. 2001. *Matrix population models*. 2d edition. Sinauer Associates, Sunderland, Massachusetts.

Caughley, G. 1994. Directions in conservation biology. *Journal of Animal Ecology* 63:215–244.

Cowan, I. M., D. G. Chapman, R. S. Hoffmann, D. R. McCullough, G. A. Swanson, and R. B. Weeden. 1974. *Report of the committee on the Yellowstone grizzlies*. National Academy of Sciences, Washington, D.C.

Craighead, J. J., J. R. Varney, and F. C. Craighead Jr. 1973. A computer analysis of the Yellowstone grizzly bear population. Montana Cooperative Wildlife Research Unit, University of Montana, Missoula, Montana.

Denniston, C. 1978. Small population size and genetic diversity: implications for endangered species. Pages 281–289 in S. A. Temple, editor, *Endangered birds: management techniques for preserving threatened species*. University of Wisconsin Press, Madison, Wisconsin.

Dunning, J. B., B. J. Danielson, B. R. Noon, T. L. Root, R. H. Lamberson, and E. Stevens. 1995. Spatially explicit population models: current forms and future uses. *Ecological Applications* 5:3–11.

Frankel, O. H., and M. E. Soulé. 1981. *Conservation and evolution*. Cambridge University Press, Cambridge, United Kingdom.

Franklin, I. R. 1980. Evolutionary change in small populations. Pages 135–149 in M. E. Soulé and B. A. Wilcox, editors, *Conservation biology: an evolutionary-ecological perspective*. Sinauer Associates, Sunderland, Massachusetts.

Gilpin, M. E., and M. E. Soulé. 1986. Minimum viable populations: processes of species extinction. Pages 19–34 in M. E. Soulé, editor, *Conservation biology: the science of scarcity and diversity*. Sinauer Associates, Sunderland, Massachusetts.

Groom, M. J., and M. A. Pascual. 1998. The analysis of population persistence: an outlook on the practice of viability analysis. Pages 4–27 in P. L. Fiedler and P. M. Kareiva, editors, *Conservation biology for the coming decade*. Chapman and Hall, New York, New York.

Kareiva, P., D. Skelly, and M. Ruckelshaus. 1997. Reevaluating the use of models to predict the consequences of habitat loss and fragmentation. Pages 156–166 in S. T. A. Pickett, R. S. Ostfeld, M. Shachak, and G. E. Likens, editors, *The ecological basis of conservation*. Chapman and Hall, New York, New York.

Lacava, J., and J. Hughes. 1984. Determining minimum viable population levels. *Wildlife Society Bulletin* 12:370–376.

Lande, R. 1988. Genetics and demography in biological conservation. *Science* 241:1455–1460.

———. 1993. Risks of population extinction from demographic and environmental stochasticity and random catastrophes. *American Naturalist* 142:911–927.

Lebreton, J.-D., K. P. Burnham, J. Clobert, and D. R. Anderson. 1992. Modeling survival and testing biological hypotheses using marked animals: a unified approach with case studies. *Ecological Monographs* 62:67–118.

Lehmkuhl, J. F. 1984. Determining size and dispersion of minimum viable populations for land management planning and species conservation. *Environmental Management* 8:167–176.

Levins, R. 1969. Some demographic and genetic consequences of environmental heterogeneity for biological control. *Bulletin of the Entomological Society of America* 15:237–240.

———. 1970. Extinction. Pages 77–107 in M. Gerstenhaber, editor, *Some mathematical questions in biology*. American Mathematical Society, Providence, Rhode Island.

Lindenmayer, D. B., M. A. Burgman, H. R. Akçakaya, R. C. Lacy, and H. P. Possingham. 1995. A review of three models for metapopulation viability analysis: ALEX, RAMAS/Space, and VORTEX. *Ecological Modelling* 82:161–174.

Ludwig, D. 1996. Uncertainty and the assessment of extinction probabilities. *Ecological Applications* 6:1067–1076.

———. 1999. Is it meaningful to estimate a probability of extinction? *Ecology* 80:298–310.

Mann, C. C., and M. L. Plummer. 1999. A species' fate, by the numbers. *Science* 284:36–37.

McCullough, D. R. 1978. Population dynamics of the Yellowstone grizzly bear. Pages 173–196 in C. W. Fowler and T. D. Smith, editors, *Dynamics of large mammal populations*. John Wiley and Sons, New York, New York.

———. 1986. The Craigheads' data on Yellowstone grizzly bear populations: relevance to current research and management. *International Conference on Bear Research and Management* 6:21–32.

Menges, E. S. 1986. Predicting the future of rare plant populations: demographic monitoring and modeling. *Natural Areas Journal* 6:13–25.

Mertz, D. B. 1971. The mathematical demography of the California condor population. *American Naturalist* 105:437–453.

Miller, R. S., and D. B. Botkin. 1974. Endangered species models and predictions. *American Scientist* 62:172–181.

Nichols, J. D. 1992. Capture-recapture models. *BioScience* 42:94–102.

Oreskes, N., K. Shrader-Frechette, and K. Belitz. 1994. Verification, validation, and the confirmation of numerical models in the earth sciences. *Science* 263:641–646.

Pfister, C. A. 1998. Patterns of variance in stage-structured populations: evolutionary predictions and ecological implications. *Proceedings of the National Academy of Sciences* (USA) 95:213–218.

Pimm, S. L. 1991. *The balance of nature?* University of Chicago Press, Chicago, Illinois.

Pulliam, H. R. 1988. Sources, sinks, and population regulation. *American Naturalist* 132:652–661.

Ralls, K., and B. L. Taylor. 1997. How viable is population viability analysis? Pages 228–235 in S. T. A. Pickett, R. S. Ostfeld, M. Shachak, and G. E. Likens, editors, *The ecological basis of conservation*. Chapman and Hall, New York, New York.

Reed, J. M., P. D. Doerr, and J. R. Walters. 1988. Minimum viable population size of the red-cockaded woodpecker. *Journal of Wildlife Management* 52:385–391.

Reed, J. M., D. D. Murphy, and P. F. Brussard. 1998. Efficacy of population viability analysis. *Wildlife Society Bulletin* 26:244–251.

Sæther, B.-E., and Ø. Bakke. 2000. Avian life-history variation and contribution of demographic traits to population growth rate. *Ecology* 81:642–653.

Salwasser, H., S. P. Mealey, and K. Johnson. 1984. Wildlife population viability: a question of risk. *Transactions of the North American Wildlife and Natural Resources Conference* 1:421–439.

Samson, F. B. 1983. Minimum viable populations: a review. *Natural Areas Journal* 3:15–23.

Samson, F. B., F. Perez-Trejo, H. Salwasser, L. F. Ruggiero, and M. L. Shaffer. 1985. On determining and managing minimum population size. *Wildlife Society Bulletin* 13:425–433.

Schonewald-Cox, C. M., S. M. Chambers, B. MacBryde, and W. L. Thomas, editors. 1983. *Genetics and conservation: a reference for managing wild animal and plant populations*. Benjamin/Cummings, Menlo Park, California.

Shaffer, M. L. 1981. Minimum population sizes for species conservation. *BioScience* 31:131–134.

Shaffer, M. L., and F. B. Samson. 1985. Population size and extinction: a note on determining critical population sizes. *American Naturalist* 125:144–152.

Shefferson, R. P., B. K. Sandercock, J. Proper, and S. R. Beissinger. 2001. Estimating dormancy and survival of a rare herbaceous perennial using mark-recapture models. *Ecology* 82:145–156.

Silvertown, J., M. Franco, I. Pisanty, and A. Mendoza. 1993. Comparative plant demography: relative importance of life-cycle components to the finite rate of increase in woody and herbaceous perennials. *Journal of Ecology* 81:465–476.

Soulé, M. E. 1980. Thresholds for survival: maintaining fitness and evolutionary potential. Pages 151–169 in M. E. Soulé and B. A. Wilcox, editors, *Conservation biology: an evolutionary-ecological perspective*. Sinauer Associates, Sunderland, Massachusetts.

———. 1987a. Introduction. Pages 1–10 in M. E. Soulé, editor, *Viable populations for conservation*. Cambridge University Press, Cambridge, United Kingdom.

———, editor. 1987b. *Viable populations for conservation*. Cambridge University Press, Cambridge, United Kingdom.

———. 1989. Risk analysis for the Concho water snake. *Endangered Species Update* 6:19–25.

2

Incorporating Stochasticity in Population Viability Analysis
Russell Lande

ABSTRACT

This chapter reviews recent advances in understanding how stochastic demographic and genetic factors affect population viability, defined by the probability of extinction during a given time interval. Stochastic fluctuations in population growth rate not only cause chance extinctions; they also produce a cumulative deterministic component that tends to decrease populations and drive them toward extinction. Environmental stochasticity decreases the long-run growth rate of a population when it is below carrying capacity, and demographic stochasticity can create a type of Allee effect or unstable equilibrium at small population size below which most population trajectories decline toward extinction. In fragmented habitats, the regional dynamics of a metapopulation interacts with the stochastic dynamics of local populations. For territorial species, there exists an extinction threshold or minimum proportion of suitable habitat necessary for metapopulation persistence. For nonterritorial species, habitat occupancy and metapopulation persistence depend strongly on the "rescue effect" and the "establishment effect," whereby immigrants to local populations decrease the rate of local extinction and increase the probability of successful colonization. Positive temporal and spatial autocorrelations of population fluctuations increase the risk of extinction. These autocorrelations depend on the temporal and spatial scales of environmental stochasticity and on the species' life history and dispersal pattern. Genetic stochasticity due to finite population size also produces deterministic or average reduction of genetic variance and adaptive potential, and loss of fitness through inbreeding depression and accumulation of new mildly deleterious mutations. The probability distribution of extinction times has an initial lag before a characteristic rate of extinction is achieved, which suggests that population viability analyses should consider time frames longer than the typical 100-year limit dictated by political and legal considerations, especially for species with long generations.

I thank I. Hanski for discussion, and S. Engen and B.-E. Sæther for comments on the manuscript. This work was supported by NSF grant DEB 9806363.

INTRODUCTION

All populations fluctuate stochastically, with coefficients of variation in annual census sizes usually in the range of about 20 to 80% (Pimm 1991). Stochastic fluctuations can drive a population or species to extinction even when its expected growth rate is positive at population sizes below carrying capacity. Following Shaffer (1981), population viability is generally defined in terms of the probability of extinction within a specified time interval. Shaffer described stochastic factors of demography and genetics that contribute to extinction risk. "Demographic stochasticity" is caused by random variation in individual fitness that is independent among individuals. This produces random fluctuations in mean fitness or population growth rate that are inversely proportional to population size. "Environmental stochasticity" caused by changes in physical or biological factors affects the fitness of all individuals in a population in a similar fashion. This produces random fluctuations in population growth rate regardless of population size. Catastrophes are sudden collapses in population size, caused by extreme environmental events such as droughts, floods, fires, and epidemics, often with a substantial random component in time occurrence (Young 1994), as well as possible periodic components (Beissinger 1995). Shaffer (1987) later included random catastrophes as the upper tail of a distribution of environmental stochasticity (cf. Erb and Boyce 1999).

Genetic stochasticity in finite populations, also known as random genetic drift, entails random changes in gene frequencies caused both by variance in family sizes and by Mendelian segregation of alleles (Wright 1969; Crow and Kimura 1970). Particularly in small populations, inbreeding due to mating between related individuals produces a random loss of alleles and a reduction of genetic variance required for adaptive evolution. Inbreeding on average increases the homozygosity of preexisting, partially recessive deleterious mutations, but by chance some can achieve appreciable frequencies in small populations. This causes "inbreeding depression," which is manifested as decreased individual fitness and population growth rate (Ralls and Ballou 1983; Falconer and Mackay 1996; Husband and Schemske 1996).

Here I review recent developments in understanding and modeling stochastic factors affecting the risk of population extinction, including (1) the relationship between stochastic demography and Allee effect (i.e., the reduction of expected growth rate in small populations), (2) the role of stochastic local dynamics in metapopulation persistence for territorial and nonterritorial species, (3) the temporal and spatial scales of environmental stochasticity and the synchrony of population fluctuations, and (4) genetic stochasticity, loss of adaptive potential and reduced fitness in small populations from fixation of both old and new mutations. I show that stochastic demographic and genetic factors have deterministic components or average effects with significant impacts on population viability. Finally, I discuss the probability distribution of extinction times and appropriate time spans for population viability analysis (PVA).

Despite the importance of stochastic factors, especially in small populations, it is important to realize that most populations initially become threatened or endangered because of deterministic human activities caused by human population growth and economic development, primarily habitat destruction and fragmentation, overexploitation, introduced species, and pollution (Groombridge 1992; Caughley 1994).

STOCHASTIC DEMOGRAPHY AND ALLEE EFFECTS

Stochastic fluctuations in population growth rate contribute to extinction risk for two reasons. Stochasticity not only causes random encounters with the "absorbing boundary" of extinction from which species cannot return; it also has a cumulative deterministic tendency to decrease populations and drive them toward extinction. This occurs because population growth is fundamentally a multiplicative process, and the long-run dynamics of population size are governed by the geometric mean growth rate (or expected rate of increase of the natural logarithm of population size), which is always less than the arithmetic mean growth rate (or expected per capita population growth rate). For example, under density-independent growth, when the population is well below carrying capacity but large enough to neglect demographic stochasticity, deterministic population dynamics in a constant environment are described by a continuous time model in which the rate of change of population size, N, with time, t, is given by $dN/dt = rN$, where r is the per capita growth rate or mean Malthusian fitness in the population. Environmental stochasticity causes r to fluctuate randomly in time with a mean $\bar{r}$ and environmental variance σ_e^2. The expected rate of increase of $\ln N$ (or "long-run growth rate") is $\bar{r} - \sigma_e^2/2$ (Tuljapurkar 1982; Lande and Orzack 1988; Lande 1993). Thus, environmental stochasticity creates a deterministic (or average) decrement in the long-run growth rate of a population. This is not an artifact of using the log scale, since simulation of stochastic discrete-time models demonstrates that surviving populations tend to grow more slowly than the deterministic rate that would occur in a constant, average environment $\bar{r}$ (fig. 2.1).

Demographic stochasticity produces similar effects in small populations. Denoting the demographic variance in individual fitness as σ_d^2, the

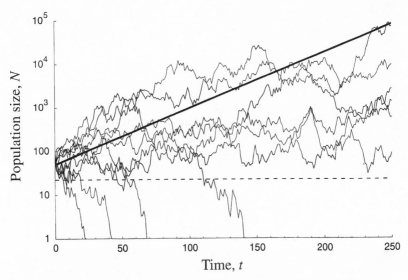

Fig. 2.1 Ten simulated trajectories of a population with initial size of 50 individuals subject to demographic and environmental stochasticity. Dynamics obey the simple discrete-time model $N_{t+1} = \lambda_t N_t$, where N_t is the population size in year t. At a given population size the finite rate of increase, λ_t, is (approximately) lognormally distributed with mean $\bar{\lambda} = 1.03$ and variance $\sigma_e^2 + \sigma_d^2/N_t$, where $\sigma_e^2 = 0.04$ is the environmental variance and $\sigma_d^2 = 1.0$ is the demographic variance. The *heavy line* gives the deterministic dynamics of geometric growth at the mean rate $\bar{\lambda}$. The *dashed line* marks the unstable equilibrium size N^* below which population trajectories tend to decrease rapidly toward the extinction boundary at $N = 1$ (see text).

variance in mean fitness or population growth rate, r, caused by demographic stochasticity is σ_d^2/N. Under both demographic and environmental stochasticity at a given population size well below carrying capacity, r in the above model fluctuates with a mean $\bar{r}$ and variance $\sigma_e^2 + \sigma_d^2/N$ (Leigh 1981; Lande 1993; Sæther et al. 2000). The expected rate of increase of ln N is $\bar{r} - \sigma_e^2/2 - \sigma_d^2/(2N)$ (cf. Lande 1998a). Thus, in addition to causing random fluctuations in population size, particularly in small populations, demographic stochasticity also creates a deterministic decrement in the long-run growth rate that is inversely proportional to population size. With sufficient demographic stochasticity, the long-run growth rate can become negative in small populations.

With both demographic and environmental stochasticity, a generalized scale transformation that resembles ln N for large populations and $\sqrt{N}$ for small populations is necessary to analyze the probabilistic tendencies of population trajectories (Lande 1998a). On this transformed scale, demographic stochasticity creates an unstable equilibrium at a value corresponding to the population size

$$N^* = \frac{\sigma_d^2/4}{\bar{r} - \sigma_e^2/2}.$$

Below a population size of N^*, most population trajectories tend to decrease toward extinction. Again, this is not an artifact of the scale transformation, as simulations of stochastic discrete-time models demonstrate that populations below N^* tend to decrease and become rapidly extinct (fig. 2.1).

The existence of an unstable equilibrium on this transformed scale bears a close resemblance to the classical Allee effect (Allee et al. 1949). Allee effects usually are defined as a deterministic decrease in individual fitness (and hence a decrease in mean fitness or population growth rate) due to a failure of cooperative interactions among individuals in small or sparsely distributed populations. Some common mechanisms for Allee effects include group foraging, group defense against predators, cooperative breeding, chemical or physical conditioning of the environment (e.g., huddling for warmth during winter), and the difficulty of finding mates (Courchamp et al. 1999). Genetic stochasticity in small populations also produces similar effects through inbreeding depression and through random fluctuations in sex ratio in species with genetic sex determination. Both classical Allee effects and the deterministic components of demographic and genetic stochasticity can cause populations below a certain size to decline rapidly to extinction.

Thus, classical Allee effects, demographic stochasticity, and genetic stochasticity may be indistinguishable in terms of their effects on the dynamics of small populations. Distinguishing them generally will require detailed studies of the behavioral, ecological, and genetic factors affecting fitness in small populations. Statistical methods for joint estimation of demographic and environmental stochasticity and uncertainty in population parameters are described in Engen et al. (1998), Kendall (1998), and Sæther et al. (1998, 2000).

METAPOPULATIONS WITH STOCHASTIC LOCAL DYNAMICS

Metapopulation concepts have become popular for analyzing the effects of habitat fragmentation on populations in which regional persistence is maintained by a balance between local extinction and colonization (Levins 1969, 1970; Hanski and Gilpin 1997). Assuming that equivalent patches of suitable territory are either occupied or unoccupied by a species, and that the local extinction rate (e) and the colonization rate (c) are constant, Levins showed that the proportion of suitable habitat patches occupied at equilibrium is $\hat{p} = 1 - e/c$. Thus, regional persistence of a metapopulation is possible ($\hat{p} > 0$) only when the colonization rate exceeds the local extinction rate ($c > e$). This and subsequent metapopulation models reveal that a metapopulation may become extinct in the presence of suitable habitat and that currently unoccupied suitable habitat may be critical for long-term persistence of a species.

However, most metapopulation models still make several of the same simplifying assumptions as in the original model of Levins (1969, 1970). Ignoring the internal dynamics within local populations fails to consider any coupling between local and global dynamics, which is known to be important through the "rescue effect" of immigrants reducing the local extinction rate (Brown and Kodric-Brown 1977). Most metapopulation models contain no description of the amount of suitable versus unsuitable habitat, which precludes their use in predicting effects of either continued habitat destruction and fragmentation or habitat improvement. This section reviews two metapopulation models that relax these simplifying assumptions for territorial species and nonterritorial species.

For territorial species in which individual females or mated pairs have exclusive or largely nonoverlapping territories or home ranges, Lande (1987) developed a metapopulation model that incorporated life history, individual dispersal behavior, and an explicit description of the amount of suitable habitat in a region. Patches of habitat the size of individual territories are assumed to be either suitable or unsuitable for survival and reproduction, and suitable habitat patches are randomly or evenly distributed over a large region such that suitable habitat is not clumped on a scale larger than the typical individual dispersal distance. The proportion of suitable habitat in the region is h. Because individual territories are identified as the unit of suitable habitat, local extinction corresponds to the death of an individual female, and colonization corresponds to settlement of a dispersing juvenile on an unoccupied suitable territory. The most basic model incorporates classical female-biased demography with age structure, assuming that all females are successfully mated. Juveniles disperse prior to reproduction, and their survival is density-dependent, based on the probability of finding a suitable unoccupied territory among a maximum number of potential territories they can search before dying from starvation or predation. The proportion of suitable habitat occupied at equilibrium takes the simple form

$$\hat{p} = 1 - (1 - k)/h.$$

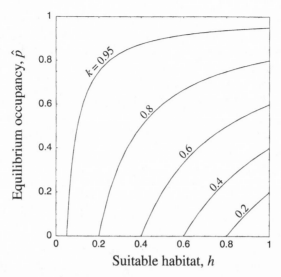

Fig. 2.2 Equilibrium occupancy of suitable habitat, $\hat{p}$, in a metapopulation model for a territorial species as a function of the proportion of suitable habitat, h, randomly or evenly distributed in a region, for different values of the demographic potential, k (modified from Lande 1987).

All information on the life history and dispersal behavior is incorporated in the composite parameter k, termed the "demographic potential" because it gives the maximum proportion of habitat occupied at equilibrium in a completely suitable region ($\hat{p} = k$ when $h = 1$). Even in a completely suitable region, not all habitat is occupied, because some time elapses before a territory vacated by the death of a resident female is settled by a dispersing juvenile. As the amount of suitable habitat decreases in the region, the equilibrium occupancy declines, eventually reaching an "extinction threshold" or minimum proportion of suitable habitat necessary to sustain the population ($h = 1 - k$). For species with high demographic potential, the equilibrium occupancy declines precipitously as the amount of suitable habitat decreases toward the extinction threshold. The equilibrium population size is proportional to the product of the amount of suitable habitat (h) and the equilibrium occupancy of suitable habitat ($\hat{p}$). Hence, the equilibrium population size declines faster than the rate of habitat destruction until the extinction threshold is reached (fig. 2.2). When habitat destruction and fragmentation occur on the same time scale as the generation time of a species, there may be little warning that the extinction threshold is being approached if the decline in population size lags behind the demographic equilibrium.

This metapopulation model for a territorial species was originally applied to data on the demography and habitat of the northern spotted owl (*Strix occidentalis caurina*) in the Pacific Northwest of the United States. It suggested that plans by the U.S. government to preserve this subspecies were seriously inadequate due to excessive habitat destruction and fragmentation. This became critical scientific evidence in litigation that eventually led to greatly increased protection of old-growth forests on which this subspecies depends (Lande 1988). Subsequent analyses using more detailed spatial information confirmed the generality of extinction thresholds in models of habitat fragmentation (Doak 1989; Nee and May 1992; Lamberson and Carroll 1993; McKelvey et al. 1993; Bascompte and Sole 1996; Hill and Caswell 1999; With and King 1999).

Incorporating stochastic dynamics of local populations into metapopulation models of nonterritorial species has proven much more difficult. Some initial results were derived by Lande et al. (1998), who modeled demographic and environmental stochasticity, and stochastic dispersal among a finite number of local populations. This approach allows local extinction and colonization rates to be derived from local population dynamics and permits analysis of the coupling between local population dynamics and metapopulation dynamics. Coupling of local and global dynamics occurs because increasing occupancy of suitable habitat in the metapopulation increases the rate of immigration into local populations. This produces the well-known "rescue effect" (Brown and Kodric-Brown 1977), whereby immigration decreases the rate of local extinction. It also produces an "establishment effect" (Lande et al. 1998), whereby continued immigration can greatly increase the probability of successful colonization during the critical initial phase, when a few individuals struggle to overcome demographic and environmental stochasticity (fig. 2.3). These effects can combine to create multiple equilibria for habitat occupancy. This includes a kind of metapopulation Allee effect or unstable equilibrium at low habitat occupancy below which the metapopulation cannot persist, as suggested by Hanski and Gyllenberg (1993), based on simple phenomenological models.

Infinite metapopulation models with an unlimited number of local populations produce deterministic dynamics of habitat occupancy (Levins 1969, 1970; Hanski and Gyllenberg 1993). In contrast, finite metapopulation models allow estimation of the risk of metapopulation extinction by stochastic local extinction and colonization. Accounting for stochastic dynamics within local populations and the coupling of local and global dynamics by the rescue and establishment effects can greatly increase metapopulation viability compared to classical metapopulation

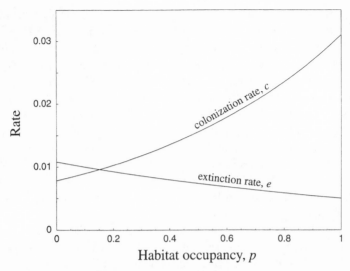

Fig. 2.3 Local extinction and colonization rates, *e* and *c,* as a function of the occupancy of suitable habitat, *p,* in a metapopulation model for a nonterritorial species. Decrease in the local extinction rate and increase in the colonization rate with increasing *p* (and higher immigration into local populations) are, respectively, the rescue effect and the establishment effect. Expected local dynamics are logistic with carrying capacity $K = 50$, mean intrinsic rate of increase $\bar{r} = 0.01$, environmental and demographic variances $\sigma_e^2 = 0.01$ and $\sigma_d^2 = 1.0$, and a low individual migration rate $m = 0.005$ (modified from Lande et al. 1998).

models that assume constant rates of local extinction and colonization (Lande et al. 1998).

TEMPORAL AND SPATIAL SCALES OF ENVIRONMENTAL STOCHASTICITY

Temporal and spatial autocorrelations in environmental stochasticity can have major impacts on population viability but often are ignored both because of the difficulty of estimating them from limited data usually available on endangered species and because these topics have only recently received attention by ecologists. Autocorrelations begin at unity for populations separated by zero time and distance, generally decline with increasing time or distance, and may be negative at some times or distances. The temporal and spatial scales of autocorrelation in environmental stochasticity determine appropriate methods for modeling and incorporating them in PVA.

Positive temporal autocorrelation in environmental stochasticity increases extinction risk, as demonstrated by several theoretical studies

(Turelli 1977; Foley 1994, 1997; Lande et al. 1995; Ripa and Lundberg 1996; Palmqvist and Lundberg 1998). This occurs because positive autocorrelation produces runs of years with consistently high or low population growth rates. Runs of years with negative growth rates can drive a population extinct, whereas runs of years with positive growth rates are damped by density dependence. The simplest method for modeling temporal autocorrelation is to multiply the environmental variance by the sum or integral of the autocorrelation function (Turelli 1977; Foley 1994, 1997; Lande et al. 1995). For positive autocorrelation, this effectively increases environmental stochasticity and the corresponding extinction risk. This approach is accurate only if the temporal scale of environmental stochasticity is not very long; otherwise, it is necessary to explicitly model the temporal autocorrelation in environmental stochasticity.

Empirical studies of autocorrelation usually deal directly with population size rather than growth rate (Pimm 1991; Halley 1996). Temporal autocorrelation in population size can also result from lags caused by age structure of populations (Lande and Orzack 1988). Studies of temporal autocorrelation in adult or total population size typically do not separate the contributions from environmental autocorrelation and life history (for an exception see Ratner et al. 1997).

Positive spatial autocorrelation in environmental stochasticity also increases extinction risk, as demonstrated in simulations of classical metapopulations with spatially correlated local extinctions (Harrison and Quinn 1989; Gilpin 1990), and simulations of metapopulations with spatially correlated environmental stochasticity in local dynamics (Heino et al. 1997; Heino 1998). This occurs because synchronized fluctuations in the sizes of local populations cause them to become extinct simultaneously, which increases the risk of regional or global extinction.

Spatial autocorrelation in population size, or "population synchrony," results from a combination of spatial autocorrelation in environmental stochasticity and localized individual dispersal. Several studies have attempted to clarify the relative contributions of environmental stochasticity and individual dispersal to population synchrony (Ranta et al. 1995, 1997a,b). Comparisons of related species have shown that population synchrony is higher over short distances for species with greater powers of dispersal. There often is a long-distance component of population synchrony on a scale much greater than the individual dispersal distance, which usually is attributed to environmental autocorrelation (Hanski and Woiwod 1993; Sutcliffe et al. 1996; Lindstrom et al. 1996).

Lande et al. (1999) analyzed a population continuously distributed in space, with environmental stochasticity caused by temporal fluctuations in the intrinsic rate of increase or carrying capacity of local populations, assuming the environmental stochasticity was spatially (but not temporally) autocorrelated. Employing the standard deviation of a function in a given direction as a measure of scale, for small or moderate fluctuations in local population size the spatial scale of population synchrony (l_p) is related to the spatial scales of environmental correlation (l_e) and individual dispersal (l) by the simple general formula $l_p^2 = l_e^2 + ml^2/\gamma$, where m is the individual dispersal rate and γ is the strength of population density regulation (or rate of return to equilibrium, $\bar{r}$, in the logistic model). Relative to environmental autocorrelation, the contribution of individual dispersal to the spatial scale of population synchrony is magnified by the ratio of the individual dispersal rate to the strength of density regulation. Even when the scale of individual dispersal is less than that of environmental autocorrelation, dispersal can substantially increase the spatial scale of population synchrony for weakly regulated populations with low $\bar{r}$. This happens because weaker density regulation allows fluctuations in local population size to build up and spread farther by individual dispersal before they are damped by density dependence.

Many threatened and endangered species are characterized by diminished values of $\bar{r}$ due to overexploitation, habitat degradation, introduced species, and pollution (Groombridge 1992; Caughley and Gunn 1996). This not only contributes to extinction risk by making local populations more susceptible to stochastic declines, but also increases population synchrony and the risk of regional or global extinction.

GENETIC STOCHASTICITY

Incorporating genetic stochasticity into PVA requires that its effects be expressed in terms of population dynamics and extinction risk. Population geneticists have developed models of stochastic evolution in finite populations of constant size, but work on the interactions between stochastic genetics and demography has barely begun. The great complexity of the genetics of finite populations makes this a daunting task because thousands of genes affect fitness. Realistic genetic models are, therefore, far more complex than most demographic models and difficult to accurately parameterize for particular species. This section reviews some recent progress in understanding the impacts of genetic stochasticity on population viability.

Genetic Variance, Adaptive Evolution, and Population Persistence in Changing Environments

Two major mechanisms of population persistence in response to major environmental change are evolution and change of geographic range.

In unfragmented habitats, theory suggests that change of geographic range is the primary mechanism of population persistence in a changing environment. Even though local populations at a fixed spatial location may evolve rapidly, the areas of highest population density move through space and time to track the environmental conditions to which a species is already adapted, so that the population as a whole maintains a nearly constant phenotype (Pease et al. 1989). This occurs through (1) active habitat selection, by individual movement along environmental gradients toward the optimal microenvironments for a species, and (2) passive habitat selection, in which local populations grow in areas to which they are well adapted and decline in areas where they are poorly adapted. Paleontological observations confirm that species often changed their geographic range in the past in response to glacial cycles while maintaining a relatively constant phenotype except for body-size evolution (Coope 1979; Smith et al. 1995).

Habitat destruction and fragmentation restrict dispersal and reduce or eliminate the ability of species to alter their geographic range (Peters and Lovejoy 1992). Species restricted to isolated habitat fragments and reserves can rely only on their limited physiological tolerances, or on evolutionary adaptation in situ, to survive rapid global warming and other environmental challenges in the coming centuries. The persistence of many species will depend increasingly on maintaining sufficient genetic variance for adaptive evolution.

Finite population size produces stochastic changes in gene frequencies known as random genetic drift, due to Mendelian segregation and variance in family size, which tends to reduce genetic variation. In the absence of natural selection, a fraction of $1/(2N_e)$ of the genetic variance (either heterozygosity or the heritable component of variance in quantitative characters) is expected to be lost from a population per generation, where N_e is the effective population size. The effective size of wild populations generally is substantially less than the actual size because of variance in family size, unequal sex ratio, and temporal fluctuations in population size (Wright 1969; Waples, chap. 8 in this volume). The ratio of effective to actual size of wild populations is often on the order of 0.1 to 0.2 (Frankham 1995; Waples, chap. 8 in this volume). To be expected to lose a large fraction of its genetic variance, a population reduced to a small N_e must remain small for at least $2N_e$ generations (Wright 1969). The genetic effects of such a population bottleneck are similar to those produced by frequent local extinction and colonization, which can reduce N_e of a metapopulation orders of magnitude below its actual size (Wright 1940; Maruyama and Kimura 1980; Hedrick 1996).

Based on estimates of mutability in quantitative characters (Lande 1976; Lynch 1988), Franklin (1980) and Soulé (1980) recommended a minimum N_e of 500 to maintain typical levels of heritable variance. Recent experiments indicate that a large fraction of the mutational variance in quantitative characters is associated with recessive lethal and semi-lethal side effects such that the quasi-neutral, potentially adaptive fraction of mutational variance is about one-tenth as large as previously thought (Mackay et al. 1992; Lopez and Lopez-Fanjul 1993a,b). Lande (1995) suggested that the Franklin-Soulé number should be increased by a factor of ten (but see Franklin and Frankham 1998). Much larger populations may be required to maintain rare alleles with major effects on disease resistance (Lande 1983; Roush and McKenzie 1987). In a relatively constant environment, however, there may be little need for adaptive evolution. Several examples exist of populations or species that recovered after reduction to small numbers, such as the northern elephant seal (*Mirounga angustirostris*; Hoelzel et al. 1993) and American bison (*Bison bison*; Miller 1990, 38–39). Following a population bottleneck and recovery to large population size, genetic variance can be replenished by spontaneous mutation, which occurs more rapidly for quantitative characters than for single-locus molecular polymorphisms (Lande 1976, 1995).

Much of adaptive evolution is based on quantitative (continuously varying) polygenic characters of morphology, behavior, and physiology. Quantitative characters usually are subject to stabilizing natural selection toward an intermediate optimum phenotype that may fluctuate with time, with phenotypes that deviate from the optimum having reduced fitness. Heritable variance in quantitative characters, therefore, imposes a fitness decrement or "genetic load" on a population, which like deleterious mutation is an inevitable cost of maintaining adaptive potential (Crow and Kimura 1970; Lande and Shannon 1996). Thus, there is an optimal level of genetic variance for maintaining both current fitness and future adaptability. When environmental change is partially predictable (i.e., when the optimal phenotype undergoes prolonged directional change, long-period high-amplitude cycles, or substantial temporal autocorrelation), then genetic variance in quantitative characters increases mean fitness and promotes population persistence (Lande and Shannon 1996). Even for very large populations, however, there is a maximum rate of directional or random environmental change to which a population can adapt without becoming extinct, depending on the amount of genetic variability maintained (Lynch and Lande 1993; Bürger and Lynch 1995; Gomulkiewicz and Holt 1995; Lande and Shannon 1996).

Inbreeding Depression and Fixation of New Mildly Deleterious Mutations

Matings between related individuals tend to reduce offspring viability and fertility due to the homozygous expression of (partially) recessive deleterious mutations, which is known as inbreeding depression in fitness. In historically large, outcrossing populations, a 10% increase in the inbreeding coefficient typically reduces fitness by a few to several percent. This applies for domesticated species as well as experimental populations of fruit flies (*Drosophila melanogaster*) and other species recently isolated from the wild (Falconer and Mackay 1996). Data on captive populations of many wild animal species suggest similar values (Ralls and Ballou 1983). Continued brother-sister mating in domesticated animals generally results in extinction of a high proportion of lines within five or ten generations (Soulé 1980). Substantial heterogeneity exists among species and populations in the magnitude of inbreeding depression (Soulé 1980; Lacy et al. 1993; Husband and Schemske 1996).

The genetic basis of inbreeding depression is best understood in *Drosophila* species, with roughly equal contributions from nearly recessive lethal mutations and from more nearly additive mildly deleterious mutations (Simmons and Crow 1977). Both types of mutations arise at thousands of loci throughout the genome in eukaryotic species (Simmons and Crow 1977). Gradual inbreeding allows natural selection to purge recessive lethal mutations from a population as they become expressed in homozygotes, but inbreeding has little or no effect on the efficiency of selection against nearly additive or additive, mildly deleterious mutations (Lande and Schemske 1984; Charlesworth and Charlesworth 1987). For populations with extremely high inbreeding depression, such as some tree species and gynodioecious plants, it may be difficult for close inbreeding to purge recessive lethals because, if nearly all the inbred offspring die before reproduction, the population is effectively outcrossed (Lande et al. 1994).

Sudden reduction to very small population size generally produces substantial inbreeding depression, unless the population quickly grows to a large size that allows natural selection to reverse the short-term effects of inbreeding and random genetic drift (Keller et al. 1994). The more gradual the reduction in population size, the greater the opportunity for purging recessive lethal mutations and avoiding a large part of the inbreeding depression. Therefore, inbreeding depression is not simply proportional to the standard inbreeding coefficient for selectively neutral genes, as was assumed in recent models of the interaction of stochastic demography with inbreeding (e.g., Mills and Smouse 1994). The rule suggested by Franklin (1980) and Soulé (1980), based on extensive data from animal and plant breeding, is that most inbreeding depression can be avoided in populations with $N_e > 50$. However, inbreeding depression may be more severe in natural than in artificial environments (Jiménez et al. 1994), and more severe in stressful than in optimal environments (Keller et al. 1994; Bijlsma et al. 1997).

Saccheri et al. (1998) showed that in a butterfly metapopulation the rate of extinction of local populations consisting of only a few families was more closely correlated with local heterozygosity than with local population size. This, combined with previous experiments demonstrating a very high inbreeding depression in the species (Saccheri et al. 1996), was used to suggest that genetics was of greater importance than demography in contributing to local extinctions. Such analyses should be viewed with caution, because heterozygosity may be a better indicator of effective population size and the tendency for local population fluctuation over time scales of $2N_e$ generations than direct observations of recent population sizes. However, in this particular case the conclusion is likely to be valid, as the mean persistence time of local populations is only a few years (I. Hanski, personal communication).

Inbreeding depression due to fixation of deleterious partially recessive mutations can be reversed, at least temporarily, by introducing genes from unrelated individuals into an inbred population, which allows natural selection to eliminate the deleterious mutations. It can be permanently prevented by continued immigration every one or two generations of a single unrelated individual into each local breeding population regardless of its size (Lande and Barrowclough 1987). Such a plan was recently implemented for the endangered Florida panther, motivated by strong circumstantial evidence of inbreeding depression and its low genetic divergence from other conspecific populations. Such genetic augmentation may be sufficient to reverse inbreeding effects and not too high to swamp possible local adaptations (Hedrick 1995). Nevertheless, the population still faces demographic threats from small size caused by past habitat destruction.

In contrast to recessive lethal mutations that generally are restricted to low frequencies by natural selection, random genetic drift can fix mildly deleterious mutations in a small population. Weakly selected genes become effectively neutral if the magnitude of selection on them is much less than $1/(2N_e)$ (Wright 1969). In the long run, nearly neutral mutations, with selection coefficients close to $1/(2N_e)$, do the most damage to a population, because strongly selected mutations rarely become fixed and selectively neutral mutations are easily fixed by random genetic drift but have no impact on fitness (Lande 1994; Lynch et al. 1995a,b). The total genomic rate of mildly deleterious mutations is on the order of one per generation in a variety of organisms. Such mutations reduce fitness on average by a few to several percent and are only partially dominant (nearly additive). After fixation of enough mildly deleterious mutations, the population becomes genetically inviable ($\bar{r} \leq 0$), and extinction rapidly ensues. For a population at carrying capacity in a constant environment with no demographic stochasticity, the mean time to reach genetic inviability from fixation of new deleterious mutations is (asymptotically) proportional to K^{1+1/c^2}, where K is the carrying capacity and c is the coefficient of variation of selection against new mutations (Lande 1994, 1995). Realistic distributions of selection on mildly deleterious mutations have a value for c on the order of one, as for an exponential distribution (Keightley 1994).

For populations with initially high mean fitness, even after reduction to very small numbers, hundreds of generations must elapse before fixation of new mildly deleterious mutations causes extinction (Lande 1994). In large populations, advantageous, compensatory, and reverse mutations can completely prevent the erosion of fitness from fixation of deleterious mutations (Lande 1994, 1998b; Schultz and Lynch 1997). Extinction from fixation of new deleterious mutations is, therefore, a serious concern within the typical 100-year time scale of conservation planning only for small populations with initially low mean fitness. For populations of moderate size, however, with N_e up to a few hundred or a few thousand, fixation of new mutations could substantially erode their mean fitness and decrease their long-term viability (Lande 1995, 1998b).

TIME FRAMES FOR PVAs

The acceptable level of extinction risk in terms of the time frame and the corresponding probability of extinction is ultimately a social (Shaffer 1981) or practical (Goodman, chap. 21 in this volume; Ludwig and Walters, chap. 24 in this volume), rather than scientific, decision. For legal and political reasons, the smallest extinction risk usually considered in classifying endangered species is a 10% chance of extinction within 100 years corresponding to the "vulnerable" category of the Red Lists of the World Conservation Union (IUCN 1994). If all species of conservation concern were managed to this minimum viability level, then within the next millennium a fraction of $1 - 0.9^{10}$, or about 65% of them, would likely go extinct. This exceeds by a factor of thousands the normal background rate of extinction for abundant species that appear in the fossil record, which typically persist for millions of years (Van Valen 1973; Jablonski 1986).

Stochastic fluctuation in population size creates a probability distribution of extinction times. Theory indicates that for populations with a positive long-run growth rate below carrying capacity, the distribution of extinction times has a lag period until a quasi-stationary distribution of population sizes is established, after which there is a relatively long phase with a nearly constant rate of extinction. Thus, after some lag, the distribution of extinction times is nearly exponential (Nobile et al. 1985; Goodman 1987; Ludwig 1996). If the initial population size is near carrying capacity, the lag period has a relatively low extinction rate. If the initial population size is far below carrying capacity, however, the lag period may begin with a short interval of low extinction rate followed by a burst of high extinction rate. Even populations with a negative long-run growth rate generally take some time to become extinct. Such lags are especially problematic for species with long generations if the standard population viability criteria of a 10% chance of extinction within 100 years is applied blindly. For example, Sukumar and Santiapillai (1993) suggested that a population of 30 Asian elephants (*Elephas maximus*) with demographic parameters producing a negative long-run growth rate should be considered viable because it satisfied these standard criteria. However, 100 years is only a few elephant generations. Using the same demographic parameters, extending the time frame to 200 years or longer beyond the lag period, it was shown that the cumulative probability of extinction is rather high (Armbruster et al. 1999). Thus, PVAs should always consider a range of time frames, including some much longer than those dictated by political and legal considerations. Although serious statistical difficulties exist in making long-term projections (Ludwig 1996, 1999; Fieberg and Ellner 2000), uncertainties are likely to accumulate more slowly with time for species with long generations than for species with short generations.

If we are to have any lasting effect in reducing the ongoing mass extinction that is expected to rival the effects of the major asteroid impact 65 million years ago, or a full-scale nuclear war, conservation plans must encompass longer time frames and lower probabilities of extinction. Conservationists should increasingly be concerned not only with threatened and endangered species and establishment of reserves, but also with maintaining and restoring abundant, healthy populations and functional ecosystems in the matrix between reserves. The proliferation of threatened and endangered species makes it increasingly obvious that effective conservation and restoration plans must be done on a landscape and ecosystem level (Scott et al. 1993; Kiester et al. 1996). In addition to monitoring ecosystem function and species diversity, large-scale conservation and restoration plans should always incorporate PVA of ecologically important, sensitive, or indicator species to confirm and monitor the efficacy of large-scale plans. PVA should therefore remain an important endeavor as long as conservation biologists exist.

LITERATURE CITED

Allee, W. C., A. E. Emerson, O. Park, T. Park, and K. P. Schmidt. 1949. *Principles of animal ecology*. Saunders, Philadelphia, Pennsylvania.

Armbruster, P., P. Fernando, and R. Lande. 1999. Time frames for population viability analysis of species with long generations: an example with Asian elephants. *Animal Conservation* 2:69–73.

Bascompte, J., and R. V. Sole. 1996. Habitat fragmentation and extinction thresholds in spatially explicit models. *Journal of Animal Ecology* 65:465–473.

Beissinger, S. R. 1995. Modeling extinction in periodic environments: Everglades water levels and snail kite population viability. *Ecological Applications* 5:618–631.

Bijlsma, R., J. Bundgaard, A. C. Boerema, and W. F. van Putten. 1997. Genetic and environmental stress, and the persistence of populations. Pages 193–207 in R. Bijlsma and V. Loeschcke, editors, *Environmental stress, adaptation, and evolution*. Birkhäuser Verlag, Basel, Switzerland.

Brown, J. H., and A. Kodric-Brown. 1977. Turnover rates in insular biogeography: effect of immigration on extinction. *Ecology* 58:445–449.

Bürger, R., and M. Lynch. 1995. Evolution and extinction in a changing environment: a quantitative-genetic analysis. *Evolution* 49:151–163.

Caughley, G. 1994. Directions in conservation biology. *Journal of Animal Ecology* 63:215–244.

Caughley, G., and A. Gunn. 1996. *Conservation biology in theory and practice*. Blackwell Scientific, Cambridge, Massachusetts.

Charlesworth, D., and B. Charlesworth. 1987. Inbreeding depression and its evolutionary consequences. *Annual Review of Ecology and Systematics* 18:237–268.

Coope, G. R. 1979. Late Cenozoic fossil Coleoptera: evolution, biogeography, and ecology. *Annual Review of Ecology and Systematics* 10:247–267.

Courchamp, F., T. Clutton-Brock, and G. Grenfell. 1999. Inverse density dependence and the Allee effect. *Trends in Ecology and Evolution* 14:405–409.

Crow, J. F., and M. Kimura. 1970. *An introduction to population genetics theory*. Harper and Row, New York, New York.

Doak, D. 1989. Spotted owls and old growth logging in the Pacific Northwest. *Conservation Biology* 3:389–396.

Engen, S., Ø. Bakke, and A. Islam. 1998. Demographic and environmental stochasticity: concepts and definitions. *Biometrics* 54:840–846.

Erb, J. D., and M. S. Boyce. 1999. Distribution of population declines in large mammals. *Conservation Biology* 13:199–201.

Falconer, D. S., and T. F. C. Mackay. 1996. *Introduction to quantitative genetics*. 4th edition. Longman, London, United Kingdom.

Fieberg, J., and S. P. Ellner. 2000. When is it meaningful to estimate an extinction probability? *Ecology* 81:2040–2047.

Foley, P. 1994. Predicting extinction times from environmental stochasticity and carrying capacity. *Conservation Biology* 8:124–137.

————. 1997. Extinction models for local populations. Pages 215–246 in I. Hanski and M. E. Gilpin, editors, *Metapopulation biology*. Academic Press, London, United Kingdom.

Frankham, R. 1995. Effective population size/adult population size ratios in wildlife: a review. *Genetical Research* 66:95–107.

Franklin, I. R. 1980. Evolutionary change in small populations. Pages 135–149 in M. E. Soulé and B. A. Wilcox, editors, *Conservation biology: an evolutionary-ecological perspective*. Sinauer Associates, Sunderland, Massachusetts.

Franklin, I. R., and R. Frankham. 1998. How large must populations be to retain evolutionary potential? *Animal Conservation* 1:69–70.

Gilpin, M. E. 1990. Extinction of finite metapopulations in correlated environments. Pages 177–186 in B. Shorroks and I. R. Swingland, editors, *Living in a patchy environment*. Oxford University Press, Oxford, United Kingdom.

Gomulkiewicz, R., and R. D. Holt. 1995. When does evolution by natural selection prevent extinction? *Evolution* 49:201–207.

Goodman, D. 1987. The demography of chance extinction. Pages 11–34 in M. E. Soulé, editor, *Viable populations for conservation*. Cambridge University Press, Cambridge, United Kingdom.

Groombridge, B., editor. 1992. *Global biodiversity: status of the earth's living resources*. Chapman and Hall, London, United Kingdom.

Halley, J. M. 1996. Ecology, evolution, and $1/f$–noise. *Trends in Ecology and Evolution* 11:33–37.

Hanski, I., and M. E. Gilpin, editors. 1997. *Metapopulation biology*. Academic Press, London, United Kingdom.

Hanski, I., and M. Gyllenberg. 1993. Two general metapopulation models and the core-satellite hypothesis. *American Naturalist* 142:17–41.

Hanski, I., and I. Woiwod. 1993. Spatial synchrony in the dynamics of moth and aphid populations. *Journal of Animal Ecology* 62:656–668.

Harrison, S., and J. F. Quinn. 1989. Correlated environments and the persistence of metapopulations. *Oikos* 56:293–298.

Hedrick, P. W. 1995. Gene flow and genetic restoration: the Florida panther as a case study. *Conservation Biology* 9:996–1007.

————. 1996. Bottleneck(s) or metapopulation in cheetahs. *Conservation Biology* 10:897–899.

Heino, M. 1998. Noise colour, synchrony, and extinctions in spatially structured populations. *Oikos* 83:368–375.

Heino, M., V. Kaitala, E. Ranta, and J. Lindstrom. 1997. Synchronous dynamics and rates of extinction in spatially structured populations. *Proceedings of the Royal Society of London*, series B, Biological Sciences, 264:481–486.

Hill, M. F., and H. Caswell. 1999. Habitat fragmentation and extinction thresholds on fractal landscapes. *Ecology Letters* 2:121–127.

Hoelzel, A. R., J. Halley, S. J. O'Brien, C. Campagna, T. Arnbom, B. LeBoeuf, K. Ralls, and G. A. Dover. 1993. Elephant seal genetic variation and the use of simulation models to investigate historical population bottlenecks. *Journal of Heredity* 84:443–449.

Husband, B. C., and D. W. Schemske. 1996. Evolution of the magnitude and timing of inbreeding depression in plants. *Evolution* 50:54–70.

International Union for Conservation of Nature (IUCN). 1994. *IUCN Red list categories*. IUCN, Gland, Switzerland.

Jablonski, D. 1986. Background and mass extinctions: the alternation of macroevolutionary regimes. *Science* 231:129–133.

Jiménez, J. A., K. A. Hughes, G. Alaks, L. Graham, and R. C. Lacy. 1994. An experimental study of inbreeding depression in a natural habitat. *Science* 266:271–273.

Keightley, P. D. 1994. The distribution of mutation effects on viability in *Drosophila melanogaster*. *Genetics* 138:1315–1322.

Keller, L. F., P. Arcese, J. N. M. Smith, W. M. Hochachka, and S. C. Stearns. 1994. Selection against inbred song sparrows during a natural population bottleneck. *Nature* 372:356–357.

Kendall, B. E. 1998. Estimating the magnitude of environmental stochasticity in survivorship data. *Ecological Applications* 8:184–193.

Kiester, A. R., J. M. Scott, B. Csuti, R. F. Noss, B. Butterfield, K. Sahr, and D. White. 1996. Conservation prioritization using GAP data. *Conservation Biology* 10:1332–1342.

Lacy, R. C., A. Petric, and M. Warneke. 1993. Inbreeding and outbreeding in captive populations of wild animal species. Pages 352–374 in N. W. Thornhill, editor, *The natural history of inbreeding and outbreeding: theoretical and empirical perspectives*. University of Chicago Press, Chicago, Illinois.

Lamberson, R. H., and J. Carroll. 1993. Thresholds for persistence in territorial species. Pages 55–62 in I. Barbieri, E. Grassi, G. Pallotti, and P. Pettazzoni, editors, *Topics in biomathematics: Proceedings of the 2d international conference on mathematical biology*. World Scientific Publishing, Singapore.

Lande, R. 1976. The maintenance of genetic variability by mutation in a polygenic character with linked loci. *Genetical Research* 26:221–235.

———. 1983. The response to selection on major and minor mutations affecting a metrical trait. *Heredity* 50:47–65.

———. 1987. Extinction thresholds in demographic models of territorial populations. *American Naturalist* 130:624–635.

———. 1988. Demographic models of the northern spotted owl (*Strix occidentalis caurina*). *Oecologia* 75:601–607.

———. 1993. Risks of population extinction from demographic and environmental stochasticity and random catastrophes. *American Naturalist* 142:911–927.

———. 1994. Risk of population extinction from fixation of new deleterious mutations. *Evolution* 48:1460–1469.

———. 1995. Mutation and conservation. *Conservation Biology* 9:782–791.

———. 1998a. Demographic stochasticity and Allee effect on a scale with isotropic noise. *Oikos* 83:353–358.

———. 1998b. Risk of population extinction from fixation of deleterious and reverse mutations. *Genetica* 102/103:21–27.

Lande, R., and G. F. Barrowclough. 1987. Effective population size, genetic variation, and their use in population management. Pages 87–124 in M. E. Soulé, editor, *Viable populations for conservation*. Cambridge University Press, Cambridge, United Kingdom.

Lande, R., S. Engen, and B.-E. Sæther. 1995. Optimal harvesting of fluctuating populations with a risk of extinction. *American Naturalist* 145:728–745.

————. 1998. Extinction times in finite metapopulation models with stochastic local dynamics. *Oikos* 83:383–389.

————. 1999. Spatial scale of population synchrony: environmental correlation versus dispersal and density regulation. *American Naturalist* 154:271–281.

Lande, R., and S. H. Orzack. 1988. Extinction dynamics of age-structured populations in a fluctuating environment. *Proceedings of the National Academy of Sciences* (USA) 85:7418–7421.

Lande, R., and D. W. Schemske. 1984. The evolution of self-fertilization and inbreeding depression in plants: 1, genetic models. *Evolution* 39:24–40.

Lande, R., D. W. Schemske, and S. T. Schultz. 1994. High inbreeding depression, selective interference among loci, and the threshold selfing rate for purging recessive lethal mutations. *Evolution* 48:965–978.

Lande, R., and S. Shannon. 1996. The role of genetic variability in adaptation and population persistence in a changing environment. *Evolution* 50:434–437.

Leigh, E. G., Jr. 1981. The average lifetime of a population in a varying environment. *Journal of Theoretical Biology* 90:213–239.

Levins, R. 1969. Some demographic and genetic consequences of environmental heterogeneity for biological control. *Bulletin of the Entomological Society of America* 15:237–240.

————. 1970. Extinction. Pages 77–107 in M. Gerstenhaber, editor, *Some mathematical questions in biology*. American Mathematical Society, Providence, Rhode Island.

Lindstrom, J., E. Ranta, and H. Lindén. 1996. Large-scale synchrony in the dynamics of capercaillie, black grouse, and hazel grouse populations in Finland. *Oikos* 76: 221–227.

Lopez, M. A., and C. Lopez-Fanjul. 1993a. Spontaneous mutation for a quantitative trait in *Drosophila melanogaster*: 1, response to artificial selection. *Genetical Research* 61:107–116.

————. 1993b. Spontaneous mutation for a quantitative trait in *Drosophila melanogaster*: 2, distribution of mutant effects on the trait and fitness. *Genetical Research* 61:117–126.

Ludwig, D. 1996. The distribution of population survival times. *American Naturalist* 147:506–526.

————. 1999. Is it meaningful to estimate a probability of extinction? *Ecology* 80: 298–310.

Lynch, M. 1988. The rate of polygenic mutation. *Genetical Research* 51:137–148.

Lynch, M., J. Conery, and R. Bürger. 1995a. Mutation accumulation and the extinction of small populations. *American Naturalist* 146:489–518.

————. 1995b. Mutational meltdown in sexual populations. *Evolution* 49:1067–1080.

Lynch, M., and R. Lande. 1993. Evolution and extinction in response to environmental change. Pages 234–250 in P. M. Kareiva, J. G. Kingsolver, and R. B. Huey, editors, *Biotic interactions and global change*. Sinauer Associates, Sunderland, Massachusetts.

Mackay, T. F. C., R. F. Lyman, and M. S. Jackson. 1992. Effects of P element insertion on quantitative traits in *Drosophila melanogaster*. *Genetics* 130:315–332.

Maruyama, T., and M. Kimura. 1980. Genetic variability and effective population size when local extinction and recolonization of subpopulations are frequent. *Proceedings of the National Academy of Sciences* (USA) 77:6710–6714.

McKelvey, K., B. R. Noon, and R. H. Lamberson. 1993. Conservation planning for species occupying fragmented landscapes: the case of the northern spotted owl. Pages 424–450 in P. M. Kareiva, J. G. Kingsolver, and R. B. Huey, editors, *Biotic interactions and global change*. Sinauer Associates, Sunderland, Massachusetts.

Miller, G. T., Jr. 1990. *Living in the environment*. 6th edition. Wadsworth Publishing, Belmont, California.

Mills, L. S., and P. E. Smouse. 1994. Demographic consequences of inbreeding in remnant populations. *American Naturalist* 144:412–431.

Nee, S., and R. M. May. 1992. Dynamics of metapopulations: habitat destruction and competition coexistence. *Journal of Animal Ecology* 61:37–40.

Nobile, A. G., L. M. Ricciardi, and L. Sacerdote. 1985. Exponential trends of first-passage time densities for a class of diffusion processes with steady-state distributions. *Journal of Applied Probability* 22:611–618.

Palmqvist, E., and P. Lundberg. 1998. Population extinctions in correlated environments. *Oikos* 83:359–367.

Pease, C. M., R. Lande, and J. J. Bull. 1989. A model of population growth, dispersal, and evolution in a changing environment. *Ecology* 70:1657–1664.

Peters, R. L., and T. E. Lovejoy. 1992. *Global warming and biological diversity*. Yale University Press, New Haven, Connecticut.

Pimm, S. L. 1991. *The balance of nature? ecological issues in the conservation of species and communities*. University of Chicago Press, Chicago, Illinois.

Ralls, K., and J. D. Ballou. 1983. Extinction: lessons from zoos. Pages 164–184 in C. M. Schonewald-Cox, S. M. Chambers, B. MacBryde, and W. L. Thomas, editors, *Genetics and conservation: a reference for managing wild animal and plant populations*. Benjamin/Cummings, Menlo Park, California.

Ranta, E., V. Kaitala, J. Lindstrom, and E. Helle. 1997a. The Moran effect and synchrony in population dynamics. *Oikos* 78:136–142.

Ranta, E., V. Kaitala, J. Lindstrom, and H. Lindén. 1995. Synchrony in population dynamics. *Proceedings of the Royal Society of London*, series B, Biologicial Sciences, 262:113–118.

Ranta, E., V. Kaitala, and P. Lundberg. 1997b. The spatial dimension in population fluctuations. *Science* 278:1621–1623.

Ratner, S., R. Lande, and B. B. Roper. 1997. Population viability analysis of spring chinook salmon in the South Umpqua River, Oregon. *Conservation Biology* 11:879–889.

Ripa, J., and P. Lundberg. 1996. Noise colour and the risk of population extinctions. *Proceedings of the Royal Society of London*, series B, Biological Sciences, 263:1751–1753.

Roush, R. T., and J. A. McKenzie. 1987. Ecological genetics of insecticide and acaricide resistance. *Annual Review of Entomology* 32:361–380.

Saccheri, I. J., P. M. Brakefield, and R. A. Nichols. 1996. Severe inbreeding depression and rapid fitness rebound in the butterfly *Bicyclus anynana* (Satyridae). *Evolution* 50:2000–2013.

Saccheri, I., M. Kuussaari, M. Kankare, P. Vikman, W. Fortelius, and I. Hanski.

1998. Inbreeding and extinction in a butterfly metapopulation. *Nature* 392:491–494.

Sæther, B.-E., S. Engen, A. Islam, R. McCleery, and C. Perrins. 1998. Environmental stochasticity and extinction risk in a population of a small songbird, the great tit. *American Naturalist* 151:441–450.

Sæther, B.-E., S. Engen, R. Lande, J. M. N. Smith, and P. Arcese. 2000. Estimating time to extinction in an island population of song sparrows. *Proceedings of the Royal Society of London,* series B, Biological Sciences, 267:621–626.

Schultz, S. T., and M. Lynch. 1997. Mutation and extinction: the role of variable mutational effects, synergistic epistasis, beneficial mutations, and degree of outcrossing. *Evolution* 51:1363–1371.

Scott, J. M., F. Davis, B. Csuti, R. Noss, B. Butterfield, C. Groves, H. Anderson, S. Caicco, F. Derchia, T. C. Edwards, J. Ulliman, and R. G. Wright. 1993. GAP analysis: a geographic approach to protection of biological diversity. *Wildlife Monographs* 123:1–41.

Shaffer, M. L. 1981. Minimum population sizes for species conservation. *BioScience* 31:131–134.

———. 1987. Minimum viable populations: coping with uncertainty. Pages 69–86 in M. E. Soulé, editor, *Viable populations for conservation.* Cambridge University Press, Cambridge, United Kingdom.

Simmons, M. J., and J. F. Crow. 1977. Mutations affecting fitness in *Drosophila* populations. *Annual Review of Genetics* 11:49–78.

Smith, F. A., J. L. Betancourt, and J. H. Brown. 1995. Evolution of body size in the woodrat over the past 25,000 years of climate change. *Science* 270:2012–2014.

Soulé, M. E. 1980. Thresholds for survival: maintaining fitness and evolutionary potential. Pages 151–169 in M. E. Soulé and B. A. Wilcox, editors, *Conservation biology: an evolutionary-ecological perspective.* Sinauer Associates, Sunderland, Massachusetts.

Sukumar, R., and C. Santiapillai. 1993. Asian elephant in Sumatra: population and habitat viability analysis. *Gajah* 11:59–63.

Sutcliffe, O. L., C. D. Thomas, and D. Moss. 1996. Spatial synchrony and asynchrony in butterfly population dynamics. *Journal of Animal Ecology* 65:85–95.

Tuljapurkar, S. D. 1982. Population dynamics in variable environments: 3, evolutionary dynamics of r-selection. *Theoretical Population Biology* 21:141–165.

Turelli, M. 1977. Random environments and stochastic calculus. *Theoretical Population Biology* 12:140–178.

Van Valen, L. 1973. A new evolutionary law. *Evolutionary Theory* 1:1–30.

With, K. A., and A. W. King. 1999. Extinction thresholds for species in fractal landscapes. *Conservation Biology* 13:324–326.

Wright, S. 1940. Breeding structure of populations in relation to speciation. *American Naturalist* 74:232–248.

———. 1969. *Genetics and the evolution of populations.* Vol. 2, *The theory of gene frequencies.* University of Chicago Press, Chicago, Illinois.

Young, T. P. 1994. Natural die-offs of large mammals: implications for conservation. *Conservation Biology* 8:410–418.

Reconciling the Small-Population and Declining-Population Paradigms

Mark S. Boyce

Ecology is not serving management well, and adaptive management is not being accepted and applied in practice.
—K. Rogers

ABSTRACT

In a pointed essay in the *Journal of Animal Ecology* shortly before his death, Graeme Caughley (1994) suggested that conservation biology was plagued by two tracks: (1) stochastic models for small populations that have few direct conservation applications, and (2) empirical investigations of declining populations that tend to be case-specific, yielding few generalizations. In the five years since Caughley's essay, several population viability analyses (PVAs) have demonstrated that conservation biologists are capable of merging theory and practice. Habitat-based PVAs for California gnatcatchers, grizzly bears, and spotted owls have combined GIS-based spatial analysis with stochastic population models to develop useful links between habitats and extinction risk. Despite progress, many of Caughley's concerns still plague conservation biology, and PVAs in particular. As Caughley pointed out, genetic models are still insufficiently developed and verified so that they cannot form more than a general basis for conservation management. Theoretical population models are poorly tied to applications because we face major hurdles in obtaining enough data for fitting models of sufficient complexity to capture the relevant ecology. The very nature of threatened and endangered species means that seldom do we have sufficient sample sizes or replication to reliably estimate parameters for complex models. Simple models present fewer problems of statistical estimation, but may poorly represent the complex dynamics of actual populations. Although adaptive management offers promise for linking science and conservation practice, institutional barriers have proven to be serious impediments. Based on a conviction that science can enhance conservation, we must give urgent priority to Caughley's charge that conservation biologists need to reconcile ecological theory and conservation practice.

INTRODUCTION

During the five years since its publication, Caughley's provocative essay "Directions in Conservation Biology" (1994) has been the most frequently cited paper in the conservation biology literature. According to *Science Citation Index*, from 1995 to 1998 the paper was cited 143 times. This was a higher rate of citation than any other paper that I could find in the conservation biology literature during those years, including relevant-content papers by M. Gilpin, R. Lande, R. Levins, M. Shaffer, D. Simberloff, and M. Soulé.

Caughley (1994) has been cited frequently because he so perceptively identified a fundamental problem with the conservation biology discipline. He claimed that conservation biology is dominated by two paradigms, which he characterized as the *small-population paradigm* and the *declining-population paradigm*. He suggested that the small-population paradigm includes theoretical models of stochastically fluctuating populations that yield a probability of extinction. The declining-population paradigm, on the other hand, involves detailed field studies that seek to identify the causes for population declines and empirical threats of extinction.

The reason that this dichotomy presents a problem for the discipline is that resolution of conservation issues requires an integration of both approaches. We cannot rely on theoretical models that may have no relevance to any particular population, but we gain little fundamental understanding of how best to approach conservation if each situation is different. Progress in the ability of conservation biologists to reduce extinction risks will occur only when we learn how to integrate the two paradigms.

Hedrick et al. (1996) criticized Caughley (1994) for inventing a false dichotomy and suggested that such a "divisive separation of approaches" could pit conservation biologists against each other. I disagree with Hedrick et al. (1996) and believe that Caughley's provocative essay has encouraged integration and the melding of the two paradigms. Indeed, this book and the conference from which it emerged were a celebration of the reconciliation between the small-population and declining-population paradigms (Mann and Plummer 1999). I would argue that there is value in recognizing weaknesses in conservation biology so that we can identify approaches that most effectively will achieve our common objective of conserving biological resources.

DICHOTOMIES IN CONSERVATION BIOLOGY

Hedrick et al. (1996) characterized Caughley's dichotomy as whether stochastic or deterministic factors were primarily responsible for extinction. But Caughley's two paradigms are multifaceted and capture much more of the structure of the science than is implied by this characterization. To clarify, I will suggest several additional dichotomies that follow a similar vein and are possibly subsumed by what Caughley intended.

Ecological Theory versus Conservation Practice

Early efforts in conservation biology were motivated by attempts to apply ecological theory to conservation, such as the theory of island biogeography (Wilson and Willis 1975). This was poorly received by many field-oriented people in the discipline of conservation. More generally, the basic theory of ecology appears to have few direct applications (Sarkar 1996). For a variety of reasons, field personnel have not been trained in theoretical ecology, and, likewise, many theoreticians have not been trained in field techniques. This dichotomy appears to reflect in part the capability and interest of scientists in the discipline. Many of my colleagues have noted that we have a difficult time finding graduate students who are skilled in quantitative science and theory as well as field skills and techniques. This dichotomy is gradually fading as we provide students in field ecology with stronger theoretical backgrounds, and as theoreticians have become more involved in real-world conservation problems.

Conservation Biology versus Wildlife Conservation

When conservation biology first emerged as a discipline about 15 years ago, there was considerable resistance by some wildlife biologists (Teer 1988). The sense was that the long-established discipline of wildlife ecology was well equipped to accommodate the conservation issues addressed by conservation biologists. However, academic ecologists were often paying no attention to conservation issues. The lack of theory in wildlife ecology motivated some of us toward the new discipline. A desire to see ecological theory applied to conservation motivated others. Conservation biologists tend to be more academic and more theoretically oriented than wildlife ecologists. The "old-school" wildlifers more often are comfortable with Caughley's declining-population paradigm and draw their legacy from Theodore Roosevelt and William T. Hornaday (Reiger 1986). In contrast, conservation biologists seem more likely to affiliate with Caughley's small-population paradigm, seeking inspiration from Robert MacArthur and Richard Levins.

Complexity versus Simplicity

Theoretical ecologists working under Caughley's small-population paradigm appear to have a passion for the complexity and interconnectedness of nature. Mathematics of nonlinear dynamics in ecology (Schaffer 1988) and debates on the role of complexity in stability (May 1973) often have developed independently of data. The other camp (declining populations) finds little use for models that are not motivated or supported by data. Statistical models developed for the analysis of ecological data rely on Occam's razor (Otis et al. 1978) or the principle of parsimony (Burnham and Anderson 1998). Clearly there must be middle ground. Theoretical ecology contains the ideas that help us understand nature. But the theory typically involves models with so many parameters that estimating all of them is virtually impossible. In contrast, simple statistical models may be tractable but cannot reasonably characterize the true complexity of ecological processes. Effective conservation modeling involves balancing complexity with statistical rigor (Starfield 1997).

Genetic Models versus Environmental Stochasticity

One of the most provocative remarks in Caughley's review (1994) noted that we cannot identify a single extinction that can be attributed to genetic malfunction. To balance this claim, I might point out that we do not have evidence to show that extinctions were not due to genetic malfunctions. Fundamentally, our interest in population viability analysis (PVA) is motivated by a desire to ensure the perpetuation of genetic diversity. Yet we are often discouraged at how difficult it can be to find theoretical or empirical support for conservation from population genetics (Ewens 1990).

Recent work has suggested an example where genetics may have contributed to the near extinction of greater prairie chickens (*Tympanuchus cupido*) in Illinois (Westemeier et al. 1998). Even in this case, however, alternative hypotheses, such as senescence, might have accounted for low productivity and viability. Other applications of genetics in conservation are reviewed by Hedrick et al. (1996) and Allendorf and Ryman (chap. 4 in this volume). Some genetic mechanisms, such as inbreeding depression, are well documented. Other mechanisms, such as mutational load (Lande 1995), are more controversial with little empirical basis (Gilligan et al. 1997). Genetic factors probably are not a driver of extinction until populations become small and isolated. Then genetics can make a bad demographic situation much worse.

There can be little question that genetics forms a foundation for conservation biology (Avise and Hamrick 1996; Landweber and Dobson 1999). Yet, in most applications, loss of habitat or environmental stochasticity is more likely to be the domain of conservation practice (Lande 1988b). This view may be due to deficiencies in our understanding of the genetic structure and dynamics of natural populations. Clearly conservation genetics is one of the hottest topics in conservation biology today.

Demography versus Habitat Ecology

Many PVAs entail age- or stage-structured models involving a projection matrix such as the Leslie matrix (Burnham and Anderson 1998). Indeed, demographic structure underlies each of the canned programs that are available for conducting PVAs, including VORTEX, RAMAS, ALEX, GAPPS, and INMAT (Boyce 1996; Mills et al. 1996; Brook et al. 2000). Seldom do we have sufficient data to do a good job of parameterizing such detailed models, however, and estimates of parameters are burdened with large sampling variance (Boyce 1992; Beissinger and Westphal 1998). In practice, such detailed models may not be necessary, especially if the primary factor driving the population is loss or alteration of habitat, as is the case in 85% of the species currently listed under the Endangered Species Act in the United States (Wilcove et al. 1998). Much more important than demographic structure is the incorporation of habitat variables into PVA models (Boyce et al. 1994).

Ultimately, habitats influence vital rates such that a fully parameterized population model would include demography as well as habitat. Having sufficient data to model the influence of habitat variables on vital rates, however, is truly exceptional. Yet we usually can expect to have sufficient data to link probability of occurrence or population density to habitats (Manly et al. 1993). An understanding of the ecology of crucial habitat can form the basis for a PVA with no information about the demographic structure of the population (Foin and Brenchley-Jackson 1991). The importance of habitat-based PVA for the northern spotted owl (*Strix occidentalis caurina;* Boyce et al. 1994) was reinforced by a sensitivity analysis that revealed that habitat variables had the greatest influence on long-term viability (Akçakaya and Raphael 1998).

Studies that Caughley (1994) would have classified under the small-population paradigm seldom have confronted details of habitat ecology. Instead they have focused on stochastic demography (Nations and Boyce 1997). Frequently, the case studies that Caughley (1994) would have classified under the declining-population paradigm have involved habitat ecology. Generally, habitat ecology is a discipline largely devoid of theory (Boyce and McDonald 1999), although I believe that the development of resource-selection functions (Manly et al. 1993) has the potential to help resolve this deficiency. This area in ecology holds great promise for contributing to the development of defensible PVAs.

MELDING THE PARADIGMS

I believe, like Hedrick et al. (1996), that we have made substantial progress in recent years in breaking down the dichotomy between the small-versus declining-population paradigms. But I would have chosen a different list of examples than those selected by Hedrick et al. For example, they tout research on the northern spotted owl as illustrating how metapopulation models gave insight into the best way to manage for persistence of spotted owls (Lande 1987, 1988a). Empirical evidence, however, contradicts this view and suggests that the metapopulation models were strictly a theoretical concept applied inappropriately to management of landscapes in the Pacific Northwest (Boyce 1994). The real issue was one of habitat loss (Boyce et al. 1994; Akçakaya and Raphael 1998; Meyer et al. 1998), not fragmentation of habitats (Fahrig 1997). Indeed, I would argue that this example highlights the risks of using theoretical models that have not been verified with field data.

Nevertheless, we have gotten much better at building PVA models that go beyond the matrix-projection approach and draw on relevant habitat data (Lindenmayer et al. 1993; Boyce et al. 1994). New models have made it easier to link extinction risk to habitats (Foley 1994). RAMAS GIS and ALEX are software packages for conducting PVA that explicitly accommodate habitat information. Furthermore, attempts to validate PVA computer software have found reasonable consistency among models and reliability in predictions (Belovsky et al. 1999; Belovsky et al., chap. 13 in this volume; Brook et al. 2000; Reed and Bryant 2000). Granted, PVA has serious weaknesses (Boyce 1992; Caughley and Gunn 1996; Beissinger and Westphal 1998; Ludwig 1999; Ludwig and Walters, chap. 24 in this volume), but it is probably the most rigorous conservation-planning tool available. Continued research and development of PVA will only enhance our ability to manage rare species in the future.

I have argued elsewhere (Boyce 1997) that PVA ought to be the basis for active adaptive management (Walters and Holling 1990). PVA forms the basis for hypotheses of how the system works and allows predictions of population response to management perturbations. Following management interventions and fieldwork to monitor the response, we evaluate new data that can be used to update or modify our PVA model to gradually increase our understanding of the system. My argument was shaken by Walters's review (1997) showing how infrequently adaptive management has been successful at large-scale ecosystem management. Attempts to implement adaptive management have failed because (1) experimental policies were viewed as too costly or risky, (2) researchers showed deplorable self-interest by viewing adaptive management as a threat, and (3) agency personnel viewed adaptive management as a threat to their management programs. Furthermore, many resource managers are not trained in hypothesis testing, monitoring is not taken seriously, and money and personnel are in short supply. Note that these are primarily social failings. Science and resource management cannot save species unless society wants them saved. The logic of adaptive management remains functional and prudent.

Developing administrative structures that would allow adaptive management to take place is beyond my expertise. But clearly, good science will conserve little if we cannot incorporate science into practice. As Michael Soulé has emphasized in the foreword to this book, learning how PVA fits into the political, societal, and economic milieu will be crucial to the effectiveness of conservation biology in the future.

LITERATURE CITED

Akçakaya, H. R., and M. G. Raphael. 1998. Assessing human impact despite uncertainty: viability of the northern spotted owl metapopulation in the northwestern USA. *Biodiversity and Conservation* 7:875–894.

Avise, J. C., and J. L. Hamrick. 1996. *Conservation genetics.* Chapman and Hall, New York, New York.

Beissinger, S. R., and M. I. Westphal. 1998. On the use of demographic models of population viability in endangered species management. *Journal of Wildlife Management* 62:821–841.

Belovsky, G. E., C. Mellison, C. Larson, and P. A. Van Zandt. 1999. Experimental studies of extinction dynamics. *Science* 286:1175–1177.

Boyce, M. S. 1992. Population viability analysis. *Annual Review of Ecology and Systematics* 23:481–506.

———. 1994. Population viability analysis exemplified by models for the northern spotted owl. Pages 3–18 in D. J. Fletcher and B. F. J. Manly, editors, *Statistics in ecology and environmental monitoring.* University of Otago Press, Dunedin, New Zealand.

———. 1996. Review of RAMAS/GIS. *Quarterly Review of Biology* 71:167–168.

———. 1997. Population viability analysis: adaptive management for threatened and endangered species. Pages 226–236 in M. S. Boyce and A. Haney, editors, *Ecosystem management: applications for sustainable forest and wildlife resources.* Yale University Press, New Haven, Connecticut.

Boyce, M. S., and L. L. McDonald. 1999. Relating populations to habitats using resource selection functions. *Trends in Ecology and Evolution* 14:268–272.

Boyce, M. S., J. S. Meyer, and L. L. Irwin. 1994. Habitat-based PVA for the northern spotted owl. Pages 63–85 in D. J. Fletcher and B. F. J. Manly, editors, *Statistics in ecology and environmental monitoring.* University of Otago Press, Dunedin, New Zealand.

Brook, B. W., J. J. O'Grady, A. P. Chapman, M. A. Burgman, H. R. Akçakaya, and R. Frankham. 2000. Predictive accuracy of population viability analysis in conservation biology. *Nature* 404:385–387.

Burnham, K. P., and D. R. Anderson. 1998. *Model selection and inference: a practical information-theoretic approach.* Springer-Verlag, New York, New York.

Caughley, G. 1994. Directions in conservation biology. *Journal of Animal Ecology* 63:215–244.

Caughley, G., and A. Gunn. 1996. *Conservation biology in theory and practice.* Blackwell Scientific, Cambridge, Massachusetts.

Ewens, W. J. 1990. The minimum viable population size as a genetic and demographic concept. Pages 307–316 in J. Adams, D. A. Lam, A. I. Hermalin, and P. E. Smouse, editors, *Convergent issues in genetics and demography.* Oxford University Press, Oxford, United Kingdom.

Fahrig, L. 1997. Relative effects of habitat loss and fragmentation on population extinction. *Journal of Wildlife Management* 61:603–610.

Foin, T. C., and J. L. Brenchley-Jackson. 1991. Simulation model evaluation of potential recovery of endangered light-footed clapper rail populations. *Biological Conservation* 58:123–148.

Foley, P. 1994. Predicting extinction times from environmental stochasticity and carrying capacity. *Conservation Biology* 8:124–137.

Gilligan, D. M., L. M. Woodworth, M. E. Montgomery, D. A. Briscoe, and R. Frankham. 1997. Is mutation accumulation a threat to the survival of endangered populations? *Conservation Biology* 11:1235–1241.

Hedrick, P. W., R. C. Lacy, F. W. Allendorf, and M. E. Soulé. 1996. Directions in conservation biology: comments on Caughley. *Conservation Biology* 10:1312–1320.

Lande, R. 1987. Extinction thresholds in demographic models of territorial populations. *American Naturalist* 130:624–635.

———. 1988a. Demographic models of the northern spotted owl (*Strix occidentalis caurina*). *Oecologia* 75:601–607.

———. 1988b. Genetics and demography in biological conservation. *Science* 241:1455–1460.

———. 1995. Mutation and conservation. *Conservation Biology* 9:782–791.

Landweber, L. F., and A. P. Dobson. 1999. *Genetics and the extinction of species.* Princeton University Press, Princeton, New Jersey.

Lindenmayer, D. B., T. W. Clark, R. C. Lacy, and V. C. Thomas. 1993. Population viability analysis as a tool in wildlife conservation policy: with reference to Australia. *Environmental Management* 17:745–758.

Ludwig, D. 1999. Is it meaningful to estimate a probability of extinction? *Ecology* 80:298–310.

Manly, B. F. J., L. L. McDonald, and D. L. Thomas. 1993. *Resource selection by animals.* Chapman and Hall, London, United Kingdom.

Mann, C. C., and M. L. Plummer. 1999. A species' fate, by the numbers. *Science* 284:36–37.

May, R. M. 1973. *Stability and complexity in model ecosystems.* Princeton University Press, Princeton, New Jersey.

Meyer, J. S., L. L. Irwin, and M. S. Boyce. 1998. Influence of habitat abundance and fragmentation on northern spotted owls in western Oregon. *Wildlife Monographs* 139:1–51.

Mills, L. S., S. G. Hayes, C. Baldwin, M. J. Wisdom, J. Citta, D. J. Mattson, and K. Murphy. 1996. Factors leading to different viability predictions for a grizzly bear data set. *Conservation Biology* 10:863–873.

Nations, C., and M. S. Boyce. 1997. Stochastic demography for conservation biology. Pages 451–469 in S. Tuljapurkar and H. Caswell, editors, *Structured population models in marine, terrestrial, and freshwater systems.* Chapman and Hall, New York, New York.

Otis, D. L., K. P. Burnham, G. C. White, and D. R. Anderson. 1978. Statistical inference from capture data on closed animal populations. *Wildlife Monographs* 62:1–135.

Reed, D. H., and E. H. Bryant. 2000. Experimental tests of minimum viable population size. *Animal Conservation* 3:7–14.

Reiger, J. F. 1986. *American sportsmen and the origins of conservation.* University of Oklahoma Press, Norman, Oklahoma.

Rogers, K. 1998. Comments on Walters. *Conservation Ecology* 2 (1). http://www.consecol.org/Journal/.

Sarkar, S. 1996. Ecological theory and anuran declines. *BioScience* 46:199–207.

Schaffer, W. M. 1988. Perceiving order in the chaos of Nature. Pages 313–350 in M. S. Boyce, editor, *Evolution of life histories of mammals.* Yale University Press, New Haven, Connecticut.

Starfield, A. M. 1997. A pragmatic approach to modeling for wildlife management. *Journal of Wildlife Management* 61:261–270.

Teer, J. G. 1988. Review of *Conservation biology: the science of scarcity and diversity,* edited by M. E. Soulé. *Journal of Wildlife Management* 52:570–572.

Walters, C. J. 1997. Challenges in adaptive management of riparian and coastal ecosystems. *Conservation Ecology* 1 (2). http://www.consecol.org/Journal/.

Walters, C. J., and C. S. Holling. 1990. Large-scale management experiments and learning by doing. *Ecology* 71:2060–2068.

Westemeier, R. L., J. D. Brawn, S. A. Simpson, T. L. Esker, R. W. Jansen, J. W. Walk, E. L. Kershner, J. L. Bouzat, and K. N. Paige. 1998. Tracking the long-term decline and recovery of an isolated population. *Science* 282:1695–1698.

Wilcove, D., D. Rothstein, J. Dubow, A. Phillips, and E. Losos. 1998. Quantifying threats to imperiled species in the United States. *BioScience* 48:607–615.

Wilson, E. O., and E. Willis. 1975. Applied biogeography. Pages 522–534 in M. L. Cody and J. M. Diamond, editors, *Ecology and evolution of communities.* Belknap Press, Cambridge, Massachusetts.

The Role of Genetics in Population Viability Analysis

Fred W. Allendorf and Nils Ryman

As some of our British parks are ancient, it occurred to me that there must
have been long-continued close interbreeding with the fallow-deer (*Cervus
dama*) kept in them; but on inquiry I find that it is a common practice to
infuse new blood by procuring bucks from other parks.
—Charles Darwin

ABSTRACT

The importance of the loss of genetic variation in decreasing the probability of persistence of isolated populations has been controversial in
spite of abundant evidence for the detrimental effects of inbreeding on
fitness-related characters. However, several recent studies have provided direct empirical evidence for the influence of genetics on population decline and recovery. Many population viability models that have
included genetics may have underestimated the effects of inbreeding
for two reasons. First, they have used estimates of inbreeding depression
derived from captive populations, and inbreeding depression is likely to
be more severe in the wild. Second, effects of inbreeding depression
usually have been incorporated in only one aspect of life history (usually
juvenile survival), whereas inbreeding depression can affect many other
attributes as well (e.g., litter size, adult survival, etc.). In addition, the
loss of phenotypic variation in small populations may reduce population
viability in the absence of inbreeding depression.

The incorporation of genetics into population viability analysis (PVA)
is problematic for many species because of unreliable estimates of de-

We thank Steve Beissinger and Dale McCullough for inviting us to contribute to this volume
and for their helpful comments on this manuscript. We are grateful to an anonymous reviewer,
J. Ballou, L. Laikre, R. Lande, G. Luikart, S. Mills, M. Schwartz, and D. Tallmon for their
helpful comments and suggestions; to M. Schwartz for allowing us to use his results in figure
4.2; and to the many authors who provided us with advance copies of their work (M. Antolin,
M. Conner, S. Daniels, P. Hedrick, J. Hogg, H. Jager, L. Keller, J. Oostermeijer, and G. White).
FWA thanks the Nature Conservancy of Montana for allowing him to stay at the Hanusa Cabin
while working on this paper. NR acknowledges support from the Swedish research program on
sustainable coastal zone management (SUCOZOMA) funded by the Foundation for Strategic
Environmental Research (MISTRA).

mographic effects and genetic parameters related to inbreeding depression. Perhaps the best way to incorporate inbreeding depression is to use a range of values likely to span how inbreeding depression will affect absolute or relative population viability. It is important that the effects of inbreeding depression be examined on many vital rates, not just juvenile survival. Some authors have suggested that it may not be important to include genetic concerns in PVA because any management option that minimizes the probability of extinction is also likely to minimize the effects of inbreeding. However, there is no way to know whether this assertion is true without testing it by incorporating genetics into the models.

Another approach is to use genetic considerations to set a lower limit to population size. We recommend retaining at least 95% of the heterozygosity in a population over 100 years. The population size required will depend on the ratio of effective population size to total population size and the average generation time. These parameters typically are easier to estimate than the effects of inbreeding depression on the vital rates needed for a PVA model. This process could be completed in conjunction with, rather than as a replacement for, PVA. The population size required to meet this genetic criterion should not be considered a goal, but rather a lower limit below which genetic considerations are likely to reduce the probability of population persistence.

Consideration of genetic effects over time frames beyond 100 years is also important for the long-term viability of populations and species. Recent considerations of this problem have led to the recommendation that an effective population size of approximately 1,000 individuals is needed to allow continued adaptive evolution and to avoid the accumulation of new harmful mutations. This recommendation would correspond to more than 5,000 individuals in many species. Such large populations will not be possible in many species except by increasing the connectivity among geographically separated populations over a wide area.

INTRODUCTION

The quote from Darwin (1896, 99) shows that both evolutionary biologists and wildlife managers have recognized for over 100 years that small isolated populations may be harmed by inbreeding. However, Darwin also was concerned that the harmful effects of inbreeding were not generally accepted: "That any evil directly follows from the closest interbreeding has been denied by many persons; but rarely by any practical breeder; and never, as far as I know, by one who has largely bred animals which propagate their kind quickly" (94). The potential harmful effects of inbreeding and the importance of genetics in the persistence of populations remains controversial to this day (Soulé and Mills 1992; Frankham and Ralls 1998; Mann and Plummer 1999).

There are a variety of reasons for this controversy. Some have suggested that inbreeding is unlikely to have significant harmful effects on individual fitness in wild populations. Others have suggested that inbreeding may affect individual fitness, but is not likely to affect population viability (Caro and Laurenson 1994). Still others have argued that genetic concerns can be ignored when estimating the viability of small populations because they are in much greater danger of extinction due to stochastic demographic effects (Lande 1988; Pimm et al. 1988). Finally, some have suggested that it may be best not to incorporate genetics into population viability analysis (PVA) because genetic and demographic "currencies" are difficult to combine, and we have insufficient information about the effects of inbreeding in most wild populations (Beissinger and Westphal 1998).

The disagreement over whether genetics should be considered in demographic predictions of population persistence has been unfortunate and misleading. Extinction is a demographic process that will be influenced by genetic effects under some circumstances. The key issue is to determine under what conditions genetic concerns are likely to influence population persistence (Nunney and Campbell 1993). There have been important recent advances in our understanding of the interaction between demography and genetics that will improve the effectiveness of our attempts to conserve endangered species (e.g., Landweber and Dobson 1999).

Perhaps most importantly, we need to recognize when management recommendations based upon strictly demographics or genetics may actually be in conflict with each other. For example, Ryman and Laikre (1991) have considered supportive breeding in which a portion of wild parents are brought into captivity for reproduction and their offspring are released back into the natural habitat where they mix with wild conspecifics. Programs like this are carried out in a number of species to increase population size and thereby temper stochastic demographic effects. Under some circumstances, however, supportive breeding may reduce effective population size and cause a reduction in genic heterozygosity that may have harmful effects on the population (Ryman and Laikre 1991).

The primary causes of species extinction today are deterministic and result from human-caused habitat loss, habitat modification, and overexploitation (Caughley 1994; Lande 1999; Lacy and Miller, chap. 23 in this volume). Reduced genetic diversity in plants and animals is generally a symptom of endangerment rather than its cause (Holsinger et al. 1999). Nevertheless, genetic effects of small populations have an important role to play in the management of many threatened species. For example, Ellstrand and Elam (1993) examined the population sizes of 743 sensitive plant taxa in California. Over 50% of the occurrences contained less than 100 individuals. In general, those populations that are the object of PVA are often small and therefore are likely to be susceptible to the genetic effects of small populations.

Genetics also plays an important and noncontroversial role in understanding the historical and current reproductive relationships among individuals and populations. Many PVA models assume a single randomly mating population or consider a metapopulation in which there is exchange among local populations. In practice, however, it is often not clear if the individuals of a particular region represent one or more reproductively isolated populations with potentially different demographic characteristics. Most species consist of multiple populations exhibiting a more or less pronounced degree of divergence with respect to historical background and life-history characteristics. Examination of the distribution of molecular genetic variation over the distribution of a species can assist in identifying what geographic units should be considered separate demographic populations (Haig 1998; Hedrick, chap. 17 in this volume).

In this chapter, we review the genetic principles underlying this controversy and provide a synthesis of the role of genetics in PVA. We first review mechanisms by which the loss of genetic variation may reduce the probability of persistence of populations. We then provide an overview of the incorporation of genetics into PVA models. Finally, we consider how genetic considerations may change the concept of what we mean by a viable population. Simply persisting over a specified period of time is not a good definition of "viable" if the demographic and genetic characteristics of the population have changed over the time period so that the population is unlikely to persist much longer.

GENETIC EFFECTS OF SMALL POPULATION SIZE

Does the loss of genetic variation in small populations decrease the likelihood that such populations will persist into the future? The classic work of Sewall Wright (1931) demonstrated that genetic drift in small populations will have two major genetic effects: changes in allele frequencies and loss of genetic variation. There are a variety of reasons why we would expect the genetic effects of small population size to reduce the persistence probability of populations (table 4.1).

Table 4.1 Overview of Genetic Effects Decreasing the Probability of Persistence of Small Populations

Source	Relative Importance	Time Frame
Inbreeding depression	High	Immediate
Mitochondrial mutations	Unknown	Immediate
Loss of phenotypic variation	Medium	Immediate
Loss of evolutionary potential	Medium	Medium/long
Mutational meltdown	Low	Long

Inbreeding Depression

Inbreeding depression is the reduction in fitness of individuals resulting from matings between related individuals. "Inbred" individuals have increased homozygosity and decreased heterozygosity over their entire genome. The pedigree inbreeding coefficient (F) is the expected increase in homozygosity for inbred individuals; it is also the expected decrease in heterozygosity throughout the genome.

Inbreeding can occur in both large and small populations. In large populations, inbreeding may occur by nonrandom mating because of self-fertilization or by a tendency for related individuals to mate with each other. For example, nearby individuals tend to be related and are likely to mate with each other because of geographic proximity. However, inbreeding will occur even in randomly mating small populations simply because all or most individuals within a small population will be related. This has been called the inbreeding effect of small populations (Crow and Kimura 1970).

Inbreeding depression may result from either increased homozygosity or reduced heterozygosity (Crow 1948). That is, a greater number of deleterious recessive alleles will be expressed in inbred individuals because of their increased homozygosity. In addition, fitness of inbred individuals will be reduced at loci at which the heterozygotes have a selective advantage over all homozygous types (heterozygous advantage or overdominance). Both of these mechanisms are likely to contribute to inbreeding depression, but it is thought that increased expression of deleterious recessive alleles is the more important mechanism (Charlesworth and Charlesworth 1987).

Three conditions must hold for inbreeding depression to reduce the viability of populations. (1) Inbreeding must occur. (2) Inbreeding depression must occur. (3) The traits affected by inbreeding depression must reduce population viability. Conditions 1 and 2 will hold to some extent in all small populations. As discussed below, matings between relatives must occur in small populations, and some deleterious recessive alleles will be present in all populations. However, condition 3 is the crux of the controversy. There is little empirical evidence that tells us when inbreeding depression will affect population viability and how important that effect will be.

Inbreeding will occur in small populations. The effective population size (N_e) has been defined by Wright (1969) as whatever has to be substituted in the basic formula

(1) $$\Delta F = \frac{1}{2N_e}$$

to describe the actual increase in homozygosity per generation in a population. N_e is generally much smaller than the actual or "census" population size (N). Frankham (1995a) concluded in a review of available data that the effective population size is usually on the order of 10% of the actual adult population size for many wild populations. However, Waples (chap. 8 in this volume) concluded that Frankham (1995a) overestimated the contribution of temporal changes to this effect by computing the N_e/N ratio as a harmonic mean divided by an arithmetic mean. The other empirical estimates of N_e that do not include the effect of temporal changes (Frankham 1995a) suggest that 20% of the adult population size is a better general value to use for N_e. Therefore, in a typical population of 50 adults, the average inbreeding coefficient (F) of individuals is expected to increase by approximately 5% per generation.

Some inbreeding depression is expected in all species (Hedrick and Kalinowski 2000). Deleterious recessive alleles are present in the genome of all species because they are continually introduced by mutation, and natural selection is inefficient in removing them because most copies are "hidden" in heterozygotes that do not have reduced fitness. We therefore would expect all species to show some inbreeding depression due to the increase in homozygosity of recessive deleterious alleles. The effects of inbreeding depression on survival is often measured by the mean number of "lethal equivalents" per gamete (per haploid genome). One lethal equivalent may be a single allele that is lethal when homozygous, two alleles each with a probability of 0.5 of causing death when homozygous, or ten alleles each with a probability of 0.10 of causing death when homozygous.

Most studies of inbreeding depression have been made in captivity or under controlled conditions, but experiments with both plants (e.g., Dudash 1990) and animals (e.g., Jiménez et al. 1994) found that inbreeding depression is more severe in natural environments. Estimates of inbreeding depression in captivity may be severe underestimates of the true effect of inbreeding in the wild. Crnokrak and Roff (1999) recently reviewed the available empirical literature of inbreeding depression for wild species. They concluded that in general "the cost of inbreeding under natural conditions is much higher than under captive conditions" (262). They tested this for mammals by comparing traits directly related to survival in wild mammals to the findings of Ralls et al. (1988) for captive species. They found that the cost of inbreeding for survival was much higher in wild mammals than in captive.

In addition, there is evidence that inbreeding depression is more severe under environmental stress and challenge events (extreme weather, pollution, or disease; Bijlsma and Loeschcke 1997; Bijlsma et al. 1999; O'Brien 2000). Bijlsma et al. (1997) found a synergistic interaction between stress and inbreeding with laboratory *Drosophila*, so that the effect of environmental stress was greatly enhanced with greater inbreeding. These conditions may occur only occasionally, and thus it will be difficult to measure their effect on inbreeding depression. For example, Coltman et al. (1999) found that individual Soay sheep (*Ovis aries*) that were more heterozygous at microsatellite loci had greater survival rates in harsh winters, apparently due to greater resistance to nematode parasites. This effect disappeared when the sheep were treated with antihelminthics.

Some authors have suggested that some species or populations are unaffected by inbreeding (e.g., Shields 1993). However, lack of statistical evidence for inbreeding depression does not demonstrate the absence of inbreeding depression. This is especially true because of the low power to detect even a substantial effect of inbreeding in many studies because of small sample sizes and confounding factors (Kalinowski and Hedrick 1999). In a recent comprehensive review of the evidence for inbreeding depression in mammals, Lacy (1997, 331) was "unable to find statistically defensible evidence showing that any mammalian species is unaffected by inbreeding."

PURGING. Some have suggested that small populations may be "purged" of deleterious recessive alleles by natural selection (Templeton and Read 1984). Deleterious recessive alleles may reach substantial frequencies in large random-mating populations because most copies are present in heterozygotes and are therefore not affected by natural selection. For example, over 5% of the individuals in a population in Hardy-Weinberg proportions will be heterozygous for an allele that is homozygous in only one out of 1,000 individuals. Such alleles will be exposed to natural selection in inbred or small populations and thereby be reduced in frequency or eliminated. Thus, populations with a history of inbreeding because of nonrandom mating (e.g., selfing) or small N_e (e.g., a population bottleneck) may be less affected by inbreeding depression because of the purging of deleterious recessive alleles.

The lack of a difference in fitness between inbred and noninbred individuals within a population that has gone through a bottleneck is not evidence for purging. Many nonlethal deleterious alleles may become fixed in such populations. The fixation of these alleles will cause a reduction in fitness of all individuals after the bottleneck, relative to the individuals in the population before the bottleneck. However, inbreeding depression will appear to be reduced following fixation of deleterious alleles because of depressed outcrossed fitness rather than increased inbred fitness (Byers and Waller 1999). To test for purging, the fitness of individuals in the postbottleneck population must be compared to the fitness of individuals in the prebottleneck population.

A recent review found little evidence for purging in plant populations (Byers and Waller 1999). Only 38% of the 52 studies included found evidence of purging. And when purging was found, it removed only a small proportion of the total inbreeding depression (roughly 10%). These authors concluded that "purging appears neither consistent nor effective enough to reliably reduce inbreeding depression in small and inbred populations" (505).

A recent review of evidence for purging in animals came to similar conclusions. Ballou (1997) found evidence for a slight decline in inbreeding depression in neonatal survival among descendants of inbred animals in a comparison of 17 captive mammal species. However, he found no indication for purging in weaning survival or litter size in these species. He concluded that the purging detected in these species is not likely to be strong enough to be of practical use in eliminating inbreeding depression. In addition, inbreeding depression can be substantial in Hymenoptera in which males are haploid, so that deleterious recessive alleles are exposed to natural selection every generation (Antolin 1999).

Failure of purging to decrease inbreeding depression can be explained by several mechanisms. Purging is expected to be most effective in the case of lethal or semilethal recessive alleles (Lande and Schemske 1985; Hedrick 1994); however, when inbreeding depression is very high (more than ten lethal equivalents), even lethals may not be purged except under very close inbreeding (Lande 1994). The lack of evidence for purging is consistent with the hypothesis that inbreeding depression is caused by many recessive alleles with minor deleterious effects. Husband and Schemske (1996) found that inbreeding depression for survival after early development and reproduction and growth was similar in selfing and nonselfing plant species. They suggested that this inbreeding depression is due primarily to mildly deleterious mutations that are not purged over long periods of time. Willis (1999) found that most inbreeding depression in the wildflower *Mimulus guttatus* is due to alleles with small effect, and not to lethal or sterile alleles. Bijlsma et al. (1999) found that purging in experimental populations of *Drosophila* is effective only in the environment in which the purging occurred, because additional deleterious alleles were expressed under changing environmental conditions.

Ballou (1997) suggested that associative overdominance also may be instrumental in maintaining inbreeding depression. Associative overdominance occurs when heterozygous advantage or deleterious recessive alleles at a selected locus result in apparent heterozygous advantage at linked loci (Ohta 1971; Pamilo and Pålsson 1998). Kärkkäinen et al. (1999) recently provided evidence that most of the inbreeding depression in the self-incompatible herb *Arabis petraea* is due to overdominance or associative overdominance.

Inbreeding depression due to heterozygous advantage cannot be purged. However, it is unlikely that heterozygous advantage is a major mechanism for inbreeding depression (Charlesworth and Charlesworth 1987). Nevertheless, there is recent strong evidence for heterozygous advantage at the major histocompatibility complex (MHC) in humans. Black and Hedrick (1997) found evidence for strong heterozygous advantage (nearly 50%) at both *HLA-A* and *HLA-B* in South Amerindians. Carrington et al. (1999) found that heterozygosity at *HLA-A*, *-B*, and *-C* loci was associated with extended survival of patients infected with the human immunodeficiency virus. Strong evidence for the selective maintenance of MHC diversity in vertebrate species comes from other approaches as well (see references in Carrington et al. 1999).

Speke's gazelle (*Gazella spekei*) has been cited as an example of the effectiveness of purging in reducing inbreeding depression in captivity (Templeton and Read 1984). Willis and Wiese (1997), however, concluded that this apparent purging may have been due to the data analysis rather than purging itself; this interpretation has been disputed by Templeton and Read (1998). Ballou (1997), in his reanalysis of the Speke's gazelle data, found that purging effects were minimal and nonsignificant. Perhaps most importantly, he found that the inbreeding effects in Speke's gazelle were the greatest in the 17 mammal species he examined. Kalinowski et al. (2000) recently concluded that the apparent purging in Speke's gazelle is the result of a temporal change in fitness and not a reduction in inbreeding depression.

INBREEDING AND POPULATION VIABILITY. For inbreeding depression to affect population viability it must affect traits that influence population viability. For example, Leberg (1990) found that eastern mosquito fish (*Gambusia holbrooki*) populations founded by two siblings had a slower growth rate than populations founded by two unrelated founders. However, it has been difficult to isolate genetic effects in the web of interactions that affect viability in wild populations (Soulé and Mills 1998). Laikre (1999) noted that many factors interact when a population is driven to extinction, and it is generally impossible to single out "the" cause.

Some authors have asserted that there is no evidence for genetics affecting population viability (Caro and Laurenson 1994, 485): "Although inbreeding results in demonstrable costs in captive and wild situations, it has yet to be shown that inbreeding depression has caused any wild population to decline. Similarly, although loss of heterozygosity has detrimental impact on individual fitness, no population has gone extinct as a result." This observation has prompted several recent papers that tested for evidence of the importance of genetics in population declines and extinction.

Newman and Pilson (1997) founded a number of small populations of *Clarkia pulchella* by planting individuals in a natural environment. All populations were founded by the same number of individuals; however, in some populations the founders were related and in some they were unrelated. All populations were demographically equivalent (N) but differed in the effective population size (N_e) of the founding population. A significantly greater proportion of the populations founded by unrelated individuals persisted.

Saccheri et al. (1998) found that extinction risk of local populations of the Glanville fritillary butterfly (*Melitaea cinxia*) increased significantly with decreasing heterozygosity at seven allozyme loci and one microsatellite locus after accounting for the effects of environmental factors. Larval survival, adult longevity, and hatching rates of eggs all were reduced by inbreeding, and were thought to be the fitness components responsible for the relationship between heterozygosity and extinction.

Westemeier et al. (1998) monitored greater prairie chickens (*Tympanuchus cupido pinnatus*) for 35 years and found that egg fertility and hatching rates of eggs declined in Illinois populations after these birds became isolated from adjacent populations during the 1970s. These same characteristics did not decline in adjacent populations that remained large and widespread. These results suggested that the decline of birds in Illinois was at least partially due to inbreeding depression. This conclusion was supported by the observation that fertility and hatching success recovered following translocations of birds from the large adjacent populations.

Madsen et al. (1999) studied an isolated population of adders (*Vipera berus*) in Sweden that declined dramatically some 35 years ago and has since suffered from severe inbreeding depression. The introduction of 20 males from a large and genetically variable population of adders resulted in a dramatic demographic recovery of this population. This recovery was brought about by increased survival rates, even though the number of litters produced by females per year actually declined during the initial phase of recovery.

Some have argued that the existence of species and populations that have survived bottlenecks is evidence that inbreeding is not necessarily harmful (Simberloff 1988; Caro and Laurenson 1994). However, we need to know how many similar populations went extinct following such bottlenecks to interpret the significance of such observations. For example, creation of inbred lines of mice usually results in the loss of many of the lines (Bowman and Falconer 1960; Lynch 1977). This argument is similar to using the existence of 80-year-old smokers as evidence that cigarette smoking is not harmful. Only populations that have survived a bottleneck can be observed after the fact. Soulé (1987) has termed this the "fallacy of the accident."

Nevertheless, this does not mean that populations that have lost substantial genetic variation because of a bottleneck are somehow doomed or are not capable of recovery. That is, increase in frequency of some deleterious alleles and loss of genomewide heterozygosity is inevitable following a bottleneck. However, the magnitude of these effects on fitness-related traits (survival, fertility, etc.) may not be large enough to constrain recovery. For example, the tule elk (*Cervus elaphus nannodes*) of the Central Valley of California has gone through a series of bottlenecks since the 1849 gold rush (McCullough et al. 1996). Simulation analysis was used to estimate that tule elk have lost approximately 60% of their original heterozygosity (McCullough et al. 1996). Analysis of allozymes (Kucera 1991) and microsatellites (D. R. McCullough, personal communication) has confirmed relatively low genetic variation in tule elk. Nevertheless, the tule elk has shown a remarkable capacity for population growth, and today there are 22 herds totaling over 3,000 animals (McCullough et al. 1996). Tule elk still may be affected by the genetic effects of the bottleneck in the future if they face some sort of a challenge event (e.g., disease).

Loss of Phenotypic Variation

Inbreeding depression is not necessary for the loss of genetic variation to affect population viability. Reduction in variability itself, even without a reduction in individual fitness, may reduce population viability (e.g., Conner and White 1999). Individual differences in life history (age at first sexual maturity, clutch size, etc.) that have at least a partial genetic basis occur in populations of plants and animals. Many of these differences may have little effect on individual fitness because of a balance or trade-off between advantages and disadvantages. Nevertheless, the loss of this life-history variability among individuals may reduce the likelihood of persistence of a population.

For example, Pacific salmon return to fresh water from the ocean to spawn and die after spawning (Groot and Margolis 1991). In most species, there are individual differences in age at reproduction that often have a substantial genetic basis (Hankin et al. 1993). Chinook salmon (*Oncorhynchus tshawytscha*) usually become sexually mature at age three, four, or five. The greater fecundity of older females (because of their greater body size) is balanced by their lower probability of survival to maturity. These different life-history types have similar fitnesses. Pink salmon (*Oncorhynchus gorbuscha*) are exceptional in that all individuals become sexually mature and return from the ocean to spawn in fresh water at two years of age (Heard 1991). Therefore, pink salmon within a particular stream comprise separate odd- and even-year populations that are reproductively isolated (Aspinwall 1974).

Consider a hypothetical comparison of two streams for purposes of illustration. The first stream has separate odd- and even-year populations, as is typical for pink salmon. In the second stream, there is phenotypic (and genetic) variation for the time of sexual maturity so that approximately 25% of the fish become sexually mature at age 1 and 25% of the fish become sexually mature at age 3; the remaining 50% of the population become mature at age 2.

All else being equal, we would expect the population with variability in age of return to persist longer than the two reproductively isolated populations. The effective population size (N_e) of the odd- and even-populations would be one-half the N_e of the single reproductive population with life-history variability (Waples 1990). Thus, inbreeding depression would accumulate twice as rapidly in the two reproductively isolated populations as in the single variable population. The two smaller populations also would each be more susceptible to extinction from demographic, environmental, and catastrophic stochasticity. For example, a catastrophe that resulted in complete reproductive failure for one year would cause the extinction of one of the populations without variability. The occurrence of separate genders or mating types is another case where the loss of phenotypic variation can cause a reduction in population viability without a reduction in the fitness of inbred individuals. Approximately 50% of flowering plant species have genetic incompatibility mechanisms (Nettancourt 1977). In one of these self-incompatibility systems, individuals are of different mating types that possess different genotypes at a self-incompatibility (S) locus (Richards 1986). Pollen grains can fertilize only plants that do not have the same S allele as carried by the pollen. Homozygotes cannot be produced at this locus, and the minimum number of alleles at this locus in a sexually reproducing population is three. Smaller populations are expected to maintain many fewer S alleles than larger populations at equilibrium (Wright 1960).

Les et al. (1991) considered the demographic importance of maintaining a large number of S alleles in plant populations. A reduction in the number of S alleles because of a population bottleneck will reduce the frequency of compatible matings and may result in reduced levels of seed set. Demauro (1993) reported that the last Illinois population of the lakeside daisy (*Hymenoxys acaulis* var. *glabra*) was effectively extinct even though it consisted of approximately 30 individuals, because all plants apparently belonged to the same mating type. Reinartz and Les (1994) concluded that some one-third of the remaining 14 natural populations of *Aster furcatus* in Wisconsin had reduced seed sets because of a diminished number of S alleles.

Loss of Evolutionary Potential
The loss in genetic variation caused by a population bottleneck may cause a reduction in a population's ability to respond by natural selection to future environmental changes. Bürger and Lynch (1995) predicted, on the basis of theoretical considerations, that small populations (N_e less than 1,000) are more likely to go extinct due to environmental change because they are less able to adapt than are large populations. The ability of a population to evolve is affected both by heterozygosity and the number of alleles present. Heterozygosity is relatively insensitive to bottlenecks in comparison to allelic diversity (Allendorf 1986). Heterozygosity is proportional to the amount of genetic variance at loci affecting quantitative variation (James 1971). Thus, heterozygosity is a good predictor of the potential of a population to evolve immediately following a bottleneck. Nevertheless, the long-term response of a population to selection is determined by the allelic diversity either remaining after the bottleneck or introduced by new mutations (Robertson 1960; James 1971).

The effect of small population size on allelic diversity is especially important at loci associated with disease resistance. Small populations are vulnerable to extinction by epidemics, and loci associated with disease resistance often have an exceptionally large number of alleles. For example, Gibbs (1991) described 37 alleles at the MHC in a sample of 77 adult blackbirds (*Agelaius phoeniceus*). Allelic variability at MHC is thought to be especially important for disease resistance (Slade 1992; Edwards and Potts 1996; Black and Hedrick 1997). For example, Paterson et al. (1998) found that certain microsatellite alleles within the MHC of Soay sheep are associated with parasite resistance and greater survival.

Accumulation of Harmful Mutations

Wright (1931) first suggested that small populations would continue to decline in vigor slowly over time because of the accumulation of deleterious mutations that natural selection would not be effective in removing because of the overpowering effects of genetic drift. Recent papers have considered the expected rate and importance of this effect for population persistence (Lynch and Gabriel 1990; Gabriel and Bürger 1994; Lande 1995). As deleterious mutations accumulate, population size may decrease further and thereby accelerate the rate of accumulation of deleterious mutations. This feedback process has been termed "mutational meltdown."

Lande (1995, 789) concluded that the risk of extinction through this process "may be comparable in importance to environmental stochasticity and could substantially decrease the long-term viability of populations with effective sizes as a large as a few thousand." The expected time frame of this process is hundreds or thousands of generations. Experiments designed to detect empirical evidence for this effect have had mixed results (Lynch et al. 1999).

Mitochondrial Mutations

Recent results have suggested that mutations in mitochondrial DNA (mtDNA) may decrease the viability of small populations (Gemmell and Allendorf 2001). Mitochondria are generally transmitted maternally so that deleterious mutations that affect only males are not subject to natural selection. Sperm are powered by a group of mitochondria at the base of the flagellum, and even a modest reduction in power output caused by a mitochondrial mutation may reduce male fertility yet have little effect on females. A recent study of human fertility has found that mtDNA haplogroups are associated with sperm function and male fertility (Ruiz-Pesini et al. 2000). In addition, the mitochondrial genome has been found to be responsible for cytoplasmic male sterility, which is widespread in plants (Hanson 1991).

The viability of small populations may be reduced by an increase in the frequency of mtDNA genotypes that lower the fitness of males. Since females and males are haploid for mtDNA, it has not been recognized that mtDNA may contribute to the increased genetic load of small populations. The effective population size of the mitochondrial genome is generally only one-quarter that of the nuclear genome, so that mtDNA mutations are much more sensitive to genetic drift and population bottlenecks than nuclear loci. Whether an increase in mtDNA haplotypes that reduce male fertility will affect population viability depends on the mating system and reproductive biology of the particular population. However, it seems likely that reduced male fertility may decrease the number of progeny produced under a wide array of circumstances. At a minimum, the presence of mtDNA genotypes that reduce the fertility of some males would increase the variability in male reproductive success and thereby decrease effective population size. This would increase the rate of loss of heterozygosity and other effects of inbreeding depression that can reduce population viability.

GENETICS AND PVA

Most published PVAs have not included genetics. Groom and Pascual (1998) reviewed PVAs published between 1987 and 1996 in *Conservation Biology, Biological Conservation, Ecological Applications*, and the *Journal of Wildlife Management*. Only 4 of 58 PVAs (7%) included genetic effects. Not including genetic effects may result in overly optimistic predictions of population persistence in an absolute sense and lead to incorrect management recommendations. Under what circumstances is it important to include genetics in a PVA for a particular species? That is, when would we expect the inclusion of genetics in PVA to have a strong effect on predicting the probability of a population's persistence?

In general, we would expect genetics to make a difference in species in which the effective population size is small enough for a long enough time for inbreeding depression or loss of genetic and phenotypic variability to affect viability. Thus, population size and effective population size are primary considerations. In addition, since most PVAs are done for some specified time period (e.g., 100 years), genetics would have a greater effect in species with shorter generation intervals. For example, 100 years may represent hundreds of generations for some rodents, but only ten or so generations for bears.

Genetics is more of a concern in PVAs for populations that are completely reproductively isolated from other populations. Even small amounts of gene flow may be genetically effective in connecting populations that are demographically independent (e.g., Keller et al. 2001). Mills and Allendorf (1996) have suggested that one to ten migrants per generation into a population is sufficient to reduce the loss of genetic variation within a population. Allendorf et al. (1991) found in a simulation study that the introduction of one male bear per year (approximately ten per generation) resulted in a reduced accumulation of inbreeding equivalent to an increase of the mean effective population size from 31 to 194 in a demographically stable population of 100 brown bears (*Ursus arctos*).

Additional insight into the importance of genetics in PVA is provided by Mills and Smouse (1994) in their consideration of the effects of inbreeding depression on the persistence of three idealized mammal life histories: low growth rate "ungulates," medium growth rate "felids," and high growth rate "rodents." They found that both genetic and nongenetic stochastic factors are critically important to extinction probability for all three cases. They concluded that deviations in the sex ratio accelerate the effects of inbreeding on extinction rate because of the effect of sex ratio on effective population size. They also concluded that low growth rate populations were extremely vulnerable to the effects of even minor inbreeding depression.

Species with Small N_e/N

All else being equal, we would expect genetics to be a greater concern with species that have a lower ratio of effective population size to adult population size (Nunney and Campbell 1993). Frankham (1995a) in a recent review of this topic concluded that the most important factors reducing effective size are fluctuations in population size, followed by variance in reproductive success, and unequal sex ratio. Fluctuations in population size would be especially important in low growth rate populations (e.g., bears) because it would take them much longer to recover from a bottleneck than high growth rate populations (e.g., rodents; Mills and Smouse 1994).

It is also important to remember that one-half N_e is the expected rate of change due to drift *per generation*. Therefore, harmful genetic effects may accumulate more rapidly over time in species or populations with larger N_e if they have a smaller generation interval (fig. 4.1). There is often a trade-off between N_e and generation interval. That is, species with larger effective population sizes tend to have shorter generation intervals (e.g., rodents versus bears). This same trade-off also can occur between different populations of the same species. For example, increasing the average life span of individuals will often increase variance

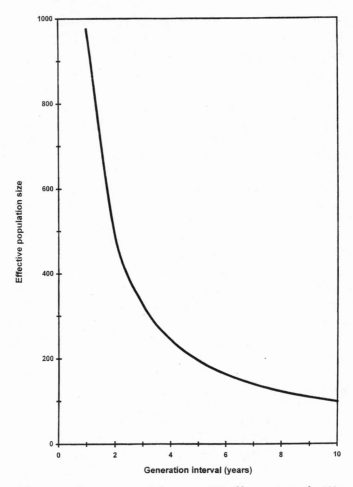

Fig. 4.1 Effective population size required to retain 95% of heterozygosity for 100 years in populations with different generation intervals.

in individual reproductive success and reduce N_e (e.g., Ryman et al. 1981).

Demographically Stable Populations

In many situations, demographic, environmental, and catastrophic uncertainty may require much larger population sizes for persistence than those based upon genetic considerations in the short or middle term (Shaffer 1987; Lande 1988; Harwood 1992). In these cases, genetic considerations may be unimportant because they will not come into play until the population is already so small that it will be unlikely to persist because of demographic stochasticity. However, small populations of some demographically stable species have been predicted to persist for long periods of time when the effects of genetics are disregarded.

Shaffer and Samson (1985) found that almost all their simulated populations beginning with 50 brown bears persisted for 100 years. Sæther et al. (1998) reported similar results for even smaller starting populations in their assessment of the viability of the Scandinavian brown bear. We would expect, however, that the vital rates of such small populations would be affected by inbreeding depression.

A population of 50 bears is likely to have an N_e of approximately 10 bears (Allendorf et al. 1991), so that inbreeding depression would accumulate rather quickly. Moreover, Sæther et al. (1998, 403) found in their PVA for Scandinavian brown bears that even "a relatively small increase in the mortality rate will strongly reduce the viability of even relatively large brown bear populations." It has been shown that inbreeding results in reduced litter size in Scandinavian brown bears (Laikre et al. 1996). Therefore, we would expect a reduction of population viability because of inbreeding depression within these small brown bear populations.

This can be seen in simulations that test the sensitivity of extinction to the amount of inbreeding depression (fig. 4.2). The life-history and demographic data for these simulations are from an analysis of brown

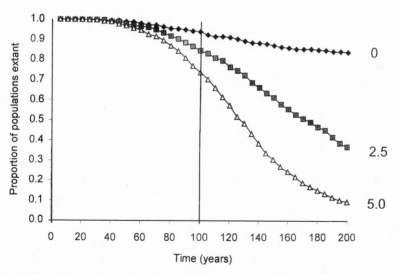

Fig. 4.2 Sensitivity analysis of the effects of different number of lethal equivalents (0, 2.5, and 5.0) on the persistence of grizzly bear populations, using VORTEX 6.2. Each point represents the proportion of 1,000 simulated populations beginning with 50 bears with a carrying capacity of 280 that did not go extinct during the specified time period ($\lambda = 1.01$).

bear recovery in the Rocky Mountains of the United States (M. K. Schwartz, personal communication). In the absence of any inbreeding depression (zero lethal equivalents), the persistence probability is similar in the first and second hundred years of the simulations. However, even moderate inbreeding depression (2.5 lethal equivalents per genome) results in a greatly decreased probability of persistence in the second hundred years.

Similar results have been reported for the tiger (*Panthera tigris*) in a spatially explicit model based on Royal Chitwan National Park, Nepal (Kenney et al. 1995). Both small (25-female territories) and large (45-female territories) populations were found to have 100% survival for 100 years. However, even a small increase in mortality caused by poaching greatly increased the probability of extinction. Moreover, poaching also nearly tripled the rate of accumulation of inbreeding in the large population. Incorporation of inbreeding depression into this model has a major effect on the probability of persistence of tiger populations (J. S. Kenney, unpublished data).

Incorporation of Genetics into PVA Models

Those PVAs that have included genetics generally incorporate inbreeding depression into only one aspect of the life cycle. For example, versions of VORTEX prior to 1999 modeled the effects of inbreeding depression only by a reduction in survival during the first year of life among inbred animals (Lacy 1993b). However, empirical evidence indicates that inbreeding depression is likely to affect many different vital rates (Lynch 1977; Ellstrand and Elam 1993; Lacy 1993a; Ballou 1997; Laikre 1999; Daniels and Walters 2000). For example, Keller (1998) found similar reductions in survival from egg to independence and from independence to first breeding associated with inbreeding (1.44 and 1.09 lethal equivalents) in a natural population of song sparrows (*Melospiza melodia*). However, inbreeding depression in one phase of the life cycle may be compensated in later stages (e.g., van Noordwijk and Scharloo 1981).

Mills and Smouse (1994) included effects of inbreeding on both survival and fecundity. They found that vulnerability to extinction is affected much more by survival depression than by fecundity depression. Mills and Smouse (1994) used a Leslie matrix projection model rather than an individual-based model such as used by VORTEX. One important advantage of the Leslie matrix approach is that it does not require as much data from the population being modeled. In some regards, however, the matrix approach is less suitable for modeling genetic effects because it does not include differences among individuals within an age or stage cohort. In a real population, individuals may have widely different inbreeding coefficients. Thus, the effect of inbreeding on population viability is more realistically incorporated into PVA by individual-based models (Letcher et al. 1998).

Some authors have suggested that genetics should not be included in PVAs until more data are available on the relationship between inbreeding depression and demographic variables (e.g., Boyce 1992; Beissinger and Westphal 1998). We agree that more empirical information on the effects of inbreeding depression on vital rates in natural populations is needed. Nevertheless, we will rarely have the data that we need for a specific population in order to include genetics in PVAs. Estimations of the effects of inbreeding depression in captivity are not likely to be accurate estimates of the effects in the wild. Furthermore, it is extremely difficult to estimate the effects of inbreeding in wild populations of animals except under experimental conditions that are not likely to be available for most populations of concern. Survival, for example, is often the most difficult vital rate to estimate in wild populations (Beissinger and Westphal 1998); it follows that it is even more difficult to estimate the effect of different inbreeding coefficients on survival. In addition, the available empirical data suggest that there may be substantial differences in the effects of inbreeding depression in different populations of the same species (Brewer et al. 1990; Lacy et al. 1996; Lacy and Ballou 1998).

It will be necessary to use values of inbreeding depression that are based upon observations from a large number of species because of the difficulties involved in obtaining reasonably adequate estimates of inbreeding depression for specific natural populations. Ralls et al. (1988) estimated the effect of inbreeding on juvenile survival in 40 populations from 38 mammalian species and found that survival declined with increasing inbreeding (fig. 4.3). However, their estimates of inbreeding depression varied greatly within and between species (e.g., see large range of values at $F = 0.25$ in fig. 4.3). Thus, Ballou (1998) found that carnivores, as a group, have less severe inbreeding depression in captivity than other mammal species. Therefore, the best approach may be to do a sensitivity analysis with a range of values that span the likely effects of inbreeding depression (e.g., fig. 4.2) to determine whether inclusion of genetic concerns affects the probabilities of extinction.

Another approach would be to assume that there is no inbreeding depression and track the expected reduction in heterozygosity or increase in mean inbreeding coefficient. If there is little loss in heterozygosity, or increase in mean inbreeding coefficient, then this would be strong testimony that genetics is not likely to affect population persistence for the time frame examined. In contrast, a substantial loss in

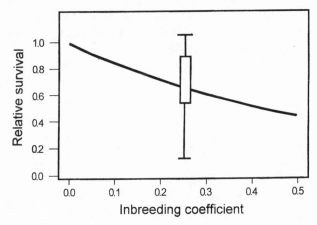

Fig. 4.3 Mean reduction in relative juvenile survival observed by Ralls et al. (1988) in 40 captive mammal populations. The box at $F = 0.25$ encompasses the middle 50% of the values; the lines at $F = 0.25$ encompass all but the highest and lowest 5% of the values.

heterozygosity (say 10% or more), or the equivalent increase in mean inbreeding coefficient, would provide evidence for the need to incorporate and consider the effects of inbreeding depression in the PVA.

Beissinger and Westphal (1998) contrasted two major uses of PVA: absolute and relative. In the former, the *absolute* probabilities of extinction resulting from PVA are used to make decisions (e.g., listing under the Endangered Species Act). Alternatively, the *relative* probabilities of extinction from PVA can be compared to help choose among alternative management actions and policies.

We share the concerns of Beissinger and Westphal (1998), who recommended against using absolute values of extinction from PVA to make policy or management decisions, because of their lack of accuracy. Nevertheless, we do not share their confidence in the "other useful criteria" that they suggest may be used alone, such as decision analysis, rules of thumb, and so forth, to make decisions, and which are also discussed by Ralls et al. (chap. 25 in this volume). We believe that it is best to use PVA in conjunction with these other criteria. Dunham et al. (1999) have considered how patterns of neutral genetic variability can be used to complement risk assessments.

Some have suggested that it may not be important to include genetic concerns in relative applications of PVA because any management option that minimizes the probability of extinction would also likely minimize the effects of inbreeding (Ralls et al., chap. 25 in this volume). However, there is no way to know whether this assertion is true unless we do incorporate genetics into the models. For example, inbreeding may be increased by management actions that increase population growth by supportive breeding or sex-biased harvest. Furthermore, the absence of accurate empirical estimates of inbreeding depression for a particular population should not preclude the inclusion of genetics into a PVA to compare relative rates of extinction.

GENETIC GOALS

The goal of many PVAs is to assess the population size necessary for a reasonably high probability of persistence over some specified period of time. For a population to be viable, however, the mere persistence of a population over the stipulated time period is not enough.

Time Scale

Many PVAs consider a probability of persistence over a period of 100 years. However, this measure may be very misleading for long-lived species (Armbruster et al. 1999). For example, the initial analysis of Shaffer and Samson (1985) found that 98% of all brown bear populations with an initial population size of 50 bears persisted for 100 years. Thus, a population of 50 bears was considered to constitute a minimum viable population (MVP) because it had a greater than 95% probability of persistence for 100 years. This estimate was changed to 50 to 90 bears based on results of sensitivity analysis. One of the early brown bear recovery plans used these results to set recovery targets for four of the six populations between 70 and 90 bears (Allendorf and Servheen 1986).

A much different picture emerged, however, when the time frame of the simulations was extended until most of the populations became extinct (Shaffer and Samson 1985). Nearly all (94%) of the populations with an initial size of 50 became extinct within 300 years. Even more surprisingly, the average extinction time for a population of 50 bears was only 114 years, and 56% of all populations became extinct in less than 114 years. Thus, application of the simple criterion of 95% probability of persistence for 100 years resulted in using an MVP at which most populations were expected to become extinct within 114 years!

This problem becomes even more serious when genetic concerns, as well as demographic, are incorporated. Populations that have persisted beyond a standard time period may have experienced a reduction in population size or genetic variation that reduces the probability of persistence over the next time period. For example, Sæther et al. (1998) recently defined a viable population of brown bears as one that has a greater than 90% chance of not going extinct (i.e., having less than two bears) during a period of 100 years. Clearly, it is imperative to use a time frame and a definition of "viable" that ensures that a population persisting over the stipulated time period is large enough and genetically variable enough to have a reasonable chance of surviving for an "adequate" number of additional generations. One way to address this problem is to use quasi-extinction risks that quantify the probability of crossing some small population threshold at which genetic, behavioral, or management-induced changes become likely to occur (Burgman et al. 1993).

Short-Term Persistence

Inbreeding depression is the most important problem with PVAs that ignore the genetic aspects of survival. Populations may be considered "viable" although they have accumulated inbreeding to an extent that they have already entered an extinction vortex (Soulé and Mills 1998) or are likely to do so in the immediate future. There is no single threshold, however, at which inbreeding is likely to reduce the viability of a population. Observations from a variety of animal species indicate that, in the early stages of the process, a 10% increase of inbreeding typically results in 5 to 10% decrease of fitness-related traits (Wright 1977; Frankel and Soulé 1981; Falconer 1989), but there is considerable variation in this value. Factors to consider include the inbreeding history of the population, the rate of inbreeding, and the outcome of random events particular to each population that determine the frequency of detrimental alleles and their mode of phenotypic effect. There are also some empirical indications that the relationship between inbreeding and extinction involves a threshold effect such that the reduction of population viability does not begin until a certain amount of inbreeding has accumulated (Frankham 1995b).

For any particular population, the ultimate evaluation of its susceptibility to inbreeding depression is necessarily a matter of trial and error, and such experimentation is generally not possible. Assessment of what can be considered a reasonably safe level of inbreeding in the future must be extrapolated, with a great degree of caution, from observations on other populations.

Soulé et al. (1986) argued that a 10% reduction of average heterozygosity ($F = 0.10$) should be considered the largest acceptable loss over 200 years. Hosack (1996) arrived at a similar conclusion suggesting a maximum of 5% loss over 100 years. By necessity, these kind of guidelines are somewhat arbitrary. Nevertheless, the genetic goal of no more than a 5 to 10% accumulation of inbreeding over 100 to 200 years seems reasonable in the context of PVA, considering both the magnitude of the observed effect of inbreeding on fitness and the lack of opportunities to predict the outcome for a particular population.

In addition to the requirement that the cumulative level of inbreeding should be kept at minimum, it is often recommended that the rate of inbreeding (ΔF) should not be allowed to accumulate too fast between any two generations in order to permit natural selection to reduce the effects of inbreeding depression. A generally accepted rule of thumb suggests that $\Delta F = 1/2N_e = 0.01$ should be considered an upper limit, which corresponds to a lower limit of $N_e = 50$ (Franklin 1980).

This recommendation of $N_e \geq 50$ is sometimes regarded as a general minimum requirement for short-term conservation. However, any such recommendation should also take into consideration the generation interval. For example, a population of $N_e = 50$ with a generation interval of two years is expected to lose nearly one-half of its initial heterozygosity in 100 years. Only a population of $N_e = 50$ with a generation interval of more than 20 years will retain more than 95% of initial heterozygosity after 100 years (fig. 4.1).

Long-Term Persistence

When considering longer periods than those of a typical PVA, avoiding the loss of genetic variation is not enough for persistence. Environmental conditions are likely to alter over time, and a viable population must be large enough to maintain sufficient genetic variation for adaptation to such changes. Evolutionary response to natural selection is generally thought to involve a gradual change of quantitative characters, and discussions on the population sizes necessary to uphold "evolutionary potential" have focused on retention of additive genetic variation of such traits.

There is current disagreement among geneticists regarding how large a population must be to maintain "normal" amounts of additive genetic variation for quantitative traits (Franklin and Frankham 1998; Lynch and Lande 1998). The suggestions for the effective sizes needed to retain evolutionary potential range from 500 to 5,000. The logic underlying these contrasting recommendations is somewhat arcane and confusing. We therefore review some of the mathematical arguments used to support the conflicting views.

Franklin (1980) was the first to make a serious attempt to provide a direct estimate of the effective size necessary for retention of additive genetic variation (V_A) of a quantitative character. He argued that, for evolutionary potential to be maintained in a small population, the loss of V_A per generation must be balanced by new variation due to mutations

(V_m). V_A will be lost at the same rate as heterozygosity ($1/2N_e$) at selectively neutral loci, so the expected loss of additive genetic variation per generation is $V_A/2N_e$. Therefore,

(2) $$\Delta V_A = V_m - \frac{V_A}{2N_e}$$

(see also Lande and Barrowclough 1987; Ewens 1990; Franklin and Frankham 1998). At equilibrium between loss and gain ΔV_A is zero, and

(3) $$N_e = \frac{V_A}{2V_m}.$$

Using abdominal bristle number in *Drosophila* as an example, Franklin (1980) also noted that $V_m \approx 10^{-3}V_E$, where V_E is the environmental variance (i.e., the variation in bristle number contributed from environmental factors). Furthermore, assuming that V_A and V_E are the only major sources of variation, the heritability (h^2, the proportion of the total phenotypic variation that is due to additive genetic effects) of this trait is $h^2 = V_A/(V_A + V_E)$, and $V_E/(V_A + V_E) = 1 - h^2$. Thus, equation 3 becomes (cf. Franklin and Frankham 1998)

(4) $$N_e = \frac{V_A}{2 \times 10^{-3}V_E} = 500 \frac{V_A}{V_E} = 500 \frac{h^2}{1 - h^2}.$$

The heritability of abdominal bristle number in *Drosophila* is about 0.5. Therefore, the approximate effective size at which loss and gain of V_A are balanced (i.e., where evolutionary potential is retained) would be 500.

Lande (1995) reviewed the recent literature on spontaneous mutation and its role in population viability. He concluded that the approximate relation between mutational input and environmental variance observed for bristle count in *Drosophila* ($V_m \approx 10^{-3}V_E$) appears to hold for a variety of quantitative traits in several animal and plant species. He also noted, however, that a large portion of new mutations seem to be detrimental, and that only about 10% are likely to be selectively neutral (or nearly neutral), contributing to the potentially adaptive additive variation of quantitative traits. Consequently, he suggested that a more appropriate value of V_m is $V_m \approx 10^{-4}V_E$, and that Franklin's (1980) estimated minimum N_e of 500 necessary for retention of evolutionary potential should be raised to 5,000.

In response, Franklin and Frankham (1998) suggest that Lande

(1995) overemphasized the effects of deleterious mutations and that the original estimate of $V_m \approx 10^{-3}V_E$ is more appropriate. They argued that empirical estimates of V_m typically have been obtained from long-term experiments where a large fraction of the harmful mutations have had the opportunity of being eliminated, such that a sizable portion of those mutations have already been accounted for. They also pointed out that, in most organisms, heritabilities of quantitative traits are typically smaller than 0.5, and that this is particularly true for fitness-related characters. As a result, the quotient $h^2/(1 - h^2)$ in equation 4 is typically expected to be considerably smaller than unity, which reduces the necessary effective size. Franklin and Frankham (1998) concluded that an N_e of the order 500 to 1,000 should be generally appropriate.

Lynch and Lande (1998) criticized the conclusions of Franklin and Frankham (1998) and argued that much larger effective sizes are justified for maintenance of long-term genetic security. They maintain that the problems with harmful mutations must be taken seriously. An important point is that a considerable fraction of new mutations are expected to be only mildly deleterious with a selective disadvantage of less than 1%. Such mildly deleterious mutations behave largely as selectively neutral ones and are not expected to be "cleansed" from the population by selective forces even at effective sizes of several hundred individuals. In the long run, the continued fixation of mildly deleterious alleles may reduce population fitness to the extent that it enters an extinction vortex (or so-called mutational meltdown; Lynch et al. 1995).

According to Lynch and Lande (1998) there are several reasons why the minimum N_e for long-term conservation should be at least 1,000. At this size, at least the *expected* (average) amount of additive genetic variation of quantitative traits is of the same magnitude as for an infinitely large population, although genetic drift may result in considerably lower levels over extended periods of time. Furthermore, Lynch and Lande (1998) considered populations with $N_e > 1,000$ highly unlikely to succumb to the accumulation of unconditionally deleterious alleles (i.e., alleles that are harmful under all environmental conditions) except on extremely long time scales. They also stressed, however, that many single-locus traits, such as disease resistance, require much larger populations for maintenance of adequate allele frequencies (Lande and Barrowclough 1987), and suggested that effective target sizes for conservation should be on the order of 1,000 to 5,000.

Discussion of the population sizes adequate for long-term persistence of populations from a genetics perspective will continue. Regardless of the precise value of this figure, there is agreement that the long-term

goal for *actual* population sizes to ensure viability should be thousands
of individuals, rather than hundreds.

Geographic Scale

There is sometimes confusion regarding when to apply short- or long-
term genetic goals, and how they relate to the conservation of local populations versus entire species. The short-term goals are appropriate for
conservation of local populations. As indicated above, those goals are
aimed at keeping the rate of inbreeding at a tolerable level. The effective
population sizes at which this may be achieved, however, are typically
not large enough for new mutations to compensate for the loss of genetic
variation through genetic drift. Some gene flow from neighboring populations is necessary to provide reasonable levels of genetic variation for
quantitative traits to ensure long-term persistence.

The long-term goal, where the loss of variation is balanced by new
mutations, refers primarily to a global population, which may coincide
with a species or subspecies that cannot rely on input of novel genetic
variation from neighboring populations. This global population may consist of one more or less panmictic unit, or it may be composed of multiple subpopulations that are connected by some gene flow, either naturally or through translocations (Mills and Allendorf 1996). It is the total
assemblage of interconnected subpopulations that forms a global population that must have an effective size meeting the criteria for long-term
conservation (e.g., $N_e \geq 500$ to 1,000). The actual size of this global
population will vary considerably from species to species, depending on
the number and size of the constituent subpopulations and on the pattern for gene flow between them (Waples, chap. 8 in this volume).

CONCLUSIONS

The effects of inbreeding depression on population viability generally
have been underestimated for two reasons: (1) the harmful effects of
inbreeding have been underestimated because they have been based
on captive populations, and (2) inbreeding depression has usually been
assumed to affect only juvenile survival rate in PVAs. In addition, even
in the absence of inbreeding depression on fitness, the reduction of phenotypic variation of individuals in a population may reduce population
persistence.

Inclusion of genetic factors is important for the prediction of the absolute and relative persistence times of small populations under a wide
variety of circumstances. Unfortunately, accurate estimates of the effect
of inbreeding depression in specific wild populations will not be avail-

able in the foreseeable future. We agree with Beissinger and Westphal (1998) that PVA models may be a waste of resources when reasonable data are not available. One approach is to use a range of values that span the likely effects of inbreeding depression to investigate whether it will affect absolute or relative population viability. It is important that the effects of inbreeding depression be examined for many vital rates, not just juvenile survival.

We also suggest that population viability should be predicted on both short (say 10 generations) and long (more than 20 generations) time frames. Shorter periods have been recommended because errors are propagated each time step in longer time periods (Beissinger and Westphal 1998). However, we should also be concerned with more than just the immediate future for which we can provide reliable predictions of persistence. The analogy of the distance that we can see into the "future" using headlights while driving at night is appropriate here. We can see only as far as our headlights reach, but we need to be concerned about what lies beyond their reach (Shaffer et al., chap. 7 in this volume).

Beissinger and Westphal (1998) recommended against using absolute extinction probabilities from PVA models to make policy and management decisions and suggested that other useful criteria could be used. (See Ralls et al., chap. 25 in this volume, for alternatives to PVA.) One such approach is to use the expected rate of loss of genetic variation to set a lower limit of population size. We recommend retaining at least 95% of the heterozygosity in a population over 100 years. The population size required to meet this criterion will depend on the ratio of effective to total population size and the average generation time. These parameters are easier to estimate than the effects of inbreeding depression on the vital rates needed for a PVA model. The population size required to meet this genetic criterion should not be considered a goal, but rather a lower limit below which genetic considerations are likely to reduce the probability of population persistence.

For example, brown bears in the Yellowstone ecosystem are currently completely reproductively isolated from other populations. The generation interval of brown bears is approximately ten years (Harris and Allendorf 1989). An N_e of nearly 100 bears would be needed to retain 95% of the heterozygosity for ten generations (fig. 4.1). The N_e/N ratio in brown bears has been estimated to be approximately 20% (Allendorf et al. 1991). Thus, a minimum of 500 bears would be needed before an isolated population of brown bears would be expected to persist for 100 years or more without experiencing loss of fitness due to inbreeding depression.

It is more difficult to develop general recommendations for a longer time frame to avoid the loss of evolutionary potential or to avoid the accumulation of new harmful mutations. An N_e of approximately 1,000 individuals in order "to maintain the long-term genetic security of a species" is the upper end of the recommendation of Franklin and Frankham (1998) and the lower end of the recommendation of Lynch and Lande (1998). This would correspond to some 5,000 brown bears, using an N_e/N ratio of 20%. Such large numbers of individuals will not be possible in many species without greatly increasing the connectivity among populations over a wide geographic area.

LITERATURE CITED

Allendorf, F. W. 1986. Genetic drift and the loss of alleles versus heterozygosity. *Zoo Biology* 5:181–190.

Allendorf, F. W., R. B. Harris, and L. H. Metzgar. 1991. Estimation of effective population size of grizzly bears by computer simulation. Pages 650–654 in E. C. Dudley, editor, *The unity of evolutionary biology*. Proceedings of the Fourth International Congress of Systematic and Evolutionary Biology. Dioscorides Press, Portland, Oregon.

Allendorf, F. W., and C. Servheen. 1986. Conservation genetics of grizzly bears. *Trends in Ecology and Evolution* 1:88–89.

Antolin, M. F. 1999. A genetic perspective on mating systems and sex ratios of parasitoid wasps. *Research in Population Ecology* 41:29–37.

Armbruster, P., P. Fernando, and R. Lande. 1999. Time frames for population viability analysis of species with long generation: an example with Asian elephants. *Animal Conservation* 2:69–73.

Aspinwall, N. 1974. Genetic analysis of North American populations of the pink salmon (*Oncorhynchus gorbuscha*), possible evidence for the neutral mutation–random drift hypothesis. *Evolution* 28:295–305.

Ballou, J. D. 1997. Ancestral inbreeding only minimally affects inbreeding depression in mammalian populations. *Journal of Heredity* 88:169–178.

———. 1998. General problems of small (carnivore) populations: minimum viable population size and inbreeding. Pages 27–39 in C. Breitenmoser-Würserm, C. Rohner, and C. Breitenmoser, editors, *The re-introduction of the lynx into the Alps*. Proceedings of the First SCALP Conference, Environmental Encounters no. 38. Council of Europe Publishing, Strasbourg, Germany.

Beissinger, S. R., and M. I. Westphal. 1998. On the use of demographic models of population viability in endangered species management. *Journal of Wildlife Management* 62:821–841.

Bijlsma, R., J. Bundgaard, A. C. Boerema, and W. F. van Putten. 1997. Genetic and environmental stress, and the persistence of populations. Pages 193–207 in R. Bijlsma and V. Loeschcke, editors, *Environmental stress, adaptation, and evolution*. Birkhäuser Verlag, Basel, Switzerland.

Bijlsma, R., J. Bundgaard, and W. F. van Putten. 1999. Environmental dependence of inbreeding depression and purging in *Drosophila melanogaster*. *Journal of Evolutionary Biology* 12:1125–1137.

Bijlsma, R., and V. Loeschcke. 1997. *Environmental stress, adaptation, and evolution.* Birkhäuser Verlag, Basel, Switzerland.

Black, F. L., and P. W. Hedrick. 1997. Strong balancing selection at HLA loci: evidence from segregation in South Amerindian families. *Proceedings of the National Academy of Sciences* (USA) 94:12452–12456.

Bowman, J. C., and D. S. Falconer. 1960. Inbreeding depression and heterosis of litter size in mice. *Genetical Research* 1:262–274.

Boyce, M. S. 1992. Population viability analysis. *Annual Review of Ecology and Systematics* 23:481–506.

Brewer, B. A., R. C. Lacy, M. L. Foster, and G. Alaks. 1990. Inbreeding depression in insular and central populations of *Peromyscus* mice. *Journal of Heredity* 81: 257–266.

Bürger, R., and M. Lynch. 1995. Evolution and extinction in a changing environment: a quantitative-genetic analysis. *Evolution* 49:151–163.

Burgman, M. A., S. Ferson, and H. R. Akçakaya. 1993. *Risk assessment in conservation biology.* Chapman and Hall, London, United Kingdom.

Byers, D. L., and D. M. Waller. 1999. Do plant populations purge their genetic load? Effects of population size and mating history on inbreeding depression. *Annual Review of Ecology and Systematics* 30:479–513.

Caro, T. M., and M. K. Laurenson. 1994. Ecological and genetic factors in conservation: a cautionary tale. *Science* 263:485–486.

Carrington, M., G. W. Nelson, M. P. Martin, T. Kissner, D. Vlahov, J. J. Goedert, R. Kaslow, S. Buchbinder, K. Hoots, and S. J. O'Brien. 1999. HLA and HIV-1: heterozygote advantage and *B*35-Cw*04* disadvantage. *Science* 283:1748–1752.

Caughley, G. 1994. Directions in conservation biology. *Journal of Animal Ecology* 63:215–244.

Charlesworth, D., and B. Charlesworth. 1987. Inbreeding depression and its evolutionary consequences. *Annual Review of Ecology and Systematics* 18:237–268.

Coltman, D. W., J. G. Pilkington, J. A. Smith, and J. M. Pemberton. 1999. Parasite-mediated selection against inbred Soay sheep in a free-living, island population. *Evolution* 53:1259–1267.

Conner, M. M., and G. C. White. 1999. Effects of individual heterogeneity in estimating the persistence of small populations. *Natural Resources Modeling* 12:109–127.

Crnokrak, P., and D. A. Roff. 1999. Inbreeding depression in the wild. *Heredity* 83: 260–270.

Crow, J. F. 1948. Alternate hypotheses of hybrid vigor. *Genetics* 33:477–487.

Crow, J. F., and M. Kimura. 1970. *An introduction to population genetics theory.* Burgess Publishing, Minneapolis, Minnesota.

Daniels, S. J., and J. R. Walters. 2000. Inbreeding depression and its effects on natal dispersal in red-cockaded woodpeckers. *Condor* 102:482–491.

Darwin, C. 1896. *The variation of animals and plants under domestication.* Vol. 2. D. Appleton, New York, New York.

Demauro, M. M. 1993. Relationship of breeding system to rarity in the lakeside daisy (*Hymenoxys acaulis* var *glabra*). *Conservation Biology* 7:542–550.

Dudash, M. R. 1990. Relative fitness of selfed and outcrossed progeny in a self-compatible, protandrous species, *Sabatia angularis* L. (Gentianaceae): a comparison in three environments. *Evolution* 44:1129–1140.

Dunham, J., M. Peacock, R. C. Tracy, and J. Nielsen. 1999. Assessing extinction risk: integrating genetic information. *Conservation Ecology* 3 (2). http://www. consecol.org/vol3/iss1/art2/.

Edwards, S. V., and W. K. Potts. 1996. Polymorphism of genes in the major histocompatibility complex (MHC): implications for conservation genetics of vertebrates. Pages 214–237 in T. B. Smith and R. K. Wayne, editors, *Molecular genetic approaches in conservation*. Oxford University Press, New York, New York.

Ellstrand, N. C., and D. R. Elam. 1993. Population genetic consequences of small population size: implications for plant conservation. *Annual Review of Ecology and Systematics* 24:217–242.

Ewens, W. J. 1990. The minimum viable population size as a genetic and demographic concept. Pages 307–316 in J. Adams, D. A. Lam, A. I. Hermalin, and P. E. Smouse, editors, *Convergent issues in genetics and demography*. Oxford University Press, Oxford, United Kingdom.

Falconer, D. S. 1989. *Introduction to quantitative genetics*. 3d edition. Longman, New York, New York.

Frankel, O. H., and M. E. Soulé. 1981. *Conservation and evolution*. Cambridge University Press, Cambridge, United Kingdom.

Frankham, R. 1995a. Effective population size/adult population size ratios in wildlife: a review. *Genetical Research* 66:95–107.

———. 1995b. Inbreeding and extinction: a threshold effect. *Conservation Biology* 9:792–799.

Frankham, R., and K. Ralls. 1998. Conservation biology: inbreeding leads to extinction. *Nature* 392:441–442.

Franklin, I. R. 1980. Evolutionary change in small populations. Pages 135–149 in M. E. Soulé and B. A. Wilcox, editors, *Conservation biology: an evolutionary-ecological perspective*. Sinauer Associates, Sunderland, Massachusetts.

Franklin, I. R., and R. Frankham. 1998. How large must populations be to retain evolutionary potential? *Animal Conservation* 1:69–70.

Gabriel, W., and R. Bürger. 1994. Extinction risk by mutational meltdown: synergistic effects between population regulation and genetic drift. Pages 69–84 in V. Loeschcke, J. Tomiuk, and S. K. Jain, editors, *Conservation genetics*. Birkhäuser Verlag, Basel, Switzerland.

Gemmell, N. J., and F. W. Allendorf. 2001. Mitochondrial mutations may decrease population viability. *Trends in Ecology and Evolution* 16:115–117.

Gibbs, A. 1991. Detection of a hypervariable locus in birds by hybridization with a mouse MHC probe. *Molecular Biology Evolution* 8:415–433.

Groom, M. J., and M. A. Pascual. 1998. The analysis of population persistence: an outlook on the practice of viability analysis. Pages 4–27 in P. L. Fiedler and P. M. Kareiva, editors, *Conservation biology for the coming decade*, 2d edition. Chapman and Hall, New York, New York.

Groot, C., and L. Margolis. 1991. *Pacific salmon life histories*. University of British Columbia Press, Vancouver, British Columbia.

Haig, S. M. 1998. Molecular contributions to conservation. *Ecology* 79:413–425.

Hankin, D. G., J. W. Nicholas, and T. W. Downey. 1993. Evidence for inheritance of age of maturity in chinook salmon (*Oncorhynchus tshawytscha*). *Canadian Journal of Fisheries and Aquatic Sciences* 50:347–358.

Hanson, M. R. 1991. Plant mitochondrial mutations and male sterility. *Annual Review of Genetics* 25:461–486.

Harris, R. B., and F. W. Allendorf. 1989. Genetically effective population size of large mammals: an assessment of estimators. *Conservation Biology* 3:181–191.

Harwood, J. 1992. Introduction. *Biological Conservation* 61:77–79.

Heard, W. R. 1991. Life history of pink salmon (*Oncorhynchus gorbuscha*). Pages 119–120 in C. Groot and L. Margolis, editors, *Pacific salmon life histories*. University of British Columbia Press, Vancouver, British Columbia.

Hedrick, P. W. 1994. Purging inbreeding depression and the probability of extinction: full-sib mating. *Heredity* 73:363–372.

Hedrick, P. W., and S. T. Kalinowski. 2000. Inbreeding depression in conservation biology. *Annual Review of Ecology and Systematics* 31:139–162.

Holsinger, K. E., R. J. Masongamer, and J. Whitton. 1999. Genes, demes, and plant conservation. Pages 23–46 in L. F. Landweber and A. P. Dobson, editors, *Genetics and the extinction of species*. Princeton University Press, Princeton, New Jersey.

Hosack, D. A. 1996. Population viability analysis workshop for the endangered Sonoran pronghorn (*Antilocapra americanus sonoriensis*) in the United States. Defenders of Wildlife, Phoenix, Arizona.

Husband, B. C., and D. W. Schemske. 1996. Evolution of the magnitude and timing of inbreeding depression in plants. *Evolution* 50:54–70.

James, J. W. 1971. The founder effect and response to artificial selection. *Genetical Research* 12:249–266.

Jiménez, J. A., K. A. Hughes, G. Alaks, L. Graham, and R. C. Lacy. 1994. An experimental study of inbreeding depression in a natural habitat. *Science* 266:271–273.

Kalinowski, S. T., and P. W. Hedrick. 1999. Detecting inbreeding depression is difficult in captive endangered species. *Animal Conservation* 2:131–136.

Kalinowski, S. T., P. W. Hedrick, and P. S. Miller. 2000. A close look at inbreeding depression in the Speke's gazelle captive breeding program. *Conservation Biology* 14:1375–1384.

Kärkkäinen, K., H. Kuittinen, R. Vantreuren, C. Vogl, S. Oikarinen, and O. Savolainen. 1999. Genetic basis of inbreeding depression in *Arabis petraea*. *Evolution* 53:1354–1365.

Keller, L. F. 1998. Inbreeding and its fitness effects in an insular population of song sparrows (*Melospiza melodia*). *Evolution* 52:240–250.

Keller, L. F., K. J. Jeffery, P. Arcese, M. A. Beaumont, W. M. Hochachka, J. N. M. Smith, and M. W. Bruford. 2001. Immigration and the ephemerality of a natural population bottleneck: evidence from molecular markers. *Proceedings of the Royal Society of London*, series B, Biological Sciences, 208:1387–1394.

Kenney, J. S., J. L. D. Smith, A. M. Starfield, and C. W. McDougal. 1995. The long-term effects of tiger poaching on population viability. *Conservation Biology* 9:1127–1133.

Kucera, T. E. 1991. Genetic variability in the tule elk. *California Fish and Game* 77:70–78.

Lacy, R. C. 1993a. Impacts of inbreeding in natural and captive populations of vertebrates: implications for conservation. *Perspectives in Biology and Medicine* 36: 480–496.

———. 1993b. VORTEX: a computer simulation model for population viability analysis. *Wildlife Research* 20:45–65.

———. 1997. Importance of genetic variation to the viability of mammalian populations. *Journal of Mammalogy* 78:320–335.

Lacy, R. C., G. Alaks, and A. Walsh. 1996. Hierarchical analysis of inbreeding depression in *Peromyscus polionotus. Evolution* 50:2187–2200.

Lacy, R. C., and J. D. Ballou. 1998. Effectiveness of selection in reducing the genetic load in populations of *Peromyscus polionotus* during generations of inbreeding. *Evolution* 52:900–909.

Laikre, L. 1999. Conservation genetics of Nordic carnivores: lessons from zoos. *Hereditas* 130:203–216.

Laikre, L., R. Andrén, H. O. Larsson, and N. Ryman. 1996. Inbreeding depression in brown bear *Ursus arctos. Biological Conservation* 76:69–72.

Lande, R. 1988. Genetics and demography in biological conservation. *Science* 241: 1455–1460.

———. 1994. Risk of population extinction from fixation of new deleterious mutations. *Evolution* 48:1460–1469.

———. 1995. Mutation and conservation. *Conservation Biology* 9:782–791.

———. 1999. Extinction risks from anthropogenic, ecological, and genetic factors. Pages 1–22 in L. F. Landweber and A. P. Dobson, editors, *Genetics and the extinction of species*. Princeton University Press, Princeton, New Jersey.

Lande, R., and G. F. Barrowclough. 1987. Effective population size, genetic variation, and their use in population management. Pages 87–124 in M. E. Soulé, editor, *Viable populations for conservation*. Cambridge University Press, Cambridge, United Kingdom.

Lande, R., and D. W. Schemske. 1985. The evolution of self-fertilization and inbreeding depression in plants: 1, genetic models. *Evolution* 39:24–40.

Landweber, L. F., and A. P. Dobson. 1999. *Genetics and the extinction of species*. Princeton University Press, Princeton, New Jersey.

Leberg, P. L. 1990. Influence of genetic variability on population growth: implications for conservation. *Journal of Fish Biology* 37:193–195.

Les, D. H., J. A. Reinartz, and E. J. Esselman. 1991. Genetic consequences of rarity in *Aster furcatus* (Asteraceae), a threatened, self-incompatible plant. *Evolution* 45:1641–1650.

Letcher, B. H., J. A. Priddy, J. R. Walters, and L. B. Crowder. 1998. An individual-based, spatially-explicit simulation model of the population dynamics of the endangered red-cockaded woodpecker, *Picoides borealis. Biological Conservation* 86:1–14.

Lynch, C. B. 1977. Inbreeding effects upon animals derived from a wild population of *Mus musculus. Evolution* 31:526–537.

Lynch, M., J. Blanchard, D. Houle, T. Kibota, S. Schultz, L. Vassilieva, and J. Willis. 1999. Perspective: spontaneous deleterious mutation. *Evolution* 53:645–663.

Lynch, M., J. Conery, and R. Bürger. 1995. Mutation accumulation and the extinction of small populations. *American Naturalist* 146:489–518.

Lynch, M., and W. Gabriel. 1990. Mutation load and the survival of small populations. *Evolution* 44:1725–1737.

Lynch, M., and R. Lande. 1998. The critical effective size for a genetically secure population. *Animal Conservation* 1:70–72.

Madsen, T., R. Shine, M. Olsson, and H. Wittzell. 1999. Conservation biology: restoration of an inbred adder population. *Nature* 402:34–35.

Mann, C. C., and M. L. Plummer. 1999. A species' fate, by the numbers. *Science* 284:36–37.

McCullough, D. R., J. K. Fischer, and J. D. Ballou. 1996. From bottleneck to metapopulation: recovery of the tule elk in California. Pages 375–403 in D. R. McCullough, editor, *Metapopulations and wildlife conservation*. Island Press, Covelo, California.

Mills, L. S., and F. W. Allendorf. 1996. The one-migrant-per-generation rule in conservation and management. *Conservation Biology* 10:1509–1518.

Mills, L. S., and P. E. Smouse. 1994. Demographic consequences of inbreeding in remnant populations. *American Naturalist* 144:412–431.

Nettancourt, D. de. 1977. *Incompatibility in angiosperms*. Springer-Verlag, New York, New York.

Newman, D., and D. Pilson. 1997. Increased probability of extinction due to decreased genetic effective population size: experimental populations of *Clarkia pulchella*. *Evolution* 51:354–362.

Nunney, L., and K. A. Campbell. 1993. Assessing minimum viable population size: demography meets population genetics. *Trends in Ecology and Evolution* 8:234–239.

O'Brien, S. J. 2000. Adaptive cycles: parasites selectively reduce inbreeding in Soay sheep. *Trends in Ecology and Evolution* 15:7–9.

Ohta, T. 1971. Associative overdominance caused by linked detrimental mutations. *Genetical Research* 18:277–286.

Pamilo, P., and S. Pålsson. 1998. Associative overdominance, heterozygosity, and fitness. *Heredity* 81:381–389.

Paterson, S., K. Wilson, and J. M. Pemberton. 1998. Major histocompatibility complex variation associated with juvenile survival and parasite resistance in a large unmanaged ungulate population (*Ovis aries* L.). *Proceedings of the National Academy of Sciences* (USA) 95:3714–3719.

Pimm, S. L., H. L. Jones, and J. M. Diamond. 1988. On the risk of extinction. *American Naturalist* 132:757–785.

Ralls, K., J. D. Ballou, and A. Templeton. 1988. Estimates of lethal equivalents and the cost of inbreeding in mammals. *Conservation Biology* 2:185–193.

Reinartz, J. A., and D. H. Les. 1994. Bottleneck-induced dissolution of self-incompatibility and breeding system consequences in *Aster furcatus* (Asteraceae). *American Journal of Botany* 81:446–455.

Richards, A. J. 1986. *Plant breeding systems*. Allen and Unwin, London, United Kingdom.

Robertson, A. 1960. A theory of limits in artificial selection. *Proceedings of the Royal Society of London*, series B, Biological Sciences, 153:234–249.

Ruiz-Pesini, E., A. C. Lapena, C. Diez-Sanchez, A. Perez-Martons, J. Montoya, E. Alvarez, M. Diaz, A. Urries, L. Montoro, M. J. Lopez-Perez, and J. A. Enriquez.

84   *Fred W. Allendorf and Nils Ryman*

2000. Human mtDNA haplogroups associated with high or reduced spermatozoa motility. *American Journal of Human Genetics* 67:682–696.

Ryman, N., R. Baccus, C. Reuterwall, and M. H. Smith. 1981. Effective population size, generation interval, and potential loss of genetic variability in game species under different hunting regimes. *Oikos* 36:257–266.

Ryman, N., and L. Laikre. 1991. Effects of supportive breeding on the genetically effective population size. *Conservation Biology* 5:325–329.

Saccheri, I., M. Kuussaari, M. Kankare, P. Vikman, W. Fortelius, and I. Hanski. 1998. Inbreeding and extinction in a butterfly metapopulation. *Nature* 392:491–494.

Sæther, B.-E., S. Engen, J. E. Swenson, Ø. Bakke, and F. Sandegren. 1998. Assessing the viability of Scandinavian brown bear, *Ursus arctos*, populations: the effects of uncertain parameter estimates. *Oikos* 83:403–416.

Shaffer, M. L. 1987. Minimum viable populations: coping with uncertainty. Pages 69–86 in M. E. Soulé, editor, *Viable populations for conservation*. Cambridge University Press, Cambridge, United Kingdom.

Shaffer, M. L., and F. B. Samson. 1985. Population size and extinction: a note on determining critical population sizes. *American Naturalist* 125:144–152.

Shields, W. M. 1993. The natural and unnatural history of inbreeding and outbreeding. Pages 143–169 in N. W. Thornhill, editor, *The natural history of inbreeding and outbreeding*. University of Chicago Press, Chicago, Illinois.

Simberloff, D. 1988. The contribution of population and community biology to conservation science. *Annual Review of Ecology and Systematics* 19:473–511.

Slade, R. W. 1992. Limited MHC polymorphism in the southern elephant seal: implications for MHC evolution and marine mammal population biology. *Proceedings of the Royal Society of London*, series B, Biological Sciences, 249:163–171.

Soulé, M. E. 1987. Introduction. Pages 1–10 in M. E. Soulé, editor, *Viable populations for conservation*. Cambridge University Press, Cambridge, United Kingdom.

Soulé, M. E., M. E. Gilpin, W. Conway, and T. Foose. 1986. The millennium ark: how long a voyage, how many staterooms, how many passengers? *Zoo Biology* 5:101–113.

Soulé, M. E., and L. S. Mills. 1992. Conservation genetics and conservation biology: a troubled marriage. Pages 55–65 in O. T. Sandlund, K. Hindar, and A. H. D. Brown, editors, *Species and ecosystem conservation*. Scandinavian University Press, Oslo, Norway.

———. 1998. No need to isolate genetics. *Science* 282:1658–1659.

Templeton, A. R., and B. Read. 1984. Factors eliminating inbreeding depression in a captive herd of Speke's gazelle. *Zoo Biology* 3:177–199.

———. 1998. Elimination of inbreeding depression from a captive population of Speke's gazelle: validity of the original statistical analysis and confirmation by permutation testing. *Zoo Biology* 17:77–94.

van Noordwijk, A. J., and W. Scharloo. 1981. Inbreeding in an island population of the great tit. *Evolution* 35:674–688.

Waples, R. S. 1990. Conservation genetics of Pacific salmon: 2, effective population size and the rate of loss of genetic variability. *Journal of Heredity* 81:267–276.

Westemeier, R. L., J. D. Brawn, S. A. Simpson, T. L. Esker, R. W. Jansen, J. W.

Walk, E. L. Kershner, J. L. Bouzat, and K. N. Paige. 1998. Tracking the long-term decline and recovery of an isolated population. *Science* 282:1695–1698.

Willis, J. H. 1999. The role of genes of large effects on inbreeding depression in *Mimulus guttatus*. *Evolution* 53:1678–1691.

Willis, K., and R. J. Wiese. 1997. Elimination of inbreeding depression from captive populations: Speke's gazelle revisited. *Zoo Biology* 16:9–16.

Wright, S. 1931. Evolution in Mendelian populations. *Genetics* 16:97–159.

———. 1960. On the number of self-incompatibility alleles maintained in equilibrium by a given mutation rate in a population of given size: a re-examination. *Biometrics* 16:61–85.

———. 1969. *Evolution and the genetics of populations*. Vol. 2, *The theory of gene frequencies*. University of Chicago Press, Chicago, Illinois.

———. 1977. *Evolution and the genetics of populations*. Vol. 3, *Experimental results and evolutionary deductions*. University of Chicago Press, Chicago, Illinois.

Metapopulations of Animals in Highly Fragmented Landscapes and Population Viability Analysis
Ilkka Hanski

ABSTRACT

This chapter reviews the application of metapopulation concepts and patch occupancy models to population viability analysis (PVA) for animals in highly fragmented landscapes. I briefly review the historical development of metapopulation ideas, how they have been applied to conservation problems, and when the metapopulation paradigm has less utility. The role of local extinction of occupied patches and colonization of unoccupied patches is considered in relation to the dispersal capabilities of species and two useful approaches used for modeling patch occupancy dynamics, the incidence function model (IFM) and the logistic regression model (LRM). IFM and LRM are illustrated by studies of the metapopulation dynamics of butterfly populations. I extend the Levins model to incorporate spatially realistic landscape structure to allow evaluation of patch-specific alterations to landscapes. Metapopulation models are best used for making short-term predictions of the effects of alternative management scenarios rather than for attempting to predict long-term extinction risk for particular populations. Metapopulations of common species with fast turnover of local populations in highly fragmented landscapes offer the best prospects for rigorously parameterizing patch occupancy models.

INTRODUCTION

Population biology has supplied two important paradigms about the spatial dynamics of species for conservation biology, the dynamic theory of island biogeography (MacArthur and Wilson 1963, 1967) in the 1970s and early 1980s and the metapopulation theory (Hanski 1989, 1996; Merriam 1991; Hanski and Simberloff 1997) in the 1990s. This succession of paradigms, or conceptual frameworks, is somewhat surprising for its timing, because the basic models for both paradigms were established at about the same time in the late 1960s when Levins (1969) published his by now well-known metapopulation model. So why did the MacArthur-Wilson model attract so much attention from conservation biologists in the 1970s, whereas the Levins model attracted none? The replacement of one model with another is also surprising, because the two models are not particularly different. The MacArthur-Wilson model for island communities can be viewed as a composite of models for independent species, each obeying the mainland-island version of the Levins model (Hanski 1991). Both models share key conceptual underpinnings, notably the vision of nature divided into discrete fragments of habitat, unstable local populations, and the possibility of long-term persistence at the regional scale. So why have conservation biologists considered the two models to be so different?

There are many possible answers to these questions, but the following considerations seem especially noteworthy (Hanski 1996; Hanski and Simberloff 1997). The initial success of the MacArthur-Wilson model partly stemmed from high esteem for the authors, but also from their ability to relate the theory to general empirical patterns (the species-area relationship) that appeared to be critical for conservation (especially for nature reserves of different kinds). In contrast, the Levins model stimulated theoretical studies but remained detached from empirical research (Hanski 1999b). In my opinion, the renaissance of the spirit (if not the letter) of the Levins model is largely due to the recent incorporation of a realistic landscape structure into the model, which has helped bridge the gap between modeling and empirical studies in metapopulation ecology (Hanski 1998, 1999b). Interest in metapopulation models has also been fueled by our awareness of ever-increasing habitat loss and fragmentation, making it practically impossible to ignore the potential of metapopulation models to predict the dynamics of populations in fragmented landscapes.

The widespread view that the island model and the metapopulation model are fundamentally different is largely mistaken (Gotelli 1991; Hanski 1996; Hanski and Simberloff 1997). The reasons for this misconception are many. Conservation biology—along with the rest of population biology—turned from communities to species and to populations in the 1980s, possibly giving the impression that the island theory with its focus on communities is somehow outdated in comparison with single-species metapopulation theories. It became fashionable to emphasize a nonequilibrium view of nature instead of the traditional equilibrium view (Wiens 1977, 1984; Chesson and Case 1986). Though there is really no difference in this respect between the island model and classical metapopulation models, an illusory difference may have been created by the emphasis in the former on the equilibrium number of species, to the extent that the theory received the nickname "equilibrium" theory. In contrast, metapopulation theories have emphasized population turnover, possibly creating a sense of a nonequilibrium theory (Hanski and Simberloff 1997). The island theory assumes a permanent mainland community that is sufficient to prevent global extinction, whereas there is no such refuge—and there is a definite possibility of global extinction—in Levins's metapopulation model. The latter is, therefore, more appropriate than the island model for endangered species with a perilous existence in fragmented landscapes. And finally, the island theory was originally developed to explain patterns at large spatial scales, whereas the metapopulation concept is associated with fragmentation of ordinary landscapes. Not many of us are familiar with oceanic islands, which supplied many key initial examples for the theory of island biogeography, but all of us are familiar with the loss and fragmentation of some natural habitats.

At present, the metapopulation concept is frequently used in conservation (Soulé 1987; Soulé and Kohm 1989; Western and Pearl 1989; Falk and Holsinger 1991; Fiedler and Jain 1992; Harrison 1994; McCullough 1996; Hanski and Simberloff 1997). Enthusiasm has reached such a level that Doak and Mills (1994), Harrison (1994), Thomas (1994), and Hanski and Simberloff (1997) perceive a potential danger in the widespread application of the metapopulation approach to species that may not be spatially structured in the way assumed by the models. There is indeed an opportunity for misapplication of metapopulation models, as for any other models. One should also recognize, however, that there is already ample support for proper use of metapopulation models of one kind or another in better understanding spatial population dynamics of many species (Hanski and Simberloff 1997; McCullough 1996). The metapopulation paradigm has also served and continues to serve the useful function of stimulating conservation biologists to gather data that are critical for the development of effective conservation strategies for many species: movement rates between populations, reproduction and mortality rates that may vary among populations, population size-dependent extinction risk, and the like (Hanski and Simberloff 1997).

Harrison (1994) called attention to substantial differences between different types of metapopulations, namely classical metapopulations, mainland-island metapopulations, nonequilibrium metapopulations, and patchy populations. Even if there are no truly distinct types in nature but rather a continuum of "metapopulation structures," Harrison's classification is useful in highlighting the point that different sorts of processes may largely underpin the dynamics and persistence of species in different environmental settings. A corollary of this observation is that different types of metapopulation models are likely to be most useful for markedly different species and landscapes.

This chapter is primarily concerned with *classical metapopulations in highly fragmented landscapes*. In highly fragmented landscapes, the habitat fragments are so small that they account for only a small fraction of the total area of the landscape. Small habitat fragments typically harbor small populations with a substantial risk of extinction; hence, the regional dynamics are dominated by local extinctions and colonizations. In such cases, Levins-type patch occupancy models that ignore local dynamics represent a practical modeling approach. In the case of landscapes with larger habitat fragments and larger local populations, it is questionable whether one is justified in ignoring local dynamics and more complex models should be used instead (for a review see Lindenmayer et al. 1995). Such models may be of value especially in comparisons between alternative scenarios, even if including local dynamics typically leads to models with many structural assumptions that are difficult to verify and to a large number of parameters that are practically impossible to estimate rigorously (Hanski and Gilpin 1991; Beissinger and Westphal 1998). In this chapter, I focus on highly fragmented landscapes and on relatively simple models that can be properly parameterized with empirical data that are not too hard to obtain.

The next section summarizes the modeling approach for highly fragmented landscapes with selected examples. The third section discusses the type of comparative predictions for which metapopulation models, and probably most models used for population viability analysis (PVA), are most appropriate. In the final section, I suggest that the domain of PVA should be expanded from the severely endangered species to "ordinary" species living in "ordinary" landscapes. This is the context in which the practical application of metapopulation models is likely to be most useful in the future.

METAPOPULATIONS IN HIGHLY FRAGMENTED LANDSCAPES

The two characteristic features of highly fragmented landscapes are that individual habitat patches are so small that the small local populations have a high risk of extinction and that the suitable habitat covers only a relatively small fraction of the total landscape. This second assumption allows us to simplify the description of the habitat patch network by considering only patch areas and their spatial locations while ignoring patch shape and many other patch characteristics. Between-patch distances can be approximated by the distances between the centers of gravity of the respective patches, though if some more appropriate distance measure is available, it can be used instead of the straightline distance. Spatial variation in habitat quality can be incorporated by appropriately adjusting patch areas, which are used as surrogates for the

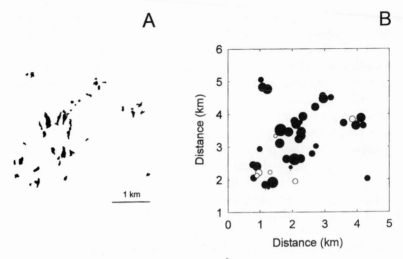

Fig. 5.1 A, A network of 50 habitat patches in the Åland Islands occupied by the Glanville fritillary butterfly. B, The idealized version of the network, as used in modeling. Dot size is proportional to the logarithm of patch area; *filled dots* are occupied patches and *open circles* represent empty patches during one survey of the distribution of the butterfly.

expected population sizes. Figure 5.1 gives an example of a real patch network of the Glanville fritillary butterfly (*Melitaea cinxia*) and the corresponding idealized network as used in the modeling.

Three Principal Processes

The key idea of classical metapopulation dynamics is to focus on the two processes that lead to a change in the size of the metapopulation in terms of the number of extant populations: local extinction and colonization of empty patches. The latter may be divided into two components, migration (dispersal) of individuals from existing local populations and the successful establishment of new populations by immigrants. In patch (occupancy) models, to which this chapter is restricted, only the presence or absence of local populations in the habitat patches is considered, while population sizes are ignored. The simplest patch models, including the familiar Levins model (Levins 1969), are completely nonstructured, meaning that all patches and hence all existing local populations are assumed to be identical. These models have been occasionally used in conservation applications (e.g., Doncaster et al. 1996; Kindvall 1996; Ås et al. 1997), but this is not to be recommended: it is plainly too great a simplification to assume identical patches if the model is applied to real metapopulations.

The modeling approaches described below allow for differences in patch areas and connectivities, which are used as surrogates for the expected sizes of existing local populations and the expected rates of immigration to particular patches. Now, as the rates of local extinction and establishment of new populations are largely determined by the sizes of local populations and by immigration, respectively, we arrive at simple "spatially realistic" metapopulation models that consider spatial variation in population sizes and distance-dependent migration without actually modeling individuals. In fact, the essential difference between the Levins model and the spatially realistic models (below) is that the latter include the effects of patch area and connectivity on local extinctions and colonizations.

The increasing risk of population extinction with decreasing population size is well established in both theoretical (Leigh 1981; Lande 1993; Foley 1994; and many others) and empirical (Williamson 1981; Diamond 1984; Hanski 1999b; and many others) literatures. Apart from truly catastrophic extinctions, small populations tend to have a higher risk of extinction than large ones regardless of the actual mechanism of extinction. It is equally well established that the probability of colonization of empty patches increases with the numbers of immigrants (Crowell 1973; Ebenhard 1987, 1991; Veltman et al. 1996). A few considerations about migration and colonization deserve to be highlighted.

Metapopulation models tend to assume unconditional emigration, which furthermore has no consequences for the source population. Though for many species these are fair assumptions, in other cases emigration is a more complex phenomenon. For instance, density-dependent emigration may contribute to population regulation. The common wisdom is that this is especially true of territorial species, but density-dependent emigration is widespread also in nonterritorial taxa (Lance and Barbosa 1979; Joern and Gaines 1990; Denno and Peterson 1995). At the lowest densities, emigration as well as immigration may become inversely density-dependent (Kuussaari et al. 1996), leading to an Allee effect that may create an unstable equilibrium below which the population goes deterministically extinct. Density-independent emigration and immigration will not directly influence population regulation, but such migration may lead to lowered and elevated population densities, respectively, with potential consequences for the persistence of populations. In extreme cases, species may occur in low-quality sink habitats only because of migration from high-quality source habitats (Pulliam 1988).

Migration distances determine the scale of recolonization of empty habitat and thereby set the spatial scale of metapopulations. The simplest model of animal movements, based on the assumption that individuals perform a random walk, predicts a normal distribution for migration distances, with the variance of the movement distances $(2Dt)$ increasing linearly with time (t) and coefficient of diffusion (D) (Okubo 1980). Empirical studies have demonstrated, however, that the distribution of migration distances is typically leptokurtic (Dobzhansky and Wright 1943; Johnson and Gaines 1990); that is, more individuals move very short, and very long, distances than predicted by simple random walk (normal distribution). One simple mechanism that leads to a leptokurtic distribution is a random walk with a constant probability per unit time that an individual will settle down and stop moving. Leptokurtic distributions may also arise from differences in individual migration tendency (Dobzhansky and Wright 1943; Skellam 1951) and from more complex movement behavior in a homogeneous population of individuals (Okubo 1980). In metapopulation models and analyses, it is commonly assumed that the distribution of migration distances is exponential, which is a simple leptokurtic distribution. The exact shape of the distribution of migration distances is unlikely to be critical for short and medium distances, but it becomes critical for the longest distances, which are what really matter in the spread of a species into a currently vacant area. Unfortunately, finding exactly how "fat" the tail of the migration distances is—that is, the fraction of individuals that moves very long distances—is very difficult using any method that is commonly available.

Turning to the establishment of new populations at empty habitat patches, an important question for metapopulation dynamics is whether colonization involves an element of inverse density dependence, such that the per immigrant probability of establishing a new population will increase with propagule size up to some threshold. Such inverse density dependence would further amplify the significance of propagule size in colonization, and would lead to a sigmoid relationship between the probability of colonization and propagule size (as commonly assumed in the incidence function model; Hanski 1994). Empirical studies on this relationship are badly needed.

Two Modeling Approaches

Two approaches have been widely used to model the dynamics of real metapopulations in highly fragmented landscapes, the incidence function model (IFM; Hanski 1994, 1999b) and the logistic regression model (LRM; Sjögren-Gulve and Ray 1996). These two models are prime examples of spatially realistic patch models; in other words these models allow for spatial variation in patch areas, connectivities, and other patch attributes, but consider only the presence or absence of local populations in the habitat patches.

In the IFM, structural assumptions are made about how increasing patch size decreases the probability of local extinction and how increasing connectivity to existing populations increases the probability of colonization (see Hanski 1999b for a description of the model and Moilanen et al. 1998 for a representative case study). Following Hanski (1994), particular functional forms have been used in most applications of the IFM, but some other forms could be assumed equally well; the choice should be made on the basis of the biology of the focal species. In principle, the model can be parameterized with just a single snapshot of occupancy data from a patch network like the one shown in figure 5.1, although, naturally, more reliable parameter estimates can be obtained if several patterns of patch occupancy from multiple surveys are available (Moilanen 1998, 1999). The effects of area and connectivity on occupancy reflect, by assumption, the above-mentioned effects of patch area and connectivity on extinction and colonization. If data from multiple surveys are available, the dependence of extinction and colonization probabilities on area and connectivity can be assessed directly in parameter estimation. Indeed, a useful feature of the IFM is that it can be rigorously parameterized with "spatial" data on the pattern of patch occupancy, with "temporal" data on extinctions and colonizations, or with a combination of the two.

One important assumption underlying parameter estimation in the IFM must be emphasized: the metapopulation occurs in a stochastic steady state. In other words, there is no increasing or decreasing trend in patch occupancy (Hanski 1994, 1999b; Moilanen 2000). Testing this assumption is difficult, unfortunately, even if data are available for a few years; formal statistical tests would require data for some dozens of years (Moilanen 2000). The best an ecologist or manager can do in practice is to use any supplementary information that might be available to judge whether the steady state assumption is reasonable or not. For instance, if the landscape has recently experienced severe habitat loss and fragmentation, the steady state assumption is dubious, as the metapopulation may be tracking a changing landscape structure with a time delay (Hanski et al. 1996; Hanski 1999b; Moilanen 2000). To some extent, the uncertainty in model predictions due to the steady state assumption can be assessed by simulations (Moilanen 2000).

In the LRM, the idea is to parameterize a generic statistical model (logistic regression) with the extinction and colonization events as two separate response variables and patch area, connectivity, and other variables as explanatory variables. Because parameter estimation is based exclusively on population turnover events, at least two snapshots of patch occupancy are needed to parameterize the model, and in practice the applicability of this approach may be limited by the amount of turnover data available. It has been suggested that an advantage of the LRM is that no steady state assumption is necessary (Sjögren-Gulve and Ray 1996), but in practice this is not so. If parameters are estimated from metapopulations with a long-term trend in patch occupancy, the LRM is likely to extrapolate the trend into the future (Hanski 1999b; Moilanen 2000). From the point of view of model application, the essential differences between the IFM and the LRM are that, first, the former makes assumptions about how patch area and connectivity influence extinction and colonization, while the latter employs the standard logistic model for this purpose; and second, the IFM but not the LRM allows one to use spatial pattern information on patch occupancy in parameter estimation.

Below, I illustrate the use of the IFM and the LRM with examples of butterfly metapopulations. Butterfly biologists have been especially quick to turn to the metapopulation approach, no doubt because butterflies often have the sort of spatial population structure assumed in the classical metapopulation model (Thomas 1994; Thomas and Hanski 1997). Other metapopulation studies that have used the IFM are reviewed in ter Braak et al. (1998) and Hanski (1998, 1999b), while the applications of the LRM are reviewed by Sjögren-Gulve and Ray (1996).

Examples in Finnish Checkerspot Butterflies

The IFM has been parameterized for a metapopulation of the Glanville fritillary butterfly (*Melitaea cinxia*) using data collected in two years from the patch network shown in figure 5.1 (Hanski 1994). This network represents only a small subset of the entire study area in the Åland Islands, southwest Finland, with about 1,600 habitat patches (Hanski 1999b). Hanski et al. (1996) tested the IFM by using the parameter values estimated from this subset (fig. 5.1) to predict the pattern of patch occupancy in the rest of the Åland Islands (fig. 5.2). For one part of the study area, the prediction failed, perhaps because of some environmental differences or perturbations away from the equilibrium (Hanski et al. 1996; Moilanen and Hanski 1999), but in most of the Åland Islands the observed fraction of occupied patches matched well the predicted one for patch networks with >15 patches (fig. 5.2). Predictions are generally expected to be worse for networks with a small number of patches because the dynamics in such networks are greatly influenced by extinction-colonization stochasticity (Hanski 1991, 1999b).

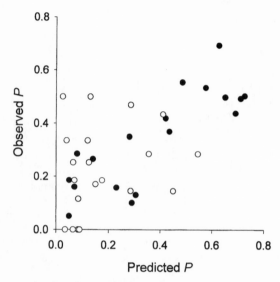

Fig. 5.2 Comparison between the predicted and observed fraction of occupied patches (*P*) in the Glanville fritillary metapopulations in western Åland Islands. *P* values were calculated for 4 × 4 km squares. *Open circles* are for squares with <15 habitat patches, *filled dots* for squares with ≥15 patches (from Hanski et al. 1996).

Another test of the model relates to the introduction in 1991 of the butterfly to the island of Sottunga in the Åland archipelago. The island has 16 meadows suitable for the Glanville fritillary within an area of about 10 km², but the butterfly was not found on the island prior to the introduction in 1991. Using the parameter values estimated from the network in figure 5.1, the species was predicted to have a lifetime of roughly 50 to 100 years on this island (fig. 5.3 shows ten predicted trajectories). The butterfly has survived up to the present time, and it has indeed functioned as a classical metapopulation with fast turnover of local populations (fig. 5.4), just as in the main Åland Island (Hanski et al. 1995; Hanski 1999b).

The third example is on the rarest of the six species of checkerspot butterflies (Melitaeini) in Finland, the false heath fritillary (*Melitaea diamina*), which is a close relative of the Glanville fritillary. *Melitaea diamina* has only two remaining metapopulations in Finland, of which one is located in one small area in south-central Finland. The caterpillars feed exclusively on *Valeriana sambucifolia*, and the butterflies occur only on the moist meadows where this plant grows (Wahlberg 1997). Having the previously estimated parameter values for the Glanville fritillary, as well as data on the patch areas and spatial locations in the *M.*

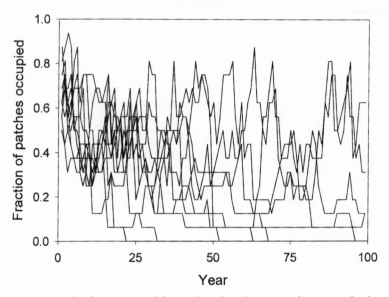

Fig. 5.3 Ten simulated trajectories of the number of patches occupied over time for the metapopulation of the Glanville fritillary butterfly on the island of Sottunga with 16 habitat patches within an area of 2 × 4 km. The predictions were made with the IFM parameterized with data from the network shown in figure 5.1 (regional stochasticity was set at 0.4; see Hanski et al. 1996).

diamina network, Wahlberg et al. (1996) used the IFM to predict the pattern of patch occupancy in the latter species, using the parameter values for the Glanville fritillary. This prediction was subsequently tested with empirical data on the actual pattern of patch occupancy in the *M. diamina* metapopulation. The match between the IFM-predicted incidences (long-term probability of patch occupancy) and the observed snapshot of patch occupancy was good (fig. 5.5), suggesting that if the opportunity exists it may be worthwhile to consider using data for a common related species to parameterize a model for an endangered species.

The Silver-Spotted Skipper Butterfly in England

The silver-spotted skipper butterfly (*Hesperia comma*) occupies heavily grazed calcareous grasslands in Britain. Female skippers lay eggs on the leaves of the grass *Festuca ovina* that are 1 to 5 cm tall and partially surrounded by bare ground. Before the 1950s, the silver-spotted skipper inhabited most of the grasslands in the southern British chalk hills (Thomas et al. 1986). In the mid-1950s, myxomatosis caused a severe decline in rabbit populations, and many grasslands became overgrown;

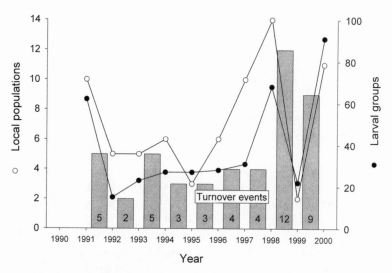

Fig. 5.4 The dynamics of a metapopulation of the Glanville fritillary butterfly introduced to the island of Sottunga in 1991. Metapopulation size is measured in two ways, by the number of local populations and by the total number of larval groups. Many turnover events (extinctions and colonizations) have occurred in the metapopulation, and no local population has persisted throughout the period 1991–2000 (M. Kuussaari and I. Hanski, unpublished data).

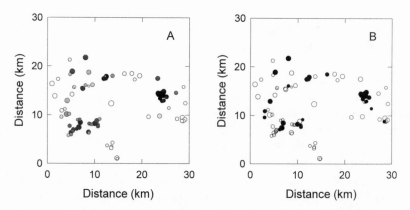

Fig. 5.5 *A*, A map of the patch network occupied by *Melitaea diamina*, showing the relative patch sizes, their spatial locations, and the predicted incidence (long-term probability of occupancy) based on parameter values estimated for the Glanville fritillary butterfly. Higher incidences are shown by *darker shading*. *B*, A snapshot of patch occupancy by *M. diamina* in 1995 (from Wahlberg et al. 1996), with *filled dots* showing the occupied patches and *open circles* the empty ones.

the skipper declined to a mere 46 sites in ten regions (Thomas and Jones 1993). Following rabbit recovery and a habitat restoration program in the subsequent years, some 150 sites had become suitable by 1982, though most of them still remained uncolonized in 1991 when a detailed butterfly survey was conducted.

Thomas and Jones (1993) used LRM to analyze the pattern of habitat occupancy, extinctions, and colonizations based on data collected in surveys conducted in 1982 and 1991. Using a simulation procedure with patch-specific colonization and extinction probabilities derived from the LRM, they examined whether the observed pattern of extinctions and colonizations in 1982–1991 could explain why the skipper had recovered only partially and why so many restored habitat patches remained vacant. Thomas and Jones (1993) found that the patches occupied in 1991 were larger and less isolated from the neighboring local populations than the unoccupied patches. Patches that had been recolonized in 1982–1991 ($n = 29$) were significantly larger and less isolated from the occupied patches than those that remained unoccupied ($n = 109$). Variation in habitat quality was not a significant determinant of patch occupancy, but this was not really surprising since patches were identified and delimited based on habitat-quality criteria. The function describing colonizations explained 69% of the observed variation in colonization. Sites at which a population went extinct ($n = 10$) were significantly smaller and more isolated from the neighboring populations than sites where a population persisted ($n = 50$), although the function describing extinctions explained only 29% of the observed variation in local extinction. Given the previous rate of spread to unoccupied but suitable patch networks (0.73 km per year in areas with abundant habitat), Thomas and Jones (1993) predicted that the presently vacant areas would become repopulated within 55 to 70 years. In contrast, simulations of the LRM predicted much more limited spread over the next 100 years, basically because zones with >10 km of unsuitable habitat acted as barriers that prevented recolonization of suitable patches.

A reanalysis of the same data by Hanski and Thomas (1994) using simulation of a structured metapopulation model and by Hanski (1994) using the IFM yielded similar results. Both the simulation model and the IFM were parameterized using data from one patch network in which the skipper had remained at a relatively steady state throughout the decline that occurred elsewhere. Using these parameter values, the spread of the species into a 75-patch network was predicted. Only 3 patches were occupied in 1982, whereas the skipper had spread to 21 patches by 1991. Both models underestimated the rate of spread. In the simulation model, only 8.6 patches on average were occupied after nine years (median 9, minimum 1, maximum 11 in 100 replicates), and 14 patches were occupied after 50 generations. In the IFM, 11 populations were predicted to exist after nine years (Hanski 1994). One possible reason for the slow predicted rate of spread is that the metapopulation used to estimate parameter values was, perhaps, not quite at equilibrium but may have been expanding (Hanski and Thomas 1994), which would result in parameter values underestimating the species' colonization potential. Alternatively, the models may have genuinely underestimated long-distance colonization, which is hard to describe correctly. During the observed expansion from 1982 to 1991, three patches isolated by up to 8.7 km from the nearest occupied patch were predicted by the models to have zero probability of colonization even after 50 years but were nonetheless colonized. Finally, it is impossible, unfortunately, to exclude the possibility that these colonizations were artifacts of erroneous classification of the respective patches as unoccupied in the first survey.

The Message

The message from this section is that rigorous modeling of metapopulation dynamics is possible in cases where it is appropriate to focus solely on extinction-colonization dynamics, which is often a reasonable approach for metapopulations living in highly fragmented landscapes. The most important caveat that must be added is that, as explained in the previous sections, parameter estimation is based on the difficult-to-test assumption that the metapopulation used for parameter estimation occurs at a stochastic steady state. Concerning the very large *M. cinxia* metapopulation, it has been learned over the past eight years that there are large-scale correlated changes in patch occupancy (Hanski 1999b), which may explain why the original prediction failed for one part of the study area (fig. 5.2; Hanski et al. 1996): this area may have been out of equilibrium. In contrast, the smaller *M. diamina* metapopulation shown in figure 5.5 has experienced substantial population turnover but no overall change in the level of patch occupancy between 1995 and 1999 (Hanski, unpublished data), which agrees with the IFM assumptions. When model predictions are made for another environment, or for another related species (as in fig. 5.5), it is always possible that predictions fail because of some known or unknown environmental differences, or because of some unsuspected interspecific differences. Predicting metapopulation expansion into a previously unoccupied patch network, which was attempted for the silver-spotted skipper, is especially difficult because such prediction critically depends on correctly modeling long-range migration. These complications do not imply that our modeling efforts are ultimately futile, but they do mean that modeling is most useful when considered as just one source of information, based on a good ecological understanding of the relationship between populations and their environment. Unfortunately, this context is often lacking in the case of endangered species for which modeling and PVA are needed. This raises questions about the type of modeling and the type of model predictions that might, in practice, be most useful.

SHORT-TERM PREDICTIONS FOR ALTERNATIVE SCENARIOS

Predicting the future is difficult, especially if the prediction concerns a long time span, say 100 years. Nonetheless, one prediction that we population biologists can confidently make is that with the current rate of habitat loss and fragmentation many more species will be added to the list of endangered species—and ultimately to the list of extinct species. Realistically, we cannot assume that even in the best scenarios severe environmental degradation will not occur in the future. In this perspective, it is not that useful to spend much time in refining predictions about the exact probability of metapopulation survival for the next 100 years, as such predictions typically assume a constant environment or are based on wild guesses as to how the environment might change.

Model predictions can be useful if the purpose is to compare two or more alternatives, for instance, to compare the probability of extinction under two or more management scenarios. Such comparative predictions are less sensitive to variation in model structure and parameter values than are predictions about the absolute risk of extinction (Hanski 1996; Ralls and Taylor 1997; Beissinger and Westphal 1998). In this context even complex metapopulation models that include a detailed description of local dynamics and dispersal might be meaningfully applied, in spite of the problems of model construction and parameter estimation. Examples of the use of complex simulation models of metapopulations include Lindenmayer's work on Leadbeater's possum (*Gymnobelideus leadbeateri*), an Australian marsupial (Lindenmayer et al. 1993; Lindenmayer and Possingham 1995), LaHaye et al.'s study (1994) of the California spotted owl (*Strix occidentalis occidentalis*), Akçakaya et al.'s modeling (1995) of the helmeted honeyeater (*Lichenostomus melanops*), and Akçakaya and Atwood's study (1997) of the California gnatcatcher (*Polioptila californica*). See Lindenmayer et al. (1995) for more examples and a comparison of the most widely used generic simulation models (ALEX, VORTEX, and RAMAS Space). Nonetheless, my opinion is that, unless an exceptional amount of empirical data are available, and provided that the landscape structure is highly fragmented, it is preferable to use relatively simple models, such as the

IFM, with fewer unverified assumptions and with properly estimated parameter values. Simple models are simplifications, but like the many questionable assumptions of complex models, these simplifications are often less critical for comparative than for absolute predictions. But be aware: some model simplifications may become absolutely critical and can be badly misleading in certain situations (Hanski 1999b, 118–119).

For management purposes, short-term predictions may be more useful than long-term predictions, even if long-term persistence is what matters in conservation. The consequences of management actions should be made evident as soon as possible to provide feedback for further management decisions. Short-term predictions are also useful for another reason. In the long run, environments will change in a manner that we cannot accurately predict, and, hence, long-term predictions about population performance are not likely to be reliable. This problem is less acute with short-term predictions.

Finally, we may take the view that in a changing environment the highest priority should be placed on maximizing the short-term performance of species to obtain favorable initial conditions for long-term persistence. One problem that may arise is that short-term and long-term population performances are maximized under different scenarios. For example, Burkey (1995) analyzed the effect of habitat subdivision on metapopulation survival by first empirically deriving the relationship between extinction rate and fragment size for mammals, birds, and lizards, and then comparing the predicted loss of species from either a single large area or five small, independent, and disconnected areas of the same total area. He found that the cumulative extinction probabilities were ultimately higher for the subdivided systems, although in the short term, the probability of extinction might be higher for the undivided habitat. One has to be aware of such possible differences between short-term and long-term performances. Here is another instance where models can be useful.

PVA FOR ORDINARY METAPOPULATIONS IN ORDINARY LANDSCAPES

Let us recall the key messages from previous sections. Long-term predictions about the destiny of any species are likely to remain academic exercises because the predictions generally assume that the environment remains unchanged—which we know is not going to happen—and because our models will rarely be good enough to make accurate predictions anyway. Modeling the viability of many highly endangered species remains an academic exercise because we do not have sufficient data to properly construct the models or to estimate the values of model parameters. What this leaves are short-term predictions for relatively common species and the question of whether such predictions matter.

I have two reasons to suggest that these predictions do matter for conservation biology. The first reason is based on the prediction that large numbers of currently not-so-endangered species will become highly endangered in the future if current trends in habitat loss and fragmentation continue. Rational applications of PVA should cover such species: conservation biologists should look ahead, be proactive as well as reactive, and be concerned with species approaching the state of being endangered. The second reason for expanding the range of species covered by PVA is that modeling can, at least in theory, be based on solid empirical results in the case of species that are not yet exceedingly rare and that fulfill model assumptions. The butterfly examples discussed above exemplify the possibility of using information for a related common species to construct and parameterize a model for an endangered species. Admittedly, this approach involves another source of uncertainty (i.e., species differences), but for robust comparative predictions this source of uncertainty might not be any worse than many other unknowns.

For species living in highly fragmented landscapes, important insights for the consequences of habitat loss and fragmentation can be gleaned from a simple extension of the Levins model (Levins 1969) incorporating a spatially realistic landscape structure with a finite number of patches (Hanski 1999a; Hanski and Ovaskainen 2000). Essentially, this model is a deterministic counterpart of the stochastic IFM and other related models, and it is especially helpful in elucidating the threshold condition for long-term persistence. Denoting by $p_i(t)$ the probability that patch i is occupied at time t, we may write the model

(1)
$$\frac{dp_i}{dt} = C_i(t)[1 - p_i] - \mu_i p_i,$$

for the rate of change in p_i (Hanski and Gyllenberg 1997), where $C_i(t)$ are the colonization rates of empty patches and μ_i are the extinction rates of extant populations. For the purpose of illustration, let us assume that $\mu_i = e/A_i$, where A_i is the area of patch i and e is an extinction parameter. Let us also assume (much like in the IFM) that the colonization rates are given by

(2)
$$C_i(t) = c \sum_{j \neq i} \exp(-\alpha d_{ij}) p_j(t) A_j,$$

where c is a colonization parameter, $1/\alpha$ gives the average distance of migration, and d_{ij} is the distance between patches i and j. Considering the threshold condition for metapopulation persistence ($p_i^* > 0$) in this model, it can be shown that λ_M, the leading eigenvalue of the matrix with elements

(3) $$m_{ij} = \exp(-\alpha d_{ij})A_j A_i, \quad m_{ii} = 0,$$

summarizes all the necessary information about landscape structure for assessing persistence (Adler and Nürnberger 1994; Hanski 1999a; Hanski and Ovaskainen 2000). In particular, there exist positive equilibrium values p_i^* if and only if

(4) $$\lambda_M > \delta,$$

where $\delta = e/c$. Hanski and Ovaskainen (2000) have termed λ_M the metapopulation capacity of a fragmented landscape. λ_M measures the amount and spatial configuration of suitable habitat in the landscape; that is, it takes into account the influence of habitat fragmentation as specified in model assumptions about the influence of patch area on extinction and connectivity on colonization. The metapopulation capacity can be closely approximated by the weighted average of the patch-specific colonization potential R_i, defined as the sum of the matrix elements m_{ij} for patch i,

(5) $$R_i = A_i \sum_{j \neq i} \exp(-\alpha d_{ij})A_j.$$

Thus (Adler and Nürnberger 1994; Hanski 1999a; Ovaskainen and Hanski 2001),

(6) $$\lambda_M \approx R_l \equiv \frac{\sum R_i^2}{\sum R_i} = \overline{R} + \frac{\text{Var}(R)}{\overline{R}},$$

which shows that metapopulation capacity increases with increasing average "value" of habitat patches, as measured by R_i, and by increasing aggregation of the habitat fragments, which increases the overall level of connectivity (the model assumes that local dynamics are independent; see Hanski and Ovaskainen 2000).

Note that condition 4 is analogous to the familiar threshold condition for persistence in the Levins model ($1 > \delta$), but with the effects of habitat patch area and connectivity on extinction and colonization explicitly included in the value of λ_M. A useful property of condition 4 is that the measure of landscape structure is located on the left-hand side of the inequality, whereas the parameters of the species are located on the right-hand side, with the exception of α, the migration range parameter (on the left-hand side, in the definition of connectivity, equation

2, setting the scale at which connectivity should be measured for the focal species). These results have the practical implication that the quantities λ_M and R_l may be used to rank different fragmented landscapes in terms of their capacity to support a viable metapopulation, even if the model has not been parameterized (e and c unknown, but note that a reasonable guess about the value of α has to be made; Hanski 1999a; Hanski and Ovaskainen 2000). Thus, the consequences of any changes to the landscape structure can be easily evaluated. One can even derive exact mathematical expressions for the values of particular patches of given areas and spatial locations in the landscape (Hanski and Ovaskainen 2000). Comparable results can be obtained using the IFM (Hanski 1999a), but an advantage of the metapopulation capacity concept and the related theory is that landscapes and individual habitat fragments can be characterized without doing any simulations. The limitation of condition 4 is that, as it is based on a deterministic model, the consequences of stochasticity are not included.

CONCLUSIONS

This chapter has three messages. First, metapopulation models are best used for making comparative predictions—for instance, in the context of comparing alternative management scenarios—rather than for attempting to predict long-term extinction risk of particular metapopulations. The latter predictions are repeatedly discredited by lack of adequate data to construct and parameterize PVA models and by our ignorance about future environmental changes.

Second, the best prospects for successful modeling are furnished by metapopulations with fast turnover of local populations in highly fragmented landscapes. In these cases, it is generally sufficient to focus on the processes of local extinction and colonization, and data can often be collected to rigorously parameterize patch occupancy models like the IFM. Insight to the effects of landscape structure on metapopulation viability can be gleaned from deterministic threshold conditions incorporating the effects of habitat patch area and connectivity on local extinction and colonization.

Third, it would be wise to expand the domain of the PVA to "ordinary" species inhabiting "ordinary" landscapes, because these are the situations that we can hope to model and because an important task for PVA should be to prevent these species from becoming endangered. Because habitat loss and fragmentation are a main cause of population and species extinction worldwide, metapopulation theory and models have the potential to make a significant contribution to PVAs extended to ordinary species in ordinary landscapes.

LITERATURE CITED

Adler, F. R., and B. Nürnberger. 1994. Persistence in patchy irregular landscapes. *Theoretical Population Biology* 45:41–75.

Akçakaya, H. R., and J. L. Atwood. 1997. A habitat-based metapopulation model of the California gnatcatcher. *Conservation Biology* 11:422–434.

Akçakaya, H. R., M. A. McCarthy, and J. L. Pearce. 1995. Linking landscape data with population viability analysis: management options for the helmeted honey-eater *Lichenostomus melanops cassidix*. *Biology Conservation* 73:169–176.

Ås, S., J. Begtsson, and T. Ebenhard. 1997. Archipelagoes and theories of insularity. *Ecology Bulletins* 46:88–116.

Beissinger, S. R., and M. I. Westphal. 1998. On the use of demographic models of population viability in endangered species management. *Journal of Wildlife Management* 62:821–841.

ter Braak, J. F., I. Hanski, and J. Verboom. 1998. The incidence function approach to modeling of metapopulation dynamics. Pages 167–188 in J. Bascompte and R. V. Solé, editors, *Modeling spatiotemporal dynamics in ecology*. Springer-Verlag, New York, New York.

Burkey, T. V. 1995. Extinction rates in archipelagoes: implications for populations in fragmented habitats. *Conservation Biology* 9:527–541.

Chesson, P. L., and T. J. Case. 1986. Overview: nonequilibrium community theories: chance, variability, history, and coexistence. Pages 229–239 in J. Diamond and T. J. Case, editors, *Community ecology*. Harper and Row, New York, New York.

Crowell, K. L. 1973. Experimental zoogeography: introduction of mice to small islands. *American Naturalist* 107:535–558.

Denno, R. F., and M. A. Peterson. 1995. Density-dependant dispersal and its consequences for population dynamics. Pages 113–130 in N. Cappuccino and P. W. Price, editors, *Population dynamics: new approaches and synthesis*. Academic Press, London, United Kingdom.

Diamond, J. M. 1984. "Normal" extinctions of isolated populations. Pages 191–246 in M. H. Nitecki, editor, *Extinctions*. University of Chicago Press, Chicago, Illinois.

Doak, D. F., and L. S. Mills. 1994. A useful role for theory in conservation. *Ecology* 75:615–626.

Dobzhansky, T., and S. Wright. 1943. Genetics of natural populations: 10, dispersion rates in *Drosophila pseudoobscura*. *Genetics* 28:304–340.

Doncaster, C. P., T. Micol, and S. P. Jensen. 1996. Determining minimum habitat requirements in theory and practice. *Oikos* 75:335–339.

Ebenhard, T. 1987. An experimental test of the island colonization survival model: bank vole (*Clethrionomys glareolus*) populations with different demographic parameter values. *Journal of Biogeography* 14:213–223.

———. 1991. Colonization in metapopulations: a review of theory and observations. *Biological Journal of the Linnean Society* 42:105–121.

Falk, D. A., and K. E. Holsinger. 1991. *Genetics and conservation of rare plants*. Oxford University Press, Oxford, United Kingdom.

Fiedler, P. L., and S. K. Jain. 1992. *Conservation biology: theory and practice of nature conservation, preservation, and management*. Chapman and Hall, London, United Kingdom.

Foley, P. 1994. Predicting extinction times from environmental stochasticity and carrying capacity. *Conservation Biology* 8:124–137.

Gotelli, N. J. 1991. Metapopulation models: the rescue effect, the propagule rain, and the core-satellite hypothesis. *American Naturalist* 138:768–776.

Hanski, I. 1989. Does it help to have more of the same? *Trends in Ecology and Evolution* 4:113–114.

———. 1991. Single-species metapopulation dynamics: concepts, models, and observations. *Biological Journal of the Linnean Society* 42:17–38.

———. 1994. A practical model of metapopulation dynamics. *Journal of Animal Ecology* 63:151–162.

———. 1996. Habitat destruction and metapopulation dynamics. Pages 217–227 in S. T. A. Pickett, R. S. Ostfeld, M. Shachak, and G. E. Likens, editors, *Enhancing the ecological basis of conservation: heterogeneity, ecosystem function, and biodiversity.* Chapman and Hall, New York, New York.

———. 1998. Metapopulation dynamics. *Nature* 396:41–49.

———. 1999a. Habitat connectivity, habitat continuity, and metapopulations in dynamic landscapes. *Oikos* 87:209–219.

———. 1999b. *Metapopulation ecology.* Oxford University Press, New York, New York.

Hanski, I., and M. E. Gilpin. 1991. Metapopulation dynamics: brief history and conceptual domain. *Biological Journal of the Linnean Society* 42:3–16.

Hanski, I., and M. Gyllenberg. 1997. Uniting two general patterns in the distribution of species. *Science* 275:397–400.

Hanski, I., A. Moilanen, T. Pakkala, and M. Kuussaari. 1996. The quantitative incidence function model and persistence of an endangered butterfly metapopulation. *Conservation Biology* 10:578–590.

Hanski, I., and O. Ovaskainen. 2000. The metapopulation capacity of a fragmented landscape. *Nature* 404:755–758.

Hanski, I., T. Pakkala, M. Kuussaari, and G. Lei. 1995. Metapopulation persistence of an endangered butterfly in a fragmented landscape. *Oikos* 72:21–28.

Hanski, I., and D. Simberloff. 1997. The metapopulation approach, its history, conceptual domain, and application to conservation. Pages 5–26 in I. Hanski and M. E. Gilpin, editors, *Metapopulation biology.* Academic Press, San Diego, California.

Hanski, I., and C. D. Thomas. 1994. Metapopulation dynamics and conservation: a spatially explicit model applied to butterflies. *Biological Conservation* 68:167–180.

Harrison, S. 1994. Metapopulations and conservation. Pages 111–128 in P. J. Edwards, R. M. May, and N. R. Webb, editors, *Large-scale ecology and conservation biology.* Blackwell Scientific, Oxford, United Kingdom.

Joern, A., and S. B. Gaines. 1990. Population dynamics and regulation in grasshoppers. Pages 415–482 in R. F. Chapmand and A. Joern, editors, *Biology of grasshoppers.* John Wiley and Sons, New York, New York.

Johnson, M. L., and M. S. Gaines. 1990. Evolution of dispersal: theoretical models and empirical tests using birds and mammals. *Annual Review of Ecology and Systematics* 21:449–480.

Kindvall, O. 1996. En naturvårdsbiologisk bektraktelse över de österlenska gröngrodornas undergång. Pages 69–80 in U. Gärdenfors and A. Carlson, editors, *Med huvudet före—Festskrift till Ingemar Ahléns 60-årsdag*. Rapport 33. Department of Wildlife Ecology, Swedish University of Agricultural Sciences, Uppsala, Sweden.

Kuussaari, M., M. Nieminen, and I. Hanski. 1996. An experimental study of migration in the butterfly *Melitaea cinxia*. *Journal of Animal Ecology* 65:791–801.

LaHaye, W. S., R. J. Gutierrez, and H. R. Akçakaya. 1994. Spotted owl metapopulation dynamics in southern California. *Journal of Animal Ecology* 63:775–785.

Lance, D., and P. Barbosa. 1979. Dispersal of larval Lepidoptera with special reference to forest defoliators. *Biologist* 61:90–110.

Lande, R. 1993. Risks of population extinction from demographic and environmental stochasticity and random catastrophes. *American Naturalist* 142:911–927.

Leigh, E. G., Jr. 1981. The average lifetime of a population in a varying environment. *Journal of Theoretical Biology* 90:213–239.

Levins, R. 1969. Some demographic and genetic consequences of environmental heterogeneity for biological control. *Bulletin of the Entomological Society of America* 15:237–240.

Lindenmayer, D. B., M. A. Burgman, H. R. Akçakaya, R. C. Lacy, and H. P. Possingham. 1995. A review of the generic computer programs ALEX, RAMAS/Space, and VORTEX for modelling the viability of wildlife metapopulations. *Ecological Modelling* 82:161–174.

Lindenmayer, D. B., T. W. Clark, R. C. Lacy, and V. C. Thomas. 1993. Population viability analysis as a tool in wildlife conservation policy: with reference to Australia. *Environmental Management* 17:745–758.

Lindenmayer, D. B., and H. P. Possingham. 1995. Modelling the viability of metapopulations of the endangered Leadbeater's possum in southeastern Australia. *Biodiversity Conservation* 4:984–1018.

MacArthur, R. H., and E. O. Wilson. 1963. An equilibrium theory of insular zoogeography. *Evolution* 17:373–387.

———. 1967. *The theory of island biogeography*. Princeton University Press, Princeton, New Jersey.

McCullough, D. R., editor. 1996. *Metapopulations and wildlife conservation*. Island Press, Covelo, California.

Merriam, G. 1991. Corridors and connectivity: animal populations in heterogeneous environments. Pages 133–142 in D. A. Saunders and R. J. Hobbs, editors, *Nature conservation: 2, the role of corridors*. Surrey Beatty and Sons, Chipping Norton, New South Wales, Australia.

Moilanen, A. 1998. Modeling metapopulation dynamics. Ph.D. dissertation, University of Helsinki, Finland.

———. 1999. Patch occupancy models of metapopulation dynamics: efficient parameter estimation with implicit statistical inference. *Ecology* 80:1031–1043.

———. 2000. The equilibrium assumption in estimating the parameters of metapopulation models. *Journal of Animal Ecology* 69:143–153.

Moilanen, A., and I. Hanski. 1999. Metapopulation dynamics: effects of habitat quality and landscape structure. *Ecology* 79:2503–2515.

Moilanen, A., A. T. Smith, and I. Hanski. 1998. Long-term dynamics in a metapopulation of the American pika. *American Naturalist* 152:530–542.

Okubo, A. 1980. *Diffusion and ecological problems: mathematical models*. Springer-Verlag, Berlin, Germany.

Ovaskainen, O., and I. Hanski. 2001. Spatially structured metapopulation models: global and local assessment of metapopulation capacity. *Theoretical Population Biology*. In press.

Pulliam, H. R. 1988. Sources, sinks, and population regulation. *American Naturalist* 132:652–661.

Ralls, K., and B. L. Taylor. 1997. How viable is population viability analysis? Pages 228–235 in S. T. A. Pickett, R. S. Ostfeld, M. Shachak, and G. E. Likens, editors, *The ecological basis of conservation: heterogeneity, ecosystems, and biodiversity*. Chapman and Hall, New York, New York.

Sjögren-Gulve, P., and C. Ray. 1996. Using logistic regression to model metapopulation dynamics: large-scale forestry extirpates the pool frog. Pages 111–137 in D. R. McCullough, editor, *Metapopulations and wildlife conservation*. Island Press, Covelo, California.

Skellam, J. G. 1951. Random dispersal in theoretical populations. *Biometrika* 38: 196–218.

Soulé, M. E., editor. 1987. *Viable populations for conservation*. Cambridge University Press, New York, New York.

Soulé, M. E., and K. A. Kohm, editors. 1989. *Research priorities for conservation biology*. Island Press, Covelo, California.

Thomas, C. D. 1994. Extinction, colonization, and metapopulations: environmental tracking by rare species. *Conservation Biology* 8:373–378.

Thomas, C. D., and I. Hanski. 1997. Butterfly metapopulations. Pages 359–386 in I. Hanski and M. E. Gilpin, editors, *Metapopulation biology*. Academic Press, San Diego, California.

Thomas, C. D., and T. M. Jones. 1993. Partial recovery of a skipper butterfly (*Hesperia comma*) from population refuges: lessons for conservation in a fragmented landscape. *Journal of Animal Ecology* 62:472–481.

Thomas, J. A., C. D. Thomas, D. J. Simcox, and R. T. Clarke. 1986. Ecology and declining status of the silver-spotted skipper butterfly (*Hesperia comma*) in Britain. *Journal of Applied Ecology* 23:365–380.

Veltman, C. J., S. Nee, and M. J. Crawley. 1996. Correlates of introduction success in exotic New Zealand birds. *American Naturalist* 147:542–557.

Wahlberg, N. 1997. The life history and ecology of *Melitaea diamina* (Nymphalidae) in Finland. *Nota Lepidopterologica* 20:70–81.

Wahlberg, N., A. Moilanen, and I. Hanski. 1996. Predicting the occurrence of endangered species in fragmented landscapes. *Science* 273:1536–1538.

Western, D., and M. Pearl. 1989. *Conservation biology in the 21st century*. Oxford University Press, Oxford, United Kingdom.

Wiens, J. A. 1977. On competition and variable environments. *American Scientist* 65:590–597.

―――. 1984. Resource systems, populations, and communities. Pages 397–436 in P. W. Price, C. N. Slobodchikoff, and W. S. Gand, editors, *A new ecology: novel approaches to interactive systems*. John Wiley and Sons, New York, New York.

Williamson, M. 1981. *Island populations*. Oxford University Press, Oxford, United Kingdom.

Plant Population Viability and Metapopulation-Level Processes

Susan Harrison and Chris Ray

ABSTRACT

We examine whether and when plant population viability studies must consider metapopulation issues such as the large-scale spatial configuration of populations. Plants differ from animals in several important ways with respect to population and metapopulation viability. Dispersal among populations is much harder to observe directly and is probably less extensive than for many animals. Dispersal may be generally less important to populations of plants than of animals because plants may resist local extinction through seed banks or clonal propagation. Pollination provides a second route by which spatial isolation may influence plant population viability. A review shows modest evidence that the large-scale spatial distribution of plant populations may influence their persistence. We present a case study to illustrate the use of simple, patch-level models for evaluating the importance of metapopulation effects. These patch-level analyses can be used to determine how important it is to include metapopulation processes in viability models.

INTRODUCTION

Deriving useful estimates of the viability of populations is a daunting task. Typically, even after considerable data are obtained, enormous uncertainties remain about the appropriateness of model structure, the accuracy of parameter estimates, and the unpredictability of the world in which real populations live. Thus, conservation biologists have come to realize that only limited and well-nuanced conclusions can be drawn from even the most careful PVAs. Recently, the concept that species exist as metapopulations has gained great popularity in conservation biology. Unfortunately, the difficulties of conducting a population viability analysis (PVA) increase when, in addition to all the other processes involved, it is necessary to measure and model metapopulation parameters, such as rates of among-population dispersal and colonization. It is well to ask, therefore, when it is reasonable for PVAs to focus on individual populations, and when it is necessary to consider multiple populations and the spatial relationships among them.

We address this question in the context of plants, in particular plants that naturally occur as small and isolated populations. Many plants are naturally rare and patchy owing to their restriction to a specialized soil type or other microhabitat, and such species often become of conservation concern (Kruckeberg and Rabinowitz 1986). In some cases the patchy habitat is ephemeral, so that local populations appear and disappear in response to disturbance (e.g., Menges 1990). But even for plants in ephemeral habitats, as well as for others, seed dispersal usually appears to be relatively localized (Husband and Barrett 1996). Recorded seed dispersal distances seldom exceed a few tens of meters, although this may reflect the difficulty of observing rare long-distance movements. For these reasons, a number of authors have suggested that it is natural to regard many plants as forming metapopulations (Eriksson 1996; Husband and Barrett 1996). Numerous studies have examined how patchy distributions and limited dispersal in plants affect the distribution of genetic variation (e.g., Barrett and Kohn 1991; Ellstrand and Elam 1993; van Treuren et al. 1993a,b; Broyles et al. 1994) and have considered the influence of metapopulation processes on plant evolution (Olivieri et al. 1990; McCauley 1994; Husband and Barrett 1996; Giles and Goudet 1997). Evidence regarding plant metapopulations remains limited (Husband and Barrett 1996), however, especially where questions of demographic viability are concerned.

SPATIAL EFFECTS MEDIATED BY SEED DISPERSAL

As in animals, one way that metapopulation structure could affect plant population viability is through dispersal leading to recolonization and rescue effects. Seed dispersal is notoriously hard to observe, however, and direct evidence for dispersal among populations is even scarcer for plants than for animals. Most evidence for plant metapopulation dynamics has been indirect. For example, in a pioneering study, Menges (1990) analyzed the viability of Furbish's lousewort (*Pedicularis furbishiae*) on the banks of the Saint John River in Maine. A matrix model was used to examine the demography of single populations; however, it was also noted that the lousewort's riverbank habitat was subject to ice-scour events that caused local extinctions. Thus, Menges (1990) inferred that metapopulation structure and long-distance waterborne dispersal were essential to the regional survival of the species.

Several studies have inferred metapopulation dynamics from analyses of the distribution of populations across habitat patches. Husband and Barrett (1996) reported that the occurrence of populations of the tropical aquatic plant *Eichhornia paniculata* was related to the density of habitat patches, and that local extinctions were frequent (37% per year) and unrelated to population size. Quintana-Ascensio and Menges (1996) found that 40 of 62 plant species occurring in patches of the rosemary scrub plant community in Florida showed significant relationships between population occurrence and the size or isolation of scrub patches. Although many of these fire-dependent species had long-lasting seed banks, Quintana-Ascensio and Menges (1996) argued that fire could nonetheless cause local extinctions, and that dispersal among patches with different fire histories was crucial to the regional persistence of these species. Subsequently, Quintana-Ascensio et al. (1998) demonstrated the existence of suitable unoccupied habitat for the rare herb *Hypericum cumulicola*, which requires recently burned patches of rosemary scrub.

Ouborg (1993) used direct observations of extinction and recolonization to infer metapopulation dynamics in plants by resurveying 143 riparian dune sites on the Rhine River that had been surveyed 32 years previously. For 5 of 15 herb species confined to these dunes, local extinction and/or colonization probabilities were related to the isolation of sites, measured either as distance to the nearest conspecific population or as the number of conspecific populations within 1.5 km. These authors pointed out the greater power of analyses based on process (i.e., observed extinction and colonization over time) as opposed to those based on pattern (i.e., a single snapshot of species distribution across sites) for disentangling the role of metapopulation processes. The case study of five plots we describe below uses a similar analysis at two points in time.

Despite these examples, it remains unclear whether seed dispersal at the landscape scale has important effects on plant demography, especially given the limited dispersal distances observed in plants (Silvertown 1991). For disturbance-dependent species, the relevant scale for dispersal and colonization may be relatively small, such as gopher mounds within a single field. Moreover, in contrast to most animals, many plants are able to survive periods of environmental adversity through dormant seeds or resistant, long-lived adult stages, and to reproduce vegetatively. Some authors have argued that such mechanisms for "escaping in time" may be more important for the persistence of plant populations than "escaping in space" through long-distance seed dispersal (Kalisz and McPeek 1993; Eriksson 1996).

SPATIAL EFFECTS MEDIATED BY POLLINATION

Another way for the size, isolation, and landscape position of plant populations to affect viability is through rates of pollination and successful reproduction. Whether plants are pollinated by insects, birds, or wind, insufficient pollination could increase extinction risks for small and isolated populations, and inhibit colonization of vacant sites. These potential problems have been of particular interest in the context of habitat fragmentation (Rathcke and Jules 1993; Bond 1994). While effects of spatial isolation on rates of pollination have been demonstrated at relatively small scales in several studies (e.g., Groom 1998), there have been far fewer studies at the scale of landscapes and populations. One such study (Aizen and Feinsinger 1994a,b) showed that habitat fragmentation led to reduced rates of pollination, but the effects of this reduction on plant demography were relatively weak. Bond (1994) pointed out that many plants are demographically buffered against a shortage of pollinators by long-lived adult stages, seed dormancy, or self-compatibility.

A related issue is whether small population size combined with isolation may lead to inbreeding depression. Population genetic structure and inbreeding in plants are the subjects of a literature too vast to be reviewed here. The genetic effects of natural rarity on plants have been reviewed by Barrett and Kohn (1991), and the genetic consequences of habitat fragmentation for plants by Holsinger (1993). In both cases, the authors concluded there is little evidence that inbreeding depression constitutes a major concern for the viability of wild populations, just as Lande (1988) concluded on more general grounds (however, see Allendorf and Ryman, chap. 4 in this volume). Populations small enough to suffer from inbreeding are likely to disappear for demographic reasons before inbreeding depression exerts a significant impact. However, demographically significant inbreeding depression was documented in small populations of two European herbs by van Treuren et al. (1993a,b).

Many more studies have focused on the genetic consequences of small population size than on the spatial isolation of populations (reviewed by Barrett and Kohn 1991; Ellstrand and Elam 1993). However, plant reproductive success in relation to large-scale habitat structure was examined in the serpentine morning glory (*Calystegia collina;* Wolf et al. 2000). Serpentine outcrops range in size from a few square meters to many square kilometers and support many endemic species (Kruckeberg 1984). Thus, small serpentine outcrops can be used to study the effects of a naturally patchy habitat, and for some purposes very large outcrops can serve as a "control." Wolf et al. (2000) compared the reproductive success of *Calystegia* in 39 plant patches on 16 small (<5 ha) serpentine outcrops and seven large (>300 ha) outcrops in a three-county region of California. Reproductive success was much higher on large outcrops, where about 25% of marked ramets produced seed capsules, compared to 5 to 10% on small outcrops. This difference was equally well explained by the number of other plant patches within

100 m of a focal patch. Rates of visitation by pollinators did not differ between small and large outcrops. Pollen augmentation showed that *Calystegia* is self-incompatible and pollen-limited. Therefore, reproductive success of *Calystegia* populations on small outcrops is sharply curtailed by the effect of isolation from sources of compatible, nonself pollen. Genetic diversity also correlated positively with reproductive success, but was independent of outcrop size and plant population isolation.

This is one of few studies to show that isolation at relatively large spatial scales, caused by the natural spatial structure of a landscape, has a significant influence on plant reproductive success. That the results may have some general significance for species distributions across the landscape is suggested by a study of species diversity patterns in the same system. Small serpentine patches supported significantly fewer serpentine-endemic species than did equivalent-sized sampling units within large outcrops, even though there were no significant differences in soil properties or species composition of the vegetation (Harrison 1997, 1999).

CASE STUDY: FIVE RARE PLANTS INHABITING SPRING-SEEPS ON SERPENTINE

The evidence reviewed above implies that it may often be reasonable for plant PVAs to overlook metapopulation considerations. Few studies provide strong evidence for the existence of regional plant networks or critical effects of isolation on local population viability. The deficit of evidence for metapopulation effects on local viability supports the argument that local persistence depends more on seed banks than on dispersal at a landscape scale. The weakness of this argument lies in the difficulty of actually observing extinctions, colonizations, and other aspects of large-scale population dynamics. In studies such as Ouborg (1993) and the one described below, however, there is some direct evidence concerning these processes.

Harrison et al. (2000) examined the persistence of five rare plant species inhabiting spring-seep habitats on serpentine soil. Serpentine seeps remain moist into midsummer and support a specialized flora. These seeps are typically isolated by hundreds to thousands of meters, forming habitat patches within large outcrops of serpentine. In the study region there are five serpentine-seep specialists, all considered sensitive taxa by the California Native Plant Society because of their narrow habitat requirements and small geographic ranges (Skinner and Pavlik 1995). These are *Helianthus exilis* (Asteraceae), *Senecio clevelandii* (Asteraceae), *Astragalus clevelandii* (Fabaceae), *Delphinium uliginosum* (Ranunculaceae), and *Mimulus nudatus* (Scrophulariaceae). We first asked how the spatial isolation of seeps affected the chances of local extinction and recolonization in these species.

In 1981–1982, the five species' distributions were surveyed in a 4,200-ha area as part of an environmental assessment for a mine (D'Appolonia 1982). This survey found a total of 218 populations on 87 seeps. In 1997 and 1999, we resurveyed all 87 sites, finding that 17 had been destroyed by construction and another 14 lacked current seep habitat. On the remaining 56 seeps, there were 32 local extinctions (i.e., species present in 1981–1982 followed by absence in 1997–1999) and 100 nonextinctions (i.e., species present in both surveys). This subset of the data was used in logistic regressions to determine the correlates of local extinction. There were also 65 apparent colonizations (i.e., absence in 1981–1982 followed by presence in 1997–1998) and 78 noncolonizations (i.e., absence in both surveys). This subset of the data was used in logistic regressions to examine the correlates of colonization.

We recognize several limitations in these data. First, since the populations were not examined every year, we missed some turnover that occurred during the 16-year period. However, we can still validly ask whether spatial isolation affected the *net* rates of extinction and colonization over the 16-year period. Second, and more seriously, we cannot be sure that populations were truly extinct in years when no plants were observed above ground. Plants could have been present but missed, and dormant seeds could have been present. There is little reason to doubt the presence data for either 1981–1982 or 1997–1999. We are also fairly certain of absences recorded in 1997–1999 because we expended considerable search effort for three consecutive years that spanned a large range of climatic conditions. We are less certain of the reliability of absences recorded in 1981–1982 because we cannot vouch for the degree of search effort in those years. Therefore, we regard our inferences about colonization as less reliable than those about extinction.

For each seep that was found in both surveys, and for each of the five species, we measured the distances to the nearest three conspecific populations, using the harmonic mean of these distances as our measure of population isolation. We also used other isolation measures, including distance to single nearest population, arithmetic distance to three nearest populations, and logarithm of mean distance to three nearest populations, and found that the results were robust to the choice of isolation measure. We measured the downstream length of each seep on topographic maps as an index of seep size and the distance from each seep to the nearest major human-caused disturbance (e.g., new roads and mining activities).

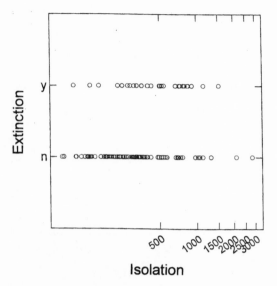

Isolation

Fig. 6.1 Isolation, defined as the harmonic mean distances in meters from the three nearest conspecific populations extant in 1997–1999, for populations found in 1981–1982 that were extinct (y) versus not extinct (n) in 1997–1999. The species were *Helianthus exilis, Astragalus clevelandii, Senecio clevelandii, Mimulus nudatus,* and *Delphinium uliginosum* on the University of California's D. and S. McLaughlin Natural Reserve in Lake and Napa Counties, California.

For all species combined, chances of local extinction increased with increasing isolation (fig. 6.1) and decreased with increasing distance from disturbance (isolation, $t = 2.19$, $P = 0.029$; distance from disturbance, $t = -1.82$, $P = 0.068$; overall model, -2 log likelihood $= 7.68$, 2 df, $P = 0.021$). Colonization also showed a significant decrease with increasing isolation ($t = -2.00$, $P = 0.046$; fig. 6.2), and the overall model was significant (2 log likelihood $= 20.29$, 5 df, $P = 0.001$). *Mimulus nudatus* ($t = 1.90$, $P = 0.057$) and *Helianthus exilis* ($t = 2.44$, $P = 0.015$) had higher rates of colonization than the other species. We used a Monte Carlo permutation method to test whether species distributions were random or clumped among seeps. All species' distributions were spatially random in 1981–1982. As a result of isolation-dependent extinction and colonization, however, the spatial distributions of two species (*Delphinium uliginosum* and *Senecio clevelandii*) were significantly clumped among seeps in 1997–1999.

We also examined correlation in population fluctuations at a regional scale. We censused population densities for three of the seep species at 50 sites across a 20 × 40 km region in 1997–1999. Between 1997

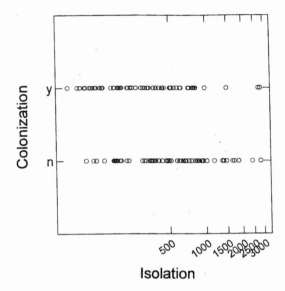

Fig. 6.2 Isolation, defined as the harmonic mean distance in meters from the three nearest conspecific populations extant in 1997–1999, for sites where species were absent in 1981–1982 that were colonized (*y*) versus not colonized (*n*) by 1997–1999. The species were *Helianthus exilis, Astragalus clevelandii, Senecio clevelandii, Mimulus nudatus,* and *Delphinium uliginosum* on the University of California's D. and S. McLaughlin Natural Reserve in Lake and Napa Counties, California.

and 1998, *Helianthus exilis* decreased strongly, *Delphinium uliginosum* increased moderately, and *Mimulus nudatus* increased strongly at virtually every site in this region. Population fluctuations within each species were so completely synchronous at a regional scale that neither population density nor change in density showed significant spatial autocorrelation under standard analyses (Cliff and Ord 1981; Isaaks and Srivastava 1989). In other words, conspecific populations behaved similarly regardless of their distance from one another. Between 1998 and 1999, *Mimulus nudatus* declined synchronously at all sites, while the other two species behaved more asynchronously; again, there was no consistent tendency for populations that were closer together to behave more similarly than those that were farther apart.

Like many plants, the species we studied were subject to regional-scale fluctuations driven by weather, and probably rely in part on long-lived dormant seeds to survive in their harsh and fluctuating environment. These considerations would appear to argue against the importance of metapopulation processes. Nonetheless, we found a detectable role for spatial isolation in determining the likelihood of local extinction and recolonization, implying a possible role for large-scale dispersal in population persistence.

Having found that spatial isolation may affect population persistence, more questions arise. First, what are the mechanisms? For one of our five seep species, *Mimulus nudatus* (an annual), we are conducting field experiments to measure seed dispersal distances, assess the effects of isolation on pollination and seed set, and test whether unoccupied sites are indeed suitable. This study should help us narrow down the list of plausible mechanisms. Another set of questions, perhaps more relevant to the present discussion, is what are the consequences for regional persistence? Do these results imply that metapopulations of serpentine-seep plants are dynamically fragile? How many populations or habitats could be removed before the species would be in danger of regional extinction?

To address the possible conservation implications of the empirical results, we used dynamic logistic regression modeling, as described by Sjögren-Gulve and Ray (1996). Metapopulation dynamics were simulated for each of the five plant species in the serpentine-seep system on a set of 56 patches with the observed spatial distribution. In these simulations, local probabilities of colonization and extinction depended on patch isolation and distance from disturbance, according to the logistic relationships determined for each species. New probabilities were calculated each time step, based on the current configuration of occupied patches. Each model was initialized with the observed distribution of populations in 1981–1982 and was run for 16 time steps (= 256 years) with 500 replicate simulations per species.

To evaluate the accuracy of the model, we asked whether the predicted number of occupied patches matched the observed values. If some of the apparent "absences" in 1981–1982 were actually populations that were missed, colonization rates in the model could have been artificially inflated. Significant errors of this type should have caused the models to predict occupancy rates higher than those observed in the 1997–1999 survey, which was never the case. In all but one case, observed occupancy rates were within the 95% confidence interval for predictions (fig. 6.3). The exception was *Delphinium uliginosum*, for which observed occupancy was always higher than that predicted by the model. This result suggests two possibilities: either the model did not capture the metapopulation dynamics of *D. uliginosum*, or the species has been declining toward a lower equilibrium occupancy than the one presently observed. Because the model predicted individual patch occupancies

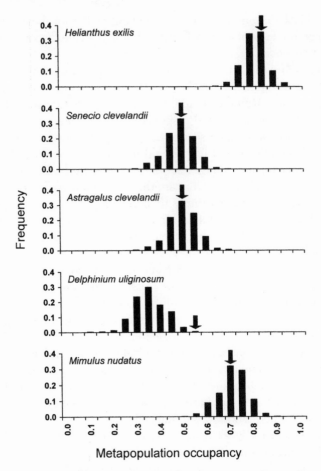

Fig. 6.3 Predicted versus observed occupancy in five plant metapopulations. Histograms show 500 model predictions for quasi-equilibrium occupancy. *Arrows* indicate recently observed occupancy values. *Delphinium uliginosum* is the species most sensitive to isolation, and may be in a period of slow decline toward lower occupancy (see text).

very well for this species ($P = 0.003$), we believe it captured the relevant metapopulation effects on local persistence. Occupancy is probably declining in a portion of this metapopulation.

To test the dynamic stability of each species under hypothetical human impacts, we performed two simulations with the models. First, we arbitrarily removed 28 patches from either the western or the eastern half of the metapopulation. Second, we extended the apparent influence of disturbance by decreasing the distance from disturbance of each population by 100 m.

The simulation results indicated a very high degree of stability in these plant metapopulations. The probability of regional extinction within 256 years was zero in nearly all cases. Even removing half of the patches almost never led to extinction for any species. Extending the influence of human-caused disturbance by 100 m brought disturbance within 0 m of 21 patches, yet never caused metapopulation extinction for any species.

Delphinium uliginosum, however, was somewhat more fragile than the other four species. Removing the eastern half of the metapopulation led to extinction in the western half with a probability of 0.24 in 256 years. But removing the western half still yielded no chance of extinction in the east. We also tried removing half the patches in a spatially random pattern, which should result in greater isolation for the remaining patches. Results were sensitive to the configuration of the randomly se- lected patches: for three different configurations of 28 patches, the chances of extinction after 256 years were 0.06, 0.07, and 0.32.

High rates of observed turnover suggest that metapopulation dynam- ics are important in this system. Extinction and recolonization appear to be only weakly dependent on patch isolation, however, probably due to the existence of seed banks and the haphazard nature of long-distance seed dispersal. Therefore, the persistence of these metapopulations does not appear to depend strongly on spatial aspects of metapopulation structure. Even in the case of the least viable species, *Delphinium uligi- nosum,* we found that it was much more likely to survive than become extinct when half the habitat patches were removed. In other words, although metapopulation processes exist in this system, they do not con- fer the fragility of local or regional persistence that has been noted in other cases, such as pool frogs in Sweden (Sjögren-Gulve and Ray 1996).

Simple, patch-level models like the one employed here, or the inci- dence function model introduced by Hanski (1994, chap. 5 in this vol- ume), can be used to determine whether metapopulation effects are important for regional persistence. Where metapopulation effects are evident, PVA models should incorporate landscape-level parameters. Of course, PVA models must also include local population processes, be- cause there are limits to our ability to extrapolate viability from snap- shots of patch occupancy or patch state transitions. This caution is espe- cially appropriate for plant metapopulations, due to the relatively poorly known role of seed banks.

CONCLUSIONS

Several lines of evidence converge on a conclusion that may be good news for PVA of plants. First, there are relatively few cases in which the spatial isolation of plant populations has been shown to affect either demographic performance or the likelihood of extinction. Second, there are several reasons to expect that plant population viability is relatively insensitive to metapopulation influences, at least in the short to medium term. These include the existence of seed banks and other extinction-resistant life stages, self-compatibility and clonal propagation, and a much poorer capacity for dispersal than most animals. Third, even when metapopulation dynamics can be successfully inferred using data on regional extinction and colonization, these dynamics may not significantly alter the likelihood of persistence. Thus, it may be reasonable in many cases for plant PVAs to continue focusing on individual populations (e.g., Menges 1998).

This conclusion must be hedged with the usual caution about the absence of evidence not equaling the evidence of absence. Clearly, conclusive data on extinctions, colonizations, and regional influences on demography are very difficult to obtain, and this may account for its scarcity. Efforts to increase our knowledge of these processes should be intensified. The availability of geographic information systems (GIS) and molecular technology should enable new advances in this area. Ironically, the importance of metapopulation effects, such as the need for rescue, recolonization, and gene flow to maintain viable populations, is likely to be greater as longer time horizons are considered. Metapopulation processes may be crucial to survival in the long term, beyond the time horizon for which PVAs are likely to provide useful answers.

When empirical evidence does exist for metapopulation processes, patch-level modeling is an excellent tool for examining its implications. Although our study is not a PVA in the traditional sense, we believe it illustrates that it is not always necessary to use age-structured or individual-based models to make predictions about population trajectories. Sometimes it can be more appropriate to use a large-scale, coarse-grained approach such as the one we have employed here.

LITERATURE CITED

Aizen, M. A., and P. Feinsinger. 1994a. Forest fragmentation, pollination, and plant reproduction in a Chaco dry forest, Argentina. *Ecology* 75:330–351.
———. 1994b. Habitat fragmentation, native insect pollinators, and feral honey bees in Argentine "Chaco Serrano." *Ecological Applications* 4:378–392.
Barrett, S. C. H., and J. R. Kohn 1991. Genetic and evolutionary consequences of small population size in plants: implications for conservation. Pages 3–30 in D. A. Falk and K. E. Holsinger, editors, *Genetics and conservation of rare plants*. Oxford University Press, Oxford, United Kingdom.
Bond, W. J. 1994. Do mutualists matter? Assessing the impact of pollinator and

disperser disruption on plant extinction. *Philosophical Transactions of the Royal Society of London*, series B, Biological Sciences, 344:83–90.

Broyles, S. B., A. Schnabel, and R. Wyatt 1994. Evidence for long-distance pollen dispersal in milkweeds (*Asclepias exaltata*). *Evolution* 48:1032–1040.

Cliff, A. D., and J. K. Ord. 1981. *Spatial processes: models and applications*. Pion, London, United Kingdom.

D'Appolonia Company. 1982. McLaughlin project: proposed gold mine and mineral extraction facility, Homestake Mining Company. Environmental report. D'Appolonia, San Francisco, California.

Ellstrand, N. C., and D. R. Elam. 1993. Population genetic consequences of small population size: implications for plant conservation. *Annual Review of Ecology and Systematics* 24:217–242.

Eriksson, O. 1996. Regional dynamics of plants: a review of evidence for remnant, source-sink, and metapopulations. *Oikos* 77:248–258.

Giles, B. E., and J. Goudet 1997. Genetic differentiation in *Silene dioica* metapopulations: estimation of spatio-temporal effects in a successional plant species. *American Naturalist* 149:507–526.

Groom, M. J. 1998. Allee effects limit population viability of an annual plant. *American Naturalist* 151:487–496.

Hanski, I. 1994. A practical model of population dynamics. *Journal of Animal Ecology* 63:151–162.

Harrison, S. 1997. How natural habitat patchiness affects the distribution of diversity in California serpentine chaparral. *Ecology* 78:1898–1906.

———. 1999. Local and regional diversity in a patchy landscape: native, alien, and endemic herbs on serpentine soils. *Ecology* 80:70–80.

Harrison, S., J. Maron, and G. Huxel. 2000. Regional turnover and fluctuation in populations of five plants confined to serpentine seeps. *Conservation Biology* 14:769–779.

Holsinger, K. S. 1993. The evolutionary dynamics of fragmented plant populations. Pages 198–216 in P. M. Kareiva, J. G. Kingsolver, and R. B. Huey, editors, *Biotic interactions and global change*. Sinauer Associates, Sunderland, Massachusetts.

Husband, B. C., and S. C. H. Barrett. 1996. A metapopulation perspective in plant population biology. *Journal of Ecology* 84:461–469.

Isaaks, E. H., and R. M. Srivastava. 1989. *An introduction to applied geostatistics*. Oxford University Press, New York, New York.

Kalisz, S., and M. A. McPeek. 1993. Extinction dynamics, population growth, and seed banks: an example using an age-structured annual. *Oecologia* 95:314–320.

Kruckeberg, A. R. 1984. *California serpentines: flora, vegetation, geology, soils, and management problems*. University of California Press, Berkeley, California.

Kruckeberg, A. R., and D. Rabinowitz. 1986. Biological aspects of endemism in higher plants. *Annual Review of Ecology and Systematics* 6:447–479.

Lande, R. 1988. Genetics and demography in biological conservation. *Science* 241:1455–1460.

McCauley, D. E. 1994. Contrasting the distribution of chloroplast DNA and allozyme polymorphism among local populations of *Silene dioica*: implications for studies of gene flow in plants. *Proceedings of the National Academy of Sciences* (USA) 91:8127–8131.

Menges, E. S. 1990. Population viability analysis for a rare plant. *Conservation Biology* 4:52–62.

———. 1998. Evaluating extinction risks in plant populations. Pages 49–65 in P. L. Fiedler and P. Kareiva, editors, *Conservation biology for the coming decade.* Chapman and Hall, New York, New York.

Olivieri, I., D. Couvet, and P.-H. Gouyon. 1990. The genetics of transient populations: research at the metapopulation level. *Trends in Ecology and Evolution* 5: 207–210.

Ouborg, N. J. 1993. Isolation, population size, and extinction: the classical and metapopulation approaches applied to vascular plants along the Dutch Rhine system. *Oikos* 66:298–308.

Quintana-Ascensio, P. F., R. W. Dolan, and E. S. Menges. 1998. *Hypericum cumulicola* demography in unoccupied and occupied Florida scrub patches with different time-since-fire. *Journal of Ecology* 86:640–651.

Quintana-Ascensio, P. F., and E. S. Menges. 1996. Inferring metapopulation dynamics from patch-level incidence of Florida scrub plants. *Conservation Biology* 10: 1210–1219.

Rathcke, B. J., and E. S. Jules. 1993. Habitat fragmentation and plant-pollinator interactions. *Current Science* 65:273–277.

Silvertown, J. 1991. Dorothy's dilemma and the unification of plant population biology. *Trends in Ecology and Evolution* 6:346–348.

Sjögren-Gulve, P., and C. Ray. 1996. Using logistic regression to model metapopulation dynamics: large-scale forestry extirpates the pool frog. Pages 111–137 in D. R. McCullough, editor, *Metapopulations and wildlife conservation.* Island Press, Covelo, California.

Skinner, M. W., and B. M. Pavlik. 1995. *California Native Plant Society's inventory of rare and endangered plants of California.* CNPS Special Publication no. 1. California Native Plant Society, Sacramento, California.

van Treuren, R., N. Bijlsma, J. Ouborg, and W. van Delden. 1993a. The effects of population size and plant density on outcrossing rates in locally endangered *Salvia pratensis. Evolution* 47:1094–1104.

———. 1993b. The significance of genetic erosion in the process of extinction: 4, inbreeding depression and heterosis effects caused by selfing and outcrossing in *Scabiosa columbaria. Evolution* 47:1669–1680.

Wolf, A. T., S. Harrison, and J. L. Hamrick. 2000. Influence of habitat patchiness on genetic diversity and spatial structure. *Conservation Biology* 14:454–463.

7

Population Viability Analysis and Conservation Policy

Mark Shaffer, Laura Hood Watchman, William J. Snape III, and Ingrid K. Latchis

ABSTRACT

This survey of the application of population viability analysis (PVA) in the policy and management arenas suggests three major lessons that have been learned but not acted upon. First, a major limitation to meaningful application of PVA to conservation problems is the lack of detailed population data for most taxa of concern. This limitation has been known for 15 to 20 years, yet little has been done by either the basic or applied research agencies to address it. Second, despite a general consensus that populations of less than a few thousand individuals are of questionable viability, the median population goal to consider a species recovered under the Endangered Species Act is about 1,500. Third, despite widespread acceptance within the scientific community that viability can only be judged relative to an explicit time horizon and probability of persistence, neither the political, management, nor legal communities have articulated, much less agreed upon, a workable definition of viability by which the adequacy of conservation efforts can be judged.

Acting on these lessons to the benefit of conservation will require five initiatives: (1) the development of standards for what constitutes an adequate PVA; (2) long-term field studies of the demography and genetics of a number of species representative of major life-history types; (3) the expansion of laboratory and field experiments on population lifetimes; (4) creation and maintenance of databases on population, PVA models, and their applications to conservation actions (e.g., listing/delisting decisions, habitat conservation plans, recovery plans, etc.); and (5) development of rules of thumb for deciding how much habitat or what population size is enough to achieve conservation in those many cases where there are neither data nor models to support an adequate PVA for the taxon in question. We suggest that three principles are important in developing rules of thumb for making viability judgments: representation, redundancy, and resiliency.

INTRODUCTION

Population viability analysis (PVA) is essentially a risk-assessment methodology applied to the issue of species extinction. It can be seen as one component of a much larger body of scientific data, information, and knowledge that could be called "extinction theory." PVA developed out of a recognition that chance factors can play a role in determining species survival that is inversely proportional to population size and thus, for many species, available habitat. The development of PVA has been fueled by the interest of society in conserving endangered species. Thus, PVA is a methodology born of a public-policy need.

The utility of PVA for meaningfully quantifying extinction probabilities is currently in question (Boyce 1992, 1993; Beissinger and Westphal 1998; Ludwig 1999). What is not in question is the need for policy and decision makers to answer the questions: How much is enough? How much habitat does any particular species need in order to maintain a population large enough to survive over the long term?

In this chapter we briefly review the many places in federal law and policy where PVA could provide potentially useful guidance. We then examine the attributes that PVA must have to be useful in the policy arena and provide an assessment of the degree to which PVA currently has achieved those attributes. Finally, we draw some broad lessons from our experience with PVA to date and offer some guidance for both improving the application of PVA to conservation policy and improving our understanding of the extinction process.

APPLICATIONS FOR PVA

Because the U.S. Endangered Species Act of 1973 (ESA; *U.S. Code,* vol. 16, secs. 1531–1544) codifies in law a national policy of avoiding the extinction of species, PVA potentially has many useful applications in environmental policy and management. It could be used to determine if a major federal action would entail an irreversible commitment of resources by causing one or more species to have an unacceptably high probability of extinction, thus helping environmental decision making under the National Environmental Policy Act of 1969 (NEPA; *U.S. Code,* vol. 42, secs. 4321–4347). PVA is a logical and potentially very useful tool in identifying critical habitats, determining jeopardy, establishing recovery goals, and designing habitat conservation plans under the ESA.

PVA can and has been used in making listing decisions under the ESA. For example, the northern spotted owl (*Strix occidentalis caurina*) was listed as threatened by the U.S. Fish and Wildlife Service after a series of lawsuits in which PVA concepts and models figured heavily (see, e.g., *Seattle Audubon v. Evans,* 952 F. 2d 297 [9th Cir. 1991], and *Northern Spotted Owl v. Hodel,* 716 F. Supp. 479 [WD Wash. 1988]). There have been at least three major PVAs conducted on the grizzly bears (*Ursus arctos*) of the Yellowstone ecosystem (Boyce 1995), and an eventual decision on delisting this population may ultimately hinge on what these models suggest about the prospects for that population's survival.

Recovery planning for endangered species can incorporate PVAs in establishing recovery goals and identifying and prioritizing management actions. Indeed, the U.S. Fish and Wildlife Service promotes the appropriate use of PVAs in recovery-planning training courses (D. Crouse, personal communication). From preliminary results of a recent study of endangered-species recovery plans sponsored by the U.S. Fish and Wildlife Service, the Society for Conservation Biology, and the National Center for Ecological Analysis and Synthesis, it appears that more recent recovery plans were more likely to use PVAs than earlier plans. Nevertheless, fewer than 20% of 86 recovery plans developed since 1992 used PVAs. The study, however, did not examine the proportion of plans for which it would be appropriate to develop a PVA. Moreover, it appears that recovery plans tended to assign tasks to collect only a portion of the data needed to develop PVAs in the future (W. Morris, personal communication). Thus, there is little preliminary evidence from this recent review to show that the recovery-planning process is likely to lead to an improvement in the availability of good data to support useful PVAs. This is unfortunate because an earlier review of recovery plans (Schemske et al. 1994) found that only 33% of endangered plant species had sufficient detailed biological information to allow a general assessment of their population dynamics, let alone for a highly detailed PVA.

By contrast, PVAs seem to be more available for species covered by a habitat conservation plan (HCP). In a sample of 50 HCPs compiled from data in Hood (1998) and Kareiva et al. (1999), 23 (46%) involved a species for which a PVA had been completed. Thirteen of those HCPs included information from the PVA in decision making for the HCP. For example, in Riverside County, California, a PVA was completed for the Stephen's kangaroo rat (*Dipodomys stephensi*) for all of the proposed core reserves as well as for individual reserves. In the Plum Creek Timber HCP involving the northern spotted owl in the central Cascade Mountains of Washington, the company modeled the population trend of spotted owls in the HCP area over the lifetime of the HCP.

PVA would also seem tailor-made for implementing the "viability" requirement of the U.S. National Forest Management Act. In the act's regulations (*Code of Federal Regulations*, vol. 36, sec. 219), it is mandated that "[f]ish and wildlife habitat shall be managed to maintain viable populations of existing native and desired nonnative vertebrate species in the planning area . . . In order to insure that viable populations will be maintained, habitat must be provided to support at least a minimum number of reproductive individuals and the habitat must be well distributed so that those individuals can interact with others in the planning area."

Clearly, applying PVAs to these forest species could serve as a tool for forest planning and management, and for the public to assess if the U.S. Forest Service is fulfilling its viability mandate. Indeed, PVA concepts and/or models have been employed in attempts to accommodate spotted owls and other old-growth-dependent species in forest planning in the Pacific Northwest, and in maintaining habitat for a cross section of avian and mammalian predators in the Tongass National Forest in southeastern Alaska.

The interpretation that maintaining diversity in national forests entails maintaining species viability is further supported by the U.S. Department of Agriculture's Committee of Scientists. In the committee's final report (COS 1999, 38), they stated that "the goals of ensuring species viability and providing for diversity are inseparable." They go on to state "a first step in providing for species viability is to assess the likelihood that a species will be viable over specified periods . . . Since viability can never be ensured with 100% certainty, whether a population is deemed viable is a decision based on an acceptable risk of extinction. Ultimately, this is a value-based, not a science-based, decision."

One promising use of PVA in the management and policy arena is the Marine Mammal Protection Act (MMPA; *U.S. Code,* vol. 16, secs. 1361–1407). The 1994 amendments to the MMPA require the secretary of commerce to consult with regional scientific review groups and draft a stock assessment for each marine mammal stock that occurs in U.S. waters. Each stock assessment is to possess the geographic range of the stock, provide a minimum population estimate and current population trend, estimate annual human-caused mortality, describe the commercial fisheries that interact with the stock, and estimate the potential biological removal for the stock, taking into account a recovery factor (*U.S. Code,* vol. 16, sec. 1386).

Thus far, there is some indication that PVA is becoming a stronger tool for supporting policy and management decisions on behalf of marine mammals through this stock-assessment process. One example is the Atlantic Large Whale Take Reduction Plan, which has aimed to reduce the level of mortality and serious injury to several endangered whale species that are frequently harmed by East Coast fisheries (Gwinn et al. 1999).

Although it is easy to find other examples of almost all these applications of PVA in the management and policy arenas, it is impossible, given readily available information, to assess either the degree to which PVA is used in these types of applications or the quality of those instances in which it has been employed.

UTILITY OF PVA

Goodman (1990) proposed that, to be useful in the policy arena, PVA would have to have five essential features: (1) PVA must provide credible predictions of population survival given management choices; (2) PVA must suggest credible interventions that would improve the prospects of population survival; (3) PVA must employ unambiguous criteria for what constitutes success and failure; (4) the policy arena itself must better define the burden of proof and how decisions are made in the absence of good information (i.e., proof) to resolve a decision; and (5) PVA must employ some acceptable standards of evidence. It is a fair question to ask to what extent PVA currently manifests these five essential features.

Credible Predictions

We can find no example of a PVA model that has been used to forecast an extinction of a wild population that actually happened within the confidence limit of the model (whether explicit or implicit). In fairness, PVA is aimed at assessing the likelihood of an outcome that is both (1) distant in the future (i.e., many decades to centuries) relative to the history of the method (only two decades), and (2) an outcome that our management system is charged with avoiding. Both of these conditions greatly confound efforts at empirical and/or experimental validation of the models. Thus, empirical evidence to corroborate PVA model predictions is likely to accrue slowly, and the lack of predicted extinctions of wild populations cannot, at this juncture, be taken as strong evidence against the credibility of PVA models.

The only experimental test of the most common models of which we are aware (Belovsky et al. 1999, chap. 13 in this volume) showed that, on a qualitative basis, the underlying precepts of PVA are correct, namely, that population lifetimes do depend on the amount of habitat provided and the variability in the quality of that habitat over time. However, these models also proved to be relatively inaccurate in terms of quantifying the expected lifetimes of the laboratory populations of brine shrimp that were the experimental system, despite having very precisely measured demographic parameters for model inputs. The more commonly used PVA models were often optimistic and overestimated the actual lifetimes of the lab populations. The best results were obtained with the simplest models.

These results indicate that it is not just the lack of good data that constrains our ability to accurately predict the lifetimes of populations. Work remains to be done on the basic structure of PVA models as well. Nonetheless, this experiment also confirms that concern for the amount and quality of habitat that is maintained for a species is well justified.

Credible Intervention

Boyce (as cited in Mann and Plummer 1999) suggested that the early PVAs on grizzly bears helped recovery efforts by highlighting the disproportionate importance of adult female mortality on the probability of persistence This, in turn, supposedly led to strong efforts to reduce mortality to this segment of the population. Given the controversy over the true status of the Yellowstone grizzly bear population (Pease and Mattson 1999), it may be premature to judge the efficacy of this intervention. In another example, the first application of the metapopulation paradigm to a conservation problem strongly suggested that retention of an adequate amount and pattern of old-growth forest were essential to assuring the survival of the northern spotted owl. More and larger blocks of old-growth-forest habitat were ultimately protected than had been planned originally. The long-term results, however, remain to be seen.

Unambiguous Criteria

One area in which it is clear that PVA has not met Goodman's admonitions is in the lack of unambiguous criteria for judging success or failure. PVAs vary in the time horizon and probability of persistence employed. Until society has expressed an acceptable level of risk and a time horizon of interest with regard to species extinction, conservation efforts are (1) not in accord with the underlying reality of the extinction process and (2) subject to endless criticism that any action taken on behalf of a species has either gone too far, or not far enough, depending on one's point of view. Although the National Marine Fisheries Service has used PVA modeling to set unambiguous criteria for judging the impacts on marine mammals of incidental take in commercial fisheries, this is a promising exception rather than the general rule. This means that, irrespective of its accuracy or precision, PVA, in most applications, is generally a weak tool for supporting policy and management decisions.

Burden of Proof

Of Goodman's five essential features for a useful PVA-based conservation policy, this is the one that has the least nexus with science. While there is no scientific reason why the burden should be placed on any particular party, both international and U.S. law presently provide authority to adopt a precautionary approach to species conservation in general, although, when applied, this usually has been interpreted by the government to be its burden. For example, the Convention on Biological Diversity (CBD; *International Legal Materials* 31 [1992]: 818) states "that where there is a threat of significant reduction or loss of biological diversity, lack of full scientific certainty should not be used as a reason for postponing measures to avoid or minimize such a threat." Similarly, the U.S. Supreme Court has observed that Congress's passage of the ESA represented its desire to "halt and reverse the trend toward species extinction, whatever the cost" (*TVA v. Hill*, 437 U.S. 153, 184 [1978]). Realization of these lofty ideals has been admittedly difficult in the face of short-term economic and political pressures (Kuhlmann 1996), but we know that civil society has been quite clear in its goal to avoid species extinction. We also know from experience that the public, and the politicians who represent the public, have a naïvely generous comprehension of how much ecologists and wildlife biologists know of the natural world, and the ease and rapidity with which that knowledge can be increased.

Of course, the *duty* to conserve species is nearly impossible without the *ability* to conserve them. So long as the burden of proof on PVAs rests with the management agencies without clearer guidance from the scientific community, the development and use of models based on whatever data are at hand will often occur. Without guidance from scientists, government agencies will often rely on PVAs that first serve political and financial interests, which in turn will elicit strong opposition from environmentalists, scientists, and many government employees. The result of this regulatory confusion will be an eventual reduction of judicial deference to agency actions. This deference to agencies was enunciated under *Chevron v. Natural Resources Defense Council* (467 U.S. 837 [1984]) and will benefit no interest in the long run. Regulatory confusion will also induce reliance on models that do not represent the best available science. If such trends continue, we predict increased litigation pursuant to *Daubert v. Merrell Dow* (509 U.S. 579 [1993]), which established a four-factor standard for admitting expert scientific evidence in federal cases. The four factors were whether the theory (1) is testable, (2) has been peer-reviewed, (3) has a known rate of error and standards for calculating it, and (4) has general acceptance in the scientific community. A PVA model would have to meet these standards to be used as scientific evidence in a federal case. The predicted increase in litigation over inappropriate or incorrect models will proceed simultaneously with continued on-the-ground declines in biodiversity.

Standards of Evidence

Here again, as currently employed in the policy arena, PVA does not meet Goodman's criteria for effectiveness. One of the principal criticisms of PVA is that the models get used, even when there are not sufficient data (Beissinger and Westphal 1998). One can view this simply as laziness or incompetence—someone else's failure to understand how things should be done properly. For example, one can see this as evidence of the degree to which the resource managers and policymakers must answer the question, How much habitat is enough? The demand for answers is outstripping the supply of information on which to base them. As with the acceptable level of risk and time horizon issues, establishing standards of evidence will be crucial to better application of PVA for the resolution of conservation issues. Nevertheless, clarification on this point will leave many situations where decisions will have to be made without data of sufficient quality to employ any PVA models.

BROAD LESSONS ON THE USE OF PVA IN CONSERVATION DECISION MAKING

The preceding review of PVA reveals how far from perfect the method is as a science-based policy and decision-making tool. It also provides a guidepost for steps to be taken, either to improve the method or to delimit the range of its applications to those situations where it provides meaningful guidance. Before turning to specific recommendations for future improvement and application of PVA, it is important to reflect on the larger, broader lessons that should have been learned by now through the many efforts to develop and apply PVAs to conservation issues. There are at least three such lessons.

Lesson One: A Major Limit to the Utility of PVA Is Lack of Suitable Data

This limitation has been known since the first PVAs were published in the early 1980s. Our response as a community has been to proliferate models and, with few exceptions, not the long-term population field studies designed to yield data suitable to such modeling. Had the latter course been chosen more often, there would now be a much better understanding of the utility these concepts have for real conservation efforts. Continuing down the current path will leave us in the same position 20 years from now, sophisticated computer models lying idle without the data to drive them.

Long-term studies of population ecology and its relationship with habitat are not only necessary to enrich our understanding of the extinction process; they are also mandated by law. Implementing regulations for the U.S. National Forest Management Act (*Code of Federal Regulations*, vol. 36, sec. 219.19) explicitly state that "population trends of the management indicator species will be monitored and relationships to habitat changes determined." Furthermore, section 219.26 states that "inventories shall include quantitative data making possible the evaluation of diversity in terms of its prior and present condition." In at least one case, *Sierra Club v. Martin* (no. 98-8358, slip op. [11th Cir., February 18, 1999]), the court's interpretation is that "taken together the two regulations require the Forest Service to gather quantitative data on management indicator species and use it to measure the impact of habitat changes on the forest's diversity." Although population monitoring, as it is usually practiced by management agencies, doesn't collect the kinds of detailed demographic data required for the more sophisticated PVA models, the ruling in this case could certainly justify the U.S. Forest Service gathering the kind of detailed, long-term population data that is necessary to conduct a useful PVA, which is lacking for most species.

Lesson Two: We Are Not Using What We Have Already Learned from PVA

Rules of thumb have had a checkered history in PVA. They are desired, promulgated, promoted, and then debunked. The 50/500 rule of conservation genetics (Franklin 1980; Frankel and Soulé 1981; Allendorf and Ryman, chap. 4 in this volume) has been generally discarded with little to take its place, save Soulé's observation (1987) that, for vertebrates at least, populations of less than a few thousand individuals either are in trouble or are likely to be in trouble in the near future. Although the scientific community's capacity for self-correction by rejecting rules of thumb, such as 50/500, is laudable, the policy process requires rules for decisions. Without good models, or at least rules of thumb that represent a broad scientific consensus, too often the decision made with regard to species conservation is that either there is no problem, or that the population of a species and its habitat remaining after the proposed action will be sufficient. After all, Noah got by on two of each.

A telling manifestation of this situation can be found in Wilcove et al.'s assessment (1993) of population sizes at the time of listing under the ESA, and Tear et al.'s review (1993) of recovery plans. Wilcove et al. (1993) found that the median population size at the time of listing was about 1,000 individuals for animals and about 100 for plants. Tear et al. (1993) found that median population size for a taxon to be considered "recovered" was about 1,552. Taking Soulé's rule of thumb (1987) of "a few thousand" at its literal minimum (2,000), most species are listed well below this level, and many of those species will be considered secure when they are only 75% of what one expert considered the minimum.

Eighty-five percent of species that are listed as threatened or endangered under the ESA are in that condition, at least in part, because of the loss or alteration of habitat (Wilcove et al. 1998). Much habitat loss is likely irreversible, because the habitat conversion either is essentially permanent (e.g., roadways, buildings, or other developments) or is long-term with respect to the species' lifetime (e.g., ancient forests converted to clear-cuts). The relatively modest population goals found in recovery plans are likely a reflection of this fact. In other words, it is the best that can be done with the habitat that remains. If this is true, then the current implementation of the ESA is actually creating a large cadre of permanently endangered species that will never be recovered in any meaningful way. Without credible models based on good data, or at least rules of thumb that represent a broad-based consensus of the scientific community, this dynamic is unlikely to change.

Lesson Three: There Are No Accepted Criteria by Which to Judge the Viability of a Species or Population

Neither scientists nor regulators have converged on a meaningful time frame or probability of survival—or any alternative formulation—that allows the rational allocation of resources across a spectrum of species in need of attention. IUCN's recently revised status criteria (1996) and the National Marine Fisheries Service's standards for marine mammal population viability embody many of the principles of uncertainty and relative risk, and thus are promising initiatives. Nevertheless, a general agreement on time frame and probability of survival is still lacking over the broad range of resource management programs that affect biodiversity conservation. So long as this situation persists, any major conservation or development decision can be viewed as right or wrong, depending on one's perspective. Our experience is that in such situations the conservation perspective seldom prevails.

EPISTEMOLOGY AND MANAGEMENT OF ENDANGERED SPECIES

Before suggesting recommendations on future directions for improving the state of affairs in PVA and conservation policy, it is worth a moment to consider whether there is anywhere to go. At the heart of the PVA issue are three old and troubling questions: Can we know the future? What will it take to do so? Is the past a reliable guide? These have been vexing questions for science and society throughout our history, and probably always will be. The answers, of course, depend on what it is about the future that we wish to know. The fates of individuals and species (death and extinction) are much more certain than are the timing and circumstances under which those fates are realized. It takes much less effort to correctly predict that the earth will have a climate in 1,000 years than to predict what that climate will be. Thus, the future is much more easily knowable in a qualitative sense than it is predictable in a quantitative sense. Yet the conservation issue PVA seeks to address can be meaningfully influenced only through quantification.

It is clear that many life-forms are at risk of extinction, often because we are usurping their habitat. And it is equally clear that all species require some critical amount of space and resources in order to survive, the dimensions of which are influenced by the biology of the species and the inherent variability of nature. Thus, the right response to the issue of habitat loss is already known, qualitatively. To maximize the chances of a species' survival, the habitat available to it should not be diminished. Unfortunately, in a world of fixed area and a growing human population, this management response is not possible, and thus, a qualitative understanding of the issue is insufficient to produce a solution. For that it is necessary to quantify how a species' chances of survival change in response to the amount of habitat provided for its use. To sustain a life-form, we must attempt to know the future. But we must also be duly circumspect with regard to our ability to do so.

Much of the current debate over the use of PVA revolves around the issues of error and the propagation of error in PVA simulation models or time-series projections (Boyce 1992; Beissinger and Westphal 1998; Ludwig 1999). Parameters describing the population dynamics of wild species, particularly animals, can be extremely difficult to measure with any kind of accuracy and precision. Data with wide confidence limits used in long-term simulations mean that the resulting forecasts of population outcomes are so uncertain as to be relatively useless. This has led some authors to conclude that reliable risk evaluations over even a 100-year time scale are probably impossible (Goldwasser et al. 2000). One rational alternative to this situation is to determine the time horizon over which available data can produce useful forecasts (Goldwasser et al. 2000), which is likely to leave us with PVAs that make forecasts on the order of several decades. This would not necessarily be a problem until the issues of error symmetry and the irreversibility of some management decisions are considered.

If short-term PVA forecasts turn out to be pessimistic with regard to the habitat requirement of species for long-term survival, some of that habitat can always be reallocated to other uses. Unfortunately, there is no reason, a priori, to believe that short-term PVA forecasts might not turn out to be optimistic with regard to habitat needed for long-term viability. Indeed, this was one of the most important results of Belovsky et al.'s experiments (1999) with brine shrimp. If extra habitat becomes necessary but has already been allocated to essentially irreversible alternative uses, then we are left with a difficult, perhaps hopeless, situation.

Another rational alternative to uncertain predicted outcomes from PVAs is to use PVAs to assess relative risks rather than to predict absolute outcomes (Beissinger and Westphal 1998; Groom and Pascual 1998). By exploring the sensitivity of persistence to different parameters, managers can focus on activities that would have the most impact on overall population persistence. Using models of different management scenarios, PVA can also provide decision makers with better information about the relative risks associated with alternatives. This approach acknowledges the valid criticism that PVA is not as effective in predicting the absolute risk of extinction, especially over long time periods (Beissinger and Westphal 1998; Ludwig 1999). Using PVA for individual management decisions and for short time frames may be the wave of the future for this tool in public policy.

However, some policy issues require absolute answers. How much old-growth habitat needs to be maintained in order to save the spotted owl? How big does the grizzly bear population of the Yellowstone ecosystem need to be in order to be removed from the list of threatened species? Even modest population sizes are likely to appear robust with regard to short time frames, and it is quite likely that PVAs justified on the narrow confidence intervals of their outcomes are likely to underestimate the amount of habitat that is actually required for long-term population viability. Thus, a normally justified emphasis on precision may actually lead to ineffective conservation over the long term.

FUTURE DIRECTIONS FOR PVA

We agree with Lindenmayer et al. (1993) that the "benefits of PVA far outweigh its limitations." These authors provide an excellent discussion of the values and limitations of PVA in the policy process. From our perspective PVA has already made three particularly important contributions to conservation efforts. First, it has forced managers and policymakers to face the fact that size does matter. We believe the general result has been that more populations and/or habitat have been conserved than would otherwise be the case, although this is a difficult conjecture to test statistically across all applications. The spotted owl is a prime example of one species for which a concerted effort to understand its viability led to much greater conservation effort than originally planned. Perhaps the current review of recovery plans by the Society for Conservation Biology will shed some light on this issue.

Second, PVA has also provided a framework for organizing what is known and what is not known about the population and habitat dynamics of species at risk. Despite the great variability from one species to another in models and the data available, the underlying principles of PVA are precipitating a trend toward comparability in management's decision making across species.

Third, despite its complexity and heavy use of jargon (e.g., stochasticity), PVA is actually making conservation decisions more transparent. If one can wade through the muddle of models and data, it is generally possible to have a fair idea of what is known about a species, what remains to be learned, and the consequences of not knowing. From the standpoint of public oversight, it is much better contesting a model whose sensitivity to assumptions can be explored, than contesting "expert judgment." This aspect of PVA model building will result in better conservation programs.

Whether PVA can expand on these positive contributions will depend in large part on how we, as a profession, use and develop both the theory and practice of PVA. Below are five recommendations that we believe are key to the credible application of PVA and to fostering the evolution of a better understanding of the fates of species.

Development of Standards for PVA

As a community, we have now sufficiently explored the use of PVA to know that not all applications have been warranted, either in terms of model structure or suitability of data. Goodman (1987) and Lande (1993) provide a very good sense of the relative hierarchy and functional form of the relationship between three categories of chance events (demographic, environmental, and catastrophic) and the likelihood of extinction for populations limited to various sizes. It is still not known how best to integrate genetic considerations into PVA models. Ferson and Ginzburg (1996) have teased apart the difference between variability and ignorance in data, the different ways they propagate, and how to treat them in PVA models. There is a growing sense in the scientific community of both the power and the limits of our models and data, and their proper role in management and conservation (Boyce 1992; Beissinger and Westphal 1998). Now, it seems, is the right time for the community to develop some standards for what constitutes an acceptable PVA in terms of the underlying data, analyses, and modeling. We applaud the efforts of Ralls et al. (chap. 25 in this volume) in beginning this process.

Long-Term Field Studies of Population Dynamics

Nearly nothing is known, quantitatively, about the population ecology of the vast majority of species. Only a handful of species have been studied for significantly long periods of time. There are, in the United States alone, over 1,000 taxa currently listed as threatened or endangered. Another 5,000 to 6,000 species are thought to be at some significant risk of extinction (Stein and Flack 1997). As a practical matter, the vast majority of these species will never be studied in sufficient detail, or for sufficient periods of time to adequately parameterize a PVA. By the same token, much could be learned by selecting a cross section of species representing major life-history types, and undertaking detailed, long-term population-ecology and population-genetics research programs aimed at understanding the interplay among population dynamics, habitat, and environmental variability. Whether or not these species could legitimately serve as models for other, unstudied species of similar life history in PVA models, as some have suggested (Goodman 1990; Hanski, chap. 5 in this volume), may bear further consideration. However, long-term studies of reference species would doubtless go a long way toward enriching extinction theory and providing better guidance to management efforts.

Our recommendations for further studies can build upon government programs that specifically fund long-term demographic research. The National Science Foundation's (NSF) programs in long-term ecological research, including the system of long-term ecological research sites, seek to fund research that promotes understanding of ecological phenomenon over long temporal and large spatial scales. Part of the research taking place at the 21 long-term research sites in the United States is investigation into the spatial and temporal distribution of populations selected to represent trophic structure. The Long-Term Research in Population Biology Initiative would also be useful in this regard if it could be funded more aggressively, and with some logical selection of "representative" taxa. The professional societies (Society for Conservation Biology, Ecological Society of America, The Wildlife Society, and others) could work together to help NSF pick a suite of "representative" taxa. NSF's recent proposal to create a biodiversity observatory network where long-term data would be collected at 20 sites across the country could provide demographic and time-series data on specific populations of the suite of taxa chosen. This proposal, and others under the recent Biocomplexity and the Environment Initiative, could provide needed funding and infrastructure for long-term demographic studies.

In addition, the Biological Resources Division (BRD) in the Department of the Interior funds various research programs on population trends, including their Bird Banding Laboratory and Breeding Bird Survey, which have tracked bird populations since 1966. They also support monitoring programs that provide tools for forecasting future trends based on alternative policy and management decisions. Because part of BRD's mission is to support the Fish and Wildlife Service and other Department of the Interior agencies, it is a logical entity to support expanded efforts at long-term population studies of benchmark or reference species to collect the kinds of data used in PVA models (Beissinger and Westphal 1998).

Experiments and Validation of PVA Models

For PVA truly to be a scientific tool, the predictions of PVA models must be tested against real populations. Belovsky et al. (1999) illustrate how to test predictions of PVA theory with well-designed laboratory experiments. Much more such work is needed, both in the lab and in the field, with nonthreatened, nonendangered species to determine how well models perform and how well they can be tested. Opportunities for testing PVA models and concepts might be found in the arena of pest control, especially where the control measures are habitat based. Such work could serve both conservation and control efforts.

Databases

Throughout this review, we have repeatedly lamented the lack of easily accessible databases, either centralized or distributed, on (1) wild populations: which ones have been studied, the quantitative depth of the studies, their duration, and their location; (2) PVAs: what species have been modeled, using what data, incorporating what factors, using what model, producing what results, and where applied; (3) listing decisions: the population characteristics at the time of listing (e.g., number, size, and trend); and (4) recovery plans: goals for population number, size, and trend to be delisted.

The best example of databases for wild populations is accessible at http://cpbnts1.bio.ic.ac.uk/gpdd. This Global Population Dynamics Database consists of over 5,000 time series of population indices on species from all over the world. The different time series come from many different sources, including annual counts of mammals or birds at individual sampling sites, and weekly counts of zooplankton and other marine fauna. The database was constructed by researchers at the National Environment Research Council's Centre for Population Biology (funded by the United Kingdom Natural Environment Research Council), in collaboration with the National Center for Ecological Analysis and Synthesis, Santa Barbara, California (funded by the NSF), and the

Department of Ecology and Evolution, University of Tennessee. In addition, Ransom Myers at Dalhousie University in Canada has brought together data on over 600 fish populations around the world, which are accessible at http://www.mscs.dal.ca/~myers/welcome.html.

Although scientists in government and research institutions are making progress in gathering other types of information and making it centralized and accessible, we still found it difficult to get information on long-term demographic data that are individual-based (as opposed to time series of population counts), PVAs, listing decisions, and recovery plans. If the scientific community is to do its job of auditing the use of the knowledge it produces and of developing that knowledge as rapidly as possible to serve society's needs, then those institutions with responsibility for the resources in question (i.e., the U.S. Fish and Wildlife Service and the National Marine Fisheries Service) have a responsibility to make information on the application of science in listing decisions, recovery plans, and HCPs readily available and analytically tractable, preferably over the Internet. Although the National Marine Fisheries Service is providing stock-assessment reports to fishermen and other interested parties, the larger body of conservation decisions under the ESA, such as listing decisions and recovery plans, still requires periodic and heroic efforts to collect, summarize, and analyze. It seems the information age has yet to arrive in our resource management agencies.

Rules of Thumb

As appealing as PVA models are for doing rigorous risk assessments of species extinction, our collective experience as a community of scientists and managers seems to be that, in most cases, we are probably overdriving our headlights. Management actions based on current models have a momentum that transcends the field of vision the data actually provide. This situation is unlikely to improve quickly. In the meantime, we have only two options for influencing the debate over how much habitat is enough to accomplish conservation: evolution of the status quo, or development of rules of thumb.

Evolution of the status quo requires the kinds of steps listed above: (1) developing standards for data, data analyses, and modeling that constitute an acceptable PVA, (2) validating the models through laboratory and field experiments, and (3) conducting long-term research to better understand how populations actually fluctuate. But what should be done when there are no data, or when we must prepare for a time horizon that exceeds the ability of the data and models to produce credible forecasts? It seems to us an inescapable conclusion that the conservation biology community must, in fact, suggest rules of thumb, or guidelines, that will lead managers to make defensible judgments about how much habitat or what population size is enough to consider a species conserved.

In developing rules of thumb to use in making viability judgments without adequate data and/or models, the following three principles should be considered: representation, redundancy, and resiliency (Stein et al. 2000).

REPRESENTATION. Harkening back to G. E. Hutchinson's classic (1965), *The Ecological Theater and the Evolutionary Play*, a species should not be considered fully conserved unless there are populations playing in all appropriate theaters. Take the gray wolf (*Canis lupus*), for example. Its pre-Colombian distribution in North America extended from the low Arctic of Canada and Alaska to the high plateau of northern Mexico. In between, it occupied a variety of habitats and preyed on a variety of species: caribou (*Rangifer tarandus*) in the Arctic, moose (*Alces alces*) in the taiga, elk (*Cervus canadensis*) and mule deer (*Odocoileus hemionus*) in the Rocky Mountains, and bison (*Bison bison*) on the prairie. A narrow definition of species conservation could be achieved by maintaining large, healthy populations of the wolf in Canada and Alaska. But to achieve biodiversity conservation, the wolf is needed in as many of its original environments as is practicable. Although many theaters have already closed, some permanently, society should still strive to maximize the "ecological representation" of species as a fundamental component of their viability and the conservation of overall biodiversity.

REDUNDANCY. The major limiting factors to the forecasting ability of PVA stem from environmental and genetic uncertainty, and catastrophes. Catastrophes are by definition unpredictable, and it has been previously noted (Shaffer 1987 and others) that the prudent strategy is for management to maintain multiple (i.e., redundant) populations of all species of concern. It follows that, if our unit of focus is the species in each of the ecological settings it originally occupied, then we should want to have several such populations in each setting as a hedge against catastrophic loss. We recognize that the larger and more widespread the range of a population, the less likely it is that any single catastrophe would eradicate the population.

RESILIENCY. How large should each population be? As noted above, the major limitations to the forecasting ability of PVAs are environmental and genetic uncertainty, and catastrophes. And, as discussed earlier, the useful horizon for PVAs that take into account the difficulties of partitioning sampling variance (sensu Ferson and Ginzburg 1996; White et al., chap. 9 in this volume) from true system variability is likely on the order of several decades. Perhaps the best approach, given that catastrophic uncertainty already necessitates multiple populations, is to assure that each population is large enough so that it is not swamped by the effects of demographic uncertainty and/or short-term inbreeding. In other words, populations need to be large enough so that they can respond to good environmental conditions when they happen, and not be trapped in the demographic and genetic vortices that result from very low numbers. In achieving the goal of redundancy, the necessary multiple, resilient populations should, to the extent possible, be located to assure they are experiencing different regimes of environmental variability as a further hedge against catastrophic loss.

Employing these principles would acknowledge both what we do know about the determinants of long-term persistence *and* the limitations of our forecasting ability. They are not meant as replacements for good models based on good data, but as defaults to use in those many instances where there are no data, or models are not adequate. They are not meant to establish objectives for the ideal distribution and abundance of a taxon to maximize its ecological or evolutionary value, but as fail-safe levels below which they should never be allowed to fall by our conservation efforts. Indeed, many species may already be so reduced that they could not meet all of these criteria (Stein et al. 2000), but without the application of such standards many more will become so reduced. We believe that these principles could be translated into real numbers by working groups of scientists knowledgeable about particular taxonomic groups. We also believe that the time is now.

LITERATURE CITED

Beissinger, S. R., and M. I. Westphal. 1998. On the use of demographic models of population viability in endangered species management. *Journal of Wildlife Management* 62:821–841.

Belovsky, G. E., C. Mellison, C. Larson, and P. A. Van Zandt. 1999. Experimental studies of extinction dynamics. *Science* 286:1175–1177.

Boyce, M. 1992. Population viability analysis. *Annual Review of Ecology and Systematics* 23:481–506.

———. 1993. Population viability analysis: adaptive management for threatened and endangered species. *Transactions of the North American Wildlife and Natural Resources Conference* 58:520–527.

———. 1995. Population viability for grizzly bears (*Ursus arctos horribilis*): a critical review. A report to the Interagency Grizzly Bear Committee.

Committee of Scientists (COS). 1999. Sustaining the people's lands: recommendations for stewardship of the national forests and grasslands into the next century. U.S. Department of Agriculture, Washington, D.C.

Ferson, S., and L. R. Ginzburg. 1996. Different methods are needed to propagate ignorance and variability. *Reliability Engineering and System Safety* 54:133–144.

Frankel, O. H., and M. E. Soulé. 1981. *Conservation and evolution.* Cambridge University Press, Cambridge, United Kingdom.

Franklin, I. R. 1980. Evolutionary change in small populations. Pages 135–149 in M. E. Soulé and B. A. Wilcox, editors, *Conservation biology: an evolutionary-ecological perspective.* Sinauer Associates, Sunderland, Massachusetts.

Goldwasser, L., L. Ginzburg, and S. Ferson. 2000. Variability and measurement error in extinction risk analysis: the northern spotted owl on the Olympic Peninsula. Pages 169–187 in S. Ferson and M. Burgman, editors, *Quantitative methods for conservation biology.* Springer-Verlag, New York, New York.

Goodman, D. 1987. The demography of chance extinction. Pages 11–34 in M. E. Soulé, editor, *Viable populations for conservation.* Cambridge University Press, Cambridge, United Kingdom.

———. 1990. Keynote address: Research. Pages 12–23 in F. D. Dottavio, P. Brussard, and J. McCrone, editors, *Protecting biological diversity in the national parks: workshop recommendations.* U.S. Department of the Interior, National Park Service, Washington, D.C.

Groom, M. J., and M. A. Pascual. 1998. The analysis of population persistence: an outlook on the practice of viability analysis. Pages 4–27 in P. L. Fiedler and P. M. Kareiva, editors, *Conservation biology for the coming decade.* Chapman and Hall, New York, New York.

Gwinn, S., A. Burns, and J. Veilleux. 1999. A review of developments in U.S. ocean and coastal law. *Ocean and Coastal Law Journal* 4:173–207.

Hood, L. C. 1998. Frayed safety nets: conservation planning under the Endangered Species Act. Defenders of Wildlife, Washington, D.C.

Hutchinson, G. E. 1965. *The ecological theater and the evolutionary play.* Yale University Press, New Haven, Connecticut.

International Union for Conservation of Nature (IUCN). 1996. *The 1996 IUCN Red List of threatened animals.* IUCN, Gland, Switzerland.

Kareiva, P., S. Andelman, D. Doak, B. Elderd, M. Groom, J. Hoekstra, L. Hood, F. James, J. Lamoreux, G. LeBuhn, C. McCulloch, J. Regetz, L. Savage, M. Ruckelshaus, D. Skelly, H. Wilbur, and K. Zamudio. 1999. *Using science in habitat conservation plans.* American Institute of Biological Sciences, Washington, D.C.

Kuhlmann, W. 1996. Wildlife's burden. Pages 189–201 in W. J. Snape III, editor, *Biodiversity and the law.* Island Press, Covelo, California.

Lande, R. 1993. Risks of population extinction from demographic and environmental stochasticity and random catastrophes. *American Naturalist* 142:911–927.

Lindenmayer, D. B., T. W. Clark, R. C. Lacy, and V. C. Thomas. 1993. Population viability analysis as a tool in wildlife conservation policy: with reference to Australia. *Environmental Management* 17:745–758.

Ludwig, D. 1999. Is it meaningful to estimate a probability of extinction? *Ecology* 80:298–310.

Mann, C. C., and M. L. Plummer. 1999. A species' fate, by the numbers. *Science* 284:36–37.

Pease, C., and D. Mattson. 1999. Demography of the Yellowstone grizzly bear. *Ecology* 80:957–975.

Schemske, D., B. Husband, M. Ruckelshaus, C. Goodwillie, I. Parker, and J. Bishop. 1994. Evaluating approaches to the conservation of rare and endangered plants. *Ecology* 75:584–606.

Shaffer, M. L. 1987. Minimum viable populations: coping with uncertainty. Pages 69–86 in M. E. Soulé, editor, *Viable populations for conservation*. Cambridge University Press, Cambridge, United Kingdom.

Soulé, M. E. 1987. Introduction. Pages 1–10 in M. E. Soulé, editor, *Viable populations for conservation*. Cambridge University Press, Cambridge, United Kingdom.

Stein, B. A., and S. R. Flack. 1997. Species report card: the state of U.S. plants and animals. Nature Conservancy, Arlington, Virginia.

Stein, B. A., L. S. Kutner, and J. S. Adams, editors. 2000. *Precious heritage: the status of biodiversity in the United States*. Oxford University Press, New York, New York.

Tear, T., J. M. Scott, P. H. Hayword, and B. Griffith. 1993. Status and prospectus for success of the Endangered Species Act: a look at recovery plans. *Science* 262: 976–977.

Wilcove, D. S., M. McMillan, and K. C. Winston. 1993. What exactly is an endangered species? An analysis of the U.S. endangered species list: 1985–1991. *Conservation Biology* 7:87–93.

Wilcove, D., D. Rothstein, J. Dubow, A. Phillips, and E. Losos. 1998. Quantifying threats to imperiled species in the United States. *BioScience* 48:607–615.

PART **2**

ISSUES IN THE
PARAMETERIZATION
AND CONSTRUCTION
OF PVA MODELS

The development of PVA models that can provide forecasts of the future trajectories of populations useful for conservation requires the careful construction and parameterization of models. Several issues have risen to prominence in recent years. First, what are the best statistical approaches to estimate vital rates and other model parameters accurately? Second, how can uncertainty in the values of vital rates and in the structure of the model be incorporated into the construction of the model?

Developing unbiased and accurate estimates of model parameters is challenging both for genetic models of effective population size and for demographic population models. Robin Waples presents an analysis and critique of different approaches to estimating effective population size. Genetic, demographic, or both kinds of information can be used to estimate effective population size, with varying levels of precision and accuracy. An important issue he raises is that the mixture of different measures of central tendency (i.e., arithmetic and geometric means) have been used to make generalizations about effective versus census population size, and this has led to erroneous conclusions. Gary White and colleagues show how demographic population models that estimate vital rates from marked animals can benefit from computer software that can calculate maximum likelihood estimates and test for significant differences among classes of animals. They discuss how lumping variation caused by sampling error with variation created by the process of interest has often confounded the estimation of environmental stochasticity and the variance of demographic rates. White et al. present approaches to estimate and discard sampling variation and show how to develop robust estimates of vital rates from marked individuals.

Even after robust parameter estimates are developed, significant uncertainties are likely to remain in the construction of most PVA models. For example, we are unlikely to know the types of distributions for many variables, or the frequency and magnitude of rare events that might have important effects on population dynamics. Bernt-Erik Sæther and Steinar Engen present one approach to incorporating uncertainty, using the concept of a population prediction interval. They demonstrate how this approach can be applied to systems that have a time series of population-size measurements. Another approach to incorporating uncertainty in parameters is using Bayesian methods to construct the PVA model. Bayesian techniques are an alternative to the frequentist methods that characterize parametric and nonparametric approaches that most students are taught in statistics classes. Paul Wade presents an overview to Bayesian thinking and shows how it can be used in PVA. Barbara Taylor and colleagues apply a Bayesian approach to classifying the category of threat that a species faces. They compare Bayesian models to standard PVA approaches and other methods for predicting extinction rates.

Definition and Estimation of Effective Population Size in the Conservation of Endangered Species

Robin S. Waples

ABSTRACT

Effective population size (N_e) is one of the most fundamental evolutionary parameters of biological systems, and it affects many processes that are relevant to biological conservation. N_e is difficult to estimate in natural populations, however, and a variety of factors complicate its application to the conservation of endangered species in general and population viability analyses in particular. Complexities include (1) ambiguity arising from multiple ways of defining N_e, (2) difficulty in adequately accounting for spatial and metapopulation structure and their effects on N_e, (3) sampling biases and other difficulties involved in evaluating N_e in species with overlapping generations, (4) confusion regarding the magnitude (and meaning) of the ratio N_e/N, and (5) concerns regarding precision and bias of indirect genetic methods for estimating N_e. Greater attention to these issues will allow effective-size considerations to be more fully integrated into endangered-species management.

INTRODUCTION

Allendorf and Ryman (chap. 4 in this volume), Hedrick (chap. 17 in this volume), and Haig and Ballou (chap. 18 in this volume) have stressed the importance of genetic considerations in risk analysis and recovery planning for endangered species. In this chapter, I focus more narrowly on what is arguably the most important genetic parameter in evolutionary biology—effective population size (N_e). N_e is the primary factor responsible for the rate of genetic drift, the rate of loss of diversity of neutral alleles, and the rate of increase in inbreeding experienced by a population. Effective size also determines the relative evolutionary importance of directional (migration and selection) and stochastic (random genetic drift) factors. Migration and selection are deterministic in large populations but can be overwhelmed by random processes in small

This paper benefited from discussions with and comments from Fred Allendorf, Steven Kalinowski, Paul McElhany, Len Nunney, Alexander Pudovkin, and Chris Ray. Part of this research was funded through contract DE-A179-89BP0091 with Bonneville Power Administration.

ones, in which favorable alleles can be lost and deleterious alleles can become fixed by chance alone—with profound effects on fitness and population viability.

N_e is difficult to estimate in natural populations, and a variety of factors related to its definition and interpretation complicate its application to the conservation of endangered species in general, and population viability analyses (PVAs) in particular. Complexities include (1) ambiguity arising from multiple ways of defining N_e, (2) difficulty in adequately accounting for spatial and metapopulation structure and their effects on N_e, (3) sampling biases and other difficulties involved in evaluating N_e in species with overlapping generations, (4) confusion regarding the magnitude (and meaning) of the ratio N_e/N, and (5) concerns regarding precision and bias of indirect genetic methods for estimating N_e. Discussion of these issues below is intended to allow considerations of effective size to be more fully integrated into endangered-species management. We shall see that, although the basic concept is elegantly simple, almost everything else involving effective population size is much more complicated.

DEFINITIONS OF N_e

In finite populations (i.e., in all biological populations), random genetic drift leads to changes in allele frequency and loss of genetic variability. In "ideal" populations (Fisher 1930; Wright 1931)—those with random mating, an equal sex ratio, discrete (nonoverlapping) generations, and random variation in reproductive success—genetic drift occurs at a rate described by an inverse function of population size (N). Because real populations almost never satisfy the conditions of an "ideal" population, Wright (1931, 1938) developed the concept of "effective population size" (N_e), which describes the size of an ideal population that would have the same rate of genetic change as the population under consideration. Departures from the "ideal" conditions generally cause N_e to be less than N, and often a good deal less.

Several different effective sizes have been identified in the literature (Ewens 1979; Crow and Denniston 1988; Caballero 1994), the two most common being variance effective size (N_{eV}), which is related to the rate of allele-frequency change, and inbreeding effective size (N_{eI}), which is related to the rate of increase in inbreeding. Typically, inbreeding is measured as the probability that the two alleles in an individual can be traced back to the same gene in a common ancestor, in which case the alleles are considered to be *identical by descent*. Formulas for N_{eI} and N_{eV} are

(1)
$$N_{el} = \frac{N\bar{k} - 2}{(\bar{k} - 1) + V_k/\bar{k}}$$

and

(2)
$$N_{eV} = \frac{N\bar{k} - \bar{k}}{1 + V_k/\bar{k}}$$

(Crow and Kimura 1970, eq. 7.6.2.17, p. 359, respectively), where N is the population size in the parental generation, $\bar{k}$ is the mean number of offspring produced per individual, and V_k is the variance of k across individuals. If reproductive success is random and population size is constant, then $V_k = \bar{k} = 2$ and the population is "ideal" ($N_e \approx N$ except for small adjustments due to the mating scheme; see Crow and Denniston 1988 for details). In natural populations, however, V_k is usually greater than $\bar{k}$, leading to $N_e < N$. High V_k values can be caused by reproductive differences among individuals within sexes or by an unequal sex ratio, which leads to differences between the sexes in mean reproductive output per individual.

In stable populations the two effective sizes are similar, and it is not important to distinguish between them for conservation purposes. Populations of conservation interest often are changing rapidly in size, however, and in this situation N_{eV} and N_{el} can be very different. Table 8.1A shows how N_{el} and N_{eV} vary over a number of generations in an otherwise ideal population with cyclical variations in N. When population size drops sharply (from 100 to 10 between generations 2 and 3), N_{eV} reflects the drop immediately, while the change in N_{el} lags by a generation (see Ryman et al. 1995). The bottleneck has an immediate effect on allele frequencies (and hence N_{eV}) because the amount of allele-frequency drift depends entirely on the number of individuals in the progeny generation. Thus, the magnitude of drift will, on average, be much higher if only 10 progeny are produced than if the population had remained constant at 100. In contrast, the bottleneck will have no immediate effect on inbreeding because the probability of identity by descent in the progeny generation is determined by the number of parents; a sample of 10 individuals produced by 100 parents will on average be no more inbred than if 100 or 1,000 progeny were produced. The effects of the bottleneck on N_{el} do not appear until the following generation when progeny of the 10 parents, which have a high probability of having two alleles that are identical by descent, mate with each other.

Note that over a full cycle (five generations), the harmonic mean

Table 8.1   Changes over Time in Inbreeding (N_{eI}) and Variance (N_{eV}) Effective Population Sizes

Generation	N	$\bar{k}$	N_{eI}	N_{eV}
A. Cyclical fluctuations				
1	100	2	99	99
2	100	0.2	90	10
3	10	200	10	900
4	1000	0.2	990	100
5	100	2	99	99
6	100			
Harmonic mean			38	38
B. Declining population				
1	1000	1	998	500
2	500	0.8	498	200
3	200	1	198	100
4	100	1	98	50
5	50	0.4	45	10
6	10			
Harmonic mean			123	36

Notes: Changes are for a hypothetical population with variable N that is "ideal" in other respects (discrete generations, random mating, equal sex ratio, random variation in reproductive success). N_{eI} and N_{eV} were computed using equations 1 and 2, respectively. $\bar{k}$ is the mean number of progeny per parent, V_k is the variance of k, and V_k is assumed to equal $\bar{k}$.

effective size is the same for the two measures ($N_e \approx 38$). This suggests that even with variable population size it may not always be important to distinguish inbreeding and variance effective sizes for conservation purposes. However, there are several situations in which the distinction can be important. First, if there is concern that low short-term N_{eI} may cause significant inbreeding depression, then conservation measures to avoid this situation must be applied at the appropriate time. Frankham (1995a) and Hedrick and Kalinowski (2000) reviewed evidence for inbreeding depression in natural and captive populations and provided several examples demonstrating that this is a real conservation concern. Second, most natural or captive populations do not exhibit such tidy cycles in abundance as shown in table 8.1A. Many populations of conservation concern are declining, and in this case N_{eI} can be substantially higher than N_{eV} for a considerable period of time (as illustrated in table 8.1B). Conversely, in recovering populations N_{eV} will recover faster than N_{eI} to prebottleneck levels.

Finally, Ryman et al. (1995) showed that variance and inbreeding effective sizes behave differently under supportive breeding, in which part of the gene pool is taken into captivity and enhanced relative to the wild component. See Nomura (1999) and Ryman et al. (1999) for more debate on this complex issue. As captive breeding is used in recovery programs for many threatened and endangered species, it is important to clarify and resolve this issue. Hedrick et al. (1995) and Hedrick (chap. 17 in this volume) provided an example in which both N_{eV} and N_{eI} were estimated for an endangered species subject to a captive-breeding program. In their example, it does not appear that the captive-breeding program had a significant impact on the effective size of the larger population, a result that can be attributed to relatively high N_e in the captive phase and relatively low contribution of captive fish to natural spawning. Waples and Do (1994) considered the genetic effects of supportive breeding in some detail and showed that whether increased population size due to captive breeding persists after the program is terminated has a profound effect on N_e and levels of inbreeding in the postsupplementation population.

N_e IN SPATIALLY STRUCTURED POPULATIONS

Spatial structure of a population is important to consider in evaluating N_e because in general the effective size of the overall population (N_{eT}) is not equal to the sum of the N_e's of the subpopulations (Nunney 1999). This is because population structure affects the variance in reproductive success of individuals within and among subpopulations. The nature of these effects depends heavily on the details of the model under consideration.

Wright (1943) was the first to consider the effective size of structured populations. In his finite island model, each of n subpopulations has the same fixed number of individuals (N). Each subpopulation is "ideal" as described above, so subpopulation $N_e = N$. The subpopulations ("islands") are connected by migration; each subpopulation contributes a fraction m of its individuals to a global migrant pool every generation, and each subpopulation receives the same fraction of migrants drawn randomly from the migrant pool. Under these conditions,

$$(3) \qquad N_{eT} \approx \frac{N_T}{1 - F_{ST}},$$

where $N_T = nN$ is the total number of individuals in the metapopulation and F_{ST} is Wright's measure of genetic differentiation among subpopulations. Inspection of equation 3 indicates that, if there is any population subdivision ($F_{ST} > 0$), N_{eT} will always be larger than N_T. Since $N_e = N$ within each subpopulation, equation 3 also implies that metapopulation N_{eT} is greater than the sum of the subpopulation N_e's. I will consider this result in some detail because understanding why population subdivision increases N_{eT} in Wright's model is critical to understanding the complex way in which many factors can affect the relationship between subpopulation and metapopulation effective size.

The result in equation 3 is due to two factors, both of which are a direct consequence of the island model's assumptions that each subpopulation is of constant and equal size and that each contributes exactly the same number of propagules to the migrant pool in the next generation. To see the effect of these assumptions on N_{eT}, consider as a point of reference an ideal population of size N_T. Within this population one could imagine hypothetical groupings of individuals into arbitrary subunits, and at any given point in time the collective reproductive success of the different groups would vary randomly—some groups contributing more to the next generation and some less, just by chance, driven by random variation among individuals in reproductive success. Now consider that these subunits are not imaginary but real and reflect permanent partitioning of the global population. Because each subunit is of fixed size N, the subunits can no longer vary among themselves in reproductive success. This constrains the degree to which individuals within and among subpopulations can differ in reproductive success, and the effect is greater the smaller N becomes. In the extreme case where each of the n subpopulations has exactly two individuals, every individual in the global population will have equal reproductive success, so $V_k = 0$ and $N_{eT} \gg N_T$ (in agreement with equations 1 and 2). The first factor operating in the island model is thus an increasing constraint on variance in reproductive success as the number of subpopulations increases and the size of each decreases.

The second, and related, factor that leads to the result in equation 3 is that the island model does not allow extinction of subpopulations, since all are constant in size. As subpopulations become more isolated, F_{ST} increases as they tend to become fixed for different alleles. Although genetic diversity within subpopulations declines as they become more isolated and cannot receive new genes through migration, genetic variation within the global metapopulation is frozen in place because different alleles, by chance, become fixed in different subpopulations. Again, this effect is enhanced as the number of subpopulations increases and the size of each decreases.

The interplay of these two factors in causing the effect Wright detected can be illustrated by substituting in equation 3 the equilibrium value of F_{ST} in the island model ($F_{ST} \approx 1/(1 + 4mN_e)$) and rearranging to yield a convenient way of expressing the effects of subdivision on the ratio N_{eT}/N_T (after Nunney 2000):

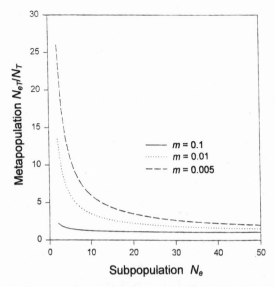

Fig. 8.1 The ratio of metapopulation effective to total size (N_{eT}/N_T) in Wright's island model as a function of subpopulation effective size (N_e) and migration rate (m; see equation 4). This figure assumes that N_T is fixed, so subpopulation N and N_e decrease as the number of subpopulations increases.

$$\textbf{(4)} \qquad \frac{N_{eT}}{N_T} \approx 1 + \frac{1}{4mN_e}.$$

Figure 8.1 illustrates this result. With N_T held constant, for any given migration rate, metapopulation N_{eT} (and hence the ratio N_{eT}/N_T) increases with increasing fragmentation of the global population (higher n and lower N_e in each subpopulation). For any given subpopulation size, metapopulation N_{eT} increases as migration rate declines and more and more of the genetic variation is "frozen" into different states in different subpopulations.

Since the validity of equations 3 and 4 depends on the extent to which forces exist that cause N to be equal and constant across subpopulations, it is important to consider whether such forces are in fact likely to occur. Ray (2001) suggested one possible scenario: if suitable habitat patches are well isolated, if the patches are similar in local carrying capacity, and if local population regulation is effective in stabilizing population size, subdivision can constrain the variance in reproductive success, leading to higher N_{eT}. The design of nature reserves and captive-breeding programs for endangered species may also provide opportunities for equalizing reproductive success among subpopulations. However, there does not appear to be a general force that constrains natural subpopulations to be equal and constant in size, so there is no reason to expect N_{eT} to be larger than N_T merely as a consequence of population subdivision. As discussed below, other models describe conditions under which population subdivision can be expected to reduce N_{eT}.

Whitlock and Barton (1997) examined the effective size of subdivided populations in some detail. One of the models they considered differed from the island model above only in that the subpopulations were allowed to vary in their contribution to the next generation through different migration rates, with the contribution to the overall migrant pool determined by subpopulation productivity. Under these conditions,

$$
(5) \qquad N_{eT} \approx \frac{N_T}{(1 + P)(1 - F_{ST}) + NPF_{ST}n/(n - 1)},
$$

where P is the variance in productivity among subpopulations. Setting $P = 0$ leads to Wright's result in equation 3. It can be shown that, if P is larger than would be expected from random variation among individuals, population subdivision will reduce N_{eT} relative to N_T (Whitlock and Barton 1997). If $F_{ST} = 0$, then equation 5 simplifies to $N_{eT} \approx N_T/(1 + P)$. That is, even if there is no permanent subdivision, but instead individuals assort randomly into temporary "subpopulations" each generation for breeding, N_{eT} will still be less than N_T if productivity differs among sites ($P > 0$). In this case, directional differences in subpopulation productivity are overlaid on random differences among individuals in reproductive success, with the overall effect of increasing variance of reproductive success within the total population.

Nunney (1999) considered a similar model that also relaxes the assumption of the island model about equal contribution of migrants. In Nunney's "interdemic genetic drift" model, variations in subpopulation productivity (and hence contribution to the migrant pool) are generated by accumulation of random differences among individuals in reproductive success. He also allowed for the possibility of inbreeding within subpopulations, measured by Wright's inbreeding statistic F_{IS}, and obtained the following relationship:

$$
N_{eT} \approx \frac{N_T}{(1 + F_{IS})(1 + F_{ST}) - 2F_{IS}F_{ST}},
$$

If mating is random within subpopulations, then the expectation of $F_{IS} = 0$ and this reduces to

$$
(6) \qquad N_{eT} \approx \frac{N_T}{1 + F_{ST}},
$$

which is exactly the opposite of Wright's result: with population subdivision ($F_{ST} > 0$), N_{eT} of the overall population is less, not greater, than total population size, provided that the subpopulations are not constrained to contribute equally to the migrant pool.

In the classical metapopulation (Levins 1970), extinction and recolonization of subpopulations (or patches) is common. Maruyama and Kimura (1980) and Gilpin (1991) showed that the extinction/recolonization process can dramatically reduce N_{eT}, particularly if the number of occupied patches is so low that the overall metapopulation is in danger of extinction. Whitlock and Barton (1997) showed that, if variance in island productivity in their model was due to subpopulation extinction and recolonization, then the following relationship approximately holds (see Nunney 2000):

(7) $$N_{eT} \approx \frac{n}{4(m + e)F_{ST}},$$

where e is the rate of subpopulation extinction. Notably, in this situation, N_{eT} of the metapopulation is proportional to the number of subpopulations rather than to N or N_T.

To examine some of the many factors that can influence the effective size of a metapopulation that is not at significant short-term extinction risk, Hedrick and Gilpin (1997) used computer simulations to monitor changes in heterozygosity (H) as an index of N_{eT}. In their model, individual patches could be vacant for one or more time periods, after which the probability of recolonization depended on the number of extant source patches and their colonization ability. Hedrick and Gilpin found that, under these conditions, N_{eT} of the metapopulation depends more on metapopulation dynamics than on the carrying capacity of the patches. The rate of patch extinction and the characteristics of the founders that recolonized vacant patches were found to be particularly important. If patch extinction probability was high, entire lineages were regularly lost, thus reducing H and N_{eT}, often by one or two orders of magnitude or more. Similarly, if new patches were colonized by just a few founders, or if the founders came from the same subpopulation rather than from the metapopulation at large, H and N_{eT} were also greatly reduced.

Hedrick and Gilpin (1997) concluded that the relatively high levels of allozyme and DNA heterozygosity found in most species suggest that they have not spent their evolutionary history functioning as classical Levins-type metapopulations. This conclusion, however, should be tempered by consideration of factors not included in their model. For example, although spatial subdivision often increases the probability of subpopulation extinction, it also may increase the independence of local extinctions. Ray (2001) showed that, if the probability of local extinction scales logarithmically with population size, then even large subpopulations are subject to extinction, and under these conditions, strong population subdivision can promote retention of more genetic diversity and increase N_{eT}.

In summary, N_{eT} of a metapopulation can be either greater or less than the sum of the N_e's for the component subpopulations. Results discussed above demonstrate that whether N_{eT} is increased or decreased by population subdivision depends heavily on parameters such as variance in size and productivity of subpopulations, the patterns of migration, and extinction and recolonization rates. Conservation biologists should consider these factors carefully in evaluating conservation programs for endangered species. They should also be careful to distinguish between the effective size of a subpopulation and the metapopulation as a whole, since results described above indicate that the two parameters can respond very differently to the same conditions. As migration rate increases, the concept of a subpopulation N_e has less and less meaning, since the subpopulation is not acting as an independent unit. We should also not lose sight of the fact that most population genetics theory (and many models) provide results expected at equilibrium. This will not often be a reasonable assumption, particularly for species of conservation concern. For example, Hedrick and Gilpin (1997) pointed out that anthropogenic fragmentation of the natural landscape may have formed (and may continue to be forming) unnatural, nonequilibrium metapopulations with dynamics that have dramatically reduced recent N_{eT} in many species.

Finally, it is also worth noting here that the assumptions in Wright's island model (equal and constant subpopulation size) that are responsible for his conclusions regarding population subdivision and N_{eT} are common to many other models used in conservation biology. In particular, any model that divides a global population into subsets while placing constraints on the mean or variance of subpopulation parameters can have the same effect as Wright's model—artificially reducing V_k and upwardly biasing N_e.

N_e AND OVERLAPPING GENERATIONS

Most population genetics theory deals with generations that are discrete. Although convenient as a theoretical construct, this scheme does not conform to the life history of many biological species. Discrete-generation models for N_e are fairly robust for organisms with overlapping generations, provided that the populations are demographically stable (Felsenstein 1971; Hill 1972). These models, however, have generally dealt with population parameters and have ignored sampling considerations. Since conservation biologists rarely know population parameters (and generally consider themselves fortunate to have at least some demographic or genetic data for their organism of interest), and because opportunities for sampling bias are considerably greater in organisms with overlapping generations, this topic deserves mention here.

One fundamental problem is that almost all sampling theory assumes a random sample from the population, and for effective population size this means sampling from an entire generation. This can be easy to achieve in organisms with discrete generations, because all the individuals in a generation may co-occur in time and space, but it is rarely simple for organisms with overlapping generations. In such species, at a given time individuals in any particular life stage (e.g., juveniles or mature adults) may constitute only a fraction of a generation. To sample an entire generation, therefore, it is often necessary to obtain an array of temporally spaced samples. Apart from the logistic challenges this presents, it may be difficult to determine the appropriate weightings to place on the individual samples, which ideally should reflect the (generally unknown) relative contribution of the component sampled to the generation as a whole.

Another complication is that V_k, which has a profound effect on N_e (see equations 1 and 2), represents the *lifetime* variance in reproductive success among individuals. Unfortunately, even when data on reproductive success are available, they typically apply to offspring produced in one season rather than in an entire lifetime, making it difficult to evaluate cumulative reproductive success of individuals.

What can be done to alleviate these difficulties, which can result in substantial bias in estimating effective population size and other genetic parameters in species with anything but the simplest life histories? One option is to ignore the sampling effects under the assumption that they will be minor. This may be reasonable if the number of intervening generations is large enough, in which case the bias from sampling error may be small relative to the signal from genetic drift. Miller and Kapuscinski (1997) made this assumption in using a discrete-generation model to estimate effective size over a period of more than 30 years in a northern pike (*Esox lucius*) population. Alternatively, a series of stratified random samples can be combined to represent the generation as a whole. Although determining the optimum sampling and weighting scheme to accomplish this will be difficult without extensive life-history and demographic data for the species of interest, simply combining multiple samples temporally spaced across a generation could substantially reduce potential biases. Nunney and Elam (1994) showed that, under some circumstances, reasonably accurate estimates of N_e for species with overlapping generations can be made using demographic data for just one or a few seasons.

Genetic methods may be able to help determine whether a sample collected from the wild can be considered quasi-random and, if not, the magnitude of potential biases. Hansen et al. (1997) used genetic data to evaluate this issue in brown trout (*Salmo trutta*) populations and found significant allele-frequency differences among age classes. Furthermore, examination of family relationships within the youngest cohort suggested that most of that sample was derived from just three full-sib families. These data clearly illustrate potential biases that can arise from assuming that samples taken from species with overlapping generations adequately represent an entire generation. Herbinger et al. (1997) used a similar approach to evaluate a natural cohort of cod larvae sampled intensively over a 21-day period. They found no evidence of any temporal or spatial family structure and concluded that the larvae came from a relatively large, homogeneous gene pool produced by a large effective number of spawners (≥2,800).

If age data are available, it may be possible to partition the samples into cohorts and make explicit adjustments for sampling (Waples 1991; Jorde and Ryman 1995, 1996). Unless there is a way to link offspring from different cohorts to the same parents, however, it will be difficult to evaluate lifetime reproductive success and, therefore, generational N_e. The unusual life history of Pacific salmon, which have variable age structure but which all die after spawning, has features of both discrete and overlapping generation models. Waples (1990a,b) showed that, because yearly reproductive output is equal to lifetime reproductive output in these species, fairly straightforward adjustments to standard discrete-generation models can be used to obtain generational estimates of N_e in Pacific salmon.

THE RATIO OF EFFECTIVE TO TOTAL POPULATION SIZE

Much has been written in the recent literature about the N_e/N ratio in natural populations. The relevance of this debate to conservation of endangered species is readily apparent: because N is generally much easier to estimate than N_e, establishing an empirical relationship between the two parameters would allow at least crude estimates of N_e for many more species. Unfortunately, there is no consensus regarding even the approximate range for N_e/N in natural populations. At one extreme Nunney (1993) argued that N_e/N would be 0.5 or higher except

Table 8.2 Contribution of Various
Factors to the Mean Ratio N_e/N

Factor	Mean N_e/N
Sex ratio, S	0.64
Family-size variance, V	0.46
$S + V$	0.35
$S + VB$ + variable N	0.11

Notes: Estimated for 102 vertebrate, invertebrate,
and plant species using demographic and genetic
data (Frankham 1995b).

under special circumstances; at the other extreme, Hedgecock (1994) argued that, in marine organisms with high fecundities and high mortality in early life stages, "sweepstakes" survival and recruitment of related individuals can lead to very nonrandom mortality across families, with the result that N_e can be several orders of magnitude less than N.

I will not attempt to resolve this debate here, but I would like to point out that at least part of the confusion regarding N_e/N ratios can be attributed to a lack of clarity regarding (1) the treatment of variable population size and its effects on N_e/N, and (2) the relationship between generational and long-term N_e/N. Frankham (1995b) reviewed empirical estimates of N_e/N for over 100 species of vertebrates, invertebrates, and plants (table 8.2) and found that the mean N_e/N ratio varied from 0.64 (using only an adjustment for uneven sex ratio) to 0.11 for a "composite" estimate (which included adjustments for sex ratio, variance in reproductive success, and temporal changes in N). From these results Frankham concluded that variable population size is the most important factor influencing the ratio N_e/N, and that N_e/N is generally lower for natural populations than many have realized. Vucetich et al. (1997) and Vucetich and Waite (1998) examined this issue analytically and also concluded that variable N has a profound effect on N_e/N ratios.

Unfortunately, the conclusions of Frankham, Vucetich, and others regarding the importance of variable population size to the ratio N_e/N are difficult to evaluate because they are confounded by a statistical artifact in computation of the N_e/N ratio for a time series. Specifically, in these papers long-term N_e/N was computed as the harmonic mean N_e ($\tilde{N}_e$) divided by the arithmetic mean N ($\overline{N}$). The effects of mixing arithmetic and harmonic means in a ratio are illustrated in table 8.3, which shows a hypothetical series of N_e and N values for five generations. In each generation, N_e/N is 0.4, similar to the mean ratios Frankham found using adjustments for variance in reproductive success (0.46) or sex ratio and reproductive variance (0.35). The ratio over the whole time series in this example is also 0.4, provided it is computed as $\tilde{N}_e/\tilde{N}$,

Table 8.3 Computation of Long-Term N_e/N in a Hypothetical Population with Variable N but Constant N_e/N per Generation

Generation	N	N_e	N_e/N
1	1000	400	0.4
2	500	200	0.4
3	150	60	0.4
4	2000	800	0.4
5	5000	2000	0.4
Arithmetic mean ($\overline{N}$, $\overline{N_e}$, $\overline{N_e/N}$)	1730	692	0.4
Harmonic mean ($\tilde{N}$, $\tilde{N_e}$, $\tilde{N_e/N}$)	482	193	0.4
Harmonic mean N_e/arithmetic mean N ($\tilde{N_e}/\overline{N}$)			0.11

$\tilde{N_e}/\overline{N}$, or any function of the individual generation N_e/N ratios. It can also be shown that 0.4 is the expected value of the harmonic mean N_e divided by the N values each generation (Kalinowski and Waples in press). When long-term N_e/N in table 8.3 is computed as $\tilde{N_e}/\overline{N}$, however, the result is 0.11, the same mean value Frankham (1995b) found for comprehensive estimates of N_e/N. This simple example shows that a reduction in N_e/N of the magnitude attributed to variable N can also be generated entirely by the statistical artifact of mixing arithmetic and harmonic means in a ratio.

This does not mean that variations in N are not important as a factor influencing N_e. If there is a strong relationship between N and N_e within generations, variable population size can have a profound influence on long-term N_e, because effective size for a series of generations is approximately the harmonic mean (rather than the arithmetic mean) of the single-generation N_e's (Wright 1938). However, I would argue that variations in N contain no additional information about the long-term ratio N_e/N beyond that contained in the individual generation values N_e/N. Variations in N by themselves do not affect the long-term ratio N_e/N, unless that ratio is calculated using a mixture of harmonic and arithmetic means. The example in table 8.3 raises the question, "What biological information is conveyed by a long-term N_e/N ratio when it is computed as $\tilde{N_e}/\overline{N}$?" This formulation ignores the information contained in the N_e/N ratios for individual generations and produces a quantity that is difficult to interpret biologically. Kalinowski and Waples (in press) provide biological interpretations for $\tilde{N_e}/\overline{N}$ and another method ($\tilde{N_e}/\tilde{N}$) for computing the long-term N_e/N, show how the two indices are related mathematically, and evaluate their properties.

INDIRECT METHODS FOR ESTIMATING N_e

Because N_e directly controls the rate of change in numerous genetic parameters in natural populations, various researchers have reasoned that by measuring these genetic parameters one can obtain an estimate of N_e. Although these indirect genetic methods for estimating N_e typically have a large stochastic variance, they have achieved some popularity because direct demographic estimates of N_e (based on sex ratio and reproductive success) are so difficult to obtain for many species that indirect genetic estimates may be of considerable use even if they are imprecise.

Waples (1991) identified three types of genetic methods for estimating N_e: those that estimate current N_e, those that estimate long-term or historical N_e, and those that provide information about population bottlenecks (see Schwartz et al. 1998 for further discussion). Estimates of historical N_e, which represents the long-term effective size that would be expected to produce currently observed levels of neutral genetic diversity, are of limited use in conservation of endangered species because of several factors: (1) they are sensitive to errors in estimating mutation rate, (2) they can be very sensitive to violations of assumptions of selective neutrality and no immigration, and (3) they assume equilibrium conditions that may take thousands or millions of generations to achieve (Avise et al. 1988; Waples 1991). Methods for estimating current N_e and evaluating bottlenecks are more promising for practical application and are discussed below.

Methods to Estimate Current N_e

Two methods (temporal changes in allele frequency and magnitude of gametic disequilibrium) have been used to estimate current or recent N_e in natural populations (Hill 1981; Nei and Tajima 1981; Waples 1989). A third method, based on heterozygote excess, has been proposed (Pudovkin et al. 1996), but it probably has too little power for application to most realistic problems involving endangered species. Both bias and precision can be a concern for indirect estimates of N_e. Biases arise from violation of the basic assumptions (random mating, random sampling, selective neutrality, no migration or immigration). Violations of strict selective neutrality are common but may introduce little bias unless the selection is strong; see Waples (1991) for discussion. As discussed above, sampling considerations are particularly important for organisms with overlapping generations. Estimates of N_e can also be seriously biased if they are applied to a subpopulation that has significant genetic linkages to other subpopulations within a larger metapopulation.

Precision is a concern for indirect estimates of N_e because they depend on evolutionary processes that are inherently stochastic and can only be predicted in a statistical sense. Only by averaging over a large number of replicate trajectories can a reliable signal be detected.

Fortunately, two factors mitigate the inherent lack of precision of methods based on random genetic processes. First, concerns for endangered species usually center not on how large N_e is but whether it is too small. Because the genetic signal detected by indirect estimates of N_e is proportional to $1/N_e$, the strength of the signal relative to stochastic noise is larger in small populations. As a consequence, although it may be difficult using indirect genetic methods alone to distinguish between $N_e = 500$ and $N_e = 5,000$ (both appearing to be "large" because the drift signal is weak), the power to detect situations in which N_e is low is generally much higher. For example, assume that one has two juvenile samples of 100 individuals each taken five generations apart from a single population, and that analysis of these samples provides genetic data for 20 diallelic loci. If the population's true N_e is 500, the temporal method will be expected to yield a wide confidence interval for its estimate of effective size (95% confidence interval [CI] = 117 to ∞, using formulas presented in Waples 1989). In contrast, if true N_e is only 50, then the expected confidence interval is much smaller (95% CI = 22 to 100) and provides evidence that the population's effective size is small enough to warrant conservation attention. Indirect genetic methods thus have some utility in monitoring species of conservation interest and may have reasonable power to detect low N_e before serious problems result.

Second, recent advances in molecular genetics permit large numbers of genetic markers to be brought to bear on problems of conservation interest. Each independent marker increases precision of indirect estimates of N_e. Although highly polymorphic microsatellites present some thorny statistical problems regarding sample sizes and the appropriate mutational model (Chakraborty 1992; O'Connell and Wright 1997), they potentially can be very useful in the temporal method for estimating N_e, since that method is less sensitive than the disequilibrium method to biases caused by low-frequency alleles. As demonstrated by Miller and Kapuscinski (1997), microsatellites can also be used to obtain historical estimates of N_e from archived samples.

Another way to deal with the limitations posed by lack of precision in indirect estimates of N_e is to integrate information across one or more dimensions. This can be done, for example, by combining the temporal and disequilibrium methods to obtain an overall estimate of N_e (as suggested by Waples 1991), or by combining data across years, populations, or species to reveal general patterns. Figure 8.2 illustrates results of applying both of these strategies to estimation of effective population size in Snake River spring/summer chinook salmon (*Oncorhynchus tshawytscha*), which have been listed as a threatened species under the U.S. Endangered Species Act since 1992 (Waples 1995). Fifteen natural

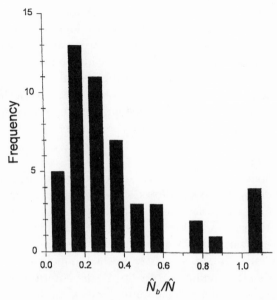

Fig. 8.2 Preliminary estimates of N_e:N (ratio of effective to total population size) for yearly samples of Snake River chinook salmon. A combination of two indirect genetic methods (temporal changes in allele frequency and gametic disequilibrium) were used to estimate the effective number of breeders each year ($\hat{N}_b$), and this was compared to an estimate of the total number of spawners ($\hat{N}$). Allozyme data (Waples et al. 1993, and unpublished data) were taken for 33 polymorphic gene loci analyzed in collections from 15 natural populations sampled in three to six years over the period 1989–1994; harmonic mean sample size was 76 individuals. Each unit in the histogram represents an estimate for one year in one population.

populations were sampled three to six times over a six-year period and analyzed for 33 polymorphic allozyme loci (Waples et al. 1993). Preliminary estimates using the temporal and disequilibrium approaches were derived using methods of Hill (1981) and Waples (1990b) and combined (unweighted harmonic mean of the two estimates) to obtain a single estimate of the effective number of breeders ($\hat{N}_b$) each year. Counts of adult fish or expansion from counts of redds (nests) allowed an estimate of the ratio $\hat{N}_b/\hat{N}$ in each year.

Two general observations can be made from the results plotted in figure 8.2. First, although they use information for a large number of individuals and gene loci and two independent methods for estimating effective size, the yearly $\hat{N}_b/\hat{N}$ ratios still show a relatively high variance. For example, $\hat{N}_b/\hat{N}$ was >1 for four samples, which is unlikely to be true for natural populations. This means that individual $\hat{N}_e$ or $\hat{N}_e/\hat{N}$

estimates based on indirect genetic methods have to be interpreted with some caution, even when they integrate a considerable amount of information. Second, in spite of this high intrinsic variance, an important general pattern emerges when data for many populations are considered: over 60% of the yearly $\hat{N}_b/\hat{N}$ estimates fall in the range 0.1 to 0.4. Given the scarcity of information about the ratio N_e/N in natural populations, this result can be of considerable use in conservation planning for at-risk salmon populations.

Methods to Detect Bottlenecks

Genetic methods for detecting population bottlenecks take advantage of the fact that changes in population size have different effects on genetic parameters such as average heterozygosity and the number of alleles per locus (or their DNA analogues, average number of nucleotide differences between a pair of DNA sequences and the number of segregating sites). Rare alleles are lost quickly from small populations, but such alleles have little effect on heterozygosity, so the latter measure is slower to reflect the effects of a population bottleneck. In theory, nonequilibrium relationships between parameters can be taken as evidence for population-size changes in the past. Cornuet and Luikart (1996) described statistical tests for this effect, and Luikart et al. (1998a) applied the tests to empirical data for a number of populations that had and had not experienced bottlenecks in size. They found that under some circumstances these tests can provide reasonable power to detect population bottlenecks. Attributes of the tests that are relevant to conservation applications include the following: (1) selective neutrality and a closed population are key assumptions, and results may be particularly sensitive to immigration from a genetically distinct population; (2) results depend on the mutational model used; (3) adequate power generally requires data for 10 to 20 gene loci, and in general increasing the number of loci is more important than increasing the number of individuals sampled; and (4) it may be possible to detect bottlenecks for up to four N_e generations, where N_e is the bottleneck effective size. Garza and Williamson (2001) recently proposed a different method for detecting bottlenecks, based on the ratio of the number of alleles to the range of allele size in microsatellites.

   The temporal method (described above) also can provide information about population bottlenecks, and Luikart et al. (1998b) showed that analysis of temporal changes in allele frequency provides much greater power to detect bottlenecks than the methods described by Cornuet and Luikart (1996). However, the temporal method requires at least two samples that span the bottleneck period, which will not be possible for many species of conservation interest, while Cornuet and Luikart's tests can be applied to data for a single sample. Richards and Leberg (1996) evaluated the usefulness of the temporal method to detect bottlenecks in simulated populations and experimentally manipulated populations of mosquito fish. They found a high variance in the estimated N_e's and an upward bias at very low bottleneck sizes. They attributed the high variance to the relatively low number of gene loci used, and they attributed the bias to extinction of low- and moderate-frequency alleles. Waples (1989) discussed ways to address these issues of precision and bias for the temporal method.

CONCLUSIONS

A fuller integration of effective size considerations into PVA and conservation of endangered species can be achieved by greater attention to some of the difficulties in definition and interpretation of N_e described above. Although the brief review in this paper considered only a few of the many factors that can influence genetic process in a metapopulation, it is clear that population subdivision can profoundly affect N_e in either direction. If Hedrick and Gilpin (1997) are correct that human influences have disrupted historical population structure in many species, greater attention to nonequilibrium population genetic processes will be increasingly important in the future. In such cases, genetic variability can be lost at rates much greater than predicted based on census numbers or traditional estimates of effective population size.

Careful attention to sampling considerations can help reduce biases that result from a failure to recognize that the population unit under consideration is closely linked in space and/or time to other units. This is particularly true for species with overlapping generations. In such cases, the effective size of the subunit sampled often will be less appropriate to consider than the effective size of the larger spatial/temporal unit as a whole.

Achieving some consensus regarding the range of N_e/N values in natural populations would facilitate the quantitative application of effective size considerations to a great many more species than at present. Improving the link between the biological interpretation of generational and long-term N_e/N ratios will be an important step in developing this consensus.

Estimating effective size in natural populations is such a difficult challenge that all feasible methods should continue to be used. Although precision is often a limiting factor for indirect genetic methods for estimating N_e, this limitation can be overcome to some extent by integrating information over multiple samples, genetic markers, and populations.

In addition, these methods are particularly well suited for use in the conservation of endangered species, since their statistical power increases dramatically as effective size decreases.

LITERATURE CITED

Avise, J. C., R. M. Ball, and J. Arnold. 1988. Current versus historical population sizes in vertebrate species with high gene flow: a comparison based on mitochondrial DNA lineages and inbreeding theory for neutral mutations. *Molecular Biology and Evolution* 5:331–344.

Caballero, A. 1994. Developments in the prediction of effective population size. *Heredity* 73:657–679.

Chakraborty, R. 1992. Sample size requirements for addressing the population genetic issues of forensic use of DNA typing. *Human Biology* 64:141–159.

Cornuet, J. M., and G. Luikart. 1996. Description and power analysis of two tests for detecting recent population bottlenecks from allele frequency data. *Genetics* 144:2001–2014.

Crow, J. F., and C. Denniston. 1988. Inbreeding and variance effective population numbers. *Evolution* 42:482–495.

Crow, J. F., and M. Kimura. 1970. *An introduction to population genetics theory.* Harper and Row, New York, New York.

Ewens, W. J. 1979. *Mathematical population genetics.* Springer-Verlag, Berlin, Germany.

Felsenstein, J. 1971. Inbreeding and variance effective numbers in populations with overlapping generations. *Genetics* 68:581–597.

Fisher, R. A. 1930. *The genetical theory of natural selection.* Clarendon Press, Oxford, United Kingdom.

Frankham, R. 1995a. Conservation genetics. *Annual Review of Genetics* 29:305–327.

———. 1995b. Effective population size/adult population size ratios in wildlife: a review. *Genetical Research* 66:95–107.

Garza, J. C., and E. G. Williamson. 2001. Detection of reduction in population size using data from microsatellite loci. *Molecular Ecology* 10:305–318.

Gilpin, M. 1991. The genetic effective size of a metapopulation. *Biological Journal of the Linnean Society* 42:165–175.

Hansen, M. M., E. E. Nielsen, and K.-L. D. Mensberg. 1997. The problem of sampling families rather than populations: relatedness among individuals in samples of juvenile brown trout *Salmo trutta* L. *Molecular Ecology* 6:469–474.

Hedgecock, D. 1994. Does variance in reproductive success limit effective population sizes of marine organisms? Pages 122–134 in A. R. Beaumont, editor, *Genetics and evolution of aquatic organisms.* Chapman and Hall, London, United Kingdom.

Hedrick, P. W., and M. E. Gilpin. 1997. Genetic effective size of a metapopulation. Pages 165–181 in I. Hanski and M. E. Gilpin, editors, *Metapopulation biology: ecology, genetics, and evolution.* Academic Press, San Diego, California.

Hedrick, P. W., D. Hedgecock, and S. Hamelberg. 1995. Effective population size in winter-run chinook salmon. *Conservation Biology* 9:615–624.

Hedrick, P. W., and S. T. Kalinowski. 2000. Inbreeding depression in conservation biology. *Annual Review of Ecology and Systematics* 31:139–162.

Herbinger, C. M., R. M. Doyle, C. T. Taggart, S. E. Lochmann, A. L. Brooker, J. M. Wright, and D. Cook. 1997. Family relationships and effective population size in a natural cohort of Atlantic cod (*Gadus morhua*) larvae. *Canadian Journal of Fisheries and Aquatic Sciences* 54 (supplement 1):11–18.

Hill, W. G. 1972. Effective size of populations with overlapping generations. *Theoretical Population Biology* 3:278–289.

———. 1981. Estimation of effective population size from data on linkage disequilibrium. *Genetical Research* 38:209–216.

Jorde, P. E., and N. Ryman. 1995. Temporal allele frequency change and estimation of effective size in populations with overlapping generations. *Genetics* 139:1077–1090.

———. 1996. Demographic genetics of brown trout (*Salmo trutta*) and estimation of effective population size from temporal change of allele frequency. *Genetics* 143:1369–1381.

Kalinowski, S. T., and R. S. Waples. 2002. The relationship of effective to census size in fluctuating populations. *Conservation Biology* 16:1–9.

Levins, R. 1970. Extinction. Pages 77–107 in M. Gerstenhaber, editor, *Some mathematical questions in biology*. American Mathematical Society, Providence, Rhode Island.

Luikart, G., J.-M. Cornuet, and F. W. Allendorf. 1998a. Empirical evaluation of a test for detecting recent historical population bottlenecks. *Conservation Biology* 12:228–237.

Luikart, G., W. Sherwin, B. Steele, and F. W. Allendorf. 1998b. Usefulness of molecular markers for detecting population bottlenecks via monitoring genetic change. *Molecular Ecology* 7:963–974.

Maruyama, T., and M. Kimura. 1980. Genetic variability and effective population size when local extinction and recolonization of subpopulations are frequent. *Proceedings of the National Academy of Sciences* (USA) 77:6710–6714.

Miller, L. M., and A. R. Kapuscinski. 1997. Historical analysis of genetic variation reveals low effective population size in a northern pike (*Esox lucius*) population. *Genetics* 147:1249–1258.

Nei, M., and F. Tajima. 1981. Genetic drift and estimation of effective population size. *Genetics* 98:625–640.

Nomura, T. 1999. Effective population size in supportive breeding. *Conservation Biology* 13:670–672.

Nunney, L. 1993. The influence of mating system and overlapping generations on effective population size. *Evolution* 47:1329–1341.

———. 1999. The effective size of a hierarchically-structured population. *Evolution* 53:1–10.

———. 2000. The limits to knowledge in conservation genetics: the value of effective population size. Pages 179–194 in M. T. Clegg, editor, *Limits to knowledge in evolutionary genetics*. Plenum Press, New York, New York.

Nunney, L., and D. R. Elam. 1994. Estimating the effective population size of conserved populations. *Conservation Biology* 8:175–184.

O'Connell, M., and J. M. Wright. 1997. Microsatellite DNA in fishes. *Reviews in Fish Biology and Fisheries* 7:331–363.

Pudovkin, A. I., D. V. Zaykin, and D. Hedgecock. 1996. On the potential for estimating the effective number of breeders from heterozygote-excess in progeny. *Genetics* 144:383–387.

Ray, C. 2001. Maintaining genetic diversity despite local extinctions: effects of population scale. *Biological Conservation* 100:3–14.

Richards, C., and P. L. Leberg. 1996. Temporal changes in allele frequencies and a population's history of severe bottlenecks. *Conservation Biology* 10:832–839.

Ryman, N., P. E. Jorde, and L. Laikre. 1995. Supportive breeding and variance effective population size. *Conservation Biology* 9:1619–1628.

———. 1999. Supportive breeding and inbreeding effective number: a comment to Nomura. *Conservation Biology* 13:673–676.

Schwartz, M. K., D. A. Tallmon, and G. Luikart. 1998. Review of DNA-based census and effective population size estimators. *Animal Conservation* 1:293–299.

Vucetich, J. A., and T. A. Waite. 1998. Number of censuses required for demographic estimation of effective population size. *Conservation Biology* 12:1023–1030.

Vucetich, J. A., T. A. Waite, and L. Nunney. 1997. Fluctuating population size and the ratio of effective to census population size. *Evolution* 51:2017–2021.

Waples, R. S. 1989. A generalized approach for estimating effective population size from temporal changes in allele frequency. *Genetics* 121:379–391.

———. 1990a. Conservation genetics of Pacific salmon: 2, effective population size and the rate of loss of genetic variability. *Journal of Heredity* 81:267–276.

———. 1990b. Conservation genetics of Pacific salmon: 3, estimating effective population size. *Journal of Heredity* 81:277–289.

———. 1991. Genetic methods for estimating the effective size of cetacean populations. *Report of the International Whaling Commission* (special issue) 13:279–300.

———. 1995. Evolutionarily significant units and the conservation of biological diversity under the Endangered Species Act. *American Fisheries Society Symposium* 17:8–27.

Waples, R. S., and C. Do. 1994. Genetic risk associated with supplementation of Pacific salmonids: captive broodstock programs. *Canadian Journal of Fisheries and Aquatic Sciences* 51 (supplement 1):310–329.

Waples, R. S., O. W. Johnson, P. B. Aebersold, C. K. Shiflett, D. M. VanDoornik, D. J. Teel, and A. E. Cook. 1993. *A genetic monitoring and evaluation program for supplemented populations of salmon and steelhead in the Snake River Basin.* Annual Report of Research. Bonneville Power Administration, Portland, Oregon.

Whitlock, M. C., and N. H. Barton. 1997. The effective size of a subdivided population. *Genetics* 146:427–441.

Wright, S. 1931. Evolution in Mendelian populations. *Genetics* 16:97–159.

———. 1938. Size of population and breeding structure in relation to evolution. *Science* 87:430–431.

———. 1943. Isolation by distance. *Genetics* 28:114–138.

Estimating Parameters of PVA Models from Data on Marked Animals

Gary C. White, Alan B. Franklin, and Tanya M. Shenk

ABSTRACT

Population viability analysis (PVA) requires knowledge of the state of a population at a given time as well as the dynamics of the population over a given time frame to be able to provide rigorous, defensible estimates of viability and/or threats to extinction. Field and statistical methods developed for the analysis of reencounters with marked animals provide biologists with techniques to study the current state of a population as well as its dynamics. Modern methods provide estimates of population size, survival, recruitment, and immigration and emigration rates, all corrected for the probability of detection, in addition to estimating the impact of density dependence on survival and recruitment. Current computer software (Program MARK) provides a common framework to estimate parameters from Cormack-Jolly-Seber models, ring or band recovery models, closed population models, and numerous extensions of these three basic models. Environmental (temporal) covariates, attribute-group covariates (e.g., spatial, gender, age), and individual-specific covariates (e.g., genetic heterozygosity) can be incorporated into models to enhance understanding of population dynamics, as well as improve precision of parameter estimates. Particularly important to PVA is that methods are available to estimate process variation of parameters across time or space with the sampling variance component removed. These estimates of process variance are the appropriate values to use in PVA models. While current thinking in PVA has concentrated on the mathematical structure of PVA models, we believe that a more constructive approach is to determine which parameters can be reliably and precisely estimated and then build PVA models around those parameters. We provide an example of such an approach with simulated data from the northern spotted owl.

INTRODUCTION

In the narrowest sense, population viability analysis (PVA) provides a methodology to estimate the probability that a biological population will persist for some defined time period. In a more general sense, PVA may provide a methodology to examine what factors are critical to persistence of a population. In either case, PVA consists of a mathematical model built to meet specific objectives (Ralls et al., chap. 25 in this volume).

Analyses of mathematical models are useful in projecting the behavior of the model as a consequence of the assumptions and parameter values used to construct the mathematical model. However, mathematical models only test hypotheses about those models; for example, will density dependence in the model cause the *modeled* population to stabilize at a carrying capacity? Too often in ecology, modelers begin to accept mathematical models as reality (or as the natural system itself), instead of as a simplified model. To provide predictive value for real populations, and hence to maximize the value for managing a biological population, a PVA model must be developed from data on the biological population of interest (or a closely related population) using appropriate statistical models. Neither mathematical nor statistical models can predict the future. They can only suggest scenarios that might take place. That is, the models themselves reflect hypotheses about biological populations and are used to project their consequences. These hypotheses and models may be of value to biologists, but only if the assumptions of the model and the parameter estimates provide a reasonable approximation to reality for the population of interest. Only data collected on the biological population can be used to test these hypotheses and corresponding models, not the behavior of the model. Thus, data are used to test hypotheses generated by models.

This chapter describes methods to estimate reliably the parameters required for PVA models to have some biological value in describing the behavior of a biological population. We discuss estimation of population size, survival, recruitment, immigration and emigration, and detecting and estimating the impact of density dependence on survival and recruitment. The focus of our methods is on the use of marked animals and the encounter histories developed from observing these animals over time. Last, we present an example to demonstrate how to use marked animals to evaluate population persistence.

PARAMETERS REQUIRED IN PVA MODELS

Ludwig (1999) suggested that PVA models developed from only observations on population size (N_t) have little value because the mechanisms driving the population's behavior cannot be deduced from the time series of population sizes. We believe that PVA models developed from more fundamental population data (i.e., survival, recruitment, immigration, and emigration) can be useful in conjecturing the persistence of a population, or at least in understanding the process more fully. The approach that we believe to be most useful is described more generally by White (2000).

Population models consist of a pool of individuals, with other individuals entering (births, immigration) and exiting (deaths, emigration) this pool. Formally, the size (N) of the population at time $t + 1$ is given in terms of the population at time t as

$$N_{t+1} = N_t + B_t + I_t - D_t - E_t,$$

where births (B) and immigration (I) contribute to the growth of the population, and deaths (D) and emigration (E) cause a reduction in the population. Typically, population models are not constructed to estimate the number of animals that are born, immigrate, die, or emigrate directly, but rather estimate these quantities as rates relative to the population size. For example, the number of animals dying in a population at time t would be estimated as $D_t = (1 - s_t)N_t$, where s_t is the survival rate of the population at time t. Similar approaches would be used for B_t, E_t, and maybe even I_t, so that the fundamental parameters of the model might be s_t, b_t, e_t, and i_t. Typically, PVA models are constructed as Leslie matrix models (Leslie 1945, 1948; Caswell 1989), where the age (or size) and possibly sex structure of the population are also modeled. As a result, the time-specific rates are now extended to age- or size- and sex-specific rates.

Variation in the fundamental model parameters can be described as *process variance* ($\sigma^2_{\text{process}}$) and variation in the true values of the parameters over time (temporal variation), space (spatial variation), and individuals (individual heterogeneity). A key requirement of PVA modeling is to examine the impact on population persistence of variation in the fundamental parameters across time, space, and individuals. For example, survival rates are not constant across time; temporal variation contributes to the process variance of s_t. All animals do not have the same survival rate at time t, so this individual heterogeneity also contributes to the process variance. The more variation in s_t across time in the model, the lower the persistence of the modeled population (Boyce 1992; White 2000). In contrast, more variation in s_t across individuals or space typically results in increased persistence (Conner and White 1999; White 2000).

Unfortunately for biologists, we can never observe the quantities s_t, b_t, e_t, or i_t and, hence, are not able to estimate directly the process variance of these parameters. Several reasons explain why the parameter values cannot be observed directly. First, even if we know exactly how many individuals were alive at time t and survived to time $t + 1$, we would only know the outcome of the stochastic survival process. For example, if a fair coin is flipped ten times and six heads are observed, we cannot claim that the coin has a probability of 0.6 of generating a head, even though that is our best estimate. The true, underlying probability of a head occurring is 0.5. Likewise, observing 65 animals surviving from a group of 100 does not mean that the survival rate governing the process was 0.65. Rather, 0.65 of the animals survived, but the true survival rate might easily have been as high as 0.7 with 0.65 observed. Second, for most populations of interest, we are not able to measure exactly how many individuals were alive at time t and survived to time $t + 1$. We are unable to count completely (i.e., census) the population of interest. As a result, we obtain only estimates of the fundamental parameters of interest. The degree of uncertainty between our estimate of a fundamental parameter and its true value is described by *sampling variance*, the variation attributable to estimating a parameter from sample data. Statistical estimators of parameters provide the sampling variance. For example, if survival is assumed to follow a binomial process, and n_l animals live out of n animals alive at the start of the interval, the estimate of survival is $\hat{s} = n_l/n$, with sampling variance $\text{var}(\hat{s}|s) = s(1 - s)/n$ that is generally estimated as $\hat{\text{var}}(\hat{s}|s) = \hat{s}(1 - \hat{s})/n$. The term $\text{var}(\hat{s}|s)$ means the variance of $\hat{s}$, given the true value s.

To make the distinction between sampling and process variance clearer, consider an example. Let the true annual survival rate for each year in a ten-year period ($t = 10$) be $s_1, s_2, \ldots, s_{10}$, and the estimates of these rates denoted as $\hat{s}_1, \hat{s}_2, \ldots, \hat{s}_{10}$. The *total variance* (process plus sampling variance) is estimated as

$$\hat{\sigma}^2_{\text{total}} = \sum_{i=1}^{t} \frac{(\hat{s}_i - \bar{s})^2}{t - 1},$$

where $\bar{s} = \sum_{i=1}^{t} \hat{s}_i/t$ (the average survival rate over ten years) and $\sigma^2_{\text{total}} = \text{process variance} + \text{sampling variance} = \sigma^2_{\text{process}} + \text{var}(\hat{s}|s)$. The typical approach to obtaining an estimate of the process variance is to subtract the sampling variance from the total variance, where the sampling variance can be approximated as the mean of the sampling variances for the individual estimates, $\hat{\text{var}}(\hat{s}|s) = \sum_{i=1}^{t} \hat{\text{var}}(\hat{s}_i|s_i)/t$. This is the basic approach suggested by Link and Nichols (1994) and Gould and Nichols (1998) and illustrated in Mills and Lindberg (chap. 16 in this volume). However, this approach ignores any sampling covariances between survival estimates across time. Burnham et al. (1987) consider the case where a more general estimator than the mean of the sampling variance is used (also described in White 2000).

A common mistake in PVA models is to use the sampling variance, or even the total variance, for a model parameter in the PVA model instead of the process variance (White 2000). For the case of temporal variation, estimates of persistence that result from this flawed model typically will predict extinction much more often than they should, because the total variance is greater than the process variance.

PARAMETER ESTIMATION FROM MARKED ANIMALS

Following marked animals through time provides a method for estimating the fundamental parameters of a biological population. A large literature has developed and been reviewed by Seber (1982, 1986, 1992) and Schwarz and Seber (1999). Most models developed to estimate fundamental population parameters from data collected on marked animals have been implemented in the software package Program MARK (White and Burnham 1999; http://www.cnr.colostate.edu/~gwhite/mark/mark.htm). Central to all of these models is the ability to estimate the probability of detection of the animal in order to estimate survival, recruitment, immigration, or emigration rate. In this section, we first describe the necessity of estimating the detection rate, and then provide an overview of Program MARK, with descriptions of specific models that are useful in estimating the fundamental parameters of a population for developing a PVA model.

Appropriate statistical treatment of the fundamental population parameters requires acknowledging that animals are not always detectable, that this detectability can vary by individual attributes, such as age and sex, and that detectability can vary across time and space. Often ecologists count animals, assume a census, and derive parameters of interest, such as abundance and survival, directly from these counts. The counts are of little use by themselves, however, because they represent some unknown sample from the population of interest. The sampling fraction needs to be estimated for parameter estimates to be unbiased, and to correct for different sampling fractions across time and/or space so that estimates are comparable.

Nichols (1992) makes the following important points concerning the need to include sampling fractions (p) when counting animals. If C_i denotes the counts of animals in some point in either space or time, p_i the sampling fraction, and N_i the true number of animals, then their relationship can be written as

$$C_i = N_i p_i \quad \text{or as} \quad \hat{N}_i = \frac{C_i}{\hat{p}_i}.$$

Often the C_i are referred to as indices and are used by themselves (e.g., $\hat{N}_i = C_i$) for comparisons, without considering the p_i. However, sampling fractions can vary over time and space in response to factors uncontrollable by the investigator. Nichols (1992) concluded that the estimation of p is a central methodological problem in the study of animal populations. In response to this problem, considerable effort has been directed to developing capture-recapture estimators for survival, recruitment, rates of population change, and abundance in demographically open and closed populations (Seber 1986; Lebreton et al. 1992; Seber 1992; Lebreton et al. 1993; Pradel 1996).

The problems encountered with ignoring detectability can be exemplified with a capture-recapture data set for the Preble's meadow jumping mouse (*Zapus hudsonius preblei*), a federally listed species. Mice were trapped for five consecutive nights in each of three seasons (posthibernation, summer, prehibernation) over the course of one year in eastern Colorado. Abundance was estimated using closed capture-recapture models in Program MARK (White et al. 1982; White and Burnham 1999) for each season and then compared with the minimum number known alive (MNA) for each season (fig. 9.1). MNA is defined as the number of different individuals captured and marked during a season and is a widely used index in the small-mammal literature (Pollock et al. 1990). Estimates of MNA lack any theoretical estimator for sampling variance and assume a capture probability of 1 for all individuals over the course of each trapping session. For these reasons, we believe MNA is an inappropriate estimator of abundance compared to the capture-recapture estimator. The two estimators give very different estimates of trends in abundance over the three seasons (fig. 9.1) and of mean abundance and $\hat{\sigma}^2_{\text{process}}$ for abundance over the three seasons (table 9.1). Estimates of mouse abundance based on MNA are lower and exhibit greater temporal process variance than those based on the capture-recapture estimator. This example is consistent with simulation work, which shows MNA is almost always negatively biased with respect to capture-recapture estimators for both abundance and survival (Pollock et al. 1990). Had the MNA estimates been used to detect trends in abundance, one would incorrectly infer an increase in abundance across seasons.

PROGRAM MARK OVERVIEW

Program MARK provides parameter estimates from marked animals when they are reencountered at a later time. Reencounters can be from dead recoveries (the animal is harvested or otherwise found dead), live recaptures (the animal is retrapped or resighted), radio tracking (the animals is marked with a radio transmitter), or from some combination

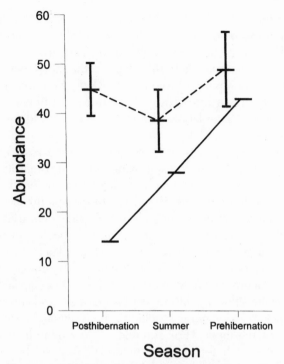

Season

Fig. 9.1 Estimates of abundance for Preble's meadow jumping mouse over three seasons. Abundance was estimated from five successive trapping occasions within each season. *Dashed line* represents estimates based on closed capture-recapture estimators, whereas *solid line* represents minimum number known alive (MNA).

Table 9.1 Estimates of Abundance for Preble's Meadow Jumping Mouse over Three Seasons

Parameter	Capture-Recapture	Minimum Number Known Alive
	Estimate from	
$\overline{N}$	43.9	28.3
$\hat{\sigma}^2_{process}$	14.6	210.3
$CV_{process}$	8.72%	51.72%

Notes: Estimates of mean abundance ($\overline{N}$), process variation over three seasons ($\hat{\sigma}^2_{process}$), and coefficient of process variation ($CV_{process}$) computed from capture-recapture estimators and minimum number known alive (MNA) for Preble's meadow jumping mouse in eastern Colorado.

of these sources of reencounters. The time intervals between reencounters do not have to be equal. More than one attribute group of animals can be modeled (e.g., treatment and control animals, or male and female), and covariates specific to the group or the individual animal can be used. The basic input to Program MARK is the encounter history for each animal. Progarm MARK can also provide estimates of population size for closed populations where capture (p) and recapture (c) probabilities are also modeled.

Parameters can be constrained to be the same across reencounter occasions, by age or by group using parameter index matrices (PIMs). The PIM provides the conceptually simplest tool to construct models that do not include covariates. Based on the parameter indices in the PIM, the design matrix allows the user to incorporate covariates into the analysis. Covariates can be time-specific, group-specific, or individual-specific. For example, a user may want to test the idea that mean annual temperature affects survival. With the design matrix, annual survival can be estimated as a function of temperature values. Other covariates might be group-specific, allowing differences in treatment and control groups, or gender-specific. Covariates specific to the individual also can be incorporated into the model. For example, to determine if survival is influenced by genetic heterozygosity, heterozygosity can be included as an individual covariate in the model.

Program MARK computes the estimates of model parameters via numerical maximum likelihood techniques. The FORTRAN program that does this computation also determines numerically the number of parameters that are estimable in the model. The number of estimable parameters is used to compute the quasi-likelihood Akaike's information criterion value (QAICc) for the model (for use in model selection). An estimate $(\hat{c})$ of the variance inflation factor provided by parametric bootstrap techniques or other goodness-of-fit tests can be used to correct estimates of parameter variance estimates and AICc to produce QAICc values. Thus, Program MARK provides a means to handle the overdispersion that is commonly found in biological populations.

Outputs for various models that the user has built and then fit are stored in a database. Input data are also stored in this database, making it a complete description of the model-building process. The database is viewed and manipulated in a Results Browser window. Summaries available from the Results Browser window include model output (estimates, standard errors, and goodness-of-fit tests), deviance residuals from the model, likelihood ratio and analysis of deviance (ANODEV) tests between models, adjustments for overdispersion, and model averaging as advocated by Burnham and Anderson (1998). Models can also be retrieved and modified to create additional models. As is described below, methods to estimate the process variance of a set of estimates are also provided.

The theory and methods used in Program MARK are described in more detail in an "electronic book" at http://www.cnr.colostate.edu/~gwhite/mark/mark.htm. The program and a set of instructions for installing it are also available from this Web site. In addition, an introduction to Program MARK is provided online at http://canuck.dnr.cornell.edu/misc/cmr/mark/docs/book/, developed by Evan Cooch and Gary White.

Estimating Survival with the Live-Recapture Model

Live recaptures are the basis of the standard Cormack-Jolly-Seber (CJS) model (Cormack 1964; Jolly 1965; Seber 1965; Lebreton et al. 1992). Animals are captured, marked, and released into the population. Marked animals are then encountered by catching them alive and rereleasing them. If marked animals are released into the population on occasion 1, then each succeeding capture occasion is one encounter occasion. Consider the following scenario:

$$\text{Release} - \phi_1 \rightarrow \text{Encounter 2} - \phi_2 \rightarrow \text{Encounter 3} \dots$$
$$p_2 \qquad\qquad\qquad p_3$$

Animals survive from initial release to the first reencounter with probability ϕ_1, and from the first encounter occasion to the second encounter occasion with probability ϕ_2. The recapture probability at encounter occasion 2 is p_2, and p_3 is the recapture probability at encounter occasion 3. At least two reencounter occasions are required to estimate the survival rate (ϕ_1) between the first release occasion and the first encounter occasion. The survival rate between the last two encounter occasions is not estimable because only the product of survival and recapture probability for this occasion is identifiable. The parameter ϕ_1 is usually termed "apparent survival," and represents the probability that an animal remains alive and on the area of interest. Animals that permanently emigrate from the study area are not available for recapture, so appear to have died in this model. Thus, $\phi_i = S_i(1 - E_i)$, where S_i is the probability of surviving interval i and E_i is the probability of emigrating from the study area during interval i.

An example of how apparent survival is estimated from an encounter history is useful in understanding the CJS model. Consider the following encounter history: 1101, with four occasions. A value of 1 represents a capture or recapture, whereas 0 means the animal was not captured.

The parameterization of this encounter history would be $\phi_1 p_2 \phi_2$ $(1 - p_3)\phi_3 p_4$. This parameterization can be translated as the animal survives the first interval with probability ϕ_1 and is captured with probability p_2, then survives the second interval with probability ϕ_2 (we know the animal is still alive and on the study area because it was captured on occasion 4) and is not captured with probability $(1 - p_3)$, and survives the third interval with probability ϕ_3 and is captured with probability p_4. When an animal is not captured on the last occasion, the parameterization of the encounter history must take into account both the possibility that the animal died or left the area, or that it was alive but not captured. For example, the encounter history 1110 would generate the parameterization $\phi_1 p_2 \phi_2 p_3[(1 - \phi_3) + \phi_3(1 - p_4)]$. The term in brackets includes the probability of not being present (dead or emigrated) and the probability of being present but not recaptured.

Estimating Survival with the Dead-Recovery Model

With dead recoveries, marked animals are released into the population and reencountered as dead animals, typically harvested. This theory was developed by Seber (1970) and Brownie et al. (1985). Estimated parameters are survival rate, S_i, and band reporting rate, r_i, following Seber (1970). If the probability of reporting the dead animal is the same off the study area where the animal was marked as on the study area, then S_i provides an unbiased estimate of "true" survival, rather than apparent survival (ϕ). Thus, this model allows estimation of survival but not emigration, under this assumption. Emigration is assumed not to affect the probability of the marked animal being reported at its death. Recovery of dead animals was used to estimate survival for Florida manatees (*Trichechus manatus latirostris*) where distinctly marked individual manatees were catalogued in a computer-based photographic system (Langtimm et al. 1998) and recovered dead animals were identified.

The primary dead-recovery model in Program MARK differs somewhat from the S_i and f_i parameterization of Brownie et al. (1985) because the f_i of Brownie et al. is reparameterized as $(1 - S_i)r_i$. The reparameterization results in better numerical estimation properties (i.e., the likelihood is easier to optimize) because there is no implicit relationship between the parameters S_i and r_i, as there is between S_i and f_i, where biologically f_i should be less than $1 - S_i$. Additionally, the S_i and r_i parameterization makes the band-recovery models consistent with the parameterization of the CJS model described above. In particular, the use of covariates with S_i and r_i is reasonable, because each parameter represents a particular process in the overall band-recovery process (unlike the f_i parameter of the Brownie et al. model, which is a function of the probability that the bird died and that the mark was reported). However, the last S_i and r_i are confounded. In addition, with the S_i and r_i parameterization, S_i is always estimated between zero and one. Both the (S_i, r_i) and (S_i, f_i) parameterizations of the band-recovery model are included in Program MARK.

Estimating Survival and Permanent and Temporary Emigration with the Joint Live and Dead Encounters Model

The joint live and dead encounters model is based on theory developed by Burnham (1993). The parameter space consists of survival rates, S_i, live recapture rates, p_i, reporting rates for dead animals, r_i, and fidelity or probability of remaining on the study area (i.e., the area where captures and recaptures occur), F_i. An extension developed by Barker (1997) that allows live resightings during the interval between live recaptures is also available. Barker's model extends the capability of Burnham's model; it allows for the option of no dead recoveries and live recaptures and live resightings.

The main advantage of these models is that the estimates of survival are separated from the probability of being present on the study area (fidelity). As noted previously, the live-recapture model estimates apparent survival, $\phi_i = S_i(1 - E_i) = S_i F_i$. The dead-recovery model estimates true survival, assuming that the probability of reporting a dead bird is the same on and off the study area. The models with both live and dead encounters further extend the capability to separate survival from emigration, with unbiased estimates of survival if the reporting probability is not a function of the study area and estimates of the probability of remaining on the study area (fidelity).

Barker's extension (1997) of Burnham's model (1993) is particularly useful for estimating parameters of PVA models because live resightings between trapping occasions can contribute considerable information in some situations, and thus improve precision of estimates. Parameters for this model in Program MARK are

S_i = probability an animal alive at occasion i is alive at occasion $i + 1$;

p_i = probability an animal at risk of capture at occasion i is captured at i;

r_i = probability an animal that dies in interval i to $i + 1$ is found dead and the band reported;

R_i = probability an animal that survives from occasion i to $i + 1$ is resighted (alive) some time during the interval i to $i + 1$;

R'_i = probability an animal that dies in interval i to $i + 1$ without being found dead is resighted alive in interval i to $i + 1$ before it died;

F_i = probability an animal at risk of capture at occasion i is at risk of capture at $i + 1$; and

F_i' = probability an animal not at risk of capture at i is at risk of capture at $i + 1$.

(Note that the definition of F_i' in Program MARK differs from the definition in Barker [1997].) By setting r_i to zero, the resulting model allows live recaptures and live resightings, but does not allow dead recoveries. Barker's model provides a more complete treatment of emigration by estimating the rate of return to the study area of animals that have left.

Estimating Survival and Temporary Emigration with the Robust Design Model

Robust design models are useful for estimating temporary emigration rates. They are a combination of the CJS live-recapture model and the closed-capture models of Otis et al. (1978) and White et al. (1982), and are described in detail by Kendall et al. (1995, 1997) and Kendall and Nichols (1995). Instead of just one capture occasion between survival intervals, multiple (>1) capture occasions are used that are close together in time, where no mortality or emigration is assumed to occur. These closely spaced encounter occasions are termed "sessions." The advantage of this model over the CJS model is that temporary emigration from the study area can be estimated, and more precise estimates of survival are obtained because of additional information on capture probabilities. Estimates of temporary emigration are useful in documenting the movement of animals on and off the study area where captures and recaptures are made. However, in this model, permanent emigration is confounded with the survival parameter, so that permanent emigration rates cannot be estimated as in the joint live and dead encounters models.

For each trapping session (j), the probability of first capture (p_{ji}) and the probability of recapture (c_{ji}) are estimated (where i indexes the number of trapping occasions within the session), along with the number of animals in the population (N_j). For the intervals between trapping sessions, parameters estimated are the probability of survival (S_j) between session j and $j + 1$; the probability of being away from the study area during trapping session j, given that the animal was previously on the study area at $j - 1$ (γ_j''); and the probability of staying away from the study area during trapping session j, given that the animal was not available for capture during session $j - 1$ (γ_j'). Indexing of these parameters follows the notation of Kendall et al. (1997). Thus, γ_2'' applies to the second trapping session, and γ_2' is not estimated because there are no marked animals outside the study area at that time. To provide identifiability of the parameters for the Markovian emigration model, Kendall et al. (1997) suggest setting $\gamma''_{k-1} = \gamma''_k$ and $\gamma'_{k-1} = \gamma'_k$, where k is the number of trapping sessions. To obtain the "no emigration" model, set all the γ parameters to zero. To obtain the "random emigration" model, set $\gamma'_j = \gamma''_j$.

The main limitation for robust design models used in Program MARK is the lack of models incorporating individual heterogeneity in the estimation of population size. Individual covariates can be used to model the parameters S_j, γ''_j, and γ'_j in the Robust Design data type. Individual covariates cannot be used with the Robust Design data type for the p_{ji}, c_{ji}, and N_j, because animals that were never captured (so their individual covariates could never be measured) are incorporated into the likelihood as part of the estimate of population size (N). However, models that can incorporate individual covariates existing in the literature (Huggins 1989, 1991) have been implemented in MARK. Estimates of population size are given for the Huggins models, but these estimates are not quite as efficient as the closed-capture data type where the statistical models are equivalent to those in Program CAPTURE (Otis et al. 1978). However, the ability to incorporate individual covariates makes the Huggins models more appropriate if individual heterogeneity exists in the data.

Estimating Survival, Recruitment, and Rates of Population Change with the Pradel Model

Pradel (1996) developed a model to estimate the proportion of the population that was previously in the population. This model, labeled the "Pradel recruitment only" model in Program MARK, estimates the seniority probability γ_i, which is the probability that an animal present at time i was already present at time $i - 1$, and the capture probability p_i, the probability of capture on occasion i. This model is seldom used because Pradel (1996) extended his recruitment-only model to also include apparent survival (ϕ_i). Parameters of the extended model are apparent survival (ϕ_i), recapture probability (p_i), and seniority probability (γ_i) as defined above.

Pradel (1996) also parameterized his model to include the parameters apparent survival (ϕ_i), capture probability (p_i), and rate of population change (λ_i), where λ_i is population size at time $i + 1$ divided by population size at time i, or N_{i+1}/N_i. The main advantage of this model is that trends in the population can be estimated directly from the encounter-history data.

An additional extension to the Pradel (1996) models is provided in

Program MARK with parameters apparent survival (ϕ_i), capture probability (p_i), and recruitment rate (f_i), defined as the number of adults at time i per adult at time $i - 1$.

All three of the Pradel models that include apparent survival and seniority probability use encounter histories that include only live recapture information. However, a subtle but important distinction exists between these models and the models discussed previously. In the previous models, the parameter estimates are conditional on the initial capture. For example, with the CJS model, no information is obtained for parameter estimation of ϕ_1, ϕ_2, p_2, and p_3 from encounter histories that start with two initial zeros (e.g., 001011). In contrast, the Pradel models estimate recruitment from these initial zeros. Because of this difference in how encounter histories are treated, an important assumption is being made with the Pradel model: unmarked and marked animals have identical parameter values. In particular, if capture probabilities change as a result of capture, then marked and unmarked animals will not have the same values of p_i. As a result, other parameter estimates from these models also may be biased.

Estimation of λ from the Pradel model offers advantages over the Leslie matrix approach (Caswell 1989); namely, λ is estimated directly and is not a function of other parameters. Further, marked animals can be specific to a particular portion of the population, such as the territorial population of spotted owls (*Strix occidentalis*), so that the rate of change of this segment of the population can be estimated from encounter-history data. However, encounter data must be collected in an appropriate fashion for estimates of λ to be useful. Specifically, if the size of the study area is expanded (e.g., boundaries extended), then estimates of λ will increase because the population being studied increased. Thus, a critical assumption is that the population from which marked animals are captured is clearly defined. A disadvantage of employing the Pradel model to estimate λ is that it does not facilitate sensitivity analysis (Mills and Lindberg, chap. 16 in this volume).

Other Models Useful for PVA

Several other classes of models are relevant to PVA. Survival of animals with known fates, such as obtained by radio tracking, can be estimated. This model is equivalent to the Kaplan-Meier method with staggered entries (Pollock et al. 1989). The multistrata model of Hestbeck et al. (1991) and Brownie et al. (1993) allows animals to move between strata with transition probabilities. Examples of strata are geographic areas and breeding status (breeding or nonbreeding). Survival and capture probabilities are estimated for each strata, along with the transition probabilities, so that metapopulation models can be developed because the movement rates between subpopulations are estimated.

The closed-capture models allow modeling of the initial capture probability (p) and the recapture probability (c) to estimate population size (N), which is necessary to assess the status of the population to begin the PVA analysis. This data type is the same as is analyzed with Program CAPTURE (White et al. 1982). All the likelihood models in Program CAPTURE can be duplicated in Program MARK. However, MARK allows additional models not available in CAPTURE, facilitates comparisons between groups, and incorporates time-specific and/or group-specific covariates into the model. The main limitation of Program MARK for closed capture-recapture models is the lack of models incorporating individual heterogeneity other than what can be explained by individual covariates.

Estimating the Effect of Density Dependence

Typically, density dependence plays an important role in PVA applications because, as the population declines, survival and recruitment rates may increase—or decrease with Allee effects. All of the types of encounter-history data described above permit survival and recruitment (if included in the model) to be modeled as a function of population density, supplied as a covariate. This relationship can then be used in the PVA. Program MARK can be used to develop such analyses.

Another possibility is to estimate both population size and survival as a function of population size at the same time in the same model, using models such as the robust design described above, which includes both population size and survival.

Estimating Process Variance

The primary purpose of variance components analysis is to separate sampling variances and covariances from process variance ($\sigma^2_{\text{process}}$, or σ^2 for short) in the set of parameter estimates of interest. Consider a set of ten annual survival estimates $\hat{S}_1, \hat{S}_2, \hat{S}_3, \ldots, \hat{S}_{10}$ that were estimated from a band-recovery model in Program MARK. If properly estimated, each $\hat{S}_t$ has a sampling variance, $\text{var}(\hat{S}_t | S_t)$, associated with it that represents the precision of the estimate, and has sampling covariance between it and the other estimates. These sources of variation must be removed to properly estimate σ^2, the variance estimate of interest for PVA models. Program MARK estimates variance components by using random-effects models with shrinkage estimators. This procedure is an extension of the one proposed by Burnham et al. (1987) and further refined by Burnham (unpublished data). First, a random-effects model can be viewed as $S_t = \mu_S + \varepsilon_t$, where μ_S is the mean of the t realizations of S, and ε_t is an independent random variable with mean 0 and variance σ^2. Thus, the parameters S_t fluctuate "randomly" around μ_S, hence the term "random"-effects model. In addition, Program MARK uses shrinkage estimators, which in our example would be denoted as $\tilde{S}_t$. These estimators are not maximum likelihood estimates but have the property $\hat{\mu}_s < \tilde{S}_t < \hat{S}_t$. Essentially, the $\hat{S}_t$ is "shrunk" toward the estimated mean $(\hat{\mu}_s)$ to provide the estimates $(\tilde{S}_t)$. The degree of shrinkage is determined by

$$\frac{\sigma^2}{[\sigma^2 + \mathrm{var}(\hat{S}_t | S_t)]},$$

where σ^2 is estimated through numerical iteration procedures. If σ^2 is small relative to the total variance, then the $\tilde{S}_t$ will be similar to the maximum likelihood estimates of S_t. However, if σ^2 is large relative to the total variance, then the $\tilde{S}_t$ will shrink much closer toward $\hat{\mu}_s$. Using this procedure, an estimate of the process variation (σ^2) of the annual estimates of S is obtained, and improved estimates of S (the $\tilde{S}_t$) are obtained by shrinking much of the sampling variation.

EXAMPLE OF AN EMPIRICALLY BASED PVA

To demonstrate the utility of marked animals to evaluate population persistence, we present a simplified example to mimic population dynamics of the northern spotted owl (*Strix occidentalis caurina*). Data were simulated based on 137 male and 130 female territorial birds for 14 occasions, and were analyzed with the Pradel (1996) model to estimate λ_i. For 14 occasions (giving 13 intervals), 11 estimates of λ_i are possible. That is, the first value and the last value are not identifiable under a fully time-specific model. Both sexes were included in the global model, resulting in six sets of parameters: apparent survival (ϕ_i) for males and females, capture probability (p_i) for males and females, and rate of population change (λ_i) for males and females.

A series of models was constructed to evaluate which combination of time- and gender-specific parameters best represented the data. The results for the ten best AICc models are shown in table 9.2. The Akaike weights (labeled "AICc Weight" in table 9.2) are defined as

$$w_i = \frac{\exp\left(-\frac{1}{2}\Delta_i\right)}{\sum\limits_{r=1}^{R} \exp\left(-\frac{1}{2}\Delta_r\right)}$$

Table 9.2 Model Selection Results with Example Northern Spotted Owl Data from Program MARK

Model	AICc	ΔAICc	AICc Weight	Number of Parameters Estimated in Model
$\{\phi_t\ p_t\ \lambda_t\}$	4505.362	0.00	0.98234	38
$\{\phi_t\ p_{g^*t}\ \lambda_t\}$	4513.862	8.50	0.01401	52
$\{\phi_{g^*t}\ p_t\ \lambda_t\}$	4517.064	11.70	0.00283	51
$\{\phi_t\ p_t\ \lambda_{g^*t}\}$	4520.499	15.14	0.00051	51
$\{\phi.\ p_t\ \lambda_t\}$	4522.976	17.61	0.00015	27
$\{\phi_g\ p_t\ \lambda_t\}$	4523.693	18.33	0.00010	28
$\{\phi_{g^*t}\ p_{g^*t}\ \lambda_t\}$	4525.162	19.80	0.00005	65
$\{\phi_t\ p_{g^*t}\ \lambda_{g^*t}\}$	4529.624	24.26	0.00001	65
$\{\phi.\ p_{g^*t}\ \lambda_t\}$	4532.108	26.75	0.00000	41
$\{\phi_g\ p_{g^*t}\ \lambda_t\}$	4532.278	26.92	0.00000	42

Notes: Pradel's model (1996) is parameterized with apparent survival (ϕ), capture probability (p), and rate of change of the population (λ). The subscripts are t for time-specific variation, g for gender-specific variation, and asterisk (*) for constant values. Akaike's information criterion (AIC), corrected for small samples (AICc), is used to order the ten models. ΔAICc is the scaled AICc, where the minimum AICc value is subtracted from each entry. AICc weights, following Burnham and Anderson (1998), suggest that only the first model is useful in representing the data.

by Burnham and Anderson (1998), where Δ_i is the difference in AICc values for the minimum-AICc model and model i. Because the minimum-AICc model had almost all of the AICc weight, we used this model (ϕ_i, p_i, λ_i; table 9.2) to make further inferences from the data. The parameter estimates shown in figure 9.2 suggested that the first estimate (λ_2) was biased high. We expect the first one or two estimates of λ to be biased high because of the field protocol used to collect the data. As researchers gained experience in locating owls, they became more proficient in locating new individuals. The result was that the population appeared to be increasing, causing the first estimates of λ in the series to be biased high. Therefore, we deleted the estimate of λ_2 from further inferences.

As shown in figure 9.2, each of the $\hat{\lambda}_i$ have an associated sampling variance that was used to construct a 95% confidence interval. This sampling variance must be removed from the total variance to obtain the process variance of λ. Program MARK provides this option, using the methods described earlier. The temporal standard deviation, σ_{process} (the square root of the temporal process variance) of λ is estimated as 0.0565 with 95% confidence interval of 0.0124 to 0.1318. The mean of $\hat{\lambda}_i$ is estimated as 0.994 with standard error (SE) = 0.021, giving 95% confidence intervals of 0.952 to 1.035. Thus, the estimated mean was not statistically different from 1.0 for $\alpha = 0.05$.

With these estimates, we can make some inferences about expected changes of this population in the future, given that the population

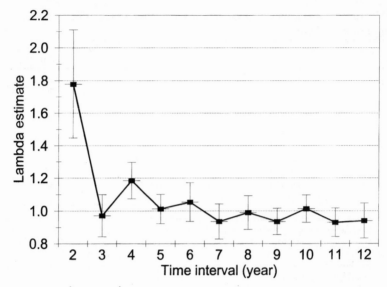

Fig. 9.2 Plot of $\hat{\lambda}_2$ through $\hat{\lambda}_{12}$ with 95% confidence intervals from the model $\{\phi_t p_t \lambda_t\}$ for northern spotted owl data. The large value of $\hat{\lambda}_2$ reflects an increase in the effective size of the study area during the first three occasions of the study as investigators gained experience in locating owls, causing a change in the size of the population being studied. Hence, $\hat{\lambda}_2$ is biased high and not used for estimating the process variance of λ.

growth rate continues to be drawn randomly from a statistical distribution with mean 0.994 and standard deviation of 0.0565. We assume a normal distribution of λ for making these projections. We asked the question, "How much will population size change during the next 20 years?" To answer it, we generated 8,000 population projections for 20 years, where we took the product of 20 randomly generated values of λ to obtain the amount of change in the population expected to occur in 21 years. A histogram of the amount of change expected over this 20-year projection is shown in figure 9.3. From this histogram we find that the probability of >30% decline ($\lambda < 0.7$) is 8.44%, of >20% decline ($\lambda < 0.8$) is 22.19%, of >10% decline ($\lambda < 0.9$) is 40.05%, and of any decline ($\lambda < 1.0$) is 57.81%. Depending on population size, we may or may not consider this population in jeopardy, and decide to take management action.

Had we mistakenly taken the estimate of temporal variation in λ as the square root of the total variance (0.0783) with the same mean of 0.994, we would have found that the probability of >30% decline ($\lambda < 0.7$) is 18.70%, of >20% decline ($\lambda < 0.8$) is 32.67%, of >10% decline ($\lambda < 0.9$) is 46.76%, and of any decline ($\lambda < 1.0$) is 59.20%. Thus, we would have concluded that the population had a greater chance of consequential declines with this inappropriate variance estimate.

The strength of the analysis is that it gives statistically rigorous estimates of the mean and process variance of λ, which are used to project the population forward in time. The projections are based on the estimated amounts of temporal and spatial variation in the observed encounter histories. The conclusions about population change are founded on the available data.

One of the critical assumptions made in this analysis is that demographic stochasticity is not included. That is, the distribution of λ does not change as the population declines to smaller values. To incorporate demographic stochasticity into the projection would require building a model with both survival and reproduction specifically represented, including their process variances. This could be accomplished from the available data, because the Pradel model in Program MARK (estimating recruitment instead of rate of population change) could be used to estimate the means and process variances of these parameters.

The second feature of the northern spotted owl example that differs from typical PVA approaches is that the probability of extinction is not

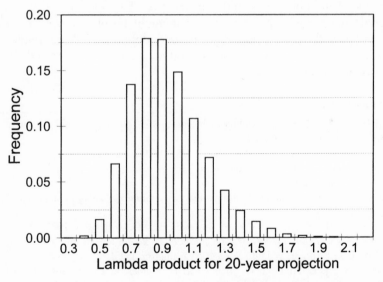

Fig. 9.3 Frequency of the amount of change expected after 20 years for a population with mean $\hat{\lambda}$ = 0.994 and process standard deviation $\hat{\sigma}_{process}$ = 0.0565. For each of the 8,000 populations simulated, values of λ for each year were randomly drawn from a normal distribution with λ = 0.994 and λ = 0.0565, and their product taken as the amount of change in the population over the 20-year period.

estimated directly. Estimating extinction may not be the only goal of a PVA, and the distribution of the predicted trend in the population may more useful for setting management options than a simplistic probability of extinction.

CONCLUSIONS

Current thinking in PVA has concentrated on the mathematical structure of PVA models as well as what parameter estimates these models can adequately incorporate (Boyce 1992; Beissinger and Westphal 1998). We believe that a constructive approach requires understanding which population parameters can be reliably and precisely estimated, and then building appropriate PVA models based on those parameters. Our example using the northern spotted owl illustrates such an approach.

PVA requires information on the dynamics of the population of interest. Reencounters of individually marked animals provide such information. Rigorous statistical methods are available in Program MARK to extract information on population size, survival, recruitment, immigration, and emigration from encounter histories of individually identifiable animals. Estimates of these parameters and their associated process variances are necessary precursors to developing population models that provide useful inferences on the persistence of a population or the projected changes in population size that might be observed over some future period. To develop rigorous and defensible PVAs, statistical methods based on marked animals need to be incorporated into the tool chests of conservation biologists.

LITERATURE CITED

Barker, R. J. 1997. Joint modeling of live-recapture, tag-resight, and tag-recovery data. *Biometrics* 53:666–677.

Beissinger, S. R., and M. I. Westphal. 1998. On the use of demographic models of population viability in endangered species management. *Journal of Wildlife Management* 62:821–841.

Boyce, M. S. 1992. Population viability analysis. *Annual Review of Ecology and Systematics* 23:481–506.

Brownie, C., D. R. Anderson, K. P. Burnham, and D. S. Robson. 1985. *Statistical inference from band recovery data: a handbook*. 2d edition. Resource Publication 156. U.S. Fish and Wildlife Service, Washington, D.C.

Brownie, C., J. E. Hines, J. D. Nichols, K. H. Pollock, and J. B. Hestbeck. 1993. Capture-recapture studies for multiple strata including non-Markovian transitions. *Biometrics* 49:1173–1187.

Burnham, K. P. 1993. A theory for combined analysis of ring recovery and recapture data. Pages 199–213 in J.-D. Lebreton and P. M. North, editors, *Marked individuals in the study of bird population*. Birkhäuser Verlag, Basel, Switzerland.

Burnham, K. P., and D. R. Anderson. 1998. *Model selection and inference: a practical information-theoretic approach*. Springer-Verlag, New York, New York.

Burnham, K. P., D. R. Anderson, G. C. White, C. Brownie, and K. H. Pollock. 1987. Design and analysis methods for fish survival experiments based on release-recapture. American Fisheries Society Monograph no. 5. American Fisheries Society, Bethesda, Maryland.

Caswell, H. 1989. *Matrix population models*. Sinauer Associates, Sunderland, Massachusetts.

Conner, M. M., and G. C. White. 1999. Effects of individual heterogeneity in estimating the persistence of small populations. *Natural Resources Modeling* 12:109–127.

Cormack, R. M. 1964. Estimates of survival from the sighting of marked animals. *Biometrika* 51:429–438.

Gould, W. R., and J. D. Nichols. 1998. Estimation of temporal variability of survival in animal populations. *Ecology* 79:2531–2538.

Hestbeck, J. B., J. D. Nichols, and R. A. Malecki. 1991. Estimates of movement and site fidelity using mark-resight data of wintering Canada geese. *Ecology* 72:523–533.

Huggins, R. M. 1989. On the statistical analysis of capture-recapture experiments. *Biometrika* 76:133–140.

———. 1991. Some practical aspects of a conditional likelihood approach to capture experiments. *Biometrics* 47:725–732.

Jolly, G. M. 1965. Explicit estimates from capture-recapture data with both death and immigration stochastic model. *Biometrika* 52:225–247.

Kendall, W. L., and J. D. Nichols. 1995. On the use of secondary capture-recapture samples to estimate temporary emigration and breeding proportions. *Journal of Applied Statistics* 22:751–762.

Kendall, W. L., J. D. Nichols, and J. E. Hines. 1997. Estimating temporary emigration using capture-recapture data with Pollock's robust design. *Ecology* 78:563–578.

Kendall, W. L., K. H. Pollock, and C. Brownie. 1995. A likelihood-based approach to capture-recapture estimation of demographic parameters under the robust design. *Biometrics* 51:293–308.

Langtimm, C. A., T. J. O'Shea, C. A. Beck, and R. Pradel. 1998. Estimates of annual survival probabilities for adult Florida manatees (*Trichechus manatus latirostris*). *Ecology* 79:981–997.

Lebreton, J.-D., K. P. Burnham, J. Clobert, and D. R. Anderson. 1992. Modeling survival and testing biological hypotheses using marked animals: a unified approach with case studies. *Ecological Monographs* 62:67–118.

Lebreton, J.-D., R. Pradel, and J. Clobert. 1993. The statistical analysis of survival in animal populations. *Trends in Ecology and Evolution* 8:91–95.

Leslie, P. H. 1945. On the use of matrices in certain population mathematics. *Biometrika* 33:183–212.

———. 1948. Some further notes on the use of matrices in population mathematics. *Biometrika* 35:213–245.

Link, W. A., and J. D. Nichols. 1994. On the importance of sampling variance to investigations of temporal variation in animal population size. *Oikos* 69:539–544.

Ludwig, D. 1999. Is it meaningful to estimate a probability of extinction? *Ecology* 80:298–310.

Nichols, J. D. 1992. Capture-recapture models: using marked animals to study population dynamics. *BioScience* 42:94–102.

Otis, D. L., K. P. Burnham, G. C. White, and D. R. Anderson. 1978. Statistical inference from capture data on closed animal populations. *Wildlife Monographs* 62:1–135.

Pollock, K. H., J. D. Nichols, C. Brownie, and J. E. Hines. 1990. Statistical inference for capture-recapture experiments. *Wildlife Monographs* 107:1–97.

Pollock, K. H., S. R. Winterstein, C. M. Bunck, and P. D. Curtis. 1989. Survival analysis in telemetry studies: the staggered entry design. *Journal of Wildlife Management* 53:7–15.

Pradel, R. 1996. Utilization of capture-mark-recapture for the study of recruitment and population growth rate. *Biometrics* 52:703–709.

Schwarz, C. J., and G. A. F. Seber. 1999. Estimating animal abundance. *Statistical Science* 14:427–456.

Seber, G. A. F. 1965. A note on the multiple recapture census. *Biometrika* 52:249–259.

———. 1970. Estimating time-specific survival and reporting rates for adult birds from band returns. *Biometrika* 57:313–318.

———. 1982. *Estimation of animal abundance and related parameters*. 2d edition. Macmillan, New York, New York.

———. 1986. A review of estimating animal abundance. *Biometrics* 42:267–292.

———. 1992. A review of estimating animal abundance 2. *Reviews of the International Statistics Institute* 60:129–166.

White, G. C. 2000. Population viability analysis: data requirements and essential analyses. Pages 288–331 in L. Boitani and T. K. Fuller, editors, *Research techniques in animal ecology: controversies and consequences*. Columbia University Press, New York, New York.

White, G. C., D. R. Anderson, K. P. Burnham, and D. L. Otis. 1982. *Capture-recapture and removal methods for sampling closed populations*. LA-8787-NERP. Los Alamos National Laboratory, Los Alamos, New Mexico.

White, G. C., and K. P. Burnham. 1999. Program MARK: survival estimation from populations of marked animals. *Bird Study* 46 (supplement): S120–S139.

10 Including Uncertainties in Population Viability Analysis Using Population Prediction Intervals

Bernt-Erik Sæther and Steinar Engen

ABSTRACT

We explore methods to incorporate into population viability analysis (PVA) forecasts the uncertainty due to unknown or inadequate parameter estimates and stochastic effects in population dynamics. The approach is based on the concept of the population prediction interval (PPI), which is the smallest time interval that includes extinction with a probability of $(1 - \alpha)$. We demonstrate the approach, using stochastic analogues of simple deterministic models without age structure. Estimates of the magnitude of demographic and environmental stochasticity are obtained from formulas that partition variance in the contributions of individuals to the next generation among individual and annual components, as well as from the demographic covariance. Demographic variance ranged from 0.27 to 0.66 in three species of small birds and from 0.155 to 0.188 in two populations of the brown bear. Environmental variance was 0.063, 0.205, and 0.411 for three species of small birds. Using the diffusion approximation to predict the time to extinction, we illustrate the application of PPI with long-term data from the white stork and song sparrow that show how forecasting the fate of populations can take into account both uncertainties in population parameters and stochastic population dynamics. Because environmental stochasticity strongly affects the risk of extinction and even the best data for endangered species may have large uncertainties in parameter estimates, we recommend that PVAs thoroughly explore different levels of environmental stochasticity and that models be kept simple to minimize the number of parameters that must be rigorously estimated.

INTRODUCTION

Assessing the risk of population extinction or decline based on the analysis of demographic models is one of the most popular and widely applied

We are grateful to R. Lande for valuable discussions, to two reviewers for comments on the manuscript, and for grants from the Research Council of Norway's program Conservation and Management of Biological Diversity and from the European Commission (project Metabird).

techniques in conservation biology. Following the pioneering work of Shaffer (1981, 1983) on the grizzly bear (*Ursus arctos*), estimates of minimum viable population have been the cornerstone of many management plans of threatened or endangered species (Burgman et al. 1993; Beissinger and Westphal 1998; Groom and Pascual 1998). The results of such analyses have also been subject to intense court examinations, such as the case of the northern spotted owl (*Strix occidentalis caurina;* Noon and McKelvey 1996).

Recently, the reliability of the predictions from population viability analysis (PVA) has been strongly questioned. This concern is based on three major problems. First, the quality of the available data is often poor, resulting in large sampling errors and biases in many important variables and parameters necessary for modeling, for instance in estimates of demographic rates and population size (Beissinger and Westphal 1998). Second, stochastic factors strongly influence the dynamics of small populations (Lande 1993; Caughley 1994). As a result, estimates of the probability of extinction become inaccurate even in "long" time series of high quality (Ludwig 1999). Third, uncertainties (Taylor et al. 1996; Akçakaya and Raphael 1998; Ludwig 1999) or biases (Walters 1985; Taylor 1995) in the estimates of important model parameters may make predictions of future population trajectories unreliable. Even in simple deterministic models, small variation in the population growth rate or in the form of the density regulation may generate large variation in the dynamics of the population and make accurate prediction of the probability of extinction difficult (May 1976; Ginzburg et al. 1990). As in population ecology in general, these factors may generate low confidence in population projections that are useful for conducting a PVA.

Here we present an approach in PVA that takes into account uncertainties in population projections due to uncertainties in parameter estimates as well as stochastic effects in population dynamics. We base our arguments on the "precautionary principle," as adopted by the World Conservation Union (IUCN 1994, see sec. II.7). It states that, in cases with uncertainties in the estimate of the extinction risk, it is legitimate to use the credible estimate that gives the highest risk of extinction. Accordingly, large uncertainties in the predictions of a PVA should favor a more cautious management approach than in cases where similar predictions can be made with greater accuracy.

Our approach is founded on the concept of the population prediction interval (PPI). This is defined as the stochastic interval that includes the unknown population size with probability $(1 - \alpha)$. If T is the smallest time in which this interval includes the extinction barrier (defined in a sexual population as population size $X = 1$), we then predict extinction in a sexual population to occur after time T. The width of the PPI increases with increasing uncertainty in the process (Heyde and Cohen 1985) as well as the estimation error. Uncertainty in the parameters does not change the extinction risk of the population, but it will affect the confidence we have in the population predictions, including the probability of extinction. For instance, if we accept the criteria of the World Conservation Union (IUCN 1994), we will consider a population of initial size N_0 to be vulnerable (an extinction risk of greater than 10% within a period of 100 years) if the 90% prediction interval includes $X = 1$ before 100 years with probability α. This is an extension of the prediction interval proposed for a Brownian-motion process by Dennis et al. (1991), because we include uncertainties in parameter estimates as well as the possibilities for extinction.

In this chapter we illustrate the use of the PPI approach in PVA. First we show how to estimate the stochastic components in the dynamics of a population without age structure. We next estimate the risk of extinction for strictly declining populations. Finally, we extend this approach to include density dependence.

We have based our analyses on stochastic analogues of simple deterministic models without age structure. A useful output from many PVA applications is an estimate of the number of individuals alive at a given point of time. Because for most threatened or endangered species the data are insufficient to support more complex models, this may involve neglecting important biological processes influencing the viability of populations, such as spatial structure, individual variation in demographic traits, and interspecific interactions. To obtain a realistic description of the population processes, however, the effect of those processes and of the environment must be taken into account. We believe that such simple models may yield realistic descriptions of the dynamics of many populations of endangered or threatened species if stochastic variables are added to the model. Thus, proper modeling and estimation of stochastic effects are crucial for obtaining meaningful results from PVAs.

DEMOGRAPHIC AND ENVIRONMENTAL STOCHASTICITY

We begin by introducing stochasticity into the dynamics of a closed population. We realize that a change in population size from one generation to the next is determined by the death or survival of each individual in addition to the number of surviving offspring. Thus, the contribution from one particular individual (c_i) will be the number of offspring recruited into the next generation plus one if the individual itself survives. For a given population size N, there will be N contributions to give the population size $N + \Delta N$ in the following generations. If $E(c)$ is the expectation of the individual contributions, this yields

(1) $$N + \Delta N = \sum_{i=1}^{N} c_i = NE(c) + \sum_{i=1}^{N} d_i,$$

where the $d_i = c_i - E(c)$ are stochastic variables with a mean of zero. Assuming that the contributions within a season are stochastically independent among individuals, we find

(2) $$\text{var}(N + \Delta N | N) = \text{var}(\Delta N | N) = N\sigma_d^2,$$

where σ_d^2 is the demographic variance. We ignore time lags by assuming that the population size in year t depends only on the population size the previous year. Hence, writing

(3) $$N_{t+1} = \lambda N_i,$$

the distribution of λ depends on N only. The mean population size can also be written as $E(N + \Delta N) = NE(c)$, which means that $E(c)$ is equivalent to the expected value of λ, while the variance of λ is

(4) $$\text{var}(\lambda) = \text{var}(\Delta N/N) = \sigma_d^2/N.$$

Thus, the variance in λ due to *demographic stochasticity* is inversely related to population size (see Lande 1993) and will most strongly affect the expected time to extinction when populations are small (Lande 1998).

Demographic variance can be estimated from data on individual variation in reproduction and survival of breeding females, following Sæther et al. (1998a). The estimate of σ_d^2 in year t,

(5) $$\sigma_{d,t}^2 = E \frac{1}{n_t - 1} \sum_{i=1}^{n_t} (C_{i,t} - \overline{C}_t)^2,$$

where $C_{i,1}, C_{i,2}, \ldots, C_{n_t,t}$ is the sample of recorded contributions to the next generation with mean $\overline{C}_t$ in year t. We then use the weighted mean across years of $\hat{\sigma}_{d,t}^2$ as an estimate of the demographic variance $\hat{\sigma}_d^2$.

In some species we have data on the magnitude of σ_d^2. In two species of small temperate passerines, the estimates of σ_d^2 were 0.66 and 0.57 for the song sparrow (*Melospiza melodia*) and the great tit (*Parus major*), respectively. A lower estimate (0.27) was found for another small passerine, the dipper (*Cinclus cinclus;* Tufto et al. 2000). For the Scandinavian brown bear (*Ursus arctos*), the estimate of $\hat{\sigma}_d^2$ based on data collected from radio-collared individuals in two populations was 0.155 and

0.180 (Sæther et al. 1998b). These data suggest that the demographic variance is likely to be less than one.

Data on individual variation in reproductive success and survival are rarely available in populations of endangered or threatened species. However, analysis of the great tit population in Wytham Wood (Sæther et al. 1998a) suggests that quite short time series may be sufficient to obtain quite reliable estimates of σ_d^2. Data from intensive individual-based studies of short duration that sometimes are made prior to conducting a PVA can provide some information about the magnitude of the demographic variance. When no information is available, demographic variance should be assumed to be less than one.

So far, we have considered only a stable environment. Realistic population models should incorporate stochastic variation in some model parameters from one year to the next. Stochasticity due to this variation is called *environmental stochasticity*. This concept has been frequently used in stochastic population models since the early 1970s (e.g., Athreya and Karlin 1971; May 1973, 1974; Turelli 1977). Here we provide precise definitions that enable estimation of the demographic and environmental components in stochastic population dynamics.

Following the approach we used for demographic stochasticity (equation 1), we again split the contribution of individual i to the next generation (c_i) into different components, but now we add an environmental effect (e) that affects all individuals in a single year similarly, but varies stochastically among years. From Engen et al. (1998) we get the relationship

(6) $$c_i = E[c(N)] + e + d_i,$$

where e, d_1, d_2, . . . , d_N are stochastic variables with zero means and $\text{cov}(e, d_i) = 0$. According to Engen et al. (1998), the conditional variance of the change in population size from one generation to the next becomes

(7) $$\text{var}(\Delta N | N) = \text{var}\left(\sum_{i=1}^{N} c_i | N \right) = N(\sigma_d^2 - \tau) + N^2(\sigma_e^2 + \tau),$$

where $\sigma_d^2 = \text{var}(d_i)$ is the demographic variance, $\sigma_e^2 = \text{var}(e)$ the environmental variance, and $\tau = \text{cov}(d_i, d_j)$, $i \neq j$. The parameter τ is created by intraspecific interactions among individuals (e.g., competition), generating a covariance between any two of the demographic contributions d_i. Engen et al. (1998) called τ the *demographic covariance*. This demographic covariance is, however, usually neglected in population modeling (Engen and Sæther 1998). For examples of the use of this concept, see Engen et al. (1998).

Using these definitions, the conditional variance in the stochastic population growth rate becomes

(8) $\mathrm{var}(\lambda|N) = (\sigma_e^2 + \tau) + (\sigma_d^2 - \tau)/N.$

Furthermore, if we assume small fluctuations in population size, we have

(9) $\mathrm{var}(\Delta \ln N|N) \approx \mathrm{var}(\lambda|N) = (\sigma_e^2 + \tau) + (\sigma_d^2 - \tau)/N.$

These definitions can be utilized to derive estimates for σ_d^2 and σ_e^2 separately. In PVA we are interested in the conditional variance of the change in population size $\mathrm{var}(\Delta N|N)$ because the extinction risk will increase with this variance (Lande, chap. 2 in this volume). Therefore, we are interested in the coefficients of N and N^2 in the expression for $\mathrm{var}(\Delta N|N)$ in equation 7. In general, these coefficients may be functions of N, say

(10) $\mathrm{var}(\Delta N|N) = \theta_1(N)N + \theta_2(N)N^2,$

where $\theta_1(N) = \sigma_d^2 - \tau$ and $\theta_2(N) = \sigma_e^2 + \tau$. If a random sample of contributions $C_1, \ldots, C_n$ is available in a given year, Sæther et al. (1998a) showed that

(11) $\mathrm{E}\left\{ \dfrac{1}{n-1} \Sigma(C_i - \overline{C})^2 \right\} = \theta_1(N),$

where $\overline{C}$ is the mean of the recorded contributions. Hence, the computation of the standard sum of squares estimate from a sample of contributions from a single year gives an unbiased estimator for the coefficient of N for the N-value that year.

If the demographic component is known, we can estimate the environmental variance from an observed time series of population sizes. Suppose we have a parametric model for $\mathrm{E}(\Delta N/N|N) = \mathrm{E}(\lambda|N) = h(\alpha, N)$, where α is some vector of unknown parameters. For instance, if we use the theta-logistic model (Gilpin and Ayala 1973; see also equation 19) for density regulation, then $\alpha = (r, K, \theta)$ and $h(\alpha)$ is the first side of equation 19. We can estimate α either by maximum likelihood or by minimizing $\Sigma[\Delta N_t/N_t - h(\alpha, N_t)]^2$ with respect to α. Since $\mathrm{var}(\Delta N/N) = \mathrm{E}[\Delta N/N - h(\alpha, N)^2] = \theta_2(N) + \theta_1(N)/N$, we may replace α by $\hat{\alpha}$ and obtain an approximately unbiased estimator for $\hat{\theta}_2(N_t)$,

(12) $\hat{\theta}_2(N_t) = [\Delta N_t/N_t - h(\hat{\alpha}, N_t)]^2 - \hat{\theta}_1(N_t)/N_t.$

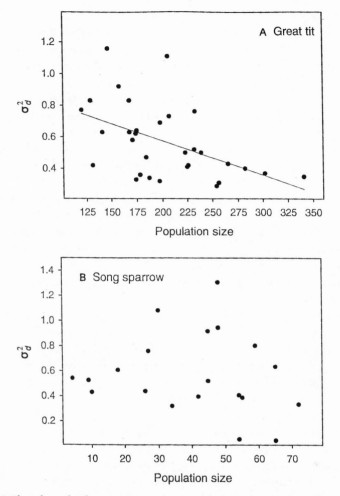

Fig. 10.1 The relationship between point estimates of demographic variance and population size in (A) the great tit in Wytham Wood, Oxfordshire, United Kingdom, and (B) the song sparrow at Mandarte Island, British Columbia.

Notice that no assumption of normality is made. Both $\hat{\theta}_1$ and $\hat{\theta}_2$ should be plotted against N to check for any density dependence in the coefficients. For the great tit, for instance, demographic variance decreased with population size (fig. 10.1A), whereas no density-dependent effects were found for the song sparrow (fig. 10.1B). In no cases have density-dependent effects in the point estimates of the environmental stochasticity been recorded (Sæther et al. 1998, 2000a,b). If there is no density dependence, θ_2 may simply be estimated as the weighted mean of the $\hat{\theta}_2(N_t)$.

This procedure has been used to obtain estimates of the demographic and environmental stochasticity in three populations of temperate passerine species (Tufto et al. 2000). The largest environmental variance was found in the song sparrow population on Mandarte Island ($\sigma_e^2 = 0.411$), where the population fluctuated between 4 and 72 pairs during the study period from 1975 to 1997 (Sæther et al. 2000a). This means that two standard deviations of the variation in λ range from 0.45 to 2.2. A high estimate of environmental variance was also found in a dipper population in southern Norway ($\sigma_e^2 = 0.205$). In this population, the stochastic variation entered the population dynamics through annual variation in the winter temperature, which affected the number of ice-free days during the winter (Sæther et al. 2000b). Smaller environmental stochasticity was found in the great tit population in Wytham Wood ($\sigma_e^2 = 0.063$). These estimates show that environmental stochasticity strongly influences the dynamics of all three populations because the effects of σ_e^2 on the variance in the conditional change in population size from one generation to the next scales with N^2 (Lande, chap. 2 in this volume).

Three problems are commonly encountered when employing these procedures to estimate demographic and environmental variances when conducting a PVA. First, long time series are necessary to obtain approximate estimates of σ_e^2 from equation 12. For instance, analyses of the great tit data from Wytham Wood showed that using subsamples of ten years resulted in very uncertain estimates of σ_e^2 (Sæther et al. 1998a). At least 15 years of data were necessary to obtain reasonable estimates of σ_e^2. Second, when data are not available to estimate σ_d^2, the estimate of σ_e^2 will be too large if $\hat{\theta}_1$ is set at zero in equation 12. Third, in cases with biased population estimates, the estimates of σ_e^2 will also be affected by the sampling variance (White et al., chap. 9 in this volume). In the latter two cases, estimates of σ_e^2 will be too large and represent an upper bound on the environmental variance in the population.

PREDICTING TIME TO EXTINCTION IN DECLINING POPULATIONS

Many populations of endangered or threatened species have shown a continuous decline over several years, so that the population size can be assumed to be far below carrying capacity. For such a population we will use a diffusion approximation to predict the time to extinction, or population sizes in the near future. Thus, we are extending the approach of Dennis et al. (1991) to include uncertainties in parameters. Gaston and Nicholls (1995) used this method to predict times to extinction of several bird species in Great Britain.

Diffusion processes are continuous in state and time and are defined by the functions called the infinitesimal mean $\mu(N)$ and variance $v(N)$ together with some boundary conditions (e.g., an extinction barrier at $N = 1$). If the fluctuations between years are not too large, diffusions may be good approximations of discrete population processes (Sæther et al. 1998a; Lande et al. 1999). Many of the theoretical results related to the dynamics of small populations (e.g., May 1973; Turelli 1977; Lande 1993, 1998) have been obtained using well-defined characteristics of such diffusion processes, as described in Karlin and Taylor (1981), for example.

Let us consider an exponentially growing population, as defined in equation 3. If the demographic stochasticity is ignored, the diffusion approximation $X_t = \ln N_t$ for has infinitesimal mean

(13) $$\mu_x(x) = E(\Delta X | X) \approx r - \frac{1}{2}\sigma_e^2$$

and variance

(14) $$v_x(x) = \text{var}(\Delta X | X) \approx \sigma_e^2.$$

This process is an example of a Brownian motion, which is a diffusion with constant infinitesimal mean and variance. We assume that environmental variance σ_e^2 is constant (i.e., density independent). In Brownian motion the parameters r and σ_e^2 are estimated using the fact that differences $X_{t+1} - X_t$ are independent and normally distributed with mean $r - \frac{1}{2}\sigma_e^2$ and variance σ_e^2 (Dennis et al. 1991). If the initial value $X(0) = x_0 > 0$, and the extinction barrier is at zero, the time to extinction follows the inverse Gaussian distribution

(15) $$f(t) = \frac{x_0}{\sigma\sqrt{2\pi t^3}} \exp\left(-\frac{(x_0 + \mu t)^2}{2\sigma^2 t}\right)$$

(Cox and Miller 1970), where $\sigma^2 = \sigma_e^2$ and $\mu = r - \frac{1}{2}\sigma_e^2$. This distribution integrates to one for $\mu \leq 0$, whereas for $\mu > 0$ the probability is $1 - \exp(-2\mu x_0/\sigma^2)$ for the process to go on increasing indefinitely and be absorbed at infinity (Karlin and Taylor 1981). The distribution of $X(t)$ for a given $t > 0$ conditional on extinction occurring later than time t is

(16) $$g(x) = \frac{1}{\sigma\sqrt{2\pi t}}\left[1 - \exp\left(-\frac{2xx_0}{2\sigma^2 t}\right)\right]\exp\left[-\left(\frac{(x - x_0 - \mu t)^2}{2\sigma^2 t}\right)\right]$$

for $x > 0$ (Cox and Miller 1970; Lande and Orzack 1988), where

(17)
$$\int_0^\infty g(x)\,dx = \int_t^\infty f(t)\,dt$$

is the probability that the process has not gone extinct before time t.

Engen and Sæther (2000) used those distributions (equations 15 and 16) to derive prediction intervals for the time to extinction and the population size at time t, based on a sequence of observations from Brownian motion. The limits of the intervals are quartiles in some normal-gamma mixtures of the above distributions. For example, this quartile defines a function T_α of the observations with the property that $P(T > T_\alpha) = 1 - \alpha$, where T is the extinction time. Then the interval $[T_\alpha, \infty]$ is a prediction interval for T with confidence $1 - \alpha$.

We illustrate the application of this method by predicting the decline of a white stork (*Cicconia cicconia*) population in Baden-Württemberg in Germany (Barlein and Zink 1979). As in many parts of western Europe (Rheinwald et al. 1989), this species has dramatically decreased in numbers during recent decades. We use data for the period 1950 to1965 to predict future fluctuations of this population. By using this limited period of time, we could compare the predictions with the population trajectory after 1965. We selected the quasi-extinction barrier (Ginzburg et al. 1982) of 20 individuals to reduce the effects of demographic stochasticity. Maximum likelihood estimates of the mean and variance of the diffusion were $\hat\mu = -0.0537$ and $\hat\sigma^2 = 0.0172$, respectively, indicating a rapidly declining population. When we compare the predicted and the actual population trajectories after 1965, we see that the population passed the quasi-extinction barrier in 1974 at $t = 9$ (fig. 10.2). This is at the lower bound of the upper 90% prediction interval for the time to quasi-extinction. If demographic stochasticity were included in the model, the lower limit of the prediction interval would be even smaller. Thus, estimates such as those provided by Gaston and Nicholls (1995) are likely to overestimate the time to extinction. In such cases we recommend that either a quasi-extinction barrier be used to reduce the influence of the demographic stochasticity (fig. 10.2), or that the process be simulated to extinction including demographic stochasticity.

TIME TO EXTINCTION IN DENSITY-REGULATED POPULATIONS

Density dependence strongly influences the risk of extinction (Ginzburg et al. 1990; Burgman et al. 1993). In spite of its importance to population dynamics, evidence for feedback of population size on the population growth rate can be difficult to verify statistically (e.g., Shenk et al. 1998).

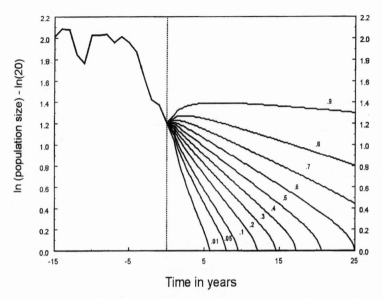

Time in years

Fig. 10.2 The recorded population fluctuation of the white stork in Baden-Württemberg from 1950 to 1965, the last year corresponding to zero on the time axis. For $t > 0$ the lower bounds of the prediction intervals are plotted against t for $\alpha = 0.01, 0.05, 0.10, 0.20,$ $\ldots, 0.90$. The corresponding lower bounds $\bar{T}$ for the prediction intervals for the time to quasi-extinction at the population size $N = 20$ are where the lines cross the horizontal axis.

Here we take another approach. We use data to estimate the parameters in a stochastic density-dependent model using a theta-logistic density regulation (Gilpin and Ayala 1973; Sæther et al. 1996; Diserud and Engen 2000),

(18)
$$E\left(\frac{\Delta N}{N}\right) = r\left[1 - \left(\frac{N}{K}\right)^{\theta}\right],$$

where the carrying capacity K is defined by $\lambda = 1$ (see equation 3). This enables us to describe a wide range of types of density regulation by varying only a single parameter, θ. For instance, when $\theta = 1$, we get the logistic model. To make this model applicable for any value of θ, including $\theta \leq 0$, we write equation 18 in the form

(19)
$$E\left(\frac{\Delta N}{N}\right) = r_1\left[1 - \frac{N^{\theta} - 1}{K^{\theta} - 1}\right]$$

for $\theta \neq 0$, where $r_1 = r(1 - K^{-\theta})$. For $\theta = 0$ we obtain the limit

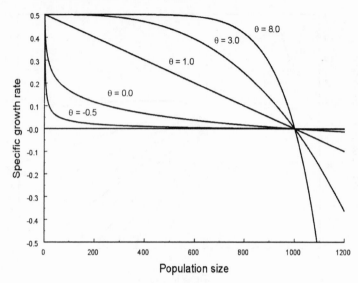

Population size

Fig. 10.3 The specific growth rate ($E(\Delta N/N)$) as a function of population size N for different θ in the theta-logistic model for density regulation (Gilpin and Ayala 1973). See text for further description.

(20)
$$E\left(\frac{\Delta N}{N}\right) = r_1\left(1 - \frac{\ln N}{\ln K}\right).$$

Note that equations 18 and 19 are two equivalent ways of expressing the same model. The formulation in equation 19 is required to cover the case $\theta = 0$, giving equation 20. Because $E(\Delta N/N) \approx E \ln N$, this is the first-order autoregressive time-series model often used in population ecology (Royama 1992).

We plot $E(\Delta N/N)$ as a function of N for different θ in figure 10.3. We also estimate the uncertainties in the different parameters (including θ) and investigate how this influences the accuracy of the predictions.

We illustrate the principles for such a PVA by analyzing the viability of a small population of song sparrows on Mandarte Island in British Columbia. In many respects this does not represent a typical population of an endangered or threatened species because it has been studied intensively for more than two decades, the population size is exactly known, and the demographic processes are understood in detail (Arcese et al. 1992). The presence of such high-quality long-term data does, however, provide the opportunity for evaluating our ability to predict the viability of small populations of this type of species.

As described previously, we obtain estimates of the demographic variance from data on individual variation in reproductive success and survival. The remaining parameters r, K, σ_e^2, and θ are estimated following Sæther et al. (1998a) by least-square techniques (for details of estimation, see Sæther et al. 2000a). Uncertainties in the estimates may be analyzed by parametric bootstrapping that simulates new data sets using the estimates rather than the true parameter values (Sæther et al. 2000a).

Density dependence of the song sparrow population was nearly logistic ($\hat{\theta} = 1.089$). However, the distributions of the bootstrap replicates of θ and r_1 suggest that the estimate of these two parameters was relatively uncertain (fig. 10.4A and B). For instance, the lower and upper 25% quartiles of θ lie at 0.6926 and 1.3383, respectively. Accordingly, a relatively wide range of models for density regulation may fit the data quite well. Less uncertainty was found in the estimates of both $\hat{\sigma}_e^2$ and $\hat{K}$ (fig. 10.4C and D). Thus, even though more than two decades of reliable data were available, reliable estimates of the density regulation and the specific population growth rate at low densities were difficult to obtain.

This song sparrow population nearly went extinct in 1989 when only four pairs were recorded (Tufto et al. 2000). Such low population sizes facilitate estimation of r_1. In many cases, however, the recorded population sizes are much larger. This often makes estimates of the growth rate at low population sizes uncertain because the estimation procedures must involve extrapolation over wide ranges of population sizes. We suggest that one useful procedure to reduce the bias in r_1 is to estimate it from other sources, such as demographic data. Then r_1 is fixed and the other parameters (K, σ_e^2, and θ) are estimated as described above.

Obviously, these uncertainties in parameter estimates are important to consider when predicting the viability of this population. We evaluate the PPIs at each point of time by stochastic simulations (Efron and Tibshirani 1993) of the process using each bootstrap replicate of the vector of parameter values. This method also takes into account the correlations between the estimates of the different parameters. The upper $(1 - \alpha)$ interval at time t then ranges from the corresponding lower quartile of the population sizes obtained from the simulations to infinity. Further stochastic simulations have indicated that this method is quite accurate for this model (Sæther et al. 2000a). The width of the prediction interval for population size increases rapidly with time (fig. 10.5) because uncertainties in the parameter estimates and stochasticity in the population dynamics (fig. 10.4) make it difficult to accurately forecast population size over long periods of time. For instance, accepting a 10% probability of extinction within a period of 100 years as suggested by Mace and Lande (1991), the extinction barrier $N = 1$ is included in the

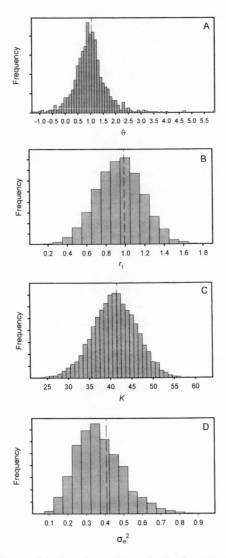

Fig. 10.4 The distribution of the bootstrap replicates for the density regulation θ (see fig. 10.2) (A), and the specific growth rate r_1 when the population size $N = 1$ (B), the carrying capacity K (C) and σ_e^2 (D) for $\hat{\theta} = 1.089$ in the song sparrow population at Mandarte Island. The *dashed lines* denote $\hat{\theta} = 1.089$ (A), $\hat{r}_1 = 0.987$ (B), $\hat{K} = 41.54$ (C), and $\hat{\sigma}_e^2 = 0.4106$ (D).

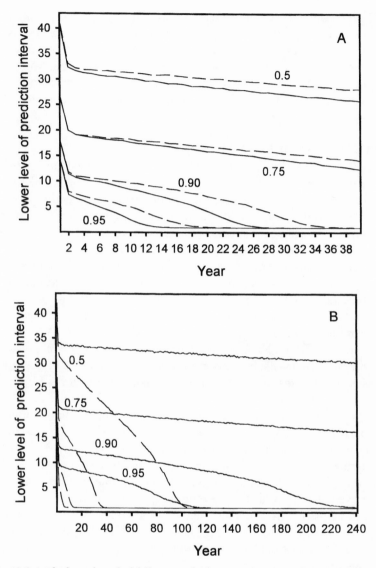

Fig. 10.5 A, The lower bound of different prediction intervals (95%, 90%, 75%, and 50%) for the future population size of the song sparrow population with demographic variance $\hat{\sigma}_d^2 = 0.66$. The *solid lines* show the prediction interval with uncertainties, whereas the *dashed lines* are the prediction intervals, assuming known parameter values equal to the estimates $\hat{r}_1 = 0.9897$, $\hat{K} = 41.54$, and $\hat{\sigma}_e^2 = 0.4106$. B, The prediction intervals using the lower (*solid line*) and higher (*dashed line*) quartile of the bootstrap distribution of θ. The values of the parameters are $\hat{r}_1 = 1.217$, $\hat{K} = 39.44$, and $\hat{\sigma}_e^2 = 0.415$ when $\theta_{25} = 0.6926$; and $\hat{r}_1 = 0.885$, $\hat{K} = 42.64$, and $\hat{\sigma}_e^2 = 0.412$ when $\theta_{75} = 1.3383$.

90% prediction interval after only 27 years (fig. 10.5A). Using only the best estimates of the parameters, this level of risk is not reached until 36 years. Thus, failure to consider uncertainties in parameter estimates will lead to a 33% overestimation of the time it takes for the extinction barrier ($N = 1$) to be included in the prediction interval for the population size. Notice that the predicted time to extinction was more than ten times longer for θ_{25} than for θ_{75} (fig. 10.5B).

DISCUSSION AND RECOMMENDATIONS

Correct predictions of future population fluctuations are the basis for a successful PVA. We have seen that such forecasting must take into account uncertainties in the estimates of population parameters, as well as various forms of stochasticity in population dynamics. A central point in future PVAs will be to estimate and correctly model stochastic factors. In particular, environmental stochasticity strongly influences the population projections and the risk of extinction. Every PVA should contain information on how environmental stochasticity is estimated and how it enters the model. Unfortunately, long data series with high precision in population estimates are necessary for obtaining reliable estimates of this parameter (Sæther et al. 1998a). We recommend that PVA should be done for different levels of environmental variance in the population growth rate that are likely to be appropriate for the species in question.

As in most cases in ecology, a demographic model for a PVA will always contain a compromise between model complexity and the number of parameters that can be estimated reliably. Here we have outlined an approach that is based on analysis of a simple stochastic population model with no age structure, but with characteristics that are reasonably well understood. Using diffusion approximations in such models, several important parameters for a PVA (e.g., expected time to extinction) can be derived (Lande et al. 1995; Sæther et al. 1998a; Ratner et al. 1997), even in age-structured models (Sæther et al. 1998b). It is important, however, that the behavior of the chosen model be checked by simulation of models that contain all necessary information (see Taylor 1995 and Sæther et al. 1998b).

Attempts to include uncertainty in the probability of extinction (Ludwig 1999) face difficulties in both interpretation and communication to managers. For example, it would be very difficult to communicate in a meaningful way to managers that the probability of extinction being smaller than some quantity is larger than some other quantity. PPI has the advantage of simplicity of interpretation. Our primary interest is not in the probability of extinction, but when extinction actually occurs. If all the population parameters are known, the probability of extinction is a useful concept that incorporates all the information we have about the extinction process. When the parameters are unknown, however, there is no reason why we should be interested primarily in the unknown probability of extinction. Rather, we should concentrate on statistical inference about the actual time of extinction, which is what prediction intervals do. They deal directly with our confidence about the actual extinction time of the population (fig. 10.5), rather than our confidence about the probability of extinction of the population before a specified time.

The frequentist concept of confidence intervals for unknown parameters is well known and commonly applied by biologists. Very few biologists seem to be aware of a parallel and purely frequentist theory on prediction, however, which is the statistical problem of predicting the value of a stochastic variable that has not been observed. Some basic textbooks in statistics do include this in chapters on confidence intervals for parameters in simple regression models. Consider for example the linear regression $E(Y) = a + bx$, where a and b are unknown parameters. Then, drawing inference about the expectation $a + bx$ for a given value of x is an estimation problem, and the uncertainty may be expressed by a confidence interval for the parameter $a + bx$, for example.

The very different problem of drawing inference about the unobserved value Y, which is a stochastic variable, is in statistical literature called prediction. Generally, confidence intervals are used to draw inference about parameters, in this case $E(Y) = a + bx$, whereas prediction intervals are used to draw inference about an unobserved stochastic variable, in this case Y. In the normal model, these two problems have different solutions. Both solutions can be expressed by quantiles of the t distribution, however, and in both cases one may completely get rid of the nuisance parameter, which in this case is the variance of the observations. For a solution of these problems, see Larsen and Marx (2001, 600). The standard solution to the prediction problem in regression given in textbooks is a purely frequentist approach. To solve the problem, there is no need to apply a Bayesian approach or Bayesian probability interpretation. One simply constructs an interval with stochastic endpoints that are functions of the observations, which cover the unobserved random variable Y with a prescribed probability $(1 - \alpha)$ in complete analogy with the common use of confidence intervals (see also Dennis 1996).

The simplest example of prediction in population dynamics is the problem of predicting a future population size when the log of the population size can be described by Brownian motion and the possibility of extinction is ignored. In this case, treated by Dennis et al. (1991), the prediction interval is also determined by some quantiles of the t distribution, the degrees of freedom depending on the number of observations of previous population sizes. When an extinction barrier is included in this model, the prediction problem gets more complicated, since one will have to consider jointly the prediction of extinction time and population sizes if a final time horizon is considered. An approximate solution to this problem was given by Engen and Sæther (2000), using the gamma-normal mixture of inverse Gaussian distributions.

When there is density regulation, we cannot construct the interval in such a way that we make the nuisance parameters (which are all the parameters in the model) disappear, as in simple linear regression and in Brownian motion without extinction barriers. We have proposed (Sæther et al. 1998b, 2000a; Engen and Sæther 2000) using the parametric bootstrap replicates to construct an interval that we expect to have an approximate coverage of $(1 - \alpha)$ regardless of the values of these parameters. One can probably easily come up with several other constructions that could also work well in practice. The point of crucial importance is, however, that one should always check the coverage of the method by doing extensive stochastic simulations from the model to observe how often the simulated intervals actually cover the simulated time to extinction. Notice that this approach is far from being Bayesian (Wade, chap. 11 in this volume). First, there is no arbitrarily or subjectively chosen prior distribution of parameters. The choice of such a distribution may represent a problem in PVA because one often has relatively short time series and often weakly density-regulated populations with long return times to equilibrium (May 1974). Second, there are few observations, and they are very strongly correlated, so they contain much less information than the same number of independent observations. As a consequence, the posterior distribution and the recommendations it implies for management may depend strongly on the choice of prior.

Our philosophy is also far from Bayesian. Applying the Bayesian approach (e.g., Ludwig 1996; Taylor et al. 1996; Tufto et al. 2000; Wade, chap. 11 in this volume), one will claim that the (Bayesian) probability that the population goes extinct before the observed 5% lower quantile of the posterior distribution of the time to extinction actually is 5%. In frequentist theory this probability remains an unknown parameter about which we admit having only very uncertain information. If one has prior information about the parameters that can be expressed by distributions without unknown parameters, and these reflect frequencies as discussed by Goodman (chap. 21 in this volume), such information should be included in the analysis through the application of Bayes's theorem, because such information in practice will lead to shorter prediction intervals. When no such information on frequencies of parameter values is available, one should try to construct prediction intervals with properties that are approximately independent of the nuisance parameters.

A central question in PVA is the inclusion of density dependence (Boyce 1992). We suggest that the Brownian-motion model is often appropriate when the stochastic growth rate is negative, $r \doteq r_0 - \frac{1}{2}\sigma_e^2 < 0$, because the population size is then so low that density regulation is not expected to occur. Furthermore, in cases where the population size is very low and the study period relatively short, density dependence need not be included (see Dennis et al. 1991; Gaston and Nicholls 1995). In other cases, we suggest that the theta-logistic model (see above) may be useful because a wide range of density regulation can be described by varying only a single parameter (fig. 10.3).

The examples provided here illustrate that large uncertainties may exist in model parameters (fig. 10.4), even when long-term data of high quality are available. Effects of such uncertainties must be evaluated before suggesting management actions based on the results from a PVA. For the brown bear, for instance, the choice of harvest strategy should depend on the level of uncertainty in the population estimates (Tufto et al. 1999). We suggest that prediction intervals created by parametric bootstrapping may be useful in such evaluations.

We also suggest that in many cases PVA should use a simplistic approach, building on analyses of simple models while neglecting biologically important processes such as age-dependent effects and spatial processes in order to reduce the number of parameters. Generally, using a large number of parameters increases the uncertainty in stochastic predictions and should be avoided (Burnham and Anderson 1998). However, we believe that uncertainties in parameter estimates and stochasticities in the population dynamics should be estimated and modeled as far as possible (see Beissinger 1995). Using the "precautionary principle" (IUCN 1994), such effects should be included when suggesting management strategies based on a PVA.

LITERATURE CITED

Akçakaya, H. R., and M. G. Raphael. 1998. Assessing human impact despite uncertainty: viability of the northern spotted owl metapopulation in the northwestern USA. *Biodiversity and Conservation* 7:875–894.

Arcese, P., J. N. M. Smith, W. Hochachka, C. M. Rogers, and D. Ludwig. 1992. Stability, regulation, and the determination of abundance in an insular song sparrow population. *Ecology* 73:805–822.

Athreya, K. B., and S. Karlin. 1971. On branching processes with random environments: extinction probabilities. *Annals of Mathematical Statistics* 42:1499–1520.

Bairlein, F., and G. Zink. 1979. Der Bestand des Weissstorchs *Cicconia cicconia* in Südwestdeutschland: eine Analyse der Bestandsentwicklung. *Journal für Ornithologie* 120:1–11.

Beissinger, S. R. 1995. Modeling extinction in periodic environments: Everglades water levels and snail kite population viability. *Ecological Applications* 5:618–631.

Beissinger, S. R., and M. I. Westphal. 1998. On the use of demographic models of population viability in endangered species management. *Journal of Wildlife Management* 62:821–841.

Boyce, M. S. 1992. Population viability analysis. *Annual Review of Ecology and Systematics* 23:481–506.

Burgman, M. A., S. Ferson, and H. R. Akçakaya. 1993. *Risk assessment in conservation biology.* Chapman and Hall, London, United Kingdom.

Burnham, K. P., and D. R. Anderson. 1998. *Model selection and inference: a practical information-theoretic approach.* Springer-Verlag, New York, New York.

Caughley, G. 1994. Directions in conservation biology. *Journal of Animal Ecology* 63:215–244.

Cox, D. R., and H. D. Miller. 1970. *The theory of stochastic processes.* Methuen, London, United Kingdom.

Dennis, B. 1996. Should ecologists become Bayesians? *Ecological Applications* 6:1095–1103.

Dennis, B., P. L. Munholland, and J. M. Scott. 1991. Estimation of growth and extinction parameters for endangered species. *Ecological Monographs* 61:115–143.

Diserud, O., and S. Engen. 2000. A general and dynamic species abundance model, embracing the lognormal and the gamma models. *American Naturalist* 155:497–511.

Efron, B., and R. J. Tibshirani. 1993. *An introduction to the bootstrap.* Chapman and Hall, New York, New York.

Engen, S., Ø. Bakke, and A. Islam. 1998. Demographic and environmental stochasticity: concepts and definitions. *Biometrics* 54:840–846.

Engen, S., and B.-E. Sæther. 1998. Stochastic population models: some concepts, definitions, and results. *Oikos* 83:345–352.

———. 2000. Predicting the time to quasi-extinctions for populations far below their carrying capacity. *Journal of Theoretical Biology* 205:649–658.

Gaston, K. J., and A. O. Nicholls. 1995. Probable times to extinction of some rare breeding bird species in the United Kingdom. *Proceedings of the Royal Society of London,* series B, Biological Sciences, 259:119–123.

Gilpin, M. E., and F. J. Ayala. 1973. Global models of growth and competition. *Proceedings of the National Academy of Sciences* (USA) 70:3590–3593.

Ginzburg, L. R., S. Ferson, and H. R. Akçakaya. 1990. Reconstructability of density dependence and the conservative assessment of extinction risks. *Conservation Biology* 4:63–70.

Ginzburg, L. R., L. B. Slobodkin, K. Johnson, and A. G. Bindman. 1982. Quasi-extinction probabilities as a measure of impact on population growth. *Risk Analysis* 21:171–181.

Groom, M. J., and M. A. Pascual. 1998. The analysis of population persistence: an outlook on the practice of viability analysis. Pages 4–27 *in* P. L. Fiedler and P.

Kareiva, editors, *Conservation biology for the coming decade*, 2d edition. Chapman and Hall, New York, New York.

Heyde, C. C., and J. E. Cohen. 1985. Confidence intervals for demographic projections based on products of random matrices. *Theoretical Population Biology* 27: 120–153.

International Union for Conservation of Nature (IUCN). 1994. *IUCN Red List categories*. IUCN, Gland, Switzerland.

Karlin, S., and H. M. Taylor. 1981. *A second course in stochastic processes*. Academic Press, New York, New York.

Lande, R. 1993. Risks of population extinction from demographic and environmental stochasticity and random catastrophes. *American Naturalist* 142:911–927.

———. 1998. Demographic stochasticity and Allee effect on a scale with isotropic noise. *Oikos* 83:353–358.

Lande, R., S. Engen, and B.-E. Sæther. 1995. Optimal harvesting of fluctuating populations with a risk of extinction. *American Naturalist* 145:728–745.

———. 1999. Spatial scale of population synchrony: environmental correlation versus dispersal and density regulation. *American Naturalist* 154:271–281.

Lande, R., and S. H. Orzack. 1988. Extinction dynamics of age-structured populations in a fluctuating environment. *Proceedings of the National Academy of Sciences* (USA) 85:7418–7421.

Larsen, R. J., and M. L. Marx. 2001. *An introduction to mathematical statistics and its applications*. 3d edition. Prentice Hall, Upper Saddle River, New Jersey.

Ludwig, D. 1996. Uncertainty and the assessment of extinction probabilities. *Ecological Applications* 6:1067–1076.

———. 1999. Is it meaningful to estimate a probability of extinction? *Ecology* 80: 298–310.

Mace, G. M., and R. Lande. 1991. Assessing extinction threats: toward a reevaluation of IUCN threatened species categories. *Conservation Biology* 5:148–157.

May, R. M. 1973. Stability in randomly fluctuating versus deterministic environments. *American Naturalist* 107:621–650.

———. 1974. *Stability and complexity in model ecosystems*. 2d edition. Princeton University Press, Princeton, New Jersey.

———. 1976. Simple mathematical models with very complicated dynamics. *Nature* 261:459–467.

Noon, B. R., and K. S. McKelvey. 1996. Management of the spotted owl: a case history in conservation biology. *Annual Review of Ecology and Systematics* 27: 135–162.

Ratner, S., R. Lande, and B. B. Roper. 1997. Population viability analysis of spring chinook salmon in the South Umpqua River, Oregon. *Conservation Biology* 11: 879–889.

Rheinwald, G., J. Ogden, and H. Schulz. 1989. *White stork: status and conservation*. Dachverbandes Deutscher Avifaunisten, Bonn, Germany.

Royama, T. 1992. *Analytical population dynamics*. Chapman and Hall, London, United Kingdom.

Sæther, B.-E., S. Engen, A. Islam, R. McCleery, and C. Perrins. 1998a. Environmental stochasticity and extinction risk in a population of a small songbird, the great tit. *American Naturalist* 151:441–450.

Sæther, B.-E., S. Engen, and R. Lande. 1996. Density-dependence and optimal harvesting of fluctuating populations. *Oikos* 76:40–46.

Sæther, B.-E., S. Engen, R. Lande, P. Arcese, and J. N. M. Smith. 2000a. Estimating time to extinction in an island population of song sparrows. *Proceedings of the Royal Society of London,* series B, Biological Sciences, 267:621–626.

Sæther, B.-E., S. Engen, J. E. Swenson, Ø. Bakke, and F. Sandegren. 1998b. Assessing the viability of Scandinavian brown bear, *Ursus arctos,* populations: the effects of uncertain parameter estimates. *Oikos* 83:403–416.

Sæther, B.-E., J. Tufto, S. Engen, K. Jerstad, O. W. Røstad, and J. E. Skåtan. 2000b. Population dynamical consequences of climate change for a small temperate songbird. *Science* 287:854–856.

Shaffer, M. L. 1981. Minimum population sizes for species conservation. *BioScience* 31:131–134.

———. 1983. Determining minimum viable population sizes for grizzly bear. *International Conference on Bear Research Management* 5:133–139.

Shenk, T. M., G. C. White, and K. P. Burnham. 1998. Sampling-variance effects on detecting density dependence from temporal trends in natural populations. *Ecological Monographs* 68:445–463.

Taylor, B. L. 1995. The reliability of using population viability analysis for risk classification of species. *Conservation Biology* 9:551–558.

Taylor, B. L., P. R. Wade, R. A. Stehn, and J. F. Cochrane. 1996. A Bayesian approach to classification criteria for spectacled eiders. *Ecological Applications* 6: 1077–1089.

Tufto, J., B.-E. Sæther, S. Engen, P. Arcese, K. Jerstad, O. W. Røstad, and J. N. M. Smith. 2000. Bayesian meta-analysis of demographic parameters in three small, temperate passerines. *Oikos* 88:273–281.

Tufto, J., B.-E. Sæther, S. Engen, J. E. Swenson, and F. Sandegren. 1999. Harvesting strategies for conserving minimum viable populations based on World Conservation Union criteria: brown bears in Norway. *Proceedings of the Royal Society of London,* series B, Biological Sciences, 266:961–967.

Turelli, M. 1977. Random environments and stochastic calculus. *Theoretical Population Biology* 12:140–178.

Walters, C. J. 1985. Bias in the estimation of functional relationships from time series data. *Canadian Journal of Fisheries and Aquatic Sciences* 42:147–149.

11 Bayesian Population Viability Analysis
Paul R. Wade

ABSTRACT

Population viability analysis (PVA) models are used to integrate various risks that a species faces into an estimate of the probability of extinction for that species. Most PVAs require estimating many parameters but allow only the specification of a single value for each parameter, and parameter uncertainty is poorly incorporated. The importance of this omission was investigated using simulations where the true extinction distribution was known, to compare PVAs using only point estimates to PVAs using Bayesian methods. Bayesian statistics uses distributions for each parameter, which allows direct incorporation of parameter uncertainty into an analysis. These distributions can be estimated from data or subjectively specified for parameters for which no data are available. This chapter introduces Bayesian parameter estimation and describes the mechanics of using the resulting probability distributions in a Bayesian PVA. I used 100 simulated data sets from a declining population to compare a Bayesian PVA with a PVA using only the maximum likelihood point estimates (MLE PVA). The resulting extinction probability distributions were compared to the true distribution of the time to extinction, as specified in the underlying simulation. The estimated Bayesian distributions were always wider and were much more likely to contain the entire true distribution than the MLE PVAs (67% versus 10% of the cases). Bayesian PVAs rarely overestimated the number of years until the first substantial chance of extinction (4% of the cases), whereas the MLE PVA commonly overestimated this quantity (32% of the cases).

INTRODUCTION

Current population viability analysis (PVA) methods rarely incorporate parameter uncertainty although such uncertainties are often very large

I thank Barbara L. Taylor for helping to provide the simulated data sets, for many discussions regarding the details of the analyses, and for substantial editing help. I also thank Douglas P. DeMaster, Gary C. White, and an anonymous reviewer for helpful comments on the manuscript, and Steven R. Beissinger and Dale R. McCullough for their substantial editorial help and considerable patience.

in ecological data. PVAs that ignore uncertainty in classifying populations according to risk (such as under World Conservation Union [IUCN] criteria) or in estimating the probability of extinction can be misleading and inaccurate (Taylor 1995; Ludwig 1996, 1999; Ralls and Taylor 1997). Because PVAs attempt to incorporate all important risk factors into the total risk of extinction, models are often complex and include many parameters (see table 1 in Beissinger and Westphal 1998). For example, in an age-structured model there are life-history parameters such as survival, fecundity, age of sexual maturity, and maximum age. The assumption may be made in using the model that some of these life-history rates vary stochastically, so there would be additional parameters controlling the variance in these rates. Often other parameters play important roles in determining risk, such as those describing density dependence, the frequency of catastrophic events, the role of breeding structure, and the effect of inbreeding depression. Regardless of the particular model used, there will always be a large number of parameters to specify in a PVA.

Uncertainty in parameters is often ignored in PVAs, with just a single value used for each parameter. For example, the software program VORTEX (Lacy 1993) takes only a single value for each of a large number of parameters. Although the user can investigate the sensitivity of the PVA results to different parameter values, in a listing decision the user must choose a single distribution to represent the best estimate of risk. This means that the user of VORTEX needs to interpret the results in a particular statistical fashion. An estimate of the distribution of time to extinction based on point estimates can itself only be considered a point estimate, and it should be understood that this estimate ignores parameter uncertainty.

Conceptually, it may seem odd to consider a distribution to be a point estimate. Because the distribution does reflect uncertainty about when a population will go extinct, it may lead users into mistakenly assuming the estimated distribution has accounted for all sources of uncertainty. However, the time to extinction has uncertainty both because of the stochastic nature of the extinction process and because our knowledge of the system being modeled is imperfect. Formally, only Bayesian statistics results in probability statements about the possible values of parameters. Bayesian approaches open the door for parameter uncertainty to be directly incorporated into the estimation of the probability of extinction in a PVA.

There can be only one true distribution for the time to extinction for a real population, and it will remain unknown. When we attempt to estimate this distribution, we understand that we do so with error, and therefore we know the estimated distribution will not exactly match the true distribution. Clearly a PVA that ignores parameter uncertainty cannot represent an accurate estimate of the true probability of extinction. Much as a point estimate has an estimated interval defining a range for the parameter, an estimate of a distribution should itself be a distribution, and it should define a range that is likely to contain the true distribution. In other words, the estimated distribution should be broader than the true distribution, and it should become narrower (and closer to the true distribution) with more precise data. A Bayesian PVA has these properties.

Here I illustrate the use of Bayesian methods by estimating population parameters from simulated data in a relatively simple example. I then use the Bayesian estimates of the parameters to perform a Bayesian PVA that incorporates parameter uncertainty. Finally, I compare the results from a series of Bayesian PVAs to results from PVAs using maximum likelihood point estimates. Because many readers may be unfamiliar with Bayesian statistics, I provide a basic explanation of how Bayesian statistics are estimated from data and how that differs from the more traditional approach scientists commonly use. Bayesian statistics are relatively new in the field of population biology, so I developed techniques for estimating probability distributions based on mark-recapture data to estimate adult survival rates. To allow the reader to follow the flow of the PVA comparison that is the main focus of this chapter, I have relegated these details to appendices.

AN INTRODUCTION TO BAYESIAN STATISTICS

There are two main schools of statistical inference. Any data can be analyzed by techniques from either school. The school of inference all scientists are taught is sometimes called frequentist statistics or Neyman-Pearson statistics. It is such a dominant school of inference that it usually goes unnamed and is simply thought of as "statistics." The other main school of statistical inference is Bayesian statistics (e.g., Press 1989; Bernardo and Smith 1994; Gelman et al. 1995). Frequentist statistics uses techniques such as maximum likelihood to estimate parameters. Statistical inference is made by calculating confidence limits, or by calculating a probability (p) value associated with a specific hypothesis test.

Bayesian statistics differ in that parameter estimates are made by calculating a probability distribution for the parameter. Statistical inference is made by making statements about the probability of different parameter values, given the observed data. Although somewhat innocuous sounding, this concept is radically different from frequentist statistics, which never leads to probability statements about the value of parameters. Frequentist methods lead only to the opposite—a statement about the probability of observing specific data, given values for parameters (e.g., a p-value is the probability of observing data as or more extreme as the data that were observed, given the null hypothesis is true). For example, a frequentist analysis of the trend of a declining population leads to an estimate of the rate of decline and confidence limits about that estimate, and to a p-value associated with the null hypothesis of no decline. A Bayesian analysis of the same data leads to a probability distribution for the rate of decline, so that statements can be made such as "there is a 0.93 probability that the population is declining," or "there is a 0.73 probability that the population is declining at more than 5% per year."

Although frequentist methods produce a distribution around the point estimate (i.e., confidence intervals), it is wrong to interpret this as a probability distribution for the parameter (DeGroot 1973; Berger and Berry 1988; Bernardo and Smith 1994). A standard confidence interval tells us only that, if one could repeatedly sample data sets from a population, the intervals calculated from each data set would include the true value of the parameter 95% of the time. We cannot use the confidence limit to state that there is a 95% probability that the true value does in fact occur within the one confidence interval we create from our data, and the confidence interval is not a probability distribution in which we expect the parameter to occur (Berger and Berry 1988). To return to the trend example, we can calculate confidence intervals around the rate of decline, but these confidence limits cannot be interpreted as a probability statement about the true rate of decline. The confidence limits are calculated from the sampling distribution centered on the point estimate. This distribution describes the probability of observing various data (i.e., point estimates), given the true rate of decline is exactly equal to the point estimate. A frequentist trend analysis never leads to a probability statement about the trend of the population. After rejecting a null hypothesis of no decline, a frequentist will often wish to conclude that the population is declining, but formally he or she can only conclude that the population is not stable—frequentists can disprove null hypotheses but they do not get to prove alternative hypotheses. In contrast, a Bayesian interval, termed a credible region or highest probability density (HPD) region, has a probability interpretation; a 0.95 HPD is the interval with a 0.95 probability of containing the true value, where no value outside the interval has a higher probability than any value inside the interval. Because many people incorrectly give frequentist confidence limits a Bayesian probability interpretation (Berger and Berry 1988; Bernardo and Smith 1994), this fundamental difference between the two schools of statistical inference is often underappreciated.

These differences become important when one is interested in incorporating parameter uncertainty into PVAs. All PVAs involve the forward projection of a population through time. To incorporate uncertainty in parameter values into a single distribution for the time to extinction, one has to think in terms of repeatedly projecting a simulated population forward with different parameter values each time to reflect the uncertainty in the estimated parameters. Therefore, parameter values need to be sampled from probability distributions for the parameters, which Bayesian methods provide. If one randomly samples values for a parameter from a distribution defined by a frequentist point estimate, such as a maximum likelihood estimate (MLE) and its estimated standard error (SE), one is simply creating the sampling distribution and calculating confidence limits. There is nothing wrong with doing this within the frequentist paradigm, but the results have to be interpreted as confidence limits. Each set of parameter values sampled from the frequentist parameter estimate leads to a distribution for the time to extinction, and therefore one has created a set of many distributions for the time to extinction. In principle, this set of distributions could be used to define a confidence limit for the entire distribution, although this is a difficult concept to consider. If one reduces the output of the PVA to a single quantity, such as the probability of extinction in three generations, then one can create a point estimate for that quantity and put confidence intervals around it by repeating PVA projections with parameter values sampled from the point estimate and its SE. This would be the correct frequentist method for incorporating parameter uncertainty into PVA results (White 2000).

The resulting difference in the Bayesian approach is the ability to incorporate parameter uncertainty into a single distribution for the time to extinction. A Bayesian PVA leads to a single probability of extinction at some specified time (e.g., 0.35 in three generations). In contrast, a frequentist PVA incorporating parameter uncertainty will lead to a best estimate of the probability of extinction, with associated confidence limits (e.g., 0.22 in three generations, with 95% confidence limits from 0.11 to 0.49). Lee and Rieman (1997) recommend Bayesian methods for estimating the probability of extinction, and assert that moving from a statistical calculation to a degree of belief about a future event in the real world is possible only using Bayesian methods.

The general form of a Bayesian analysis combines a prior probability distribution for the parameter with the data to form a posterior probability distribution for the parameter. The posterior is found by integrating the product of the likelihood function and the prior distribution across the space of the parameters:

(1)   Posterior distribution =

$c\int$(Likelihood function · Prior distribution),

where c is the normalizing constant that makes the posterior a probability distribution. Symbolically, this is written

(2) $$p(\theta|y) = c\int_{\theta} p(y|\theta) \cdot p(\theta)d\theta,$$

where θ represents the parameters and y represents the data. The likelihood function is the same function used in frequentist methods to find MLEs. The only new concept is the prior distribution, which is a probability distribution for the parameter prior to consideration of the data used in the likelihood function. Prior distributions can be estimated from previous independent data. If no direct information is available about a parameter, a prior distribution is usually specified in one of two ways. The first method is to specify a prior probability distribution for a parameter using expert opinion and judgment. This form of analysis has been termed subjective Bayesian analysis (Efron 1986); examples include Wolfson et al. (1996) and Lee and Rieman (1997).

A second method is to specify a "noninformative" prior distribution for a parameter, such that the prior distribution has little influence on the shape of the posterior distribution. This often takes the form of a uniform or flat distribution that conveys the belief that all parameter values are equally probable. For some parameters, the prior distribution can appear to be truly noninformative (e.g., rate of survival with a prior distribution that is uniform from 0.0 to 1.0). In some situations, however, there may be no clear choice for a noninformative prior distribution because a density that is uniform in one parameterization may not be exactly uniform in another. Additionally, the task of specifying a noninformative joint prior distribution for a model containing many parameters can be difficult. Gelman et al. (1995) discuss these problems, along with several possible solutions. First, they point out that with even a modest amount of data the influence of a prior distribution often becomes small, and therefore the choice between a range of relatively flat prior densities will not matter. Second, a formal method based on Jeffreys's invariance principle can be used in univariate cases to define a noninformative prior distribution that is invariant to transformations. Finally, in more complex modeling situations Gelman et al. (1995)

recommend a hierarchical approach where model parameters are themselves assumed to come from distributions. This has the advantage that noninformative prior distributions need be specified for only a limited number of hyperparameters, rather than for all the model parameters. Bernardo and Smith (1994) argue that in most cases some additional information is available that allows one to construct a minimally informative prior distribution, but it does not have to be purely noninformative across the entire parameter space. They also outline an approach for complex models where one distinguishes a priori between the parameter of interest and nuisance parameters, and one then need only construct a joint prior distribution that is noninformative for the parameter of interest. The specification of prior distributions is an important and difficult issue that cannot be fully explored here; see Bernardo and Smith (1994), Gelman et al. (1995), and Press (1989) for further discussion.

A thorough explanation of Bayesian statistics is beyond the scope of this chapter. Introductory articles and discussions about Bayesian statistics include Efron (1986), Berger and Berry (1988), and Stern (1998). Introductions to the use of Bayesian methods in ecology and conservation include Ellison (1996) and Anderson (1998). Examples of the application of Bayesian methods in ecology include Gazey and Staley (1986), Walters and Ludwig (1994), Raftery et al. (1995), and Taylor et al. (1996). Punt and Hilborn (1997) provide a review of Bayesian methods in fisheries population modeling. Comparisons between Bayesian methods and frequentist methods are in Pascual and Kareiva (1996), Omlin and Reichert (1999), Poole et al. (1999), and Wade (1999).

METHODS

The Simulated Population and Data

The population model representing the "true" population was a fairly simple age-structured model with four main parameters controlling the population's dynamics: first-year survival ($s_0 = 0.40$), second-year survival ($s_1 = 0.60$), adult survival ($s_a = 0.70$), and the birthrate ($f = 0.80$, or 0.40 females per female, as an equal sex ratio was specified; see Taylor et al., chap. 12 in this volume). The population was modeled as a birth pulse with postbreeding census (Caswell 1989). The age of first reproduction was 2 (age class 3, where newborns are age class 0). Therefore, the first nonzero fecundity was for age class 1—females give birth for the first time on their second birthday, and then immediately enter age class 3 and are available to be censused. The oldest age was 40; all animals alive at the end of their 40th year died prior to reproduction. The model contained no density dependence, no environmental stochasticity, no catastrophes, and no genetic effects, but did include demographic stochasticity. These parameters resulted in a population that was declining, on average, by 20% per year. The population started at a size of 1,087 at the beginning of the five-year study. This population size was chosen because it would leave the population, on average, at a size of 400 at the end of the five-year study. For each simulation the population was initialized stochastically to a stable age distribution. This initialization was done by randomly assigning individuals to sex and age class according to the probabilities of individuals belonging to these categories when the population was in stable age distribution with a growth rate of zero.

Data were collected from the simulated population by random sampling. Data on birthrate and first- and second-year survival rates were collected by randomly sampling from binomial distributions with the binomial parameter set to the true survival values. For the survival rates, this simulates the kind of data that might be collected by attaching radio tags to a specified number of animals and following them for a year to determine if they survived. Similarly, for the birthrate, this simulates the kind of data that could be collected from a population with a birth pulse, where adult females are inspected for the presence of a newborn immediately after births occur.

Data on adult survival were collected by randomly sampling from the simulated population to determine whether an adult animal was seen (captured) in a given year, over five years, based on a given capture probability. This represents the kind of data that might be collected if animals had unique markings that allowed individuals to be identified, and a photo-identification or mark-resighting study were carried out. To keep the mark-recapture analysis simple, the simulated population was given the "characteristic" that only adult animals could be identified. Therefore, the mark-recapture analysis was used only to estimate adult survival, and the binomial tag data described above was necessary for estimating first- and second-year survival. Two scenarios were specified: an imprecise case with fewer data and a precise case with more data. The capture probability was set at 0.25 for "imprecise" runs and at 0.60 for "precise" runs. The sample sizes for the tagging experiments were set at 15 (imprecise) and 30 (precise). Other sample sizes (such as number of adult females seen) were the result of the specified capture probability. See Taylor et al. (chap. 12 in this volume) for further explanation and justification of the simulation scenarios.

Parameter Estimation

Five parameters were estimated from the data sampled from the simulated population: s_0, s_1, s_a, f, and N_{total}. Three parameters (s_0, s_1, f) had binomial data available for them, meaning data in the form of n trials with x outcomes (e.g., the number of animals with tags, and the number that survived). In the simulation, the tags were placed on animals immediately after reproduction. Therefore, the tag data can be used to estimate first- and second-year survival rates. Similarly, the mark-recapture data were collected immediately after the birth pulse. Therefore, the number of adult females with a newborn can be used to directly estimate the birthrate. The analytic methods for calculating maximum likelihood and Bayesian estimators for the binomial probability are explained in appendix 11.1.

Adult survival was estimated using the mark-recapture model of Pradel (1996). This particular model was chosen because it represents a reparameterized form of the standard Jolly-Seber type of models and provides a flexible general model for markrecapture studies. An additional important feature was that Pradel (1996) fully described the likelihood function, which then allowed me to implement both maximum likelihood and Bayesian estimation using that model. The simplest form of the model was chosen using constant (not time dependent) parameters for survival (ϕ), capture probability (p), and seniority probability (γ; Pradel 1996), for a total of three parameters. Note that the simulated "true" population also had constant survival and capture probability, so the true model was being fit to the data, and hence there was no model selection uncertainty as is normally encountered in a real mark-recapture analysis (Burnham and Anderson 1998). In this context, with only adult animals marked, ϕ represents an estimate of adult survival, s_a. A search algorithm was used to find the MLEs. A numerical integration technique called Markov chain Monte Carlo (MCMC) was used to find the Bayesian estimates and calculate a probability distribution for each parameter (called the posterior distribution in Bayesian statistics). The numerical methods used to calculate maximum likelihood and Bayesian estimators of ϕ, p, and γ are fully described in appendix 11.2.

The abundance of a population can also be estimated from mark-recapture data. In this example, only adult abundance could be estimated because mark-recapture data were available only for adults. Adult abundance was estimated from the capture probability and the number of captures in the last year of the study. Adult abundance in the fifth year was therefore estimated as

(3)
$$\hat{N}_{\text{adults}} = \frac{n_5}{\hat{p}},$$

where n_5 was the total number of captures in the fifth year of the study and p was the estimated capture probability.

Total population size can be estimated from this estimate of adult population size by using the estimated birth and survival parameters, along with an assumption of stable age distribution. Therefore, the estimate of s_a can be used in conjunction with the estimates of s_0, s_1, and f to estimate the stable age distribution from the full Leslie matrix (Caswell 1989) associated with these life-history parameter values. Age of first reproduction (age class 3) and maximum age (40) were assumed to be known and were set to the true values used in the simulation to generate the data. From the estimated stable age distribution, the proportion of the population that is adult (age class 3 or older) was then used to prorate adult abundance to total abundance:

(4)
$$\hat{N}_{total} = \frac{\hat{N}_{adults}}{\sum_{x=3}^{w} \hat{c}_x} = \frac{\hat{N}_{adults}}{1 - \hat{c}_1 - \hat{c}_2},$$

where $\hat{c}_x$ is the estimated stable age distribution.

Bayesian PVA

A Bayesian PVA has parameters specified by probability distributions that have been estimated via a Bayesian analysis. Therefore, it was relatively simple to perform a Bayesian PVA, given that posterior distributions were available for the parameters. Posterior probability distributions were calculated for the five parameters necessary to run the model: s_0, s_1, s_a (also called ϕ), f, and N_{total}. All that was required was to randomly sample values from the joint posterior distribution of these five parameters and project the population model forward using these values. Because their sources of data were independent, 5,000 random samples were directly drawn from the marginal posterior distributions for s_0, s_1, and f. Conveniently, the MCMC approximation method had already provided a random sample of 5,000 values from the joint posterior distribution for ϕ (s_a) and p. Given values for f, s_0, s_1, s_a, and p, equations 3 and 4 were used to estimate N_{total}. This provided a set of values of the five parameters needed to run the PVA, and automatically incorporated any correlations that exist between the parameters, such as ϕ and N_{total}. The PVA was carried out, in turn, by repeatedly projecting a model using one set of values for the five parameters, then repeating for all 5,000 sets of values.

The population model was individual-based to incorporate demographic stochasticity. As previously noted, the "true" population model that generated the data contained no density dependence, no environmental stochasticity, no catastrophes, and no genetic effects. The individual-based model used in the PVA similarly contained none of these effects. The model was similar to a Leslie matrix model. However, each birth or death was treated as a random event. For mortality, each individual in an age class had a survival probability proportional to the survival rate of its age class. For example, if a survival rate was 0.5 and there were ten individuals, rather than just have five survive (as would occur in a Leslie matrix model), each individual had an independent 0.5 probability of dying. Therefore, the outcome might be that any number from zero to ten would survive on any random realization, although the most likely outcome would still be five. Births were treated in the same way. Females and males were modeled separately, although they had the same survival rates.

For each set of 5,000 parameter values, the model was projected 500 times for 100 years. Each population trajectory was potentially different because of demographic stochasticity. If the population went extinct, the year it went extinct was recorded. Extinction occurred whenever the number of females or males became zero. In this way, a probability distribution for the time to extinction was constructed. In this case, there were $5,000 \times 500 = 2,500,000$ extinction times used to build the distribution.

Comparison between a Bayesian PVA and an MLE PVA

To compare the performance of Bayesian PVAs to MLE PVAs using just point estimates, I used the true extinction distribution as a performance measure. First, the true distribution for the time to extinction was calculated by running the PVA model with the parameter values set to their true values. This provided the reference by which to judge the accuracy of the estimated time to extinction. Next, 100 separate data sets were simulated as described above, 50 using the precise parameter values and 50 using the imprecise parameter values. These 100 simulated data sets were also used in Taylor et al. (chap. 12 in this volume); they also used an additional 100 simulated data sets that had different survival rates and a different rate of decline. All 100 data sets were sampled from a model with the same population parameters, and only the sample sizes differed between the two sets of 50.

Each of the 100 data sets was analyzed twice, once using maximum likelihood and once using Bayesian methods. Then two PVAs were performed. An MLE PVA used only the MLE point estimates for each parameter. A Bayesian PVA incorporated parameter uncertainty from the Bayesian parameter estimation process described above. In each case, a single distribution for the time to extinction was calculated.

In each analysis, the lower and upper bounds of the distributions were compared to see if the estimated distribution included the true distribution (using the 2.5th and 97.5th percentiles of both the true and estimated distributions). If the estimated distribution contained the entire true distribution within the 2.5th and 97.5th percentiles, it was said to cover "both" tails of the true distribution. If it included only the lower tail within the 2.5th percentile (but not the upper tail within the 97.5th), it was said to cover the "lower" tail of the true distribution. The lower tail of the distribution is particularly important, as this represents an estimate of the first year in which extinction could happen. If the estimated distribution covered the lower tail, extinction could not occur earlier than predicted by the analysis. If the estimated distribution did not cover the lower tail, extinction could occur before the earliest predicted date for possible extinction. This would not be precautionary and could lead to poor management decisions.

RESULTS

Parameter estimates from the analysis of the first simulated data set are presented to illustrate the Bayesian results. Of the 15 "radio-tagged" newborns and one-year-olds that were tracked for one year, in the first simulated data set 5 and 11 survived, respectively. Over the course of the five-year study, 172 mature females were seen, 140 newborns were found, and a total of 416 different adults were identified. The Bayesian posterior means for ϕ, p, and γ were 0.698, 0.262, and 0.899, respectively, and the MLEs were 0.702, 0.258, and 0.905, respectively. Bayesian parameter estimates of the five main parameters necessary for performing the PVA were similar to the MLEs, and it can be seen that some of the parameters were estimated fairly imprecisely (table 11.1). The major difference between the estimation methods, in this case, is that the Bayesian methods result in probability distributions for the parameters (figs. 11.1–11.3). When results from the first simulated data set were used to perform a Bayesian PVA, the resulting time to extinction distribution had a peak at 19 years, with a posterior mean of 25 years because the distribution was skewed (fig. 11.4). The 0.95 HPD interval was from 12 to 55 years.

Results from the Bayesian and MLE PVAs were compared across the 100 data sets. To give examples of the types of outcome that resulted, the estimated distributions for the time to extinction were plotted for the first through fourth imprecise data sets (fig. 11.5). The Bayesian distribution always included the MLE distribution. This occurred because the Bayesian and MLE point estimates were similar. However, because the Bayesian PVA incorporates parameter uncertainty, it was more uncertain and thus wider at either end. Therefore, the Bayesian

Table 11.1 An Example of Bayesian and Maximum Likelihood Parameter
Estimates Using the First Simulated Data Set

Parameter	True Values	Bayesian Posterior Mean	Bayesian 0.95 HPD Interval		Bayesian Posterior Distribution	MLE
Birth rate, f	0.800	0.810	0.749,	0.865	Beta(141, 33)	0.814
First year survival, s_0	0.400	0.353	0.142,	0.574	Beta(6, 11)	0.333
Second year survival, s_1	0.600	0.706	0.495,	0.903	Beta(12, 5)	0.733
Adult survival, s_a	0.700	0.698	0.622,	0.774	Approximated by MCMC	0.702
N_{total}	~400	326	256,	409	Derived from other posterior distributions	329

Notes: The specified true values that were used to define a scenario for a hypothetical species declining at 20% per year are shown for comparison. The true value for N_{total} is shown as ~400, as the true simulated population was started at a size of 1,087 four years earlier, and therefore there is a distribution of true population sizes in that year with expected value of 400. See the text for explanations of a Bayesian posterior mean, highest probability density (HPD) interval, and posterior distribution. MCMC stands for the numerical integration routine Markov chain Monte Carlo.

PVA inevitably worked as well as or better than the MLE PVA in covering the true distribution. In some cases, both the Bayesian and MLE PVAs came close to estimating the true distribution, although the Bayesian distribution is wider (fig. 11.5A). In other cases, the Bayesian PVA covered both tails of the true distribution, while the MLE PVA covered only one tail (fig. 11.5B and C). In particular, note how the MLE PVA in figure 11.5B only slightly overlapped the true distribution. In other cases, both the Bayesian PVA and the MLE PVA covered only one tail of the true distribution (fig. 11.5D).

Overall, the Bayesian PVA provided better coverage of the true distribution (table 11.2). The estimated Bayesian distributions were always wider and were much more likely to contain the entire true distribution (67% of the time) than the MLE PVA (10% of the time). In particular, the Bayesian PVAs covered the lower tail of the distribution 96% of time, whereas the MLE PVA covered the lower tail only 68% of the time. This means that the Bayesian PVA would only rarely overestimate the number of years until there was any substantial chance of extinction, whereas the MLE PVA would overestimate this quantity 32% of time.

The MLE PVA did better in the precise case than in the imprecise case. This was expected, because with better data the point estimate is more likely to be closer to the truth. In the Bayesian PVA, the coverage in the imprecise case was nearly identical to the coverage in the precise case, although slightly better coverage actually occurred in the imprecise case. This was also expected, as the Bayesian PVA should reflect the precision of the data by incorporating parameter uncertainty. In

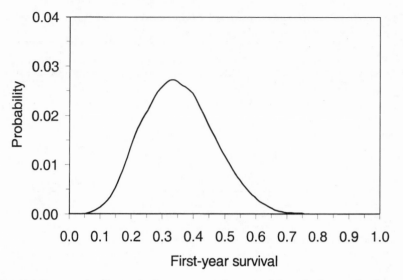

Fig. 11.1 Posterior distribution for first-year survival estimated from the first simulated data set.

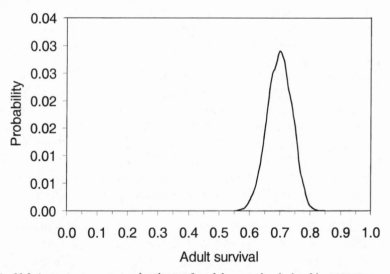

Fig. 11.2 Approximate posterior distribution for adult survival, calculated by MCMC.

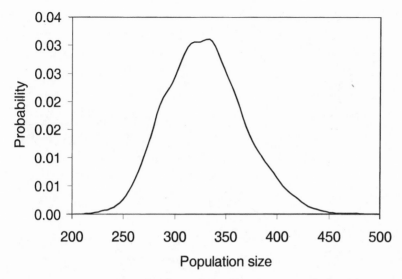

Fig. 11.3 Approximate posterior distribution for total population size, calculated from posterior distributions for s_0, s_1, s_a, f, and p.

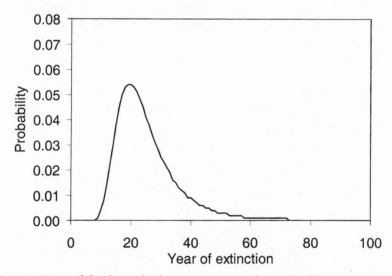

Fig. 11.4 Estimated distribution for the time to extinction for simulated data set 1, using Bayesian posterior distributions for the estimated parameters birthrate (f), first-year survival (s_0), second-year survival (s_1), adult survival (s_a), and total abundance (N_{total}) (see table 11.1).

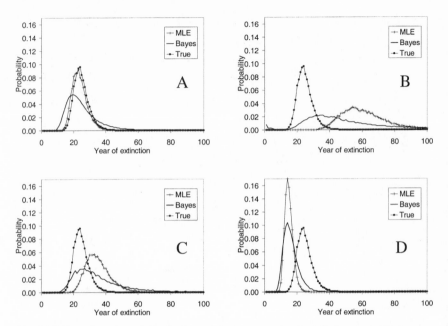

Fig. 11.5 Example estimated distributions of time to extinction for (A) data set 1, (B) data set 2, (C) data set 3, and (D) data set 4. These were the first four sample data sets out of a total of 50 for the "imprecise" scenario. MLE is the estimated distribution using the maximum likelihood point estimates. Bayes is the estimated distribution from the Bayesian PVA, incorporating parameter uncertainty. True is the true distribution calculated from the known parameter values.

Table 11.2 Percentage of Time That the Estimated Distribution for the Time to Extinction Covered the True Distribution for the Time to Extinction

Case	Bayesian PVA (%)	MLE PVA (%)
Imprecise (both)	72	4
Imprecise (lower)	98	52
Precise (both)	62	16
Precise (lower)	94	84

Notes: "Both" indicates that the estimated distribution included both the lower and upper ends of the true distribution. "Lower" indicates that the lower end of the estimated time to extinction distribution included the true value. "Imprecise" refers to a scenario using less data. "Precise" refers to a scenario using more data.

conclusion, the Bayesian PVA was much more likely to contain the true distribution within its estimated time to extinction than was the estimated distribution from the MLE PVA. This result occurred because the Bayesian PVA incorporated parameter uncertainty whereas the MLE PVA did not.

DISCUSSION

A Bayesian PVA, which incorporates parameter uncertainty, always results in wider distributions for the time to extinction than a PVA that uses only point estimates and ignores parameter uncertainty. In this analysis, the estimated Bayesian distributions for the time to extinction were more uncertain, but were more likely to include the true range of times to extinction. Ignoring parameter uncertainty can lead to an estimated range for the time to extinction that has little or no overlap with the true time to extinction (e.g., fig. 11.5B). It is obviously undesirable to make management decisions based on an estimated time to extinction distribution that is completely outside the true range of times to extinction. Most importantly, the Bayesian PVA rarely fails to cover the lower tail of the distribution. Therefore, a Bayesian PVA is unlikely to predict too long a time to the first year with any true chance of extinction. Bayesian PVAs are more conservative where our level of ignorance about a species is high and are less conservative for species with better knowledge.

Bayesian statistics provides a technique for producing a single distribution for the time to extinction that incorporates parameter uncertainty. The standard frequentist technique for incorporating uncertainty is to calculate confidence limits. Unfortunately, there is no convenient way to calculate confidence intervals around a distribution. Alternatively, a frequentist can reduce the distribution to a single quantity, such as the probability of extinction in 50 years, and calculate confidence limits around that estimated quantity (White 2000). But the frequentist cannot present a distribution of time to extinction that incorporates parameter uncertainty. In this regard, the Bayesian solution in the form of a single distribution seems simpler.

Another way in which parameter uncertainty is dealt with in some PVAs is to interpret sampling variation of parameters as true variation in the parameters over time (e.g., environmental variation). Examples include Dennis et al. (1991) and Stacey and Taper (1992). These methods tend to lead to overly pessimistic estimates of the probability of extinction because the amount of environmental variability in the population is overestimated (White 2000). A point estimate of persistence from a Bayesian PVA is not overly pessimistic. However, the lower tail of the probability distribution for extinction will be overly pessimistic with imprecise data (as will the frequentist lower confidence limit on the probability of extinction in a specified number of years), but the lower tail will become appropriately less pessimistic as more data are collected. In contrast, the methods of Dennis et al. (1991) and Stacey and Taper (1992) will continue to overestimate the probability of extinction no matter how many years of data are collected, because sampling variation will continue to be interpreted as environmental variation. Taylor et al. (chap. 12 in this volume) show how Bayesian and other methods for incorporating parameter uncertainty can be used to make precautionary decisions in the face of uncertainty, and these methods also have the property that the collection of more data leads to appropriately less precautionary decisions.

Bayesian methods are not as common as frequentist methods but are being used increasingly more often. Bayesian methods are not, in principle, more difficult to use than traditional methods. At this time, however, Bayesian methods are not as familiar to conservation scientists, and less software is available for analyses. For example, there is currently no Bayesian analogue to Program MARK (White 1998; White et al., chap. 9 in this volume) for capture-recapture analysis, although such a program would likely be no more difficult to implement.

Another difficult aspect of PVAs is that there are often some or many parameters for which few data are available. Values for such parameters often have to be determined from more indirect knowledge, such as information from other species or from biological intuition. Rather than fix such unknown parameters at single values, Bayesian methods allow a distribution to be specified, which seems a better way of reflecting the uncertainty inherent in such unknown parameters. In the absence of data, frequentist methods do not really allow for anything other than fixing a parameter at a single value. This issue has been examined in fisheries stock assessments, and an ad hoc method has been proposed for specifying a distribution for an unknown parameter in maximum likelihood analysis (Restrepo et al. 1992). It has been shown that this technique can produce biased results, however, and it generally performs no better, and sometimes much worse, than Bayesian methods (Poole et al. 1999). Beissinger and Westphal (1998) discuss the potential pitfalls of using biological intuition or data from other species to choose average vital rates, and point out PVA outcomes can range from complete extinction to no extinction if uncertainty in such rates is fully explored. Specifying a distribution for an unknown parameter seems preferable to fixing the parameter at a single value, but the outcomes may then be of little use because they are so uncertain. Conservation decisions must often be made when little information is available, however, and delay itself carries a risk (Ludwig 1999).

Purist versus Pragmatic Approaches to Incorporating Uncertainty

A statistical purist will note that the only correct way to sample parameter values based on data is by using Bayesian methods. However, many may find it natural to think of sampling from a distribution determined by MLEs of a parameter and its estimated variance. A non-Bayesian scientist might reasonably decide to sample parameters from such distributions as a way of incorporating parameter uncertainty into a PVA. As sensible as this may sound, there is no theoretical basis for sampling parameter values from frequentist parameter estimates. If a biologist performs a PVA by sampling parameter values from frequentist estimates, and then uses the single resulting distribution for the time to extinction to make probability statements, then that biologist has magically turned a sampling distribution for data into a probability distribution for a parameter. However, it is well known that the only way to go from a sampling distribution for data to a probability distribution for a parameter is through the application of Bayes's rule (i.e., equation 2). Biologists who sample parameter values from frequentist distributions and claim to be making statistically based calculations are, therefore, guilty of performing Bayesian analyses without realizing it, and furthermore have done the Bayesian part incorrectly.

A more pragmatic statistician might recognize a few other relevant points. First, the frequentist sampling distribution will sometimes approximate the Bayesian posterior distribution reasonably well, especially in simpler analyses where noninformative priors are used. Second, frequentist techniques are well known and readily available to the average conservation biologist. In my view, incorporating uncertainty by sampling parameter values from a distribution is preferable to just using point estimates, whether one does it in the technically correct way or not. It is important to address the issue of parameter uncertainty in some way. One could use non-Bayesian statistics to calculate frequentist sampling distributions, and then sample from them to calculate a PVA. For example, York et al. (1996) used this approach to incorporate parameter uncertainty into a single probability distribution for extinction risk of Steller sea lions. Although York et al. (1996) do not refer to it this way, their work can be viewed as an approximate Bayesian analysis, where the approximation is accurate to the degree that the estimated sampling distribution approximates the Bayesian posterior distribution.

This would then be the pragmatic solution if Bayesian methods were not yet available to a biologist—to sample from frequentist distributions but to label the results as an approximate Bayesian solution.

Looking Forward: Using Formal Decision Theory

Conservation biology often needs to provide scientific analyses that are used for making decisions important to the management of an endangered species. An advantage of the Bayesian approach is that it yields a result that can be immediately used in formal decision theory (Berger 1985; Goodman, chap. 21 in this volume). Most forms of decision theory are Bayesian, and it has been argued that the non-Bayesian types of decision theory do not have many good properties (Bernardo and Smith 1994). The simplest explanation is that it is necessary to know the probability of various outcomes to be able to weigh how much consideration needs to be given to them in making a decision, and the probability of outcomes can come only from a Bayesian analysis. Decision analysis requires one to specify the "cost" of making various wrong decisions. These costs are not necessarily economic, but simply describe the effect of various outcomes relative to one another (Possingham et al., chap. 22 in this volume). Using decision theory allows for the preference for over- and underprotection errors to be explicitly stated, so that the basis for the decision-making process is clear to all stakeholders. In its simplest form, decision theory would allow decisions to be made about endangered species based on their estimated time to extinction distribution, and the "costs" associated with different management actions given a specific true time to extinction. Taylor et al. (chap. 12 in this volume) provide an example of how a decision analysis would use the results of a Bayesian PVA to make a classification decision. Further investigation of such methods could be useful for listing and other management decisions.

APPENDIX 11.1. BINOMIAL ESTIMATION USING ANALYTICAL METHODS

An analytical solution can be directly calculated from the data. There are analytical solutions to both the maximum likelihood and Bayesian estimates of a binomial probability. Survival was estimated from a binomial distribution using data on the number of tagged animals that survived one year (e.g., assuming no unknown tag failures and that no animal leaves the study area). Birthrate was estimated from the number of mature females seen with newborns. In both cases, n represents the total number of trials (tagged animals or mature females seen), and x

represents the number of successes (number surviving or number of newborns).

The MLE is the value of the parameter that maximizes the likelihood function. For the binomial distribution, the MLE of the binomial parameter s (the letter p is often used, but p is already used here for capture probability in the mark-recapture analysis) is simply

(5)
$$\hat{s} = \frac{x}{n}$$

with estimated variance

(6)
$$\hat{var}(\hat{s}) = \frac{\hat{s} \cdot (1 - \hat{s})}{n}.$$

The same kind of data can be analyzed by different methods to get a Bayesian estimate. The solution will be in the form of a probability distribution for the binomial probability (the survival rate). The same likelihood function, from the binomial distribution, is used. To do a Bayesian analysis, we have to specify a prior probability distribution for the survival rate. This distribution should represent what is known about the survival rate prior to consideration of the data. In this example, nothing is known about the survival rate of this population prior to the data, and so in this case a "noninformative" prior (also called a Jeffreys prior) is chosen. Obviously, survival rates cannot be lower than 0.0, or higher than 1.0, so a uniform distribution from 0.0 to 1.0 specifies that any value in that range is equally likely, prior to collecting any data about the population.

In the same way that there is an analytical solution to the MLE, there is an analytical solution to the Bayesian binomial estimate. This analytical solution depends on the prior distribution's being in the form of a beta distribution. If the prior is a beta distribution, and the likelihood is a binomial distribution, a known result is that the posterior distribution is also a beta distribution, with parameters that are a direct function of the data and the parameters of the prior (e.g., Press 1989). This is the case because it has been shown that the integral of the product of a beta and a binomial is itself another beta distribution. In Bayesian jargon, one says that the beta is the conjugate distribution to the binomial; if a prior and likelihood are conjugate distributions, they have a posterior distribution with a known distributional form. Specifically, in this case, given that the prior is a Beta(a, b) and the likelihood is a Bin(n, x), then

(7)   Posterior distribution = Beta($x + a, n - x + b$)

and

(8)   Posterior mean $= \dfrac{(x + a)}{(n + a + b)}$.

The beta distribution serves as an ideal prior distribution for a binomial parameter estimate. With parameters equal to one, the beta distribution becomes identical to a U(0.0, 1.0) distribution. So a Beta(1, 1) can serve as a noninformative prior distribution for a parameter like survival that is essentially a probability. With a Beta(1, 1) prior, the above equations become

(9)   Posterior distribution = Beta($x + 1, n - x + 1$)

and

(10)   Posterior mean $= \dfrac{(x + 1)}{(n + 2)}$.

In Bayesian statistics, one can calculate what is termed a credible interval rather than a confidence interval. The usual Bayesian credible interval used is the HPD interval (Press 1989). There is a 0.95 probability that the true value lies within the HPD interval, and no value outside the interval has a higher probability than any value inside the interval. Credible intervals were calculated using the IMSL 3.0 routine BETDF (IMSL 1994). For the Bayesian PVA, random samples from these posterior distributions were made using IMSL 3.0 routine RNBET.

APPENDIX 11.2. MARK-RECAPTURE ESTIMATION OF SURVIVAL USING NUMERICAL METHODS

There are a few examples available of Bayesian mark-recapture analysis (e.g., Gazey and Staley 1986; Dupuis 1995; Shaughnessy et al. 1995), but none was suitable for the study in this chapter. Therefore, I developed a Bayesian analysis of the mark-recapture model of Pradel (1996). I also implemented a maximum likelihood analysis of the same model as in the original paper. I used the parameterization of the model that has survival, capture probability, and seniority probability, where ϕ is the annual apparent survival rate, p is the probability an animal is "captured" in a given year, and γ is the probability that an animal present in a given year was present in the previous year. The simplest form of this model was used with no time dependence. I specified the likelihood function directly from equation 2 in Pradel (1996). I confirmed that I had implemented the likelihood function correctly by comparing MLEs using my program with results from Program MARK (White 1998), choosing the Pradel survival and seniority option in MARK, and the $\phi(*)p(*)$ model for constant parameters (White et al., chap. 9 in this volume).

In contrast to the relatively simply binomial estimation detailed in appendix 11.1, mark-recapture analyses generally do not have analytical solutions, and numerical methods have to be used. To find the MLEs for the three mark-recapture parameters ($\phi, p,$ and γ), I used the simplex search algorithm (e.g., Press et al. 1989). A search algorithm is a numerical routine that searches for the parameters that maximize the likelihood function. The routine was started with initial values of 0.5 for all three parameters.

For the Bayesian analysis, a numerical integration technique (e.g., Tanner 1993) was necessary, rather than a search algorithm. I used the MCMC routine (Geyer 1992), specifically, the Metropolis algorithm version of MCMC (Gelman et al. 1995). The routine was used to approximate the posterior distributions for the parameters $\phi, p,$ and γ by generating a random sample of points from the posterior distributions. These sets of parameter values were accumulated into frequency distributions and were then normalized to probability distributions to approximate the posterior distribution for each parameter. The more samples that are taken, the better the approximation becomes.

The MCMC algorithm works by randomly jumping from the current set of parameter values to a new set of trial parameter values, using a specified jumping distribution, which can be any symmetric distribution (e.g., a uniform or normal distribution could be a jumping distribution). Either the old set or the new set of values is accepted as a random sample from the posterior based on the ratio of the posterior probabilities of the two sets of parameter values. The rules for accepting a new point are simple. If the new point has a higher posterior probability, it is always accepted. If it has a lower posterior probability, it is randomly accepted with a probability equal to the ratio of the new posterior probability to the old posterior probability. MCMC is relatively easy to implement; the only tools needed for MCMC are (1) the ability to calculate the prior probability of any parameter value, (2) the ability to calculate the likelihood of any parameter value given the data, and (3) the ability to generate random numbers. Practical advice on using MCMC can be found in Geyer (1992), Besag et al. (1995), Gelman et al. (1995), and Kass et al. (1998).

All three parameters represent probabilities; therefore a U(0.0, 1.0) distribution was used as the prior distribution for all three parameters. The MCMC routine was started from the maximum likelihood estimates and was run for 2,500,000 iterations. I used a uniform distribution to randomly jump to new parameter values. This jumping distribution was given a width of 10% of each initial parameter value (e.g., if the initial parameter value is 0.5, the width of the distribution is set as 0.05, so the first jumping distribution is U(0.475, 0.525)). I chose to jump all the parameter values before deciding to accept the new set of parameter values, so the routine randomly chose new values for all three parameters (ϕ, p, and γ) before deciding whether to accept the new set of values ("point"). Every 500th point was saved, for a total of 5,000. The frequency distribution of 5,000 points, normalized to a probability distribution, serve as the estimated posterior distribution (e.g., fig. 11.2).

LITERATURE CITED

Anderson, J. L. 1998. Embracing uncertainty: the interface of Bayesian statistics and cognitive psychology. *Conservation Ecology* 2 (2). http://www.consecol.org/vol2/iss1/art2.

Beissinger, S. R., and M. I. Westphal. 1998. On the use of demographic models of population viability in endangered species management. *Journal of Wildlife Management* 62:821–841.

Berger, J. O. 1985. *Statistical decision theory and Bayesian analysis*. Springer-Verlag, New York, New York.

Berger, J. O., and D. A. Berry. 1988. Statistical analysis and the illusion of objectivity. *American Scientist* 76:159–165.

Bernardo, J. M., and A. F. M. Smith. 1994. *Bayesian theory*. John Wiley and Sons, New York, New York.

Besag, J., P. Green, D. Higdon, and K. Mengersen. 1995. Bayesian computation and stochastic systems. *Statistical Science* 10:3–66.

Burnham, K. P., and D. R. Anderson. 1998. *Model selection and inference: a practical information-theoretic approach*. Springer-Verlag, New York, New York.

Caswell, H. 1989. *Matrix population models*. Sinauer Associates, Sunderland, Massachusetts.

DeGroot, M. H. 1973. Doing what comes naturally: interpreting a tail area as a posterior probability or as a likelihood ratio. *Journal of the American Statistical Association* 68:966–969.

Dennis, B., P. L. Munholland, and J. M. Scott. 1991. Estimation of growth and extinction parameters for endangered species. *Ecological Monographs* 61:115–143.

Dupuis, J. A. 1995. Bayesian estimation of movement and survival probabilities from capture-recapture data. *Biometrika* 82:761–772.

Efron, B. 1986. Why isn't everyone a Bayesian? *American Statistician* 40:1–11.

Ellison, A. M. 1996. An introduction to Bayesian inference for ecological research and environmental decision making. *Ecological Applications* 6:1036–1046.

Gazey, W. J., and M. J. Staley. 1986. Population estimation from mark-recapture experiments using a sequential Bayes algorithm. *Ecology* 67:941–951.

Gelman, A, J. Carlin, H. Stern, and D. Rubin. 1995. *Bayesian data analysis*. Chapman and Hall, London, United Kingdom.

Geyer, C. J. 1992. Practical Markov chain Monte Carlo. *Statistical Science* 7:473–511.

IMSL. 1994. IMSL Math/Library user's manual, version 3.0. Visual Numerics, Houston, Texas.

Kass, R. E., B. P. Carlin, A. Gelman, and R. M. Neal. 1998. Markov chain Monte Carlo in practice: a roundtable discussion. *American Statistician* 52:93–100.

Lacy, R. C. 1993. VORTEX: a computer simulation model for population viability analysis. *Wildlife Research* 20:45–65.

Lee, D. C., and B. E. Rieman. 1997. Population viability assessment of salmonids by using probabilistic networks. *North American Journal of Fisheries Management* 17:1144–1157.

Ludwig, D. 1996. Uncertainty and the assessment of extinction probabilities. *Ecological Applications* 6:1067–1076.

———. 1999. Is it meaningful to estimate a probability of extinction? *Ecology* 80: 298–310.

Omlin, M., and P. Reichert. 1999. A comparison of techniques for the estimation of model prediction uncertainty. *Ecological Modelling* 115:45–59.

Pascual, M. A., and P. Kareiva 1996. Predicting the outcome of competition using experimental data: maximum likelihood and Bayesian approaches. *Ecology* 77: 337–349.

Poole, D., G. H. Givens, and A. E. Raftery. 1999. A proposed stock assessment method and its application to bowhead whales, *Balaena mysticetus*. *Fisheries Bulletin* 97:144–152.

Pradel, R. 1996. Utilization of capture-mark-recapture for the study of recruitment and population growth rate. *Biometrics* 52:703–709.

Press, S. J. 1989. *Bayesian statistics: principles, models, and applications*. John Wiley and Sons, New York, New York.

Press, W. H., B. P. Flannery, S. A. Teukolsky, and W. T. Vetterling. 1989. *Numerical recipes*. Cambridge University Press, Cambridge, United Kingdom.

Punt, A., and R. Hilborn. 1997. Fisheries stock assessment and decision analysis: the Bayesian approach. *Reviews in Fish Biology and Fisheries* 7:35–63.

Raftery, A. E., G. H. Givens, and J. E. Zeh. 1995. Inference from a deterministic population dynamics model for bowhead whales. *Journal of the American Statistical Association* 90:402–416.

Ralls, K., and B. L. Taylor. 1997. How viable is population viability analysis? Pages 228–235 in S. T. A. Pickett, R. S. Ostfeld, M. Shachak, and G. E. Likens, editors, *The ecological basis of conservation: heterogeneity, ecosystems, and biodiversity*. Chapman and Hall, New York, New York.

Restrepo, V. R., J. M. Hoenig, J. E. Powers, J. W. Baird, and S. C. Turner. 1992. A simple simulation approach to risk and cost analysis, with application to swordfish and cod fisheries. *Fisheries Bulletin* 90:736–748.

Shaugnessy, P. D., J. W. Testa, and R. M. Warneke. 1995. Abundance of Australian fur seal pups, *Arctocephalus pusillus doiferus*, at Seal Rocks, Victoria, in 1991–92 from Peterson and Bayesian estimators. *Wildlife Research* 22:625–632.

Stacey, P. B., and M. Taper. 1992. Environmental variation and the persistence of small populations. *Ecological Applications* 2:18–29.

Stern, H. S. 1998. A primer on the Bayesian approach to statistical inference. *Stats* 23:3–9.

Tanner, M. A. 1993. *Tools for statistical inference: methods for the exploration of posterior distributions and likelihood functions.* Springer-Verlag, New York, New York.

Taylor, B. L. 1995. The reliability of using population viability analysis for risk classification of species. *Conservation Biology* 9:551–558.

Taylor, B. L., P. R. Wade, R. A. Stehn, and J. F. Cochrane. 1996. A Bayesian approach to classification criteria for spectacled eiders. *Ecological Applications* 6: 1077–1089.

Wade, P. R. 1999. A comparison of statistical methods for fitting population models to data. Pages 249–270 in G. W. Garner, S. C. Amstrup, J. L. Laake, B. J. F. Manly, L. L. McDonald, and D. G. Robertson, editors, *Marine mammal survey and assessment methods.* Balkema, Rotterdam, Netherlands.

Walters, C., and D. Ludwig. 1994. Calculation of Bayes posterior probability distributions for key population parameters. *Canadian Journal of Fisheries and Aquatic Sciences* 51:713–722.

White, G. C. 1998. MARK: mark and recapture survival rate estimation, version 1.0. Colorado State University, Fort Collins, Colorado.

———. 2000. Population viability analysis: data requirements and essential analyses. Pages 288–331 in L. Boitani and T. K. Fuller, editors, *Research techniques in animal ecology: controversies and consequences.* Columbia University Press, New York, New York.

Wolfson, L. J., J. B. Kadane, and M. J. Small. 1996. Bayesian environmental policy decisions: two case studies. *Ecological Applications* 6:1056–1066.

York, A. E., R. L. Merrick, and T. R. Loughlin. 1996. An analysis of the Steller sea lion metapopulation in Alaska. Pages 259–293 in D. R. McCullough, editor, *Metapopulations and wildlife conservation.* Island Press, Covelo, California.

12

Incorporating Uncertainty in Population Viability Analyses for the Purpose of Classifying Species by Risk

Barbara L. Taylor, Paul R. Wade,
Uma Ramakrishnan, Michael Gilpin,
and H. Resit Akçakaya

ABSTRACT

Population viability analysis (PVA) has been criticized as a technique for estimating absolute levels of risk because current models fail to incorporate uncertainty in many parameters or in the model structure itself. We call PVAs that use a single estimate for each parameter "point-estimate PVAs," because the point estimate or mean is commonly used for parameters estimated from data. Point-estimate PVAs have the disadvantages that they are not comparable between species with differing levels of uncertainty and that, because they ignore uncertainty, they cannot employ the precautionary principle in listing decisions. We use a simple simulation both to demonstrate the problem of using point-estimate PVAs for listing decisions and to provide alternatives that directly incorporate uncertainty. A single life history was created to mimic the type of data typically available for a PVA. To compare different types of PVAs, we limited the number of estimated parameters. Simulations were run at two levels of precision and at two population growth rates: one that would result in an IUCN listing of critically endangered and another that would result in a listing of endangered. The sampled data were sent to a single author to make likelihood estimates of the five unknown parameters. The estimates were checked for bias and distributed to the other authors, who ran different PVA models: VORTEX, RAMAS, and a custom individual-based model. All models were run using the maximum likelihood estimates as the point estimates in the PVAs. Listing decisions were made using the IUCN PVA criteria (criterion E). Uncertainty was directly incorporated in two additional ways. RAMAS was run using maximum likelihood estimates minus one standard deviation for all five variable parameters. A Bayesian PVA using the custom model was run that was identical to the individual-based model used in the point-estimate analysis, except that repeated simulations were run drawing the five parameters from the likelihood distributions (called posterior distributions by Bayesians) for those parameters.

For simulations that should have resulted in a critically endangered listing, PVAs that incorporated uncertainty performed much better than point-estimate PVAs, especially as precision decreased. Similarly, underprotection errors were much less for the PVAs that included uncertainty than for the point-estimate PVAs for the simulations that should have resulted in an endangered listing. However, improved performance came at the cost of a much higher overprotection error (incorrectly listing as critically endangered). We conclude that making decisions in a framework of formal decision analysis would better account for the costs of making over- and underprotection errors.

INTRODUCTION

Population viability analysis (PVA) was originally designed to incorporate all threats facing a species into a single estimate of risk, such as the probability of extinction over a specified time interval (Gilpin and Soulé 1986; Beissinger, chap. 1 in this volume). The technique was first developed to estimate a minimum viable population size. Thus, the estimate of risk was interpreted in absolute terms, because it was assumed that decisions regarding habitat alteration were likely to be irreversible.

Over the past decade PVAs have been used for other objectives, such as making listing decisions and deciding among different management options (Boyce 1992; Beissinger and Westphal 1998). Using PVA to decide among management options is a less stringent use of PVA, because it requires only a relative measure of risk that is comparable among the different options (Ralls and Taylor 1997; Akçakaya and Raphael 1998; Beissinger and Westphal 1998). Point-estimate PVAs develop estimates of risk based on single estimates for the means of each model parameter. Taylor (1995) questioned the use of point-estimate PVAs to evaluate absolute extinction risks. She used simulations to demonstrate poor performance of such PVAs in classifying species according to risk. Ludwig (1999) employed available data sets to show that using point estimates for parameters results in large errors in the extinction probability. He concluded that PVAs provide little or no meaningful information about extinction probabilities. Beissinger and Westphal (1998) discussed the amount of data necessary for a good PVA, and suggested that PVA be used not to classify species, but primarily to choose among various management options. In contrast, a recent study conducted a retrospective test of PVA based on 21 long-term ecological studies and found that PVA predictions were quite accurate using five PVA packages (Brook et al. 2000). For most species, however, data are poorer than for these selected studies.

Our objective here is to improve PVA as a technique to estimate absolute extinction probabilities for a wide range of species with differing amounts and qualities of data available. This chapter is offered as a step in this direction, so that PVAs can be better used in situations that require absolute extinction probabilities (Ralls et al., chap. 25 in this volume), such as listing species according to risk or determining minimum viable population size. Most listing decisions are now based on simpler measures of risk, such as a single abundance estimate, a rate of decline over a period of time, or the remaining amount of a species' range. Such measures are attractive because they match the poor quality and quantity of data available for most species at risk of extinction. Although these simple criteria allow quick decisions, such decisions could be risk-prone because they ignore many of the risks that the species faces. Only PVA explicitly attempts to account for all risks, yet it is the least frequently used World Conservation Union (IUCN) criterion for classifying species. Scientists justifiably feel insecure performing a complex analysis that may require estimates of dozens of demographic and environmental parameters when a good estimate for even the simplest parameter (e.g., abundance) does not exist. PVAs allow for incorporation of some forms of uncertainty. For example, variations in parameters over time or space (called process variation) are often incorporated as environmental stochasticity. However, the models do not incorporate the greatest uncertainty of all—our ignorance about the species. This type of variance is often called sampling variance and reflects our uncertainty in estimating the parameters of the model. There is also uncertainty in the form of the model itself, which can be very important. We do not address model structure here, but acknowledge that this form of uncertainty is important to incorporate in any real listing decision.

Without incorporating the uncertainty that results from our poor knowledge about a species, it is difficult to interpret what an extinction distribution means. For example, suppose that data are used to get an estimate of abundance with a mean of 500 individuals and a range between 200 and 1,200 individuals. Similarly, suppose a trend in abundance was available with a mean annual decline of 5%, with a range of decline between an increase of 2% and a decline of 15%. If we used the mean values, we would get an extinction distribution that appeared to be quite peaked, and the range of years when extinction could take place would probably span a few decades at most. It is clear from a quick look at the data, however, that the population could actually be relatively healthy ($N = 1,200$) and increasing ($r = +2\%$) or it could be in great peril ($N = 200$, $r = -15\%$). The distribution of extinction times using point estimates does poorly at capturing our uncertainty about the future of this population.

Taylor (1995) pointed out this shortcoming of PVAs and suggested that Bayesian techniques could be used to solve the problem. She and coauthors made progress toward incorporating uncertainty in a Bayesian PVA that used trends in abundance data for spectacled eiders (*Somateria fischeri*; Taylor et al. 1996). Although these authors assert that incorporating uncertainty will improve the performance of PVA at capturing the real risks that species face, no actual comparisons were done between PVAs that do and do not incorporate uncertainty. Publicly available PVA software uses demographic data, rather than the trend data that were used in the eider analysis. Therefore, to compare the performance of different PVA techniques, we created simulations that used demographic data.

The goals of this chapter are to strengthen PVA by promoting techniques that incorporate uncertainty directly into PVA and that incorporate the precautionary principle into listing decisions. A technique with good performance, according to the precautionary principle, should make more conservative decisions as risk increases, in order to make the level of conservatism in decisions commensurate with the level of ignorance. Decisions need to be more conservative when less is known about the species (Holt and Talbot 1978). We use simulations where the listing status of the hypothetical species is known, to evaluate the performance of the different types of PVAs against the IUCN criteria for listing according to risk. The techniques used to estimate demographic parameters (e.g., mark-recapture or survival of radio-tagged animals) are not important to the objective of this exercise, which is to compare PVA techniques that do and do not treat uncertainty. Similarly, the exact levels of precision chosen are not intended as suggestions but were merely selected to provide contrast between the point-estimate type PVAs and PVAs that treat uncertainty in decisions to list according to risk.

METHODS

Life-History Description

The life history used in the simulation emulated a mammalian strategy with a lower survival rate in the first two years of life, followed by a constant adult survival. Adults were individually identified and sexed from photographs. Females gave birth on their second birthday with an annual probability of 0.80. The sex ratio was 50:50 with a maximum age of 40. In the simulation, maximum population growth was set at $r = 0.036$, survival from birth to first birthday = 0.57, and survival at all other ages = 0.85. Humans caused two disturbances to this population: loss of habitat and poaching of adults. In normal conditions many juveniles would disperse to neighboring patches of habitat. Because no suitable habitat existed within the maximum dispersal distance, juvenile dispersers were lost to the population at a constant rate, which decreased first- and second-year survival to 0.40 and 0.60, respectively ($r = -0.05$). Poaching of adults lowered adult survival rates to 0.70 and 0.78 and resulted in population declines of $r = -0.20$ and $r = -0.12$, respectively, depending on the simulation scenario. Effects of both risks had occurred long enough that the population was in stable age distribution by the beginning of the five-year study. By then the population was so far below carrying capacity of the remaining patch of habitat that birthrates were set at the maximal rate. Thus, there was no density dependence.

Model Structure

To facilitate comparison of different types of PVAs, we wanted results to be as simple as possible, so we evaluated model performance using simulations. We deliberately chose a very simple model structure requiring estimation of only five parameters: abundance (at the end of the mark-recapture research), birthrate, first-year survival rate, second-year survival rate, and adult survival rate. We specified all other necessary parameters (see "Life-History Description," above) and the model structure, which had no density dependence, no environmental stochasticity, no catastrophes, and no genetic effects. Thus, we did not examine the problem of model choice. Although other uncertainties are very important components to address in future simulation testing, our simple model was sufficient to meet the objectives of this chapter.

To evaluate model performance, we ran simulations at two levels of precision and at two growth rates that would result in different IUCN listings. The simulations were individual-based Monte Carlo projections of the populations until extinction was reached (either no females or no males). Levels of precision were chosen to contrast species that were well known (capture probability of 0.6 each year and radio tags of 30 each of one- and two-year-old individuals) and relatively poorly known (capture probability of 0.25 each year and tags on 15 each of one- and two-year-old individuals). The rates of decline were chosen to present a difficult listing challenge, with the actual probabilities of extinction being close to the threshold criteria dividing the classification between critically endangered and endangered. The IUCN criteria for using PVA results are critically endangered (CR) = 50% chance of extinction in three generations, endangered (EN) = 20% chance of extinction in five generations, vulnerable (VU) = 10% chance in 100 years, and least concern (LC) = none of the above. A population growth rate of $r = -0.20$ results in a 63% chance of extinction in three generations, as determined by running 1,000 simulations, and thus represents a species that should be classified CR. A population growth rate of $r = -0.12$ results in a 7% chance of extinction in three generations (i.e., does not qualify as CR) and a 91% chance of extinction in five generations (i.e., qualifies for EN). Thus, although this species should be classified as EN, it is relatively close to qualifying for CR. We would expect a greater chance of misclassifying it as CR than as VU, because the actual status of the species is closer to being CR.

One author (Taylor) ran 50 simulations at each rate of decline. This number was sufficient to contrast the results of point-estimate PVAs to PVAs that treated uncertainty, which was the goal of this exercise. For each simulation, the population was initialized stochastically to a stable age distribution. This initialization was done by randomly assigning individuals to sex and age class according to the probabilities of individuals belonging to these categories when the population was in stable age distribution with a growth rate of zero. The model that projected the population into the future was individual-based (i.e., the history of each individual was tracked independently), and the population was recorded as extinct in the year that either the number of males or the number of females became zero. In the first five years, a mark-recapture study was simulated by sampling adults just after the birth pulse according to two different capture probabilities: $p = 0.60$ for the precise case and $p = 0.25$ for the imprecise case. Initial abundance was 1,087 and 729 for the $r = -0.20$ and $r = -0.12$ cases, respectively. These initial values resulted in both cases having abundances that averaged 400 when the abundance estimates were made using the mark-recapture data in the fifth year. Thus, estimated abundance in year five would not provide a clue to those implementing the different PVAs as to which scenarios might match. When a female was "captured," the presence or absence of a newborn was recorded. In the fourth year of the simulation, newborns and one-year-olds were radio-tagged: 30 for each year class for the precise case and 15 for each year class for the imprecise case. Although these numbers may not seem high, they do represent a fairly large proportion of these age classes and quite realistically portray early juvenile survival as parameters that are often difficult to estimate. The fate of these individuals was recorded in the fifth year. Thus, two data sets for the different precision levels resulted from each simulation and consisted of the capture-recapture histories, number of adult females captured with and without newborns, and number of newborns and one-year-olds that were tagged and survived the subsequent year. These data were given to another author to perform maximum likelihood estimation of the vital rates (details are in Wade, chap. 11 in this volume). The maximum likelihood and standard error estimates were returned to Taylor, who confirmed that the estimates were correct and unbiased. Then 200 sets of parameter estimates (50 each for the different precision levels and rates of decline) were distributed to other authors, who performed different types of PVA models. Each modeler had no foreknowledge of what the classifications should have been for any of the 200 mixed data sets that they received and analyzed. We evaluated three point-estimate models of PVA: VORTEX (Lacy 1993), RAMAS (Akçakaya 1998), and a custom individual-based model developed by Wade for the Bayesian analysis (Wade, chap. 11 in this volume) but run here using only the maximum likelihood estimates (MLEs) for the five variable parameters. We refer to the latter model as the "custom" PVA. Each of these techniques yielded single values from each simulation for the probability of extinction after three generations, five generations, and 100 years. Classification was made according to these values.

Two PVA methods were used that accounted for uncertainty. A second run of RAMAS (RAMAS-P) used a precautionary value for each of the five parameters, which was the mean estimate minus one standard error. The choice of one standard error was arbitrary but could be changed to make this simple technique more or less precautionary. We include this comparison to demonstrate a very simple method that treats uncertainty within the models already available for PVA. It is another form of point-estimate PVA that simply uses a lower percentile from a distribution for a parameter rather than the mean estimate. In contrast, the Bayesian technique differed fundamentally from the other models because all of the information in the parameter distributions was used to calculate the final extinction distribution. Details of the Bayesian analysis are given in Wade (chap. 11 in this volume) but are summarized here to explain how classifications were made. The Bayesian analysis was conducted as follows: (1) choose the five parameters randomly, according to their probability from the posterior distributions; (2) run a PVA and save the probability of extinction for three generations, five generations, and 100 years; (3) repeat both previous steps 500 times. Each one of the 500 replicate PVAs resulted in a probability of extinction in three generations, five generations, and 100 years. Given uncertainty in each parameter, we expect at times to draw all optimistic values from the posterior distribution and perhaps have only a 10% probability of extinction in 100 years (VU classification). At other times we may draw values leading to a high rate of decline and consequently result in a greater than 50% probability of extinction in three generations (CR classification). Thus, all 500 iterations will yield probabilities that the study population should have been classified into each of the four IUCN categories. To incorporate precaution into the listing process, we used a precautionary set of decision rules for listing: list as CR if the probability of being CR is at least 10% (e.g., ≥50 out of 500 replicates meets the CR criterion of a greater than 50% probability of extinction in three generations), list as EN if the probability of being EN is at least 20%, list as VU if the probability of being VU is at least 30%, and otherwise list as LC.

RESULTS

Performance of the point-estimate PVAs was very similar and displayed undesirable characteristics if precautionary management is desired (table 12.1). When data were precise and simulations should have classified the species as CR ($r = -0.20$), the point-estimate PVAs resulted in underprotection errors (classified as EN) that ranged between 14% and 32%. Underprotection errors increased to 34% to 46% for the imprecise case. Not only is this error rate unlikely to be acceptable, but considerably more errors were made when less was known, which is the opposite of being precautionary. To make matters worse, some simulations with imprecise data were classified as VU, indicating a relatively low risk for

Table 12.1 Percentage Classified by Risk Category for Different PVA Models for Different Growth Rates and Levels of Knowledge as Indicated by the Precision of the Data

		Rate of Decline							
		$r = -0.20$ (Should be CR)				$r = -0.12$ (Should be EN)			
Knowledge Level	PVA Method	CR	EN	VU	LC	CR	EN	VU	LC
Precise	VORTEX	68	32	0	0	32	60	8	0
	RAMAS	86	14	0	0	38	54	8	0
	Custom (MLE)	68	32	0	0	22	64	14	0
	Bayes	98	2	0	0	74	24	2	0
	RAMAS-P	98	2	0	0	80	18	2	0
Imprecise	VORTEX	60	34	6	0	34	50	14	2
	RAMAS	62	34	4	0	40	42	16	2
	Custom (MLE)	54	46	10	0	34	42	22	2
	Bayes	92	6	2	0	86	14	0	0
	RAMAS-P	96	4	0	0	84	16	0	0

Note: Risk categories are critically endangered (CR), endangered (EN), vulnerable (VU), and least concern (LC).

a population that was declining at 20% per year. Incorporating uncertainty using the Bayesian PVA greatly improved performance by reducing underprotection errors. Choosing a conservative estimate of vital rates (RAMAS-P) yielded similar results. As expected, the PVA methods that accounted for uncertainty were precautionary as they reduced underprotection errors to 6% or less regardless of data precision.

Simulations of the population that should have been classified as EN ($r = -0.12$) allowed both over- and underprotection errors. Recall that the true state of this endangered population was closer to CR than to VU, so we did not expect errors to be symmetrical. Again, for the point-estimate PVAs, underprotection errors were large and occasionally made the classification error of LC for the imprecise case. As before, the PVA methods that accounted for uncertainty substantially reduced the rate of underprotection errors. Note, however, that reduction in underprotection errors came at a substantial cost in overprotection errors. This example suggests that when under- and overprotection errors can be made, there is a clear trade-off between the two. If we desire to avoid underprotecting a species, then it comes at the expense of frequently overprotecting it when it requires a lower level of intervention.

DISCUSSION

This exercise demonstrates that PVAs using point estimates are not acceptable if precautionary management is desirable. Not accounting for uncertainty in estimates of vital rates made classification more variable. In addition, point-estimate PVAs are explicitly not precautionary because the less we know the higher the rate of underprotection errors becomes (table 12.1). Both techniques that accounted for uncertainty improved performance. Nevertheless, our results were not fully precautionary, as underprotection errors did not remain constant as knowledge decreased (table 12.1). For example, when $r = -0.20$, the Bayesian method correctly classified 98% with precise data but only 92% with imprecise data. Thus, there is still room for improvement in modifying criteria to be fully precautionary. However, maintaining a constant underprotection error rate for any quality of data will lead to greater overprotection error rates for imprecise data.

It is surprising, and probably coincidental, that the RAMAS-P and the Bayesian approaches performed so similarly. The former was an ad hoc approach that simply assumed a bad case scenario by arbitrarily subtracting one standard deviation from each parameter. This does not permit us to make any probabilistic statement about the results, but it does give us a fair idea that the real-world picture is unlikely to be as grim as our depiction. Because the level of precaution (one standard error) is not based in units that managers or stakeholders can understand, it may be difficult to support this approach even if it performs quite well and is simple to carry out. However, it is possible to make this approach comparable to the Bayesian approach, by making multiple simulations for each of the 200 sets, and sampling the parameters of each simulation from statistical distributions with the estimated standard error, representing uncertainty. (See Wade, chap. 11 in this volume, for further discussion of the comparability of Bayesian and likelihood approaches.) The "minus one standard error" approach described here was employed only to demonstrate the possibility of using a simple and fast method that, nevertheless, gave comparable results and reduced the underprotection error. Another alternative to the Bayesian approach is using fuzzy arithmetic to propagate uncertainty thorough the decision rules of the IUCN criteria. Such an approach has been developed to incorporate uncertainty into all five IUCN criteria (Akçakaya et al. 2000).

The Bayesian approach has statistical underpinnings that are well documented (Wade, chap. 11 in this volume). Although it is more complex to implement, Bayesian PVA has one very important advantage over most other types of PVA models: the resulting distribution of extinction probabilities can be used in formal decision theory. Our analysis was constrained by our decision to use the IUCN criteria as a performance measure. However, these criteria are not precautionary. To make them precautionary for our Bayesian PVA analyses, we had to impose another set of decision criteria. We arbitrarily decided that a population should be listed as CR if it has at least a 10% chance of being CR. This is a very awkward way to think about the problem because it essentially yields a probability of a probability, which is not an easy thing to contemplate.

There is another way we could have reported the Bayesian results if we were not using the IUCN criteria. Recall that the Bayesian PVA randomly chose five parameters from the posterior distributions and ran a full PVA with those parameters. Each time the population was projected into the future, an extinction time was generated. If we accumulated all the extinction times for the many different choices of five parameters into a single distribution, we would have an extinction distribution that incorporated all the parameter uncertainty. Such distributions are shown in Wade (chap. 11 in this volume). Figure 12.1 shows hypothetical distributions for a precise and an imprecise case. If we define the degree of risk by time, again using the IUCN time threshold criteria, we can see that there is a probability of extinction for each time interval that is defined as the area under the probability distribution in

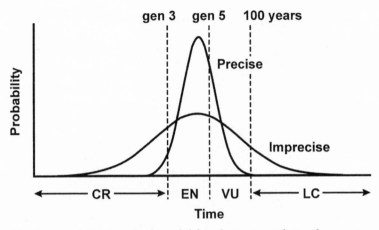

Fig. 12.1 Posterior distributions for the probability of extinction with time for a precise and an imprecise case. The cutoff points for time are determined by the criteria used. For the IUCN criteria, *CR* is any time less than three generations, *EN* between three and five generations, *VU* between five generations and 100 years, and *LC* any time exceeding 100 years. The probability of being in different risk states by different cutoff times is the area of the probability distribution included in the time span of a risk category (table 12.2).

Table 12.2 Posterior Probabilities for Being in Different Risk States According to Figure 12.1

Knowledge Level	Risk States			
	CR	EN	VU	LC
Precise	0.10	0.45	0.45	0.00
Imprecise	0.30	0.15	0.35	0.20

that interval. Table 12.2 gives the probabilities that the observed data were from populations in the different risk states.

Decision theory is a framework that integrates the cost of making various errors with the probabilities of being in various states, to make the optimal decision (Possingham et al., chap. 22 in this volume). Table 12.3 gives a cost table where the costs of making over- and underprotection errors are equal. For example, consider the case where the true state is that a population is EN. An equal cost is incurred if an under- or an overprotection error is made, with both CR and VU costing one-half point. Contrast this with a precautionary cost table (table 12.4), where the cost of an underprotection error is double the cost of an overprotection error (one for listing an EN population as VU versus one-half for listing the same population as CR). The cost table allows the users great flexibility in assigning costs. For example, we could make

Table 12.3 Cost Table for Equal Over- and
Underprotection Costs

	True State			
Decision	CR	EN	VU	LC
CR	0	½	1	2
EN	½	0	½	1
VU	1	½	0	½
LC	2	1	½	0

Notes: The table is read as follows: if the true state is that this population is critically endangered, then the cost that decisions will incur is
nothing if the decision is to list as CR, one-half if the decision is to list
as endangered, and so forth. The symmetrical nature of the table results
from equal costs being assessed for placing a population in either one
category too restrictive or one category too lenient.

Table 12.4 Precautionary Cost Table Where
Underprotection Costs Are Double Overprotection Costs

	True State			
Decision	CR	EN	VU	LC
CR	0	½	1	2
EN	1	0	½	1
VU	2	1	0	½
LC	4	2	1	0

Table 12.5 Costs of Making Various Decisions Using the Probabilities of Being
in Different States

	Equal Costs		Precautionary Costs	
Risk Categories	Precise	Imprecise	Precise	Imprecise
CR	0.675	0.825	0.675	0.825
EN	**0.275**	0.525	**0.325**	**0.675**
VU	0.325	**0.475**	0.65	0.85
LC	0.875	0.925	1.75	1.85

Notes: Table 12.2 shows the probabilities of different risk states. Tables 12.3 and 12.4 are the cost tables
for equal and precautionary decision schemes. Decisions incurring the lowest cost are in bold.

costs increase much more for underprotection errors made at high-risk
categories than for low-risk categories. This simple example uses relative
costs (i.e., the cost of making error A is twice the cost of making error
B), but costs could be assessed in monetary terms as estimated by economists.

The decision that minimizes risk of making errors for this example
is shown in bold in table 12.5. The true state is that the population
should be classified as EN. In the equal-costs approach, the correct decision is made with precise data, but an incorrect underprotective decision

is made with imprecise data. This mistake is corrected by using a precautionary cost table.

The simulation approach presented here clearly demonstrates the nonprecautionary nature of using point-estimate PVAs to make listing decisions. It also helps refine PVA techniques that incorporate uncertainty and the way that risk probabilities can be used to make decisions. When we started this exercise, we expected a simple set of comparative simulations. Along the road, we learned a number of lessons that rivaled the importance of our original goal. First, analytical techniques to incorporate uncertainty are not readily available. Even though we created a very simple scenario, we still had to limit ourselves to one of the few mark-recapture analysis methods for which the full likelihood function has been published to be able to compare maximum likelihood and Bayesian techniques. Although progress is being made to make analytical methods more available, dedicated technical work needs to continue in the area of parameter estimation (White et al., chap. 9 in this volume).

For all real PVAs there would be parameters for which no data exist. These factors, such as Allee effects, catastrophes, and inbreeding depression, are often omitted from PVAs. Although this may be acceptable for PVAs that compare relative estimates of risk, this practice is not acceptable for estimating absolute risk to the population. We feel it would be useful to develop default distributions for these parameters for different life-history strategies, based on comparative studies for PVAs that estimate absolute risk. Such default distributions would incorporate what is known about these factors from similar species. We believe they would be an improvement over either omitting these factors or requiring each PVA modeler to make arbitrary choices.

Our final conclusion concerns the importance of choosing a decision-making framework. As mentioned above, the IUCN criteria were not framed to incorporate uncertainty into decision making. The objective of the IUCN Red List (http://www.redlist.org) is to yield a crude indicator of the health of global biodiversity. The simplicity of its decision criteria and the fact that most countries do not have conservation laws or risk criteria have propelled the use of this classification system for single-species management. Our experience in this exercise, however, points out a shortcoming of using these criteria for management. The criteria do not allow the user to choose the levels of over- and underprotection errors appropriate for the magnitude of risk or for the particular political and economic situation confronting a species. The costs of making different errors can be superimposed on the IUCN framework using fuzzy arithmetic (Akçakaya et al. 2000) or by the Bayesian approach shown here. We encourage the development and testing of such classification schemes that incorporate uncertainty and account for the costs of making different types of errors.

LITERATURE CITED

Akçakaya, H. R. 1998. RAMAS Metapop: viability analysis for stage-structured meta-populations, version 3.0. Applied Biomathematics, Setauket, New York.

Akçakaya, H. R., S. Ferson, M. Burgman, D. Keith, G. Mace, and C. Todd. 2000. Making consistent IUCN classifications under uncertainty. *Conservation Biology* 14:1001–1013.

Akçakaya, H. R., and M. G. Raphael. 1998. Assessing human impact despite uncertainty: viability of the northern spotted owl metapopulation in the northwestern USA. *Biodiversity and Conservation* 7:875–894.

Beissinger, S. R., and M. I. Westphal. 1998. On the use of demographic models of population viability in endangered species management. *Journal of Wildlife Management* 62:821–841.

Boyce, M. S. 1992. Population viability analysis. *Annual Review of Ecology and Systematics* 23:481–506.

Brook, B. W., J. J. O'Grady, A. P. Chapman, M. A. Burgman, H. R. Akçakaya, and R. Frankham. 2000. Predictive accuracy of population viability analysis in conservation biology. *Nature* 404:385–387.

Gilpin, M. E., and M. E. Soulé. 1986. Minimum viable populations: processes of species extinction. Pages 19–34 in M. E. Soulé, editor, *Conservation biology: the science of scarcity and diversity*. Sinauer Associates, Sunderland, Massachusetts.

Holt, S. J., and L. M. Talbot. 1978. New principles for the conservation of wild living resources. *Wildlife Monographs* 59:1–33.

Lacy, R. C. 1993. VORTEX: a computer simulation model for population viability analysis. *Wildlife Research* 20:45–65.

Ludwig, D. 1999. Is it meaningful to estimate a probability of extinction? *Ecology* 80:298–310.

Ralls, K., and B. L. Taylor. 1997. How viable is population viability analysis? Pages 228–235 in S. T. A. Pickett, R. S. Ostfeld, M. Shachak, and G. E. Likens, editors, *The ecological basis of conservation*. Chapman and Hall, New York, New York.

Taylor, B. L. 1995. The reliability of using population viability analysis for risk classification of species. *Conservation Biology* 9:551–558.

Taylor, B. L., P. R. Wade, R. A. Stehn, and J. F. Cochrane. 1996. A Bayesian approach to classification criteria for spectacled eiders. *Ecological Applications* 6:1077–1089.

PART 3

INTEGRATING
THEORY AND
PRACTICE IN
THE USE OF
POPULATION
VIABILITY
ANALYSIS

Theory and empiricism meet when evaluating the chance that a species or population may go extinct. This part examines various aspects of PVA, largely through examples of how the theory and practice of conservation come together in PVA for a variety of taxa using a diversity of approaches.

Building and parameterizing a PVA model is often much easier than testing it, as Gary Belovsky and colleagues show. They present novel tests of simple extinction models using experimental systems in the laboratory to track the fate of scores of replicated populations of brine shrimp. Their results remind us that many attempts at predicting the fate of populations, although well meaning, may be misleading. A good example of how this problem occurs in practice is presented by David Maehr and colleagues, who trace the fate of the Florida panther based on three PVAs spanning 20 years. The panther persists in small numbers (74 individuals) in the southern third of the Florida peninsula. Maehr et al. show how the prognosis for this species changed from certain extinction to certain persistence as more data and better knowledge of the species and its ecosystem were accumulated. Dan Doak and collaborators consider the special problems that seed banks create for PVA models of plants. Small, numerous, and often undetectable, seeds not only contribute to short-term reproduction but act as a bet-hedging mechanism to enhance persistence over long periods of time. Rates of seed survival have important implications for population forecasts of a variety of plant life histories, but Doak et al. show that current technologies are unlikely to provide good estimates of seed survival.

One way that we can deal with uncertain or unknown parameter estimates in PVA models is to conduct sensitivity analyses. Sensitivity analyses can indicate which parameter values have the greatest impacts on model outcomes. There are a variety of ways to conduct these analyses, and Scott Mills and Mark Lindberg explore the implications of different approaches, both to understand the problems of small populations and to evaluate approaches to manage overabundant species.

Molecular and population genetics also have made key contributions to PVA applications. Philip Hedrick reviews the role of molecular genetics in managing endangered species, presenting several examples from his work with endangered fishes. Molecular techniques have made important contributions to the identification of stocks and population units, and to evaluating relatedness among populations. He also considers the use of genetic measures for highly variable loci (e.g., microsatellite loci). Susan Haig and Jonathan Ballou show how detailed knowledge of the pedigree of a small population can be used to develop gene drop models. These models can be used to evaluate the effects of adding or subtracting individuals on various measures of genetic diversity. They present several compelling examples of using these models with endangered birds for structuring captive-breeding programs, restoring wild populations, and evaluating small-population management options.

Although PVA is often used with small populations to evaluate risks and management options, two contributions consider the application of PVA to conservation at large scales. Large, widespread populations can decline to small numbers rapidly. It is challenging to identify the factors responsible and to determine how to reverse the trend. Stuart Pimm and Oron Bass consider how PVA models can be used to evaluate the consequences of various causes of decline for the Cape Sable seaside sparrow in the Florida Everglades. Different subpopulations of the sparrow are declining due to different causes related to ecosystem management decisions—fire and flooding. By comparing measured sparrow population trends to theoretical predictions of persistence based on the fluctuations of surrogate sparrow species, Pimm and Bass show how wrong the view of a problem from the armchair can be. PVA has played a modest role in conservation planning at bioregional scales. Fred Samson reviews how risk assessment is implemented for hundreds of species at large scales using coarse- and fine-filter approaches. He discusses the impediments to making better large-scale conservation planning decisions and suggests that wise investments of resources would place less emphasis on assessments and more on implementation. It seems likely that PVA will always have a limited but important role to play in developing and assessing bioregional plans. However, PVA could benefit greatly if more resources were used to determine the impacts of plans as part of adaptive management.

13

How Good Are PVA Models? Testing Their Predictions with Experimental Data on the Brine Shrimp

Gary E. Belovsky, Chad Mellison, Chad Larson, and Peter A. Van Zandt

ABSTRACT

We employed replicate laboratory populations of brine shrimp (*Artemia franciscana*) to develop a data set on extinction dynamics (Belovsky et al. 1999). Sets of replicate populations were established with different initial numbers of adults and with different carrying capacities (food-supply rates) that were randomly varied each day by different coefficients of variation. The experiment examined the action of demographic and environmental stochasticity on extinction. Extinction by virtue of the above stochastic processes produces a mean population persistence time with a probability density function (variance) for identical populations experiencing identical environmental conditions; therefore we measured mean population persistence and its variance for each set of replicate populations. An additional experiment with replicated laboratory populations measured brine shrimp demography (birth, death, and population growth rates) and nonlinear (oscillatory) dynamics under the same conditions as the extinction experiment. Strong density-dependent population dynamics were observed: birthrate increased, death rate decreased, and population growth rate increased as population number relative to carrying capacity decreased. Furthermore, brine shrimp population growth rates should produce intrinsic population oscillations (nonlinear dynamics). Mean persistence time for sets of replicate populations increased with carrying capacity, decreased with variability in population numbers, and increased with initial population number. Population variability increased with variation in carrying capacity, increased

The study was supported by funds to G. E. Belovsky from the National Science Foundation (DEB-9322576); the Office of the Vice-President of Research, Utah State University; and Utah Division of Wildlife Resources. We wish to thank Sincere Jackson, Courtney Olsen, Patrick Lee, and Jason Jacobson for assistance in censusing individuals in populations. Patrick Lee aided in modeling and setting up the experiments examining environmental variability. T. Crowl provided comments on the experimental design and statistics, and J. B. Slade provided comments on the manuscript.

with the intensity of intrinsic oscillations in the population, and increased as the ratio of starting population numbers to carrying capacity increased. The effect of initial population number on persistence was not as great as often hypothesized. Furthermore, variance in mean persistence time was much smaller than assumed. Our findings support qualitative perceptions about extinction and pose additional conservation concerns.

Finally, experimental demographic data were used in a variety of population viability models to predict mean population-persistence time and its variance, and predictions were compared with the results from the extinction experiments. Many commonly used population viability models provided predictions that deviated considerably from experimental results, but others were very predictive. We address the reliability of commonly employed population viability models and their utility for conservation planning.

INTRODUCTION

Population viability analysis (PVA) examines the likelihood of a population becoming extinct over a particular time period due in part to stochastic extinction processes. Stochastic extinction processes include genetic, demographic, environmental, and catastrophic factors (Gilpin and Soulé 1986; Soulé and Kohm 1989; Boyce 1992; Burgman et al. 1999). PVA has been used to predict the smallest population size necessary for the population to persist over a specified time period and a probability acceptable to society (minimum viable population: MVP), although recent applications rarely estimate MVPs (Beissinger, chap. 1 in this volume). Common PVA metrics include the expected persistence time and the probability that a population will persist for a given time period.

PVA concepts are important to ecological questions dealing with population stability, colonization ability, and island biogeography (MacArthur and Wilson 1967; Diamond 1984; Pimm 1991) and to paleontological questions dealing with the extinctions observed through the geological record (Raup 1991). PVA concepts have received the greatest attention in conservation biology, however, where they are used in planning and policy, especially for managing endangered species and preserving biodiversity (Soulé and Kohm 1989; Noss 1999). PVA models are used to project a population into the future and evaluate if it is to survive for some time period with a certain degree of success that is acceptable to society (Shaffer 1987; Soulé 1987; Thomas 1990; Boyce 1992; Murphy and Noon 1992; Lindenmayer et al. 1993; Wilcove et al. 1993; Beissinger and Westphal 1998).

PVA has its critics. First, PVA has been criticized for focusing on stochastic processes rather than the deterministic factors (e.g., over-harvesting, habitat loss) that are responsible for initial population decline (Caughley 1994; Caughley and Gunn 1996). This is not a clean dichotomy, however, because PVA addresses how extinction may occur when populations are reduced in number, even though deleterious deterministic factors have been eliminated (Boyce, chap. 3 in this volume). Second, PVA has been criticized for providing a false sense of quantitative precision in conservation planning and policy, when understanding of population dynamics may be uncertain and population data is subject to sampling error (Boyce 1992; Lindenmayer et al. 1993; Possingham et al. 1993; Beissinger and Westphal 1998). Unfortunately, even though PVA concepts have been widely applied in conservation, they have not been well validated. What constitutes validation of PVA models is debatable. Areas of debate include the following issues.

Genetic factors (e.g., inbreeding depression, genetic drift) have been demonstrated to reduce survival and reproduction in small populations (see Frankham 1995a,b, 1998). This demonstrates that genetic factors hypothesized to increase the extinction risk for small populations can operate, but this does not test the extinction predictions of PVA models.

Qualitative predictions of extinction from PVA models (e.g., the likelihood of extinction increases as population size decreases) have been tested in one laboratory experiment (Forney and Gilpin 1989) and in a number of field observational studies (e.g., Crowell 1973; Diamond 1984; Belovsky 1987; Ebenhard 1987; Pimm et al. 1988; Burgman et al. 1992; Tracy and George 1992; McCarthy et al. 1995). These studies find a negative relationship between population size and the likelihood of extinction, but they are unable to determine how well various PVA models quantitatively predict the likelihood of extinction.

Quantitative predictions of extinction from PVA models using observations of field populations are problematic. One can compare PVA model projections of expected population size with observations from a single population over time (Brook et al. 1997; Brook and Kikkawa 1998). The essence of PVA models is the likelihood of extinction due to stochastic factors, however, and a single population sheds no light on this, because there is no replication. One can also compare the proportion of populations of the same or different species that have become extinct to PVA model predictions (Brook et al. 2000). But differences between environments inhabited by populations and differences between species produce inordinate variation, which makes rigorous comparison with model predictions impossible. Finally, field studies observe the number of extinctions over a set time period but not a set of populations until all become extinct. This measures quasi-extinction probabilities (Ginzburg et al. 1982), because the extinction probability distribution is truncated. Quasi-extinction probabilities make testing of different PVA models difficult, because distributions of extinction probability have a long right tail—which strongly influences PVA predictions—and this tail is not examined by quasi-extinction measures.

Without quantitative tests of PVA models, numeric projections from these models for conservation planning and management can be questioned (Boyce 1992; Yaffee 1994; Beissinger and Westphal 1998). For example, opponents can ask why projections from one PVA model should be trusted over projections from another model that may be less dire.

Although the necessary control and range of conditions cannot be obtained naturally, a way to assess the quantitative validity of PVA projections is a controlled experiment. A robust test of PVA requires sets of replicate populations that experience identical conditions and are monitored over time until each population goes extinct, so that mean persistence time and its variance can be measured. With experimental sets of replicate populations started with different numbers of individuals, with different mean carrying capacities, and with different levels of random variation in carrying capacity, the populations' mean persistence times and their variances can be related to these factors to test PVA concepts based on genetic, demographic, and environmental stochasticity. Finally, to test more rigorously the quantitative predictions of PVA models, demographic data (birth, death, and population growth rates) for the study species can be entered into a variety of PVA computer software programs, and the models' predictions can be compared with experimentally determined mean persistence times and variances.

We explicitly designed an experiment to test various PVA models (Belovsky et al. 1999). Replicated laboratory populations of brine shrimp (*Artemia franciscana*) were maintained in environmental chambers with different initial numbers of individuals, with different mean carrying capacities (food-supply rates), and with different random variation in carrying capacity (coefficient of variation in food supply). We then used *Artemia* demographic parameters in some PVA computer software models to predict persistence times and variances to test whether they agreed with our experimental results.

METHODS

Brine Shrimp Data Set

The details of the brine shrimp experiments and their results have been published elsewhere (Belovsky et al. 1999). All brine shrimp were hatched from eggs commercially harvested from the Great Salt Lake (Utah), a noninbred population. This experiment did not reflect brine shrimp populations in nature because the temperature and food levels employed in the experiment cannot be directly related to field conditions in the source, the Great Salt Lake. Furthermore, brine shrimp reproduction was restricted to live births (ovoviviparity), because diapausing cysts (oviparity) were removed from the experimental population.

Extinction experiments examined replicate sets of 4, 10, or 20 populations experiencing the same initial population size ($I = 2, 4, 6, 8, 12, 16,$ or 20 adults at an equal sex ratio), carrying capacity (13 levels of K, ranging from <1 to 12 adult equivalents based on body mass, were obtained by varying food-supply rate), and variation in carrying capacity (coefficient of variation or $CV_K = 0, 1,$ or 3). All populations were censused until they became extinct. Censuses counted every individual, and each was identified to developmental stage and sex. Counts of known numbers of individuals by observers who were unaware of actual numbers were always ± 1 from actual values, producing an average deviation of about 0.5%. Therefore, there was virtually no sampling error. Censuses were converted to numbers of adult equivalents based on the body masses of individuals of different developmental stages, so that actual population numbers were 15 to 55 times greater than the number of adult equivalents. Finally, mean persistence time and its variance were computed for each set of replicate populations.

Demography experiments examined replicate populations at a constant density (20 individuals) of the same developmental stage (three stages: nauplii, juveniles, and adults, where adults were at equal sex ratio), and a constant food-supply rate (five levels). These populations were maintained at the same temperature and light conditions as the extinction experiment. As in the extinction experiment, all individuals were counted, and individuals were added or removed to maintain constant density; this virtually eliminated experimental error. Results from this experiment provided estimates of density-dependent (constant density/food-supply rate) changes in population birth, death, and growth rates that were independent from the population dynamics observed in the extinction experiment.

Testing PVA Models

A number of PVA models have been developed to predict mean persistence times and their variances. To examine the quantitative precision of a sample of the most commonly cited models, we calculated mean persistence times and their variance for different experimental initial numbers (I), carrying capacities (K), and variation in carrying capacity (CV_K) by parameterizing the models with the brine shrimp demographic data and comparing model predictions with the results from the brine shrimp extinction experiment. The PVA models and their characteristics are summarized below.

Model 1: MacArthur and Wilson (1967) developed the first commonly employed PVA model. This model treats populations as growing exponentially in a continuous fashion until they reach K, when population growth rate becomes 0. Mean persistence time (T) is

(1)
$$T = \frac{1}{K\lambda}\left(\frac{\lambda}{\mu}\right)^K,$$

where λ is birthrate, μ is death rate, and K is carrying capacity. This model assumes no environmental variability. Persistence times are assumed to be distributed as a negative exponential (Goodman 1987a,b; Mangel and Tier 1993).

Model 2: Richter-Dyn and Goel (1972) used the MacArthur-Wilson model (1) but allowed per capita birth (λ) and death (μ) rates to change with population density and populations to grow continuously. Their most commonly cited model produces logistic-like population dynamics, where

(2a)
$$\lambda = \lambda_{max}\left[1 - \left(\frac{N}{K^*}\right)^\beta\right],$$

and

(2b)
$$\mu = \mu_{min}\left[1 + \left(\frac{N}{K^*}\right)^\beta\right].$$

β is a constant that is commonly assumed to equal 1, and $K^* = K/[(\lambda - \mu)/(\lambda + \mu)]^{1/\beta}$ (N can never exceed K^*). Richter-Dyn and Goel (1972) provide an approximate analytical solution for a population's mean persistence time given this model. This model does not include environmental variation. Mean persistence times are assumed to be distributed as a negative exponential (Goodman 1987a,b; Mangel and Tier 1993).

Model 3: Dennis et al. (1991) developed a general stochastic calculus approach to extinction dynamics, where the population grows exponentially until K is reached. The mean persistence time (T) from a population at K is

(3)
$$T = \frac{\ln(K)}{|r - v_r/2|},$$

where r is the per capita population growth rate and v_r is the variance in r (calculated sensu Dennis et al. 1991). The persistence times are predicted to be in an inverse Gaussian distribution, rather than a negative exponential.

Model 4: May (1975) and Renshaw (1993) developed a general stochastic model for discrete population growth:

(4a) $$N_{t+1} = N_t e^r + Z_{t+1}.$$

N_{t+1} is population number at time $t + 1$ and N_t is the population number at time t. r is the population's per capita growth rate ($\ln[\lambda - \mu]$), which reflects positive and negative components of exponential population growth between time t and $t + 1$. Z_{t+1} is a random normal deviate with mean of zero and a variance (V) based upon individual birth and death processes, which are assumed to be distributed in a Poisson fashion, the same assumption employed in many PVA models (Feller 1939, 1966; Renshaw 1993).

(4b) $$V = N_{t+1}\left(\frac{\lambda + \mu}{\lambda - \mu}\right)(e^r - 1), \quad \text{if } r \neq 0.$$

(4c) $$V = 2N_t\lambda, \quad \text{if } r = 0.$$

Two versions of the difference equation model were examined: (1) Exponential population growth (λ and μ are constant) occurs until the population reaches K, when growth rate equals zero ($\lambda = \mu$). This model is a discrete equivalent of the continuous MacArthur and Wilson model (1967). (2) Density-dependent population growth occurs ($\lambda - \mu$ declines linearly with N/K). The linear function for density dependence is obtained if λ is constant and μ equals $\lambda N/K$ (May 1975; Renshaw 1993). In this model, extinction can only occur stochastically (i.e., $N_{t+1} \leq 0$ when $N_t > 0$ only by the action of Z_{t+1}; the model can produce N_{t+1} values less than or equal to zero, at which time the population is considered extinct and no further calculations are made), while some density-dependent models also can deterministically produce extinction; if N_t is sufficiently greater, then N_{t+1} may decline below zero (i.e., $N_{t+1} \leq 0$ when $N_t > 0$, even if Z_{t+1} is 0) (May 1975; Renshaw 1993). This model is similar to the continuous Richter-Dyn and Goel model (1972).

This model does not include environmental variation. We used Mathcad Plus 7.0 to solve the difference equation model (equation 4a) for mean persistence time and its variance by Monte Carlo simulation, because the model does not have an analytical solution. Solutions were based on 500 populations that were followed until they became extinct (i.e., one or no individuals). This provided the probability distribution of persistence times, rather than assuming a distribution, as was done for the above models.

Model 5: Foley (1994) presents diffusion approximations for solutions to mean persistence time when the environment varies stochastically. These approximations were found to be similar whether populations grew exponentially until they reached K, or exhibited density-dependent growth prior to reaching K (Foley 1994). The approximate solution when per capita population growth rate (r) does not equal zero and I equals K is

(5) $$T \approx \frac{\ln K^2}{\theta_r}\left[1 + \frac{2}{3}\left(\frac{r \ln K}{\theta_r}\right)\right],$$

where θ_r is the variance of r with environmental variation (calculated sensu Dennis et al. 1991). Foley (1994) argued that persistence times would be distributed as a negative exponential.

Model 6: Goodman (1987a) and Leigh (1975, 1981) developed a discrete model of exponential population growth until K is attained, when the environment stochastically varies. These models have approximate analytical solutions, which are available in the commercial package NEMESIS 1.0 (Gilpin 1993).

(6) $$T_I = \sum_{x=1}^{I}\sum_{y=x}^{K}\frac{2}{y[yV_y - r_y]}\prod_{z=x}^{y-1}\frac{V_z z + r_z}{V_z z - r_z},$$

where T_I is the expected persistence time started with I individuals, K is the population ceiling (carrying capacity), r_z is the population growth rate at population size z, r_y is the population growth rate at population size y, V_z is the variance in population growth rate at population size z, and V_y is the variance in population growth rate at population size y. Persistence times are assumed to be distributed as a negative exponential (Goodman 1987a,b; Mangel and Tier 1993).

Model 7: VORTEX 6.3 (Lacy 1993; Lacy et al. 1994), a commercial model for discrete, age/stage-structured population growth, was applied. For comparison to the simpler models presented above, genetic inbreeding, metapopulation, and age-structure options were not employed. Solutions were based on 500 populations that were followed until all became extinct (i.e., one or fewer individuals), which provided the probability distribution of persistence times.

Model 8: RAMAS Metapop 1.1 (Akçakaya 1994), a commercial model for discrete, age/stage-structured population growth, was applied. For comparison to the simpler models presented above, metapopulation and age-structure options were not employed. Solutions were based on 500 populations that were followed until they became extinct (i.e., one or no individuals), which provided the probability distribution of persistence times.

There are numerous other PVA models in the literature. For example, some PVA models based on discrete population growth account for males and females separately (Gabriel and Bürger 1992; Stephan and Wissel 1994). We did not examine these models because of their complexity and possible problems in their construction. Other models are commercially available or free on the Internet or from the author (e.g., INMAT, GAPPS, ALEX), but these usually produce similar results to VORTEX and RAMAS Metapop (Mills et al. 1996; Brook et al. 2000). Finally, better PVA models can and no doubt will be developed, but our purpose was to examine current PVA models commonly employed.

Statistical Analyses

SYSTAT 7.0 was used for all statistical analyses of the experimental data and comparison of observations with model predictions (Wilkinson 1997).

RESULTS

Brine Shrimp Data Set

The *extinction experiment* found that populations persisted from 1 to 65 weeks or 0.25 to 16 generations, with few persisting longer than 16 weeks or 4 generations. The extinction experiment provided mean persistence times (T) for replicated populations with the same I, K, and CV_K values (fig. 13.1). Results were qualitatively consistent with PVA expectations (Gilpin and Soulé 1986; Shaffer 1987; Soulé 1987; Pimm et al. 1988; Boyce 1992; Tracy and George 1992; Wissel and Zaschke 1994): T increased as I increased, T increased as K increased, and T decreased as population variability (CV_P) increased (Belovsky et al. 1999). Our experimental results represent the first case where extinction dynamics can be attributed to specific factors, because of the controlled conditions. We found, however, that the relative importance of each factor sometimes differed from expectation and that other factors often not considered in PVA were very important (Belovsky et al. 1999). (1) Demographic stochasticity accounted for 48% of the variation in mean persistence time for brine shrimp populations, with 32% due to variation in K and 16% due to variation in I. Conservationists often consider initial

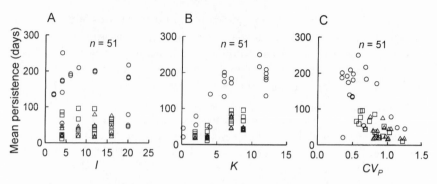

Fig. 13.1 Results from the brine shrimp extinction experiments presented as the effect of A, initial population size (*I*); *B*, carrying capacity (*K*); and *C*, coefficient of variation population numbers (CV_P) on mean persistence time. *Circles* indicate experiments without environmental variability ($CV_K = 0$), *triangles* indicate low environmental variability ($CV_K = 1$), and *squares* indicate high environmental variability ($CV_K = 3$).

population size (*I*) the most important factor, especially for species reintroduction (Gilpin and Soulé 1986; Pimm and Gilpin 1989), but the experimental results did not support this perspective. (2) Environmental stochasticity accounted for 14% of the variation in mean persistence time for brine shrimp populations. Environmental stochasticity is sometimes considered more important than demographic stochasticity (Leigh 1975, 1981; Goodman 1987a,b), but the experimental results did not support this. (3) Deterministic population oscillations caused by nonlinear dynamics arising from density-dependent population responses (Renshaw 1993) accounted for 38% of the variation in mean persistence times. This refers to intrinsic oscillations in a population that can be amplified if a population is initiated with numbers greater than *K*. This effect is rarely considered in conservation planning. (4) Thus, factors that reduced population size so that demographic stochasticity could operate were of approximately equal or greater importance (i.e., effects of environmental stochasticity and nonlinear dynamics) to demographic stochasticity (48% versus 52% of variance explained). (5) The standard deviation for mean persistence time appears to increase linearly with mean persistence time (fig. 13.2).

Extinction occurs when all individuals in a population disappear, and when either all males or all females disappear for a sexually reproducing organism, like this species of brine shrimp. The relative importance of the above modes of extinction differed with CV_P ($\chi^2 = 23.89$, df = 4, $P < 0.001$). When $CV_P = 0$, disappearance of females accounted for 67% of the extinctions, but it decreased to 38% in importance with increasing population variability ($r^2 = 0.97$, $n = 3$, $P < 0.13$). This result suggests that males may be less able to cope with conditions producing population variation.

Extinction did not appear to be due to reproductive failure arising from adult senescence. This was determined by increasing the food-supply rate for five populations that were persisting after 29 weeks. These populations contained only old adults and were near extinction due to very low reproductive output. With increased food-supply rate, these populations increased their reproduction and were still reproducing through week 58. Finally, carrying capacity is a critical determinant of population persistence. Some laboratory populations of brine shrimp have persisted for more than 25 years under similar laboratory conditions, but with higher carrying capacities ($\sim$200 adults; Browne 1983).

Demography experiments showed that per capita birth, death, and population growth rates were density dependent (fig. 13.3), where density dependence is expressed as the constant density relative to K (food-supply rate). The density-dependent response produces a linear relationship between the constant density relative to K (food-supply rate), which is indicative of logistic population dynamics (Renshaw 1993). The maximum per capita population growth rate for brine shrimp

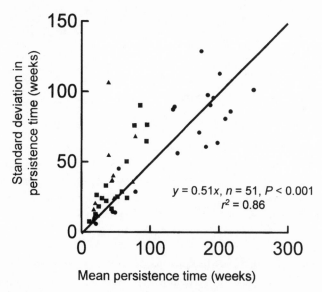

Fig. 13.2 Standard deviation for mean persistence time plotted against mean persistence time. A regression line, equations, and statistics are also presented. *Circles* indicate experiments without environmental variability ($CV_K = 0$), *triangles* indicate low environmental variability ($CV_K = 1$), and *squares* indicate high environmental variability ($CV_K = 3$).

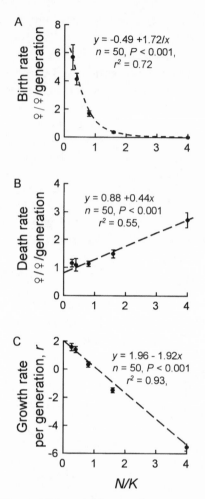

Fig. 13.3 Means (±1 SD, $n = 10$) from the demography experiments: A, production of female offspring surviving to become adults for each adult female starting a generation (adult females/adult female/generation); B, observed death rate (adult female deaths/adult female starting a generation/generation); and C, per capita population growth rate per generation (natural logarithm of female adults produced minus female deaths). Regression lines (*dashed lines*), equations, and statistics are also presented.

(r_{max}) was 1.96 female offspring produced/adult female/generation when $CV_K = 0$, 2.22 when $CV_K = 1$, and 1.45 when $CV_K = 3$ (Belovsky et al. 1999).

We observed an average generation time (nauplii to reproductive adult) of 4.0 ± 0.04 weeks (±SE, $n = 50$), an average life expectancy of adults of 3.12 ± 0.14 weeks (±SE, $n = 50$), and a food-supply rate of 0.2 mg of yeast/day when per capita population growth equaled zero. Browne (1980, 1983) observed similar generation times and adult life expectancies in food-limited laboratory populations and determined that maintenance and replacement reproduction required approximately 0.2 mg of yeast/day. Therefore, it would appear that our experimental populations were food-limited.

Testing PVA Models

PVA models were solved using the observed demographic parameters for the brine shrimp and the experimental K and I values (summarized in appendix 13.1). Models assuming exponential population growth until K is attained were solved where $\lambda = \lambda_{max}$, $\mu = \mu_{max}$, and $r = r_{max}$. Models assuming density-dependent population growth were solved assuming logistic population dynamics (r declines linearly with N/K, so that λ is constant and $\mu = \lambda N/K$; May 1975; Renshaw 1993). Age/stage-structured models (models 7 and 8) were solved with only a single stage category for comparison with the models without age/stage-structure, but similar results emerged when three stage categories (nauplii, juveniles, adults) were used. Models assuming no environmental variation (models 1–4, 7, and 8) were only compared with experimental data where $CV_K = 0$, and models assuming environmental variation (models 5–8) were only compared with experimental data where $CV_K = 1$ or 3. The ranges of mean persistence times observed in the experiments and predicted by the models are presented in figure 13.4.

With *no environmental variability*, the continuous population models (models 1 and 2) predicted mean persistence times that were three to five orders of magnitude greater than observed. The stochastic calculus model for discrete exponential growth without K (model 3) predicted mean persistence times that were less than one to one order of magnitude smaller than observed. The general discrete stochastic models with K (model 4) predicted the observed persistence very well, with the density-dependent form fitting exceptionally well. The commercial age/stage-structured models (models 7 and 8) generally overestimated persistence time, but by less than an order of magnitude.

With *low environmental variability* ($CV_K = 1$), the diffusion approximation model for exponential growth without K (model 5) overestimated mean persistence times by less than one to two orders of magnitude greater than observed. The model for exponential growth until K is attained (model 6) tended to overestimate mean persistence times by less than an order of magnitude. The commercial age/stage-structured models (models 7 and 8) overestimated persistence times at low K and I values, but came close to observed persistence times at high K values.

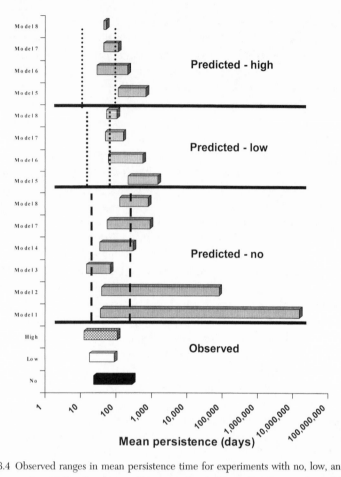

Fig. 13.4 Observed ranges in mean persistence time for experiments with no, low, and high environmental variability (*bottom panel*) compared with the predicted ranges from different PVA models (*top three panels*) for the experimental conditions. The observed range for the appropriate conditions (no environmental variability = *long-dashed vertical lines;* low environmental variability = *short-dashed vertical lines;* high environmental variability = *dotted vertical lines*) are compared with appropriate PVA models.

With *high environmental variability* ($CV_K = 3$), the diffusion approximation model for exponential growth without K (model 5) overestimated mean persistence times by less than one to two orders of magnitude greater than observed. The model for exponential growth until K is attained (model 6) generally overestimated mean persistence times by less than one order of magnitude. The commercial age/stage-structured models (models 7 and 8) tended to overestimate persistence times by less than one order of magnitude at low K and I values, but at higher K

values, model 7 came close to observed persistence times, while model 8 underestimated persistence times by less than one order of magnitude.

Mean persistence time is one output of PVA models; another is the probability distribution of persistence times. Negative exponential distributions have often been hypothesized (Goodman 1987a,b; Mangel and Tier 1993; Foley 1994), but an inverse Gaussian distribution also has been suggested (Dennis et al. 1991). Negative exponential distributions have the standard deviation in persistence times equal to mean persistence time for similar populations (I, K, CV_K). Our data indicated that the standard deviation increased linearly with mean persistence time, but with a slope significantly less ($\sim 1/2$) than 1 ($t = 17.11$, df = 49, $P < 0.001$). The observed linear relationship between the standard deviation and mean persistence time indicated a gamma distribution for which the negative exponential is a special case. An inverse Gaussian distribution is nonlinear, increasing at a decreasing rate, which was significantly different from the observations ($t = 5.0$, df = 49, $P = 0.001$).

DISCUSSION

Experimental Realism

Before discussing the implications of our experimental results, we wish to reiterate that our experimental populations do not portray brine shrimp populations in nature, nor do the results quantitatively reflect extinction dynamics for any specific endangered or threatened species. However, our experimental system is a useful test of qualitative and quantitative predictions based on PVA models. This has great utility in validating the underlying premises of extinction dynamics and conservation biology.

Genetics can be important in PVA projections (Allendorf and Ryman, chap. 4 in this volume). We did not, however, observe reduced individual survival or fecundity during the experiment that might be due to genetic effects. First, the demographic data were obtained from a large number of individuals (20 individuals plus replacements due to mortality) within a single generation (Belovsky et al. 1999), which reduces the likelihood of any negative genetic factors, and the demographic data predicted extinction dynamics very well (Belovsky et al. 1999). Second, most extinction-experiment populations became extinct within four generations, which did not provide time for substantial genetic effects. Therefore, we concluded that extinction in our experiments was primarily due to demographic and environmental stochasticity, and the PVA model (model 4) that explained observed extinction dynamics the best did not include genetic factors.

One might argue that the brine shrimp per capita population growth rate (r_{max} = 1.45–2.22 per generation) is much larger than observed growth rates for most endangered species. First, observed population growth rates even for nonendangered species and our brine shrimp can be very small (fig. 13.3) as K is approached. However, population models require maximum growth rate (r_{max}). Second, the brine shrimp r_{max} is for a generational time frame; r_{max} on a generational time frame for many vertebrates, including humans, approaches two, and it exceeds two for many invertebrates (Charnov 1993). Therefore, because endangered species exhibit low population growth rates does not imply that with better conditions (e.g., more habitat, less pollution, etc.) they would not exhibit much larger growth rates (Dennis et al. 1991). Growth potential is what the population models require.

Qualitative Findings

The low predictive power of initial population number (I) is somewhat surprising, because this is often considered to be very important in conservation planning, especially in reintroduction of species (Gilpin and Soulé 1986; Pimm and Gilpin 1989). This suggests that rapid population growth when resources are abundant (i.e., low density of conspecifics, competitors, or predators) may quickly counter the extinction-enhancing effects of low initial numbers, assuming no major genetic effects of small numbers.

Population carrying capacity (K) is important because it sets the upper ceiling to population growth. The higher the ceiling, the farther the population is from extinction and the longer it will take to become extinct. However, variability in population numbers (CV_P) will periodically reduce a population below K and increase the likelihood of extinction due to demographic stochasticity (Gilpin and Soulé 1986; Shaffer 1987; Soulé 1987).

A number of factors that can reduce a population periodically below K are included in our measure of population variation (CV_P). First, environmental variation (CV_K) can reduce K, decrease birthrates, and increase death rates, which reduces population numbers. Second, intrinsic oscillations with nonlinear dynamics due to density dependence can lead a population to overshoot K and then decrease below K. Third, intrinsic oscillations with time lags in density-dependent responses (e.g., stunted growth; Renshaw 1993) can lead a population to overshoot K and then decrease below K. Fourth, intrinsic oscillations can emerge with changes in age or stage structure (Renshaw 1993). The effects of intrinsic oscillations due to nonlinear dynamics, time lags, and age or stage structure are seldom considered for extinction dynamics. Oscillations can also emerge as a population interacts with predators, parasites, pathogens, and competitors. Therefore, the impact of population oscillations on extinction dynamics may be common.

Our experimental studies found intrinsic oscillations due to nonlinear dynamics that were particularly important for CV_P. Nonlinear dynamics can emerge in populations that grow at discrete time intervals rather than continuously over time (Renshaw 1993). Our brine shrimp populations exhibit discrete growth for several reasons. First, populations were started with similar-aged virgin adult females, and reproduction occurs by the production of broods, which requires an interbrood interval; this tends to lead to reproduction occurring periodically rather than continuously. Second, generation time approaches the life expectancy of adults, which means that adults tend to die before their offspring begin to reproduce. This will create a scenario in which the adults at any point in time tend to be individuals from a single cohort, which approaches the condition of nonoverlapping generations, an extreme case of discrete population growth. While most species may not approach nonoverlapping generations, discrete population growth may be most common in nature (Renshaw 1993). Third, initial numbers (I) greater than carrying capacity (K) will, for a period, amplify intrinsic oscillations produced by nonlinear dynamics. The greater the initial population number relative to K (I/K), the greater the amplification of oscillations.

Given nonoverlapping generations and logistic population growth, a population will exhibit greater oscillatory behavior (damped oscillations, 2, 4, 8 . . . point limit cycles, and finally chaos) as r_{max} increases above one (May 1975; Renshaw 1993). Our measures of brine shrimp r_{max} ranged between 1.45 and 2.22, suggesting that damped oscillations and multiple-point limit cycles occur (Belovsky et al. 1999). Furthermore, we observed that increasing values of I/K increased CV_P as expected with nonlinear dynamics (Belovsky et al. 1999). We have not investigated whether time lags and age/stage-structure also contribute to oscillations.

PVA Projections

Discrete-growth PVA models (models 3–8) predicted shorter mean persistence times and matched observed mean persistence times better than continuous-growth models (models 1 and 2). Similar birth and death rates produce slower population growth if births and deaths occur at discrete intervals rather than continuously, and this results in shorter mean persistence times.

In the absence of environmental variability $(CV_K = 0)$, a general discrete density-dependent model (model 4; May 1975; Renshaw 1993) did very well at predicting mean persistence times $(r^2 = 0.73, n = 19, P < 0.001)$ and their standard deviations $(r^2 = 0.57, n = 19, P < 0.001)$.

Furthermore, the slopes of the regressions between predicted and observed values were not different from 1 (0.82 and 0.97, respectively, for mean persistence time and its standard deviation). Predictions of other models without environmental variability matched observed values much more poorly ($r^2 < 0.2$), and slopes were less than 0.3. Therefore, commonly employed PVA models (models 1, 2, 7, and 8) did not do well and generally overestimated persistence times, which made them overly optimistic about a population's likelihood of not going extinct.

With environmental variability ($CV_K = 1$ or 3), none of the PVA models (models 5–8) predicted mean persistence times in the experiment very well, but they did better on average than they did without environmental variability. Predicted mean persistence times sometimes were smaller and at other times were larger than observed values, but they generally overestimated mean persistence times, which made them overly optimistic about a population's likelihood of persisting. It would be interesting to modify the general discrete density-dependent model (model 4) that did so well without environmental variability to include environmental variability.

Mean persistence time only partially defines extinction dynamics, because the probability distribution of persistence times for replicate populations is very broad and positively skewed with a very heavy right tail (Dennis et al. 1991). This makes knowledge about the standard deviation of mean persistence time critical for assessing the likelihood of extinction. None of the hypothesized probability distributions (negative exponential and inverse Gaussian) was supported by our experimental results. Instead, the observed probability distribution was not as broad as hypothesized, but still highly skewed (gamma distribution: SD $\cong$ 1/2 T), which means that populations will be more likely to approach the mean persistence time than predicted by most PVA models.

The easiest way to envision the combination of mean persistence time with its standard deviation is to use the two values to estimate how large a population needs to be to survive a given time period with a certain probability of success, a common conservation question asked of PVA. As an example, we computed how large a brine shrimp population should be if it is to persist for 100 days (about four generations) with a 95% certainty (fig. 13.5).

With *no environmental variability*, some PVA models (models 1, 2, 7, and 8) were too optimistic, predicting a 10 to 60% lower required carrying capacity (K). This means that the predicted K was 0.9 to 0.4 times smaller than needed for a 100-day persistence with 95% certainty. Model 3 was too pessimistic, predicting a 61% higher required carrying

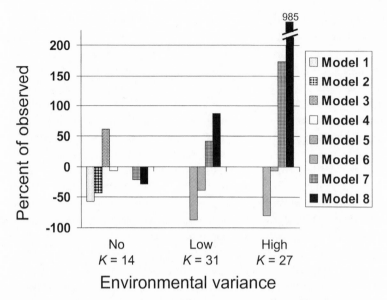

Fig. 13.5 Predicted K necessary for a population to persist 100 days with 95% certainty (I = four adults with equal sex ratio: no environmental variation for models 1–4, 7, and 8; low and high environmental variation for models 5–8) is presented as a percentage of the experimentally observed K (presented below each environmental variability). A positive value means that the model overestimates the necessary K, and a negative value means that the model underestimates the necessary K.

capacity, or a value 2.57 times larger than needed. The general discrete stochastic model (model 4) predicted the necessary K very well.

With *low environmental variability* ($CV_K = 1$), models 7 and 8 were too pessimistic, predicting a 40 to 80% larger required carrying capacity (K). This means that the predicted K was 1.42 to 1.78 times larger than needed for a 100-day persistence with 95% certainty. Model 6 was too optimistic, predicting a 40% smaller required carrying capacity, or a value 0.71 times smaller than needed. Model 5 was very optimistic, predicting a 61% smaller required carrying capacity, or a value 2.26 times smaller than needed.

With *high environmental variability* ($CV_K = 3$), models 7 and 8 were too pessimistic, predicting a 175 to 985% larger required carrying capacity (K). This means that the predicted K was 2.78 to 33.33 times larger than needed for a 100-day persistence with 95% certainty. Model 6 did very well, predicting only a 10% smaller required carrying capacity, or a value 0.9 times smaller than needed. Model 5 was very optimistic, predicting a 49% smaller required carrying capacity, or a value 2.1 times smaller than needed.

Without environmental variability, the general discrete density-dependent model (model 4) predicted the K needed for a population to persist with a degree of certainty very well. This occurred because model 4 predicted mean persistence time and its standard deviation very well. With environmental variability, another PVA model also did well (model 6 with high CV_K), but for the wrong reasons, because the model did not correctly predict either mean persistence time or its standard deviation. Although incorrect estimates of mean persistence time and its variance might counter each other to make a correct prediction, that does not make the model valid.

Our tests of PVA models address how well the models predict the observed mean persistence times and their standard deviations, when the observed values are well documented in controlled experiments where experimental and sampling error are very small. This is different from studies that have examined whether different PVA models provide similar predictions, when provided with the same parameters (Mills et al. 1996; Brook et al. 2000). These studies do not indicate PVA model validity, but similarity in model construction.

Each PVA model was developed assuming different population dynamics, usually to make the mathematics more tractable, and solutions to these models depend upon various simplifying approximations. This makes all of the models and their solutions caricatures of real populations. Particular assumptions and simplifications may be more appropriate to particular species' life histories; however, these models and their solutions are usually applied without this consideration, because it is not always apparent how particular assumptions and approximations relate to the details of species' life histories. We believe that our brine shrimp experimental system (not brine shrimp in nature with their extended diapause period, which we eliminated) is representative of species that exhibit discrete population growth and are active year round. This is common for many, if not most, species, and, as discussed above, the r values in our experiments are typical of many species, including vertebrates. Therefore, our experimental results may be typical for many species, as well as the quantitative applicability of PVA models to their populations.

Conservation Implications

The qualitative concepts emerging from PVA models were supported, but many PVA models were not quantitatively supported by experimental observations. This has important implications for the use of PVA in conservation planning and leads to a number of specific recommendations. Qualitative PVA concepts should be a major focus in conservation, because these concepts were largely validated.

Quantitative PVA predictions generally provided overly optimistic estimates of population persistence times, except for the simple discrete stochastic models (model 4). Therefore, the use of PVA models in conservation may lead to recommending a population size that is too small to achieve protection. It should not be surprising that the PVA predictions are overly optimistic, because this has been a general pattern in the history of these models. First, the earliest PVA models represented continuous density-independent population growth until K was attained (MacArthur and Wilson 1967); these models suggested that 20 individuals were adequate to buffer a population from demographic stochasticity. Second, continuous density-dependent population growth was then incorporated into PVA estimates (Richter-Dyn and Goel 1972; Ginzburg et al. 1982, 1990), which led to a conservation rule of thumb for an MVP of 50 individuals to avoid demographic stochasticity. Third, these small MVP estimates led ecologists to suggest that environmental stochasticity and catastrophes must be factors, in addition to demographic stochasticity, that reduce the likelihood of persistence and require larger populations, which might approach hundreds of individuals (Leigh 1981; Wright and Hubbell 1983; Strebel 1985; Ewens et al. 1987; Goodman 1987a,b; Stacey and Taper 1992; Wissel and Zaschke 1994). Finally, our results indicate that demographic stochasticity may produce much shorter persistence times, even without environmental stochasticity and catastrophes, than previously thought, when discrete population growth and its potential to generate population oscillations (nonlinear dynamics) are considered. Perhaps we must consider maintaining thousands of individuals in a population for adequate protection. This evolution of PVA projections reflects our general underestimate of nature's volatility.

Quantitative PVA predictions need to be used with caution, especially those based on commonly used commercial programs that did not perform well in predicting our experimental results (models 7 and 8). In fact, the simple discrete density-dependent stochastic model (model 4) did very well. This suggests that simple models may be more valuable for PVA than complex simulation models. We suggest that conservation planners and managers might be better served by adapting these simple models to their specific problem by entering them into one of the many mathematical programs, e.g., Mathcad, Mathematica (Wolfram 1996; Hanselman and Littlefield 1997) to obtain PVA projections. Finally, density-dependent dynamics and their potential to produce population oscillations (nonlinear dynamics and time lags) may need to be more frequently considered in applications of PVA to conservation planning.

Population numbers that increase in habitat fragments due to the influx of displaced individuals from the surrounding destroyed habitat could reduce persistence times. This is seldom considered, but our experiments indicated that population variability increased as I/K increased (Belovsky et al. 1999). Lovejoy et al. (1986) have suggested this process may occur for some bird and mammal populations in Amazonian forest fragments as surrounding areas are logged.

Differential survival of one sex could have important effects on persistence. Our experiments indicated that males were less capable of surviving with environmental variability. This shifted the main form of extinction from the loss of all females when the environment was constant, to the loss of all males when the environment varied. How often is this considered in conservation planning? None of the PVA models that we used considered differential survival of males and females.

CONCLUSION

Our extinction experiment with brine shrimp is the first quantitative validation of PVA concepts. It provides mixed evaluations of the utility of PVA models in conservation. Qualitatively, PVA models were instructive, but quantitatively, many PVA models did not predict our experimental observations very well. Our experiment provides precise demographic data, but often conservationists have only rudimentary demographic data for natural populations that are threatened with extinction. This further complicates the utility of PVA models. There is hope for quantitative PVA projections, however, because we found that simple discrete stochastic models predict extinction dynamics very well, and these models require the least detailed demographic data. Therefore, for conservation, we recommend that simple PVA models may be adequate and preferred.

Our extinction experiment dealt with isolated populations. This is only one aspect of PVA, because multiple populations that are linked (i.e., metapopulations) persist longer when migration between populations prevents extinction (Hanski and Gilpin 1991). However, extinction dynamics of component populations of a metapopulation occur as in isolated populations, except individuals are periodically added and lost by migration. Therefore, our results are relevant to metapopulations.

We do not claim that our experiment with brine shrimp directly extrapolates to specific endangered species, but the experiment allows us to validate PVA, models, and concepts. Without validation, current and future PVA models will be hard pressed to provide the requested conservation projections to society. Only in this way can challenges from conservation opponents be countered.

APPENDIX 13.1. EXPERIMENTAL PARAMETERS IN THE PVA MODELS

The experimental parameters used in the PVA models to predict mean population persistence times were as follows.

Model 1: $\lambda = 5.65$, $\mu = 0.88$ (fig. 13.3).

Model 2: λ and μ functions in figure 13.3.

Model 3: $r_{max} = 1.96$ per generation, $v_r = 0.5$.

Model 4: λ and μ functions in figure 13.3.

Model 5: For $CV_K = 1$, $r_{max} = 2.22$ per generation, $\theta_r = 1.13$; for $CV_K = 3$, $r_{max} = 1.45$ per generation, $\theta_r = 1.48$.

Model 6: For $CV_K = 1$, $r_{max} = 2.22$ per generation, $V_r = 2.22$; for $CV_K = 3$, $r_{max} = 1.45$ per generation, $V_r = 4.35$.

Model 7: For $CV_K = 1$, $r_{max} = 2.22$ per generation, $V_r = 2.22$; for $CV_K = 3$, $r_{max} = 1.45$ per generation, $V_r = 4.35$; density dependence from figure 13.3; one age category; maturity in one generation; equal sex ratio; no genetics; no metapopulation.

Model 8: For $CV_K = 1$, $r_{max} = 2.22$ per generation, $V_r = 2.22$; for $CV_K = 3$, $r_{max} = 1.45$ per generation, $V_r = 4.35$; density dependence from figure 13.3; one age category; maturity in a generation; equal sex ratio; no metapopulation.

LITERATURE CITED

Akçakaya, H. R. 1994. RAMAS Metapop: viability analysis for stage-structured meta-populations, version 1.0. Applied Biomathematics, Setauket, New York.

Beissinger, S. R., and M. I. Westphal. 1998. On the use of demographic models of population viability in endangered species management. *Journal of Wildlife Management* 62:821–841.

Belovsky, G. E. 1987. Extinction models and mammalian persistence. Pages 35–57 in M. E. Soulé, editor, *Viable populations for conservation.* Cambridge University Press, Cambridge, United Kingdom.

Belovsky, G. E., C. Mellison, C. Larson, and P. A. Van Zandt. 1999. Experimental studies of extinction dynamics. *Science* 286:1175–1177.

Boyce, M. S. 1992. Population viability analysis. *Annual Review of Ecology and Systematics* 23:481–506.

Brook, B. W., and J. Kikkawa. 1998. Examining threats faced by island birds: a population viability analysis on the Capricorn silvereye using long-term data. *Journal of Applied Ecology* 35:491–503.

Brook, B. W., L. Lim, R. Harden, and R. Frankham. 1997. Does population viability analysis software predict the behaviour of real populations? A retrospective study on the Lord Howe Island woodhen *Tricholimnas sylvestris* (Sclater). *Biological Conservation* 82:119–128.

Brook, B. W., J. J. O'Grady, A. P. Chapman, M. A. Burgman, H. R. Akçakaya, and R. Frankham. 2000. Predictive accuracy of population viability analysis in conservation biology. *Nature* 404:385–387.

Browne, R. A. 1980. Reproductive pattern and mode in the brine shrimp. *Ecology* 61:466–470.

———. 1983. Divergence of demographic and reproductive variables over 25 years in laboratory and natural populations of the brine shrimp, *Artemia. Crustaceana* 45:164–168.

Burgman, M., D. Cantoni, and P. Vogel. 1992. Shrews in suburbia: an application of Goodman's extinction model. *Biological Conservation* 61:117–123.

Burgman, M. A., D. A. Keith, F. J. Rohlf, and C. R. Todd. 1999. Probabilistic classification rules for setting conservation priorities. *Biological Conservation* 89:227–231.

Caughley, G. 1994. Directions in conservation biology. *Journal of Animal Ecology* 63:215–244.

Caughley, G., and A. Gunn. 1996. *Conservation biology in theory and practice.* Blackwell Scientific, Cambridge, Massachusetts.

Charnov, E. L. 1993. *Life history invariants: some explorations of symmetry in evolutionary ecology.* Oxford University Press, Oxford, United Kingdom.

Crowell, K. L. 1973. Experimental zoogeography: introductions of mice to small islands. *American Naturalist* 107:535–558.

Dennis, B., P. L. Munholland, and J. M. Scott. 1991. Estimation of growth and extinction parameters for endangered species. *Ecological Monographs* 61:115–143.

Diamond, J. M. 1984. "Normal" extinctions of isolated populations. Pages 191–246 in M. H. Nitecki, editor, *Extinctions.* University of Chicago Press, Chicago, Illinois.

Ebenhard, T. 1987. An experimental test of the island colonization survival model: bank vole (*Clethrionomys glareolus*) populations with different demographic parameter values. *Journal of Biogeography* 14:213–223.

Ewens, W. J., P. J. Brockwell, J. M. Gani, and S. I. Resnick. 1987. Minimum viable population size in the presence of catastrophes. Pages 59–68 in M. E. Soulé, editor, *Viable populations for conservation.* Cambridge University Press, Cambridge, United Kingdom.

Feller, W. 1939. Die Grundlagen der Volterraschen Theorie des Kampfes ums dasein in wahrscheinlichkeitstheoretischer Behandlung. *Acta Biotheoretica* 5:11–40.

———, editor. 1966. *An introduction to probability theory and its applications.* J. Wiley and Sons, New York, New York.

Foley, P. 1994. Predicting extinction times from environmental stochasticity and carrying capacity. *Conservation Biology* 8:124–137.

Forney, K. A., and M. E. Gilpin. 1989. Spatial structure and population extinction: a study with *Drosophila* flies. *Conservation Biology* 3:45–51.

Frankham, R. 1995a. Conservation genetics. *Annual Review of Genetics* 29:305–327.

———. 1995b. Inbreeding and extinction: a threshold effect. *Conservation Biology* 9:792–799.

————. 1998. Inbreeding and extinction: island populations. *Conservation Biology* 12:665–675.

Gabriel, W., and R. Bürger. 1992. Survival of small populations under demographic stochasticity. *Theoretical Population Biology* 41:44–71.

Gilpin, M. E. 1993. NEMESIS 1.0: conservation biology simulations. Michael Gilpin, Bozeman, Montana.

Gilpin, M. E., and M. E. Soulé. 1986. Minimum viable populations: processes of species extinction. Pages 19–34 in M. E. Soulé, editor, *Conservation biology: the science of scarcity and diversity*. Sinauer Associates, Sunderland, Massachusetts.

Ginzburg, L. R., S. Ferson, and H. R. Akçakaya. 1990. Reconstructability of density dependence and the conservative assessment of extinction risks. *Conservation Biology* 4:63–70.

Ginzburg, L. R., L. B. Slobodkin, K. Johnson, and A. G. Bindman. 1982. Quasi-extinction probabilities as a measure of impact on population growth. *Risk Analysis* 2:171–181.

Goodman, D. 1987a. Consideration of stochastic demography in the design and management of biological reserves. *Natural Resource Modeling* 1:205–234.

————. 1987b. The demography of chance extinction. Pages 11–34 in M. E. Soulé, editor, *Viable populations for conservation*. Cambridge University Press, Cambridge, United Kingdom.

Hanselman, D., and B. Littlefield. 1997. The MATLAB curriculum series: the student edition of MATLAB: version 5, user's guide. Prentice Hall, Upper Saddle River, New Jersey.

Hanski, I., and M. E. Gilpin. 1991. Metapopulation dynamics: brief history and conceptual domain. *Biological Journal of the Linnean Society* 42:3–16.

Lacy, R. C. 1993. VORTEX: a computer simulation model for population viability analysis. *Wildlife Research* 20:45–65.

Lacy, R. C., K. A. Hughes, and T. J. Kreeger. 1994. VORTEX: user's manual, version 6. Chicago Zoological Society, Chicago, Illinois.

Leigh, E. G., Jr. 1975. Population fluctuations, community stability, and environmental variability. Pages 51–73 in M. L. Cody and J. M. Diamond, editors, *Ecology and evolution of communities*. Belknap Press, Cambridge, Massachusetts.

————. 1981. The average lifetime of a population in a varying environment. *Journal of Theoretical Biology* 90:213–239.

Lindenmayer, D. B., T. W. Clark, R. C. Lacy, and V. C. Thomas. 1993. Population viability analysis as a tool in wildlife conservation policy: with reference to Australia. *Environmental Management* 17:745–758.

Lovejoy, T. E., R. O. Bierregaard, A. B. Rylands, J. R. Malcolm, C. E. Quintela, L. H. Harper, K. S. Brown, A. H. Powell, G. V. N. Powell, H. O. R. Schubart, and M. B. Hays. 1986. Edge and other effects of isolation on Amazon forest fragments. Pages 257–285 in M. E. Soulé, editor, *Conservation biology: the science of scarcity and diversity*. Sinauer Associates, Sunderland, Massachusetts.

MacArthur, R. H., and E. O. Wilson. 1967. *The theory of island biogeography*. Princeton University Press, Princeton, New Jersey.

Mangel, M., and C. Tier. 1993. A simple direct method for finding persistence times of populations and application to conservation problems. *Proceedings of the National Academy of Sciences* (USA) 90:1083–1086.

Mathsoft. 1997. Mathcad7 user's guide. Mathsoft, Cambridge, Massachusetts.

May, R. M. 1975. Biological populations obeying difference equations: stable points, stable cycles, and chaos. *Journal of Theoretical Biology* 51:511–524.

McCarthy, M. A., M. A. Burgman, and S. Ferson. 1995. Sensitivity analysis for models of population viability. *Biological Conservation* 73:93–100.

Mills, L. S., S. G. Hayes, C. Baldwin, M. J. Wisdom, J. Citta, D. J. Mattson, and K. Murphy. 1996. Factors leading to different viability predictions for a grizzly bear data set. *Conservation Biology* 10:863–873.

Murphy, D. D., and B. R. Noon. 1992. Integrating scientific methods with habitat conservation planning: reserve design for northern spotted owls. *Ecological Applications* 2:3–17.

Noss, R. 1999. Is there a special conservation biology? *Ecography* 22:113–122.

Pimm, S. L. 1991. *The balance of nature? ecological issues in the conservation of species and communities.* University of Chicago Press, Chicago, Illinois.

Pimm, S. L., and M. E. Gilpin. 1989. Theoretical issues in conservation biology. Pages 287–305 in J. Roughgarden, R. M. May, and S. A. Levin, editors, *Perspectives in ecological theory.* Princeton University Press, Princeton, New Jersey.

Pimm, S. L., H. L. Jones, and J. M. Diamond. 1988. On the risk of extinction. *American Naturalist* 132:757–785.

Possingham, H. P., D. B. Lindenmayer, and T. W. Norton. 1993. A framework for the improved management of threatened species based on population viability analysis. *Pacific Conservation Biology* 1:39–45.

Raup, D. M. 1991. *Extinction: bad genes or bad luck?* W. W. Norton, New York, New York.

Renshaw, E. 1993. *Cambridge studies in mathematical biology: 11, modelling biological populations in space and time.* Cambridge University Press, Cambridge, United Kingdom.

Richter-Dyn, N., and N. S. Goel. 1972. On the extinction of a colonizing species. *Theoretical Population Biology* 3:406–433.

Shaffer, M. 1987. Minimum viable populations: coping with uncertainty. Pages 69–86 in M. E. Soulé, editor, *Viable populations for conservation.* Cambridge University Press, Cambridge, United Kingdom.

Soulé, M. E. 1987. Introduction. Pages 1–10 in M. E. Soulé, editor, *Viable populations for conservation.* Cambridge University Press, Cambridge, United Kingdom.

Soulé, M. E., and K. A. Kohm, editors. 1989. *Research priorities for conservation biology.* Island Press, Covelo, California.

Stacey, P. B., and M. Taper. 1992. Environmental variation and the persistence of small populations. *Ecological Applications* 2:18–29.

Stephan, T., and C. Wissel. 1994. Stochastic extinction models discrete in time. *Ecological Modelling* 75/76:183–192.

Strebel, D. E. 1985. Environmental fluctuations and extinction: single species. *Theoretical Population Biology* 27:1–26.

Thomas, C. D. 1990. What do real population dynamics tell us about minimum viable population sizes? *Conservation Biology* 4:324–327.

Tracy, C. R., and T. L. George. 1992. On the determinants of extinction. *American Naturalist* 139:102–122.

Wilcove, D. S., M. McMillan, and K. C. Winston. 1993. What exactly is an endangered species? An analysis of the U.S. endangered species list: 1985–1991. *Conservation Biology* 7:87–93.

Wilkinson, L. 1997. SYSTAT 7.0. SPSS, Chicago, Illinois.

Wissel, C., and S.-H. Zaschke. 1994. Stochastic birth and death processes describing minimum viable populations. *Ecological Modelling* 75/76:193–201.

Wolfram, S. 1996. *The Mathematica book*. 3d edition. Wolfram Media, Champaign, Illinois; Cambridge University Press, Cambridge, United Kingdom.

Wright, J. S., and S. P. Hubbell. 1983. Stochastic extinction and reserve size: a focal species approach. *Oikos* 41:466–476.

Yaffee, S. L. 1994. *The wisdom of the spotted owl*. Island Press, Covelo, California.

Evolution of Population Viability Assessments for the Florida Panther: A Multiperspective Approach

David S. Maehr, Robert C. Lacy, E. Darrell Land, Oron L. Bass Jr., and Thomas S. Hoctor

ABSTRACT

We conducted a population viability analysis (PVA) for the Florida panther (*Puma concolor coryi*) and compared the results with two previous PVAs conducted in 1989 and 1992 that suggested complete extinction within 100 years. Despite a lack of full consensus regarding the inputs and results of these modeling efforts, management of this endangered subspecies moved forward first with a plan for captive breeding and more recently with planned genetic introgression. Since 1994, eight female cougars, introduced from Texas, have produced at least 40 hybrid kittens. Panther recovery has been controversial, with genetic restoration efforts questioned by analyses suggesting that demographic stability of the population may obviate the need for such radical management.

We conducted another PVA that demonstrates the value of periodically updating previous analyses. We constructed independent VORTEX models based on demographic inputs provided by each author: a federal field biologist, a state field biologist, a university conservation biologist, a university landscape ecologist, and a nongovernmental-organization population biologist. Our results indicate that the Florida panther has a high (>0.98) probability of persisting for 100 years, compared to PVAs in 1989 and 1992 that predicted complete extinction. However, an apparent capability of Florida panthers for rapid population growth is countered by potential genetic problems that might become severe beyond 100 years. Genetic erosion can be forestalled or halted by allowing for population expansion, but the conservation status of panther habitat in Florida is uncertain. We recommend an approach that emphasizes the expansion of Florida panther range, incorporates controlled genetic introgression, reconsiders rapid population growth in captivity, and embraces the long-term goal of conserving the Florida panther genome with a landscape-based strategy.

We appreciate the helpful suggestions of D. McCullough, S. Beissinger, and P. Beier on earlier drafts of this manuscript.

INTRODUCTION

Wide-ranging vertebrates, especially large carnivores, are ideally suited to serve as conservation flagships and are increasingly the subject of population viability analysis (PVA). However, there is great variation in the implementation of PVAs, whether due to the common dilemma of insufficient data (Minta and Kareiva 1994; Beissinger and Westphal 1998), or due to the inconsistency with which they are applied (Backhouse et al. 1994). At best, PVAs can be used to illustrate the linkage between populations and landscapes, and can drive adaptive management on behalf of a target species (Minta and Kareiva 1994). At worst, PVAs either are not used at all, or are relied upon as fact (Reed et al. 1998) to support agency policies in the presence of extreme uncertainty or political division among managers. Not only can the utility of PVAs be diminished by limited data, but the interpretation of available information can be influenced by the personal perspectives of the individuals conducting them (Reed et al. 1998). Because PVAs can use best guesses for model inputs, and because the collection of accurate demographic data from small populations can be a slow, painstaking process (especially for naturally rare carnivores), PVA results can easily outpace the collection of data needed to properly drive the model (Lacy et al. 1995). Caughley (1994) noted that most PVAs "are essentially games played with guesses."

Shaffer (1990) referred to the Florida panther as an example of an organism whose "survival may hinge on who mates with whom," and likened its status to that of the California condor (*Gymnogyps californianus*)—a species that has recently been repatriated to the wild with captive-raised individuals. Recovery of the panther now depends on success of a program to hybridize the Florida subspecies with individuals of another subspecies from Texas (*P. c. stanleyana*), a decision based in part on the results of a PVA generated from the program VORTEX (Lacy et al. 1995), and concerns over possible inbreeding depression.

Florida panther PVAs began in the late 1980s when little was known about the demographics of the population (Seal and Lacy 1989). The first PVA in 1989 estimated that the Florida panther had a 100% chance of becoming extinct within 100 years. Captive breeding was the recommended course of action, based on speculative data. A subsequent PVA was inconclusive, suggesting complete extinction or persistence over 100 years depending upon the values used in the model (Seal and Lacy 1992). It resulted in recommendations to abandon captive breeding and to introduce genetic material from another wild population (Seal and Lacy 1992).

Now, with two decades of supporting information, we offer a third PVA based on long-term demographic research that minimizes speculation and produces results that are very different from previous efforts. This approach is unique because it allows a retrospective analysis of two previous PVAs based on 20 years of experience and because it resulted in a consensus among participants with diverse backgrounds. We offer this case study to emphasize the dangers of applying management actions based on speculative analyses, and show that PVA can be an instructive process that becomes more realistic over time with better information.

OVERVIEW OF PANTHER STATUS

History of Distribution

The Florida panther has been a state and federally listed endangered subspecies since 1973. It was once distributed throughout the southeastern coastal plain, but now is confined to less than 5% of its original range in a hyper-peninsular distribution in extreme south Florida. Radio-telemetry studies began in 1980, but it was not until the end of that decade that sample sizes exceeded ten radio-collared individuals in any given year. This was due to the remote and hostile nature of the landscape, as well as the inherent rarity and uneven distribution of the panther. More recently, successful capture efforts have been attributed to a population increase beginning in the 1980s. R. McBride (personal communication) has systematically surveyed for panthers in south Florida since the early 1970s and found that current centers of population abundance were virtually uninhabited two decades ago. While there is no evidence of comparable growth in the 1990s, recent dispersal events suggest the potential for future population expansion beyond currently occupied range.

Basic Demographics

The panther currently inhabits an area of about 8,800 km^2 (Maehr 1990). The most recent empirically derived population estimate is 74 individuals (9 resident males, 28 resident females, 9 transient males, and 28 dependent kittens; Maehr et al. 1991b). Panther density follows a northwest to southeast gradient: the highest densities are associated with the better drained, more productive soils and private lands of the upper Big Cypress basin, whereas the lowest densities are associated with the poorer quality, frequently inundated soils associated with the Everglades (Maehr 1997a). This density gradient also follows a pattern of declining forest abundance. The population exists in two rather distinct subunits. A larger, source population encompasses the better drained, more productive portions of the Big Cypress physiographic region in southwestern Florida (Maehr 1997a). A smaller, ephemeral sink population inhabits the Everglades ecosystem in extreme southeastern Florida (Bass and Maehr 1991).

There is no known natural immigration from outside of south Florida. However, there are recent records of panthers north of the Caloosahatchee River (Layne and Wassmer 1988; Maehr et al. 1992). Dispersal away from south Florida is unusual. During the summer of 1998 a subadult male left south Florida by crossing the Caloosahatchee River and continues to inhabit south-central Florida at the time of this writing. Since then, at least two other males have crossed this landscape filter.

Mortality rates, litter sizes, social ecology, and patterns of land tenure are similar to some populations of *Puma concolor* in unhunted regions of western North America (Lindzey et al. 1988; Anderson et al. 1989). The greatest single cause of mortality is intraspecific aggression between males (Maehr et al. 1991a). Mortality of kittens during the period following parturition to 12 months of age has been reported to be less than 20% for both males and females (Maehr and Caddick 1995). Overall, mean annual mortality for all age and sex groups is also less than 20%. The highest mortality rates are for nonresident, dispersing males (Maehr 1998). Females are readily established in the population, and they conceive their first litter at about 18 months of age. Average litter size is two, but ranges from one to four. Successful establishment of breeding males is dependent upon avoiding intraspecific aggression, winning an encounter with an older, decrepit male, and/or demonstrating patience and luck in waiting for the disappearance of a resident adult. Most males do not breed before three years of age. Maehr and Caddick (1995) found that the production of kittens outpaced the death of adults.

Male panthers exhibit a high incidence of deformed spermatozoa (Barrone et al. 1994), and both sexes possess morphological characteristics that have been attributed to inbreeding (Belden 1986; O'Brien et al. 1990). Most Florida panthers exhibit reduced allozyme heterozygosity and a high rate of band sharing for DNA fingerprints, characteristics that have been used to explain testicular abnormalities, an apparent increase in atrial septal defects, and high levels of disease in the population (O'Brien et al. 1990; Roelke and Glass 1992; Hedrick 1995). However, discrepancies between field detections and subsequent necropsies suggest that atrial septal defects are fewer than has been previously suggested (Cunningham et al. 1999). In addition, while it is true that antibodies to a number of diseases have been detected in panther serum samples (Roelke et al. 1993), such sera-positive results do not necessarily arise from loss of genetic diversity. On the contrary, the formation of titers to viruses and other infectious diseases is more likely explained by a functioning immune system where an encounter with a potentially lethal disease is dealt with successfully and permanently. Although environmental contaminants may have contributed to demographic instability in the small Everglades segment of the panther population (Bass and Maehr 1991), infectious diseases and contamination play an insignificant role in total mortality (Maehr et al. 1991a).

It is widely accepted that habitat loss is the most significant short- and long-term threat to the Florida panther. For example, the recent southward retreat of the state's freeze line has placed new pressures on southwestern Florida's wildlands to produce citrus. This, in addition to other intensifying agricultural practices and urbanization, will undoubtedly reduce the ability of the region to support panthers (Maehr 1992). Subsequently, Pearlstine et al. (1995) suggested that an increase in citrus development in southwest Florida could reduce "potential panther habitat" by 25 to 60%.

Such concerns led to the development of an interagency planning document to stem panther habitat loss (Logan et al. 1993). Curiously, while such reports are remarkably detailed and accurate in their portrayal of Florida's rapidly increasing human population and the spread of agriculture, no analyses have compellingly demonstrated the influence of habitat loss on Florida panther population size. In addition, there is no evidence, despite intense study, to suggest that panther numbers have declined during two decades of increasing human population. Indeed, the popular, quasi-official 1980s estimate of "30 to 50" has been replaced by the 1990s estimate of "30 to 50 *adults.*" This counterintuitive result may be caused by several factors. First, panther numbers might actually be declining, but intensive monitoring and capture activities maintain a static study sample that may be an increasingly large percentage of the total population. In the absence of density-dependent adjustments this scenario seems unlikely, however, because there have not been concurrent declines in other demographic variables that would suggest a shrinking population. Second, because panthers reproduce relatively slowly and because there is little turnover in the adult segment of the population, there may be a lag between habitat loss and a measurable impact on the population. Furthermore, there may be sufficient prey resources in the landscape and adequate social flexibility in the subspecies to allow a temporary increased concentration of individuals in the face of contracting range. The impacts of loss of habitat would also be delayed in this scenario. This explanation also seems unlikely because the basic shapes and sizes of home ranges have not changed since the inception of telemetry studies in 1980 (Maehr 1997a).

Perhaps the most likely scenario for why panther numbers have not declined over the last 20 years is because key portions of productive, occupied range have not been lost. The more serious potential problem is that loss of habitat to development is occurring in areas where panthers live but are not documented. Alternatively, lost habitat may be potential range that could be colonized, or that could serve as an important landscape linkage to potential habitat. At best, the loss of peripheral and dispersal habitat could limit the ability of the population to increase.

Despite the lack of specific data to demonstrate a negative impact of habitat loss on panthers, it must be presumed that current patterns of expanding human presence in south Florida are not benefiting the population. For the input requirements of VORTEX, then, determining a trend in carrying capacity becomes a matter of judgment. No trend has, as yet, been empirically determined. This is one example where best guesses enter into the modeling process.

Recent Panther Management

Although a number of authors have addressed the need for panther management to occur at the landscape level (Maehr 1990; Schortemeyer et al. 1991; Harris and Cropper 1992; Cox et al. 1994; Maehr and Cox 1995; Harris et al. 1996; Maehr 1997a,b), most recovery actions have focused on the individual or population level. The two most controversial recovery-based actions have been the initiation of a captive-breeding program in 1991 and purposeful genetic introgression in 1994. Both were based, in part, on the inputs and results of earlier modeling efforts (Seal and Lacy 1989, 1992). Captive breeding was abandoned in 1994 after ten kittens were removed from south Florida during 1991 and 1992. Since 1994, eight female cougars introduced from west Texas have produced at least 40 hybrid kittens. The intent of these introductions is for 20% of the Florida gene pool to be derived from Texas cougars (Seal 1994). Although breeding between Texas cougar females and Florida panther males has been successful, no plans had been made as of early 2000 to remove the Texas animals from the population. This is a potentially important issue inasmuch as the original recommendations called for a one-time infusion of genetic material from the original eight females. Some of these females have been in south Florida now for more than four years, and have not only produced their second F_1 litters but have produced 75% pure Texas offspring as the result of back-crossing.

METHODS

The PVA exercise we present here takes advantage of this history and as such represents a mature synthesis not possible for most PVAs. Most PVA efforts are too recent to allow such a retrospective analysis of their strengths and weaknesses. We used VORTEX version 7 (Lacy et al. 1995) to model Florida panther population viability independently from the different perspectives of the five authors: a federal field biologist, a state field biologist, a university conservation biologist, a university landscape ecologist, and a nongovernmental-organization population biologist. "The VORTEX program is a Monte Carlo simulation of the effects of deterministic forces as well as demographic, environmental, and genetic stochastic events on wildlife populations" (Lacy et al. 1995, 4). Key inputs of the model are listed in table 14.1.

Each contributor had equal access to published, peer-reviewed literature on panther biology as well as to recent unpublished agency progress reports from field studies of the south Florida panther population. Relevant references included those on genetics (O'Brien et al. 1990; Roelke et al. 1993), general ecology (Belden et al. 1988; Maehr et al. 1989a,b, 1990a,b, 1991b; Maehr 1990, 1997a; Maehr and Cox 1995; Dalrymple and Bass 1996), demographics (Bass and Maehr 1991; Maehr et al. 1991a; Barrone et al. 1994; Maehr and Caddick 1995; Maehr 1997a), and disease (Forrester et al. 1985; Greiner et al. 1989; Forrester 1992; Roelke et al. 1993; Glass et al. 1994; Maehr et al. 1995). Each author independently provided inputs to drive the VORTEX model on a standardized form. Collaboration among authors was prohibited during this initial step of input preparation. Each model was then run on the same computer by the senior author.

Author inputs were used to develop persistence probabilities for each of the five models (table 14.1). Each model ran 500 iterations of simulations for 100 years. Authors were then given the results of their simulation before a series of E-mail and fax transmissions were used to develop a consensus model for the group. Each author was then able to view the inputs of the other authors and reconsider the variables entered into his original simulation. Where discrepancies were more than slight, each author was asked to justify the variable in question. When a single view did not prevail, compromise was sought by averaging the five versions of the contentious variable. Results of the consensus model were then compared (table 14.2) to the original VORTEX-based PVA that was developed in 1989 (Seal and Lacy 1989), and to a subsequent analysis in 1992 (Seal and Lacy 1992).

Six other simulations based upon the consensus model were also run. Scenario 1 excluded population supplementation. Scenario 2 retained

Table 14.1 Comparison of VORTEX Model Inputs Provided Independently by the Five Authors and the Outputs Generated from These Simulations

Model Inputs and Output	Population Ecologist (Lacy)	State Field Biologist (Land)	Federal Field Biologist (Bass)	University Landscape Ecologist (Hoctor)	University Conservation Biologist (Maehr)
Inputs					
Inbreeding depression?	Yes	No	No	No	No
Lethal equivalents	3.14	—	—	—	—
% due to recessive lethals	50	—	—	—	—
Reproduction correlated with survival?	Yes	No	No	No	No
Polygynous mating system?	Yes	Yes	Yes	Yes	Yes
Age 1st female reproduction	2	1	3	2	2
Age 1st male reproduction	4	3	2	3	3
Maximum individual age	12	12	12	9	12
Reproduction density dependent?	No	No	No	No	No
Sex ratio at birth	50:50	50:50	50:50	50:50	50:50
Maximum litter size	4	4	2	3	4
% females with litter/year	50	50	50	60	50
SD of above	20	5	10	10	5
% litter of size 1	32.5	17.5	50	20.0	10.0
% litter of size 2	40.0	50.0	50	50.0	50.0
% litter of size 3	20.0	30.0	—	30.0	30.0
% litter of size 4	7.5	2.5	—	0	10.0
Female mortality in year 1	26.5	20	0	20	20
SD in female mortality, year 1	6.625	2.0	4	10.0	5.0
Female mortality in year 2	10.1	—	0	10	20
SD in female mortality, year 2	2.5	—	4	5.0	5.0
Female mortality in year 3	—	—	0	—	—
SD in female mortality, year 3	—	—	4	—	—
Female mortality in adults	10.1	17	25	20	20
SD in female mortality, adults	2.5	3.0	4	5.0	10.0
Male mortality in year 1	26.5	20	0	20	20
SD in male mortality, year 1	6.625	10.0	6	10.0	5.0
Male mortality in year 2	21.7	15	0	20	50
SD in male mortality, year 2	5.425	3.0	6	5.0	5.0
Male mortality in year 3	21.7	15	—	20	60
SD in male mortality, year 3	5.425	3.0	—	5.0	5.0

(continued)

Table 14.1 (*continued*)

| Model Inputs and Output | Originator of Variable Estimates for the VORTEX Simulation | | | | |
	Population Ecologist (Lacy)	State Field Biologist (Land)	Federal Field Biologist (Bass)	University Landscape Ecologist (Hoctor)	University Conservation Biologist (Maehr)
Male mortality in adults	21.7	20	66	20	20
SD in male mortality, adults	5.425	3.0	6	5.0	10
Number of catastrophe types	0	0	0	2	1
Probability for catastrophe 1	—	—	—	0.05	0.02
Probability for catastrophe 2	—	—	—	0.01	—
Reproduction rate for catastrophe 1[a]	—	—	—	0.80	.98
Reproduction rate for catastrophe 2[a]	—	—	—	0.50	—
Survival for catastrophe 1[a]	—	—	—	0.80	0.95
Survival for catastrophe 2[a]	—	—	—	0.50	—
% of adult males breeding	100	50	100	50	40
Starting population size	50	50	6	60	70
Habitat carrying capacity	50	60	8	70	85
SD of above	0	5	2	10	5
Change in habitat	Lost	0	0	Lost	0
# of years of habitat loss	25	0	0	20	0
% habitat change per year	−1.0	0	0	−1.5	0
Will panthers be removed?	No	No	No	Yes	No
At what annual interval?	—	—	—	1	—
For how many years?	—	—	—	10	—
# males removed/year	—	—	—	1	—
# females removed/year	—	—	—	1	—
Population augmentation?	Yes	Yes	Yes	No	No
If yes, at what interval?	20 years	10 years	10 years	—	—
For how many years?	100	100	100	—	—
# males added per event	0	0	1	—	—
# females added per event	6	1	2	—	—
Outputs					
Expected heterozygosity	0.682	0.597	0.659	0.537	0.635
Number of extant alleles	6.38	4.58	3.89	3.58	4.68
Probability of persistence to 100 years over 500 iterations	0.998	1.00	0.0689	.998	1.00
Mean final population	34.19	59.41	5.52	50.24	83.29
Median time to extinction	—	—	7.13 years	—	—

Note: SD = standard deviation.

[a]These values represent multipliers that reduce survival and reproduction due to catastrophe.

Table 14.2 Comparison of Variables Used in the PVA Models

Model Inputs and Outputs	1989 Panther PVA	1992 Panther PVA Consensus	1992 Panther PVA Optimistic	1999 Consensus Simulation
Inputs				
Inbreeding depression?	Yes	Yes	No	Yes
Lethal equivalents	3.4	3.0	0	3.14
% due to recessive lethals	0	0	0	50
Reproduction correlated with survival?	Yes	Yes	No	No
Polygynous mating system?	Yes	Yes	Yes	Yes
Age 1st female reproduction	3	2	2	2
Age 1st male reproduction	3	2	2	4
Maximum individual age	15	12	12	12
Reproduction density dependent?	No	No	No	No
Sex ratio at birth	50:50	50:50	50:50	50:50
Maximum litter size	5	3	3	4
% females with litter/year	50	50	50	50
SD of above	1	0	0	10
% litter of size 1	10	25	25	17.5
% litter of size 2	20	50	50	50.0
% litter of size 3	40	25	25	30.0
% litter of size 4	20	—	—	2.5
% litter of size 5	10	—	—	—
Female mortality in year 1	50	50	20	20
SD in female mortality, year 1	5	0	0	6
Female mortality in year 2	30	20	20	20
SD in female mortality, year 2	3	0	0	3
Female mortality in year 3	25	—	—	—
SD in female mortality, year 3	3	—	—	—
Female mortality in adults	25	20	20	17
SD in female mortality, adults	3	0	0	3
Male mortality in year 1	50	50	50	20
SD in male mortality, year 1	5	0	0	6
Male mortality in year 2	30	20	20	30
SD in male mortality, year 2	3	0	0	5
Male mortality in year 3	25	—	—	30
SD in male mortality, year 3	3	—	—	5
Male mortality in adults	25	20	20	15
SD in male mortality, adults	3	0	0	5
Number of catastrophes	2	0	0	1
Probability for catastrophe 1	0.01	—	—	0.5
Probability for catastrophe 2	0.02	—	—	—
Reproduction rate for catastrophe 1	—	—	—	0.95
Reproduction rate for catastrophe 2	—	—	—	—
Survival for catastrophe 1	—	—	—	0.95
Survival for catastrophe 2	—	—	—	—
% of adult males breeding	100	50	50	50
Starting population size	45	50	50	60
Habitat carrying capacity	45	50	50	70
SD of above	1	0	0	5

(continued)

Table 14.2 (*continued*)

Model Inputs and Outputs	1989 Panther PVA	1992 Panther PVA Consensus	1992 Panther PVA Optimistic	1999 Consensus Simulation
Change in habitat	Lost	Lost	No	No
# of years of habitat loss	50	25	—	—
% habitat change per year	1.0	1.0	—	—
Will panthers be removed?	Yes	No	No	No
At what annual interval?	1	—	—	—
For how many years?	3	—	—	—
# males removed/year	1	—	—	—
# females removed/year	1	—	—	—
Population augmentation?	No	No	No	Yes
If yes, at what interval?	—	—	—	10 years
For how many years?	—	—	—	100
# males added per event	—	—	—	0
# females added per event	—	—	—	2
Outputs				
Expected heterozygosity	0	0	0.594	0.759
Number of extant alleles	0	0	4.02	8.91
Probability of persisting 100 years over 500 iterations	0	0	0.998	1.00
Mean final population	0	0	47.44	65.72
Median time to extinction	23.13	43.71	—	—

Note: SD = standard deviation.

supplementation and included a 25% loss in habitat. Scenario 3 combined the changes in scenarios 1 and 2 (no supplementation; 25% habitat loss), and scenario 4 combined no supplementation, 25% habitat loss, and the removal of six females over a three-year period (table 14.3). A fifth set of simulations used the consensus model to examine the effect of a growing population on heterozygosity (table 14.4). In this scenario, carrying capacity was increased gradually from 70 to 500. The sixth scenario allowed the population to grow beyond the constraints of south Florida and simulated a population that had the capacity to colonize vacant range. This was done by increasing K to 300 and leaving the other variables constant. The final scenario divided existing and potential range into a metapopulation of three subpopulations: the Everglades sink, the Big Cypress Swamp source, and the region north of the Caloosahatchee River. Dispersers were two- to four-year-old panthers of either sex with a 1% likelihood of leaving either extant population.

Table 14.3 Consensus Model Simulation Compared to Four Variations in Management Trajectory over 100 Years

Model Inputs and Outputs	Consensus Simulation	No Cats Added	25% Habitat Loss	25% Loss, No Cats Added	25% Loss, No Cats Added, Cats Removed
Inputs					
Change in habitat	No	No	Yes	Yes	Yes
# of years of habitat loss	—	—	25	25	25
% habitat change per year	—	—	1	1	1
Will panthers be removed?	No	No	No	No	Yes
At what annual interval?	—	—	—	—	1
For how many years?	—	—	—	—	3
# males removed/year	—	—	—	—	0
# females removed/year	—	—	—	—	2
Population augmentation?	Yes	No	Yes	Noy	No
If yes, at what interval?	10 years	—	10 years	—	—
For how many years?	100	—	100	—	—
# males added per event	0	—	0	—	—
# females added per event	2	—	2	—	—
Outputs					
Expected heterozygosity	0.759	0.690	0.719	0.672	0.603
Number of extant alleles	8.91	5.54	7.65	5.19	4.25
Probability of persisting 100 years over 500 iterations	1.00	1.00	1.00	0.998	0.992
Median final population	65.58	64.16	46.72	45.21	44.70

Notes: Only those variables that diverge from the consensus model are displayed. All others are identical to column 4 of table 14.2.

Table 14.4 Effects of Increasing Carrying Capacity on Genetic Heterozygosity after 100 Years, Using the Consensus VORTEX Simulation

Carrying Capacity	Predicted Heterozygosity (%)[a]
70	72.2
100	80.6
150	84.1
200	86.5
250	87.5
300	89.6
400	90.7
500	92.4

[a] As percentage of initial value of *H*.

RESULTS AND DISCUSSION

Independent Simulations

Four out of the five independent models resulted in projections of the south Florida panther population that maintained a >99% probability of persisting for 100 years (table 14.1). Mean ending population size ranged from 34 to 83 individuals. The only model that had a very low probability of persistence (<7%) reflected only the demographics of the Everglades subpopulation. The Everglades represents a large proportion (>50%) of occupied panther range, but it supports only a small portion (<10%) of the total population in south Florida. Compared to the original PVA conducted in 1989 (table 14.2), only the Everglades model (federal wildlife biologist) and the 1992 PVA consensus model (table 14.2) follow similar trajectories to extinction in less than 50 years (figs. 14.1–14.3).

Inasmuch as the ultimate results were similar, the predicted extinctions appear to be driven by different input variables. Because the Everglades inputs reflect a very small initial population as well as a low carrying capacity, environmental variation and stochastic events are likely

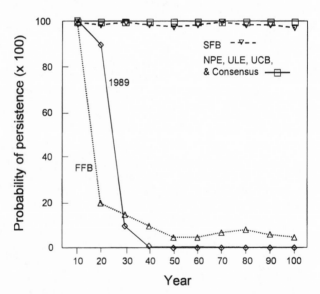

Fig. 14.1 Comparison of persistence probabilities for seven PVAs for the Florida panther: *1989*, the first PVA performed on the subspecies; *FFB*, inputs provided by a federal field biologist; *SFB*, inputs provided by a state field biologist; *NPE*, inputs provided by a non-governmental-organization population ecologist; *ULE*, inputs provided by a university landscape ecologist; *UCB*, inputs provided by a university conservation biologist; and *Consensus*, inputs resulting from the agreement among the authors.

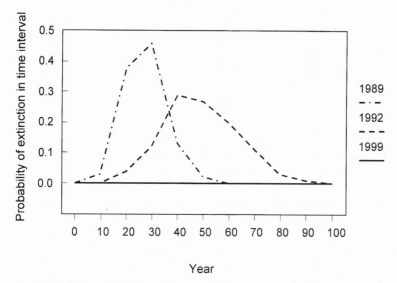

Fig. 14.2 Probabilities of Florida panther extinction for PVAs conducted in 1989 and 1992, and the consensus simulation from 1999.

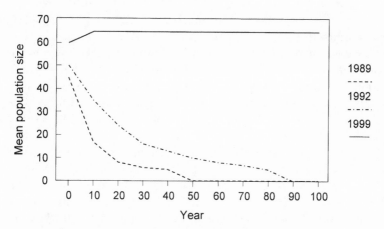

Fig. 14.3 Mean extant population sizes per ten-year intervals for Florida panther PVAs conducted in 1989 and 1992, and the consensus simulation from 1999.

powerful influences on population performance. Without the possibility of immigration from a nearby population (which is behaviorally and demographically possible and was excluded in this particular model), there is no opportunity for recolonization after an initial extinction. The other four independent models considered the panther as a single, south Florida–wide population. In the 1989 PVA, estimates based on small sample sizes and speculation suggested mortality rates of 50% for kittens and 25 to 30% for older age classes. Although litter sizes were relatively large in the 1989 PVA, age at first reproduction in females was three years, and habitat loss was predicted to be 1% a year for 50 years (table 14.2). In addition, catastrophes, which were predicted to occur infrequently, were modeled to have a greater impact on the population than in subsequent analyses. The 1992 consensus model had slightly lower mortality rates in older panthers, but did not allow population supplementation and precluded the possibility of litter size greater than three. Data collected during the late 1980s and early 1990s was not used in the 1992 consensus PVA because it was viewed as overly optimistic.

Agreement among Simulation Inputs

In the present study all independent PVAs were very similar in their basic demographic composition with the exception of the federal field biologist model (table 14.1). Only the population biologist incorporated inbreeding depression, predicted a 25% loss in habitat over 25 years, and suggested that litter sizes were somewhat skewed toward one. These factors were insufficient to drive the model to extinction, although population size declined to a mean of 34 (table 14.1). General agreement among models was reflected in similar biological inputs, such as age at first reproduction, age of senescence, sex ratio at birth, lack of density dependence in reproduction, and proportion of adults breeding. The probability of catastrophe was low or nonexistent in all independent simulations. All authors agreed that the panther has exhibited a virtual immunity against natural disasters such as hurricanes and periodic drought and flood, and unnatural disturbances such as highways and accidental mortalities at captures.

Disagreement about variable inputs was more prevalent than agreement (table 14.1). Four of us initially chose to exclude inbreeding depression from the model because there was no demographic proof of its occurrence. Distribution of litter-size probabilities and first-year mortality were highly variable, and resulted from the individual interpretation of existing data. Starting population sizes, habitat carrying capacity, and trend in habitat availability were more subjective inputs that reflected personal perspectives of each author. Variation in most of the inputs, either demographic- or habitat-related, were within reasonable limits, given the relatively small population of panthers in south Florida and the uncertainty in some variables. For example, the high mortality estimates provided by the university conservation biologist for males aged two to three years stemmed from the disappearance of uncollared animals in an age and sex cohort that experienced the highest mortality in the collared population (Maehr et al. 1991a). The greatest uncertainty was associated with trends in available habitat. None of the initial inputs agreed on carrying capacity, in part because current population estimates are based on extrapolation (Maehr et al. 1991b) and because each of us had different visions for the trajectory of future south Florida development.

Development of the Consensus Simulation

Reaching consensus among the five authors was relatively straightforward. For variables that have received research attention since earlier publications, we relied heavily on the input from active field biologists (Land and Bass) for the most current information. Despite the discrepancy between initial inputs (table 14.1), the state and federal biologists were able to resolve opposing views of a south Florida–wide panther population versus the Everglades. Thus, model inputs based on observed demography in the Everglades were not used in subsequent iterations of the model.

Demographic Parameters

Compromises on the distributions of litter sizes, survival probabilities, and age at first reproduction were quickly achieved. Starting population size was the arithmetic mean of the three different values (two of which were identical) offered for the entire south Florida population. A carrying capacity higher than previous PVAs was based on a 20-year pattern of gradual population expansion noted by R. McBride. Under this scenario it was difficult to argue that there was a decreasing trend in the area of habitat used by panthers, despite the intuitive reasoning that some forest loss must be occurring in south Florida. The consensus that population augmentation will occur and that panther removals will not be resumed reflects the facts that augmentation has occurred and that captive breeding has not.

Genetic Considerations

Inbreeding depression was the most difficult variable to reconcile. We agreed that, despite the apparent lack of demographic declines in the panther, genetic erosion is predicted by theory, and prudence dictated that we accept this potential as a default position. Without immigration or rapid population growth, some form of depression is theoretically predicted, and the population may be otherwise doomed to extinction (Diamond 1978; Allendorf and Ryman, chap. 4 in this volume; Lande, chap. 2 in this volume). Several of us were uncomfortable with this compromise inasmuch as genetic introgression has the potential to alter the Florida panther genome sufficiently to alter important local adaptations

(Maehr and Caddick 1995). Furthermore, because the population has always existed at a distributional extreme on a large peninsula that has effectively become an island, deleterious alleles may have been purged if the decline was not too rapid. Although we recognized that demographic stability does not ensure genetic stability, most of us were reluctant to model inbreeding as an important component of simulations for the reasons outlined by Caughley (1994).

Despite the low rates of extinction, all of the simulations presented in tables 14.1 through 14.3, including those that did not incorporate inbreeding depression, projected considerable loss of genetic variability over the next 100 years. The loss of heterozygosity ranged from 24% in the consensus model to as high as 46% in some of the projections using input values provided independently by some authors. These losses of heterozygosity (and concomitant accumulated inbreeding) could cause a reduction in juvenile survival of 31% to 51%, if the response of Florida panthers to inbreeding is comparable to the median effect observed in a survey of 40 mammalian populations (Ralls et al. 1988). Furthermore, reductions in adult survival and fecundity resulting from inbreeding depression are often of similar magnitude as reductions in juvenile survival (Falconer 1989), so theoretically, the much smaller panther population should be more susceptible to inbreeding effects. Although Florida panthers exhibit morphological and physiological abnormalities (e.g., deformed tail vertebrae and cryptorchidism) that may result from past inbreeding, there is no compelling evidence that links inbreeding with demographic performance. However, more than a decade's worth of intense monitoring has not verified that either kittens or adults suffer increased mortality. Like the cheetah, the Florida panther exhibits genetic impoverishment without clear impacts on demographics (Caughley 1994).

However, there may be long-term consequences of inbreeding. In some species, inbreeding depression does not appear until inbreeding has accumulated to higher levels, after which rapid fitness declines can lead to extinction (Frankham 1995; Lacy and Ballou 1998). Although the Florida panther population is projected to remain demographically viable through 100 years, the consensus population model declined over longer time periods, presumably because of a decline in survival due to inbreeding. If allowed to run for 500 years, the consensus model declines steadily with an increasing probability of extinction beyond 100 years (fig. 14.4). Even if the Florida panther is one of the few mammalian populations that is relatively unaffected by inbreeding, lack of genetic diversity would reduce adaptability in the taxon (Lacy 1997),

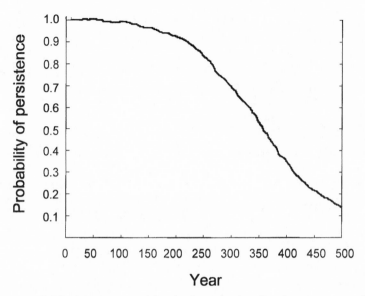

Fig. 14.4 Probability of persistence of the Florida panther based on the consensus simulation run for 500 years.

and it could be questioned whether conservation objectives would be achieved if the population retained only a small portion of the gene diversity that was present when recovery efforts began. Of course, this presumes that environmental conditions will not change, or that the population will not be challenged with disease or other stresses.

At workshops held in 1991 and 1992, it was recommended that some pumas from Texas be released in south Florida to restore genetic variability to reverse existing inbreeding (Seal and Lacy 1992). The subsequent releases of female Texas cougars in Florida will partly accomplish this. However, even in the simulations that included continued releases of unrelated females (two per decade, or about one per generation), the population lost 24% to 28% of its heterozygosity over 100 years. These models suggest that the population in south Florida is too small to avoid considerable losses of genetic variation unless there is a steady inflow of unrelated animals. If the population was allowed to grow dramatically, these individuals would, of necessity, come from a different subspecies.

To prevent significant future losses of the remaining genetic diversity in the Florida panther, the population would have to be expanded severalfold. Franklin's rule of thumb (1980) for keeping inbreeding below 1% per generation would require an effective population size of at least 50 and a doubling of panther carrying capacity. Using our consensus model, we gradually increased carrying capacity (K) in order to judge the subsequent affect of a growing population on heterozygosity (H) over 100 years. Increasing K from 70 to 500 resulted in an increase in H from 72.2% to 92.4%, respectively (table 14.4). If the genetic goal of panther recovery were to retain 90% of the population's initial heterozygosity (Soulé et al. 1986), then the population would need to be increased to more than 300.

Rapid recovery from a population bottleneck may be the most effective form of genetic management. Fortunately, demographic modeling suggests that the panther population could rapidly expand into additional habitat should it be made available. This focuses attention on the need for landscape management as part of Florida panther recovery, but also raises the question of whether captive breeding should be revisited as a population-expanding mechanism. This approach was abandoned, presumably, under the perception that future genetic problems could be circumvented with a large and rapid influx of genetic material from another subspecies. Our analyses suggest that genetic management alone may not be sufficient, because the introduction of Texas cougars will not increase the total population size unless there is enough habitat for such growth to occur. Thus, along with a needed increase in genetic variability, the Florida panther also requires a boost in carrying capacity.

Additional Variations on the Consensus Model

Whereas the consensus model indicates that genetics should play an increasingly important role in future management, we also recognize that the panther in south Florida has the capability for population growth. Relatively low mortality (Maehr et al. 1991a), low adult turnover (Maehr 1997b), high kitten survival (Maehr and Caddick 1995), recent records of the panther north of the population core (Layne and Wassmer 1988; Maehr et al. 1992), and recent dispersal of south Florida panthers to south-central Florida suggest that population expansion and colonization may occur even without human intervention. This led us to simulate the consensus model without the constraints of a finite range. To do this, we increased the carrying-capacity limit to 300 and kept all other model inputs constant. This resulted in the expansion of an initial population of 60 to 300 within 20 years. While this scenario may be unlikely under today's landownership and management patterns, the facilitation of panther movement between south Florida and south-central Florida, and the adoption of a statewide wildlife habitat conservation plan (Noss 1987; Harris and Atkins 1991; Cox et al. 1994; Carr et al. 1998; Hoctor et al. 2000) could lead to population growth that would enhance the long-term survival of the subspecies.

Alternate Futures

We examined five possible future scenarios by altering three management and habitat variables either singly or in combination (table 14.3). The single effects of no population supplementation or 25% habitat loss over 25 years had no effect on the 100-year persistence probability of the simulated population. A combination of both 25% habitat loss and no supplementation reduced the probability of persistence for 100 years from 1.00 to 0.998. Adding the removal of panthers from the population to both habitat loss and no supplementation reduced the survival probability to 0.992. Without the influence of rapid genetic deterioration in the model, simulated populations were virtually assured of persisting for at least 100 years. Although managed genetic introgression has already occurred, even the exclusion of two supplemental females per decade from another population had little impact on the persistence probabilities predicted by the consensus model. We believe that the very minor impacts of these permutations on the consensus model are due to the robust demographics that have been exhibited by the panther over the last 15 years.

Overview of Model Results

The development of our consensus model followed a progression of simulations that were overly pessimistic with regard to Florida panther demographics (Seal and Lacy 1989, 1992). The early age of first reproduction in females, resistance to natural and artificial perturbations, and low mortality in juvenile and adult age classes were population features that required a decade for research to reveal. As recently as 1992 these parameters were considered too optimistic, and another decade transpired before panther PVA was revisited. In the short term, habitat loss had little impact on survival prospects of the Florida panther. In the long term, genetics were revealed as the most important factor leading to population decline. Simulations that projected a future with significantly higher carrying capacity resulted in larger final populations and adequate genetic variability.

The iteration of panther PVA presented here has suggested that conservation of the subspecies need not be driven by a crisis mentality with respect to short-term survival. Rather, the pursuit of efforts that facilitate real or artificial population growth should be the primary focus of agencies responsible for panther recovery. Thus, the immediate threats to the panther's future are those that reduce habitat availability or sever the connections between occupied habitat and habitat that can be colonized. If any crisis exists, it is due to the reluctance of agencies to grapple with the hurdles of not only securing existing panther habitat but *increasing* its distribution. The panther population has time; its habitat does not.

Future Needs

We agree with Boyce (1992) and Reed et al. (1998) that most PVAs are useful as starting points for management, but are not in and of themselves sufficient. Although our consensus simulation predicts a high probability of survival for the panther, this future is contingent upon the continued availability of habitat. Despite the relative ease with which ecological change can be determined (Harcourt 1995), habitat trend and its influence on panther population increase are the most poorly understood aspects of Florida panther conservation. There are many satellite and high-altitude images available for Florida. Yet analyses related to panther habitat have been only snapshots in time. Analyses of panther habitat change over time have yet to be done. To more accurately predict the ability of the south Florida landscape to support a panther population in the future, comparisons between aerial images over time would allow managers to measure the rate of forest retreat or expansion. Geographic information system (GIS) technology could also create an image layer of permitted construction/agricultural activities that would allow the calculation of future changes. This would enhance the accuracy of predictions resulting from future Florida panther viability analyses, and could quantitatively link ongoing panther management activities with efforts to plan a statewide ecological reserve network (Carr et al. 1998; Hoctor et al. 2000). The panther, as a flagship species, could help drive the ongoing efforts to conserve and restore native landscapes in Florida and the southeastern United States.

Our simulations suggest that, even with genetic supplementation, under current conditions the south Florida panther population will have persistence problems in the long term. On the other hand, raising carrying capacity results in rapid growth in modeled populations. Without the initiation of extensive forest restoration, panther habitat availability in south Florida will, at best, remain static for the next 100 years. Thus, in situ population expansion will have to occur north of the Caloosahatchee River. If the landscape can be managed to facilitate panther movement across this landscape filter, or if panthers can be moved to potential habitat, the population will be able to increase naturally.

A Metapopulation Approach

Our final simulations examined the creation of a metapopulation by allowing the existing population to colonize vacant range. To do this, we divided existing range into the Everglades sink ($N = 10$ panthers) and the Big Cypress source ($N = 50$ panthers), and added a new region north of the Caloosahatchee River. We believe that the recent dispersal of male panthers across the river and their use of potentially high-quality habitat (Maehr et al. 1992) suggest that this area could be colonized through dispersal from the south. With a 1% probability that an animal will disperse from either the Everglades or the Big Cypress in any direction in a given year, the Big Cypress core remained at 50, the Everglades dropped to 5, and the new population north of the Caloosahatchee River grew to 35 in 100 years. Increasing the probability of dispersal to 6% increased the total population to 100, with additional growth ($N = 44$ panthers) in the colonized area. Probability of population persistence ranged from 72 to 97% for the new population, 98 to 100% for the Big Cypress core population, and 35 to 70% for the Everglades population. Persistence of simulated Everglades populations appears dependent upon frequent rescues by the Big Cypress core population. It is unknown what panther population the landscape north of the Caloosahatchee River can support. However, analyses by Cox et al. (1994) and Maehr and Cox (1995) clearly demonstrate the potential of this landscape to support panthers.

LESSONS FROM THE PANTHER CASE STUDY

Our collaboration emphasized the advantages of multiple perspectives in developing a credible PVA. This third iteration of a VORTEX-based panther PVA also demonstrated the need to periodically revisit previous analyses, reevaluate available data, and renew simulations in light of new information, compelling trends, and new management direction. It also demonstrated that no matter how well-intended the participants of a PVA may be, they can make mistakes during the process. The results of our analyses, with or without supplementation with Texas cougars, contrasts sharply with the results of the 1989 and 1992 PVAs. The 1989 panther PVA and the version presented here have few similarities despite four of the authors being participants in both simulations.

Differences in the results of the three PVAs conducted over a decade were primarily due to the use of fewer guesses in the model inputs. Importantly, we found that kittens were surviving at a higher rate, females began breeding 1.5 years earlier than other cougar populations, and adult survival was higher. Although one of the 1992 model runs offered a very optimistic view of panther demographics, it was discarded because the consensus of experts at the time indicated that there was no way that the population could be performing so well. The performance of the Florida panther population over the last decade suggests that the earlier optimistic run was probably correct.

The PVA model presented in this paper represents an evolution in both demographic information and conservation thinking as applied to the panther. The earliest applications of VORTEX to panther persistence occurred when there was great uncertainty about panther ecology and when ex situ conservation (captive breeding) was a more popular small-population management tool. Today, panther demographics and genetics are better understood, and the management focus is on in situ conservation (reserve creation and landscape management). The PVA presented here incorporated increased initial population size, increased carrying capacity, and reduced mortality. Whether these changes are the result of improving demographic conditions in the population or they reflect an improvement in the accuracy of field data is unknown.

Our PVA model suggests that, despite a demographically secure population in the short run, genetic considerations may become more important over time. Indeed, all of the initial independent models that considered the entire south Florida population predicted a very high probability of survival for 100 years. However, genetic complications frequently drove the consensus model to extinction if it was allowed to run for 500 years. Despite the uncertainty that pertains to short-term PVAs, this exercise helped create a consensus that genetic factors must be considered in the long-term management of the subspecies.

What sets this PVA apart from earlier panther PVAs is the availability of more current data and our consensus approach. The value of a consensus goal for PVAs should not be underestimated. While previous panther PVAs were done in a group setting, neither the inputs nor the results were unanimously agreed to by participants. The five authors of this paper have a wide range of perspectives on the Florida panther and on small-population management, but this was not an impediment to developing the consensus product. This diversity of expertise was likely a key factor leading to a better analysis than could have been done by a more narrowly focused group. Perhaps the clear ground rules for participation and compromise gave all of us a sense of ownership, involvement, and bipartisanship that is not always possible with PVAs where politics can influence the outcomes.

Both previous PVAs resulted in radical, single-pronged management actions. It is clear, however, that no single approach will assure the future of the Florida panther and the Florida panther genome. We hope the PVA presented here will compel managers to consider multiple solutions to the Florida panther's small-population problem. Genetic restoration, which dominates recovery efforts today, will be successful only if it is part of a larger plan that considers demographic and landscape concerns. Ultimately, recovery should seek to expand from the quick-fix solution of genetic introgression and embrace the long-term goal of conserving the Florida panther genome in an expanding landscape.

The subspecies' recovery plan calls for the reestablishment of at least three viable populations in the panther's historic range (U.S. Fish and Wildlife Service 1987). Thus, we recommend that future recovery efforts combine controlled genetic introgression with aggressive landscape management that provides opportunities for the panther to colonize significant new tracts of suitable habitat north of the Big Cypress source population. Should such efforts succeed and a third population be established, managers may wish to use artificial dispersal by moving individuals between populations. The movement of panthers among metapopulation patches would counter some future genetics problems by increasing the effective population size, reducing male dominance, and maintaining proper sex ratios. Alleles could wink out in some patches but be "recolonized," just as natural subpopulations are by dispersers. Expanding numbers and distribution might also justify revisiting the role of recovery technologies such as captive breeding and artificial reproduction, without the anxiety created by a crisis mentality.

LITERATURE CITED

Anderson, A. E., D. C. Bowden, and D. M. Kattner. 1989. Survival in an unhunted mountain lion (*Felis concolor hippolestes*) population in southwestern Colorado. *Mountain Lion Workshop* 3:57.

Backhouse, G. N., T. W. Clark, and R. P. Reading. 1994. The Australian eastern bandicoot recovery program: evaluation and reorganization. Pages 252–269 in T. W. Clark, R. P. Reading, and A. L. Clarke, editors, *Endangered species recovery: finding lessons, improving the process*. Island Press, Covelo, California.

Barrone, M. A., M. E. Roelke, J. Howard, J. L. Brown, A. E. Anderson, and D. E. Wildt. 1994. Reproductive characteristics of male Florida panthers: comparative studies from Florida, Texas, Colorado, Latin America, and North American zoos. *Journal of Mammalogy* 75:150–162.

Bass, O. L., and D. S. Maehr. 1991. Do recent panther deaths in Everglades National Park suggest an ephemeral population? *National Geographic Research and Exploration* 7:427.

Beissinger, S. R., and M. I. Westphal. 1998. On the use of demographic models of population viability in endangered species management. *Journal of Wildlife Management* 62:821–841.

Belden, R. C. 1986. Florida panther recovery plan implementation: a 1983 progress report. Pages 159–172 in S. D. Miller and D. D. Everett, editors, *Cats of the world: biology, conservation, and management*. National Wildlife Federation, Washington, D.C.

Belden, R. C., W. B. Frankenberger, R. T. McBride, and S. T. Schwikert. 1988. Panther habitat use in southern Florida. *Journal of Wildlife Management* 52:660–663.

Boyce, M. S. 1992. Population viability analysis. *Annual Review of Ecology and Systematics* 23:481–506.

Carr, M., P. Zwick, T. Hoctor, W. Harrell, A. Goethals, and M. Benedict. 1998. Using GIS for identifying the interface between ecological greenways and roadway systems at the state and sub-state scales. Pages 68–77 in G. Evink, P. Garrett, D. Ziegler, and J. Berry, editors, *Proceedings of the International Conference on Wildlife Ecology and Transportation*. Florida Department of Transportation, Tallahassee, Florida.

Caughley, G. 1994. Directions in conservation biology. *Journal of Animal Ecology* 63:215–244.

Cox, J., R. Kautz, M. MacLaughlin, and T. Gilbert. 1994. *Closing the gaps in Florida's wildlife habitat conservation system*. Florida Game and Fresh Water Fish Commission, Tallahassee, Florida.

Cunningham, M. W., M. R. Dunbar, C. D. Buergelt, B. L. Homer, M. E. Roelke-Parker, S. K. Taylor, R. King, S. B. Citino, and C. Glass. 1999. Atrial septal defects in Florida panthers. *Journal of Wildlife Diseases* 35:519–530.

Dalrymple, G. H., and O. L. Bass Jr. 1996. The diet of the Florida panther in Everglades National Park, Florida. *Bulletin of the Florida Museum of Natural History* 39:173–193.

Diamond, J. M. 1978. Critical areas for maintaining viable populations of species. Pages 27–40 in M. W. Holdgate and M. J. Woodman, editors, *The breakdown and restoration of ecosystems*. Plenum Press, New York, New York.

Falconer, D. S. 1989. *Introduction to quantitative genetics*. 3d edition. Longman, New York, New York.

Forrester, D. J. 1992. *Parasites and diseases of wild mammals in Florida*. University Press of Florida, Gainesville, Florida.

Forrester, D. J., J. A. Conti, and R. C. Belden. 1985. Parasites of the Florida panther (*Felis concolor coryi*). *Proceedings of the Helminthological Society of Washington* 52:95–97.

Frankham, R. 1995. Inbreeding and extinction: a threshold effect. *Conservation Biology* 9:792–799.

Franklin, I. R. 1980. Evolutionary change in small populations. Pages 135–149 in M. E. Soulé and B. A. Wilcox, editors, *Conservation biology: an evolutionary-ecological perspective*. Sinauer Associates, Sunderland, Massachusetts.

Glass, C. M., R. G. McClean, J. B. Katz, D. S. Maehr, C. B. Cropp, L. J. Kirk, A. J. McKiernan, and J. F. Evermann. 1994. Isolation of pseudorabies (Aujesky's disease) virus from a Florida panther. *Journal of Wildlife Diseases* 30:180–184.

Greiner, E. C., M. E. Roelke, C. T. Atkinson, J. P. Dubey, and S. D. Wright. 1989. *Sarcosystis* spp. in muscles of free-ranging Florida panthers and cougars (*Felis concolor*). *Journal of Wildlife Diseases* 25:623–628.

Harcourt, A. H. 1995. Population viability estimates: theory and practice for a wild gorilla population. *Conservation Biology* 9:134–142.

Harris, L. D., and K. Atkins. 1991. Faunal movement corridors in Florida. Pages

117–134 in W. E. Hudson, editor, *Landscape linkages and biodiversity*. Island Press, Covelo, California.

Harris, L. D., and W. Cropper. 1992. Between the devil and the deep blue sea: implications of climate change for Florida's flora and fauna. Pages 309–324 in R. Peters and T. Lovejoy, editors, *Global warming and biological diversity*. Yale University Press, New Haven, Connecticut.

Harris, L. D., T. S. Hoctor, D. S. Maehr, and J. Sanderson. 1996. The role of networks and corridors in enhancing the value and protection of parks and equivalent areas. Pages 173–197 in R. G. Wright, editor, *National parks and protected areas: their role in environmental protection*. Blackwell Scientific, Cambridge, Massachusetts.

Hedrick, P. W. 1995. Gene flow and genetic restoration: the Florida panther as a case study. *Conservation Biology* 9:996–1007.

Hoctor, T. S., M. H. Carr, and P. D. Zwick. 2000. Identifying a linked reserve system using a regional landscape approach: the Florida ecological network. *Conservation Biology* 14:984–1000.

Lacy, R. C. 1997. Importance of genetic variation to the viability of mammalian populations. *Journal of Mammalogy* 78:320–335.

Lacy, R. C., and J. D. Ballou. 1998. Effectiveness of selection in reducing the genetic load in populations of *Peromyscus polionotus* during generations of inbreeding. *Evolution* 52:900–909.

Lacy, R. C., K. A. Hughes, and P. S. Miller. 1995. VORTEX: a stochastic simulation of the extinction process: version 7 user's manual. IUCN/SSC Conservation Breeding Specialist Group, Apple Valley, Minnesota.

Layne, J. N., and D. A. Wassmer. 1988. Records of the panther in Highlands County, Florida. *Florida Field Naturalist* 16:70–72.

Lindzey, F. G., B. B. Ackerman, D. Barnhurst, and T. P. Hemker. 1988. Survival rates of mountain lions in southern Utah. *Journal of Wildlife Management* 52:664–667.

Logan, T., A. C. Eller Jr., R. Morrell, D. Ruffner, and J. Sewell. 1993. Florida panther habitat preservation plan. Florida Panther Interagency Committee and U.S. Fish and Wildlife Service, Gainesville, Florida.

Maehr, D. S. 1990. The Florida panther and private lands. *Conservation Biology* 4:167–170.

———. 1992. Florida panther. Pages 176–189 in S. R. Humphrey, editor, *Rare and endangered biota of Florida*, vol. 1, *Mammals*. University Press of Florida. Gainesville, Florida.

———. 1997a. The comparative ecology of bobcat, black bear, and Florida panther in south Florida. *Bulletin of the Florida Museum of Natural History* 40:1–176.

———. 1997b. *The Florida panther: life and death of a vanishing carnivore*. Island Press, Covelo, California.

———. 1998. The Florida panther in modern mythology. *Natural Areas Journal* 18:179–184.

Maehr, D. S., R. C. Belden, E. D. Land, and L. Wilkins. 1990a. Food habits of panthers in southwest Florida. *Journal of Wildlife Management* 54:420–423.

Maehr, D. S., and G. B. Caddick. 1995. Demographics and genetic introgression in the Florida panther. *Conservation Biology* 9:1295–1298.

Maehr, D. S., and J. A. Cox. 1995. Landscape features and panthers in Florida. *Conservation Biology* 9:1008–1019.

Maehr, D. S., E. C. Greiner, J. E. Lanier, and D. Murphy. 1995. Notoedric mange in the Florida panther (*Felis concolor coryi*). *Journal of Wildlife Diseases* 31:251–254.

Maehr, D. S., E. D. Land, and M. E. Roelke. 1991a. Mortality patterns of panthers in southwest Florida. *Proceedings of the Annual Conference of Southeastern Fish and Wildlife Agencies* 45:201–207.

Maehr, D. S., E. D. Land, and J. C. Roof. 1991b. Social ecology of Florida panthers. *National Geographic Research and Exploration* 7:414–431.

Maehr, D. S., E. D. Land, J. C. Roof, and J. W. McCown. 1989a. Early maternal behavior in the Florida panther (*Felis concolor coryi*). *American Midland Naturalist* 122:34–43.

―――. 1990b. Day beds, natal dens, and activity of Florida panthers. *Proceedings of the Annual Conference of Southeastern Fish and Wildlife Agencies* 44:310–318.

Maehr, D. S., J. C. Roof, E. D. Land, and J. W. McCown. 1989b. First reproduction of a panther (*Felis concolor coryi*) in southwestern Florida, USA. *Mammalia* 53:129–131.

Maehr, D. S., J. C. Roof, E. D. Land, J. W. McCown, and R. T. McBride. 1992. Home range characteristics of a panther in south central Florida. *Florida Field Naturalist* 20:97–103.

Minta, S. C., and P. M. Kareiva. 1994. A conservation science perspective: conceptual and experimental improvements. Pages 275–304 in T. W. Clark, R. P. Reading, and A. L. Clarke, editors, *Endangered species recovery: finding lessons, improving the process*. Island Press, Covelo, California.

Noss, R. F. 1987. Protecting natural areas in fragmented landscapes. *Natural Areas Journal* 7:2–13.

O'Brien, S. J., M. E. Roelke, J. Howard, J. L. Brown, A. E. Anderson, and D. E. Wildt. 1990. Genetic introgression within the Florida panther (*Felis concolor coryi*). *National Geographic Research and Exploration* 6:485–494.

Pearlstine, L. G., L. A. Brandt, W. M. Kitchens, and F. J. Mazzotti. 1995. Impacts of citrus development on habitats of southwest Florida. *Conservation Biology* 9:1020–1032.

Ralls, K., J. D. Ballou, and A. Templeton. 1988. Estimates of lethal equivalents and the cost of inbreeding in mammals. *Conservation Biology* 2:185–193.

Reed, J. M., D. D. Murphy, and P. F. Brussard. 1998. Efficacy of population viability analysis. *Wildlife Society Bulletin* 26:244–251.

Roelke, M. E., and C. M. Glass. 1992. Strategies for the management of the endangered Florida panther (*Felis concolor coryi*) in an ever shrinking habitat. Pages 38–43 in R. E. Junge, editor, *Proceedings of the American Association of Zoo Veterinarians and the American Association of Wildlife Veterinarians*. St. Louis, Missouri.

Roelke, M. E., J. S. Martenson, and S. J. O'Brien. 1993. The consequences of demographic reduction and genetic depletion in the endangered Florida panther. *Current Biology* 3:344–350.

Schortemeyer, J. L., D. S. Maehr, J. W. McCown, E. D. Land, and P. D. Manor.

1991. Prey management for the Florida panther: a unique role for managers. *Transactions of the North American Wildlife Conference* 56:512–526.

Seal, U. S. 1994. A plan for genetic restoration and management of the Florida panther (*Felis concolor coryi*). Report to the U.S. Fish and Wildlife Service. Conservation Breeding Specialist Group, SSC/IUCN, Apple Valley, Minnesota.

Seal, U. S., and R. Lacy. 1989. Florida panther population viability analysis. Report to the U.S. Fish and Wildlife Service. Captive Breeding Specialist Group, SSC/IUCN, Apple Valley, Minnesota.

———. 1992. Genetic management strategies and population viability of the Florida panther (*Felis concolor coryi*). Report to the U.S. Fish and Wildlife Service. Captive Breeding Specialist Group, SSC/IUCN, Apple Valley, Minnesota.

Shaffer, M. L. 1990. Population viability analysis. *Conservation Biology* 4:39–40.

Soulé, M. E., M. E. Gilpin, W. Conway, and T. Foose. 1986. The millennium ark: how long a voyage, how many staterooms, how many passengers? *Zoo Biology* 5: 101–113.

U.S. Fish and Wildlife Service. 1987. *Florida panther (Felis concolor coryi) recovery plan*. U.S. Fish and Wildlife Service, Atlanta, Georgia.

15

Population Viability Analysis for Plants: Understanding the Demographic Consequences of Seed Banks for Population Health

Daniel F. Doak, Diane Thomson, and Erik S. Jules

ABSTRACT

Many plants share life-history traits that greatly complicate estimation and use of demographic data for viability assessment: large and long-lived seed banks and relatively large temporal variation in demographic rates. Seed banks are especially vexing because they are governed by demographic rates that can be extremely difficult to quantify, especially over the short time periods typically available for conservation planning. Furthermore, if seed banks are demographically important, census and demographic information for adult populations alone may yield misleading information about population viability. Following a survey of how past studies have dealt with these problems, we use simulation models to explore the implications of imperfect seed bank data for conservation planning. Our results emphasize how poor or misleading data on seed banks and demographic variability can alter estimates of extinction times and population growth rates for plants, and how these problems vary with plant life history. Most worrisome is the need for good estimates of environmental variance to qualitatively assess how seed banks will influence population viability analysis results. Finally, we discuss reasonable ways to construct population viability analyses for plants, given the sparse and possibly misleading data usually available.

INTRODUCTION

Since the inception of population viability analysis (PVA), both its practice and underlying research on extinction processes largely have focused on animal (mostly vertebrate) species, rather than plants, fungi, water molds, or other major taxonomic groups. This emphasis resulted in part from simple anthropocentrism and in part from the unique legal

We thank Alison Graff, Caroline Christian, and Ingrid Parker for very helpful discussion that led to many of the ideas presented. Steven Beissinger and Dale McCullough provided helpful editorial comments. Partial support for this work came from the following NSF awards: DEB-9806722 (to Doak and Jules), DEB9424566 (to Doak) and DEB-9902269 (to Thomson and Doak).

protections enjoyed by vertebrates in many countries. Perhaps as importantly, much more demographic information is available for many rare vertebrates than exists for endangered species of other taxa. While the basic theory underlying PVAs is not taxon- or life-history-specific, ecologists typically use demographic data to tie general ideas of stochastic growth and extinction to particular populations. In this chapter, we explain the difficulties of obtaining and using adequate demographic data to conduct PVAs for many plants and explore the consequences and possible solutions to these problems.

For many plants, PVAs are not hard to do. Some herbs, and many trees and shrubs, have life histories very similar to those of the vertebrates for which the most informative PVAs are conducted. For example, the life history of the Madagascar triangle palm (*Neodypsis decaryi*; Ratsirarson et al. 1996) is essentially identical to that of a loggerhead sea turtle (*Caretta caretta*; fig. 15.1A); both have high and variable mortality during the ephemeral "newborn" stage, a prolonged juvenile phase with slow growth and increasing annual survival, and finally, a long-lived, high-survival adult period. In this case, there is no sense in which the plant is any more difficult to study than the animal, and many ways in which it is easier (e.g., most life-history stages don't move).

However, this life history is at one extreme of a continuum for plants. Other species follow a very different pattern that is less easy to study. The crucial features of these life histories are relatively stable, long-lived, dormant, or resting stages (seeds, cysts, etc.), often coupled with shorter-lived and less environmentally buffered adult stages. In addition to short-lived plants, this life-history pattern is common among many freshwater invertebrates, some marine algae, and other taxa (Hairston et al. 1996; fig. 15.1B). The importance for PVAs of the seed or cyst banks found in these groups can be thought of in two ways. First, from the most practical viewpoint, these cryptic, often buried, parts of the life cycle are hard to study and their demographic rates difficult to quantify. Second, long-lived resting stages serve to decouple the two life-cycle roles most important for population viability: reproduction and buffering against environmental variation. In most vertebrate life histories, adults are more physiologically and behaviorally buffered against the vicissitudes of their biotic and abiotic environments than other life-history stages; they also constitute the part of the life cycle that directly contributes to population growth. In contrast, for many plants these two functions are performed by very different parts of the life cycle. Dormant seeds in particular do not reproduce, but may be the crucial stage in buffering populations against environmental variability.

Our goals in this chapter are to briefly describe life histories featuring

A. Long-lived vertebrates and many trees, shrubs, and long-lived forbs

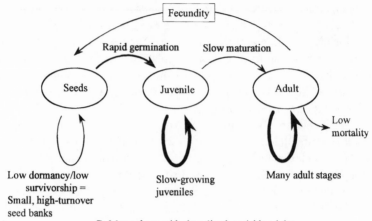

Low dormancy/low
survivorship =
Small, high-turnover
seed banks

Slow-growing
juveniles

Many adult stages

B. Many plants with short-lived, variable adults

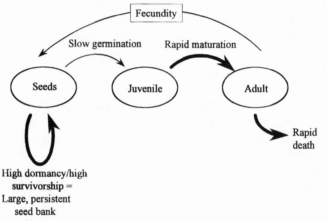

High dormancy/high
survivorship =
Large, persistent
seed bank

Fig. 15.1 Extremes of plant life-history strategies. *A,* Many plants show life histories qualitatively identical to those of many vertebrates, with long-lived, environmentally buffered adults, weak and catastrophe-prone offspring, and slow maturation. *B,* At the other extreme, many "short-lived" plants exhibit long-lived, highly buffered, and high-dormancy seeds. After breaking dormancy, the resulting juveniles quickly mature to become reproductive adults. This life history is shared by many freshwater invertebrates, fungi, and other species.

seed banks, explore the problems that they generate for PVA, and suggest some possible approaches to overcoming these difficulties in the analysis and management of rare plant species.

SEED DORMANCY AND SEED BANKS

Seed dormancy is an inactive state characterized by reduced respiration and the suspension of embryonic growth. This dormant stage can arise in two general ways: innate (intrinsic) dormancy and enforced (extrinsic) dormancy. In seeds with innate dormancy, factors such as immaturity of the embryo, impermeability of the seed coat, chemical inhibition, or the lack of required environmental cues prevent germination upon release from the parent plant (Rees 1997). Enforced dormancy operates independently of innate dormancy, arising when requirements for germination are not present (e.g., light or water; reviewed in Baskin and Baskin 1998). When seed longevity is high relative to seed production by adults, a substantial "bank" of living seeds may form in the soil. For species with such life histories, successful PVAs depend on the recognition that seeds are individuals in the population, just as are aboveground, photosynthetic plants, and that their demography is important for an understanding of population viability.

Dormant seeds can play an essential role in population growth and persistence, both by virtue of their sheer abundance and because seeds are often less subject to environmental variation than are aboveground individuals. For some species, the seed bank may contain most of a population. Epling et al. (1960) estimated that seeds made up 94 to 97% of the population of the Mojave Desert annual *Linanthus parryae* (15 to 30 seeds for every aboveground plant) and that ungerminated seeds remained viable for up to seven to ten years. Populations of annuals with sporadic reproductive failures provide the most dramatic examples of environmental buffering by seed banks. For example, a population of the winter annual *Sedum pulchellum* in north-central Kentucky that suffered complete mortality of aboveground plants prior to seed production during a drought year subsequently recovered entirely from seed bank recruitment (Baskin and Baskin 1980; see also McCue and Holtsford 1998). Adults of many perennial plants may also experience extreme variation in demographic performance through time (Steenbergh and Lowe 1977, 1983; Burgman and Lamont 1992). For some of these species, too, a seed bank may be crucial for population persistence if seeds are less subject to environmental variation than are aboveground members of the population.

Available evidence suggests that seeds of many plant species can remain viable in the soil for extremely long periods of time (Baskin and

Baskin 1998). For example, *Trifolium trichocalyx*, an endangered fire-dependent annual endemic to Monterey pine forests of California, recruited from a seed bank where it had not been observed for 86 years (U.S. Fish and Wildlife Service 1998). Extreme examples of long-term seed storage include radiocarbon-dated seeds of sacred lotus (*Nelumbo nucifera*) collected from a dry lake bed in China germinating after 1,288 ± 271 years (Shen-Miller et al. 1995) and viable seeds of *Lupinus arcticus* found in frozen silt with the skull of a collared lemming (*Dicrostonyx groenlandicus*) that had been extinct from the area for 10,000 years (Porsild et al. 1967). The frequency of such extreme seed longevities is unknown. However, these cases suggest that dormant seeds can be extraordinarily long-lived, and that understanding the survival and germination rates governing the demography of such seeds may be very difficult.

Experimental data on seed longevity in seminatural conditions are available for over 500 species from at least 33 different studies (reviewed in Baskin and Baskin 1998). Seed survivorship is generally inferred by measuring seedling emergence. This approach misses two important components of survivorship: germination without seedling establishment and seed mortality (Rees and Long 1993). Furthermore, while most of these studies examined buried seeds in natural conditions, many used greenhouse trials to determine germination rates. This almost certainly overestimates germination rates of naturally buried seeds, where enforced dormancy (e.g., by lack of light) frequently inhibits germination. Nonetheless, these studies reveal much about the size and temporal variation of seed banks. Figure 15.2 shows germination rates of known numbers of seeds over a five-year period. The negative exponential germination curve shown by two of the species is typical of many plants and is frequently assumed to describe seed survivorship over long time periods (Rees and Long 1993). However, the wide variety of other emergence patterns (fig. 15.2) suggests that such inferences are dubious. Rees and Long (1993) reanalyzed data collected by H. A. Roberts and colleagues for 145 plant species and concluded that a negative exponential pattern should not generally be assumed to predict seed bank decay.

Two basic messages emerge from this brief review of seed biology. First, understanding seed banks is essential for constructing PVAs of many plant species. While seed banks are often considered separately from the aboveground population, both their numbers and capacity to buffer populations make seeds crucial individuals to account for in plant populations. Second, as for other life-history stages, the demographic behavior of seeds can be summarized with a few conceptually simple, but probably age-dependent, demographic rates. In general, to describe

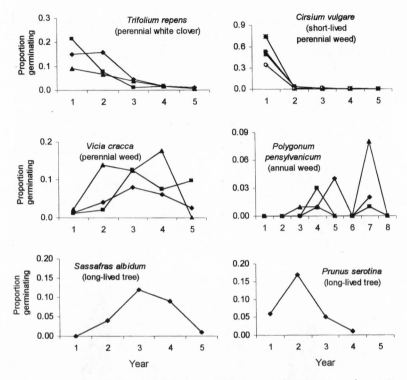

Fig. 15.2 Experimentally determined seed germination rates with increasing seed ages. Germination curves for *Cirsium vulgare* and *Trifolium repens* show classic negative exponential declines in emergence, indicating roughly constant seed germination and survivorship rates. In contrast, data for four other species show evidence of prolonged dormancy and/or variable survivorship. Different lines for each species correspond to different cohorts. Data from Roberts and Boddrell (1985), Toole and Brown (1946), Wendel (1977), and Roberts and Chancellor (1979).

seed demography one must estimate three sets of parameters: age-dependent germination rates of seeds, age-dependent survival rates of nongerminating seeds, and production of new seeds by reproductive plants (fig. 15.1). With this simple framework for incorporating seed banks into PVAs, we next look for precedents in the study of seed bank demography.

HOW HAVE SEED BANKS BEEN INCORPORATED IN PVAs AND OTHER DEMOGRAPHIC STUDIES?

Given that seed banks can be large, long-lived, complex, and difficult to quantify, how have plant demographers typically incorporated seeds into their analyses? We surveyed 70 demographic studies of herbaceous plants and relatively short-lived shrubs, for both rare and nonrare taxa, to assess how they addressed seed banks. Our sample included 34 studies of herbaceous plants compiled by Silvertown et al. (1993) and an additional 36 located through literature searches. Of the 70 total studies, 22 addressed at least one rare species (a list of these studies is available from the authors).

We first determined which studies provided information on seed dormancy of the focal species. If the study mentioned dormancy, we then assessed whether data were available or were collected on the seed bank. Of the 70 studies, 25 did not mention seed banks (fig. 15.3A). Six of the studies stated, or cited evidence, that seed banks were unlikely to be important for the focal species. However, the majority of studies that mentioned seed banks also collected some form of data on dormancy;

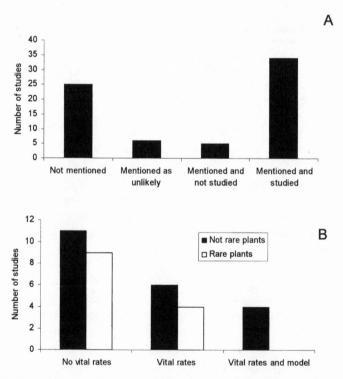

Fig. 15.3 Treatment of seed demography in 70 demographic studies of rare and common plants. *A,* While many studies do not mention the existence of a persistent seed bank, most that do mention it report some data collected to quantify seed demography. *B,* Of the 34 studies involving study of seed banks, the majority did not estimate seed vital rates and, thus, would not allow inclusion of dormant seeds in a demographic analysis.

in only five cases were seed banks mentioned as potentially important and not studied.

We next categorized the 34 papers with information on seed banks by the type of data gathered and how the data were incorporated into demographic analyses. Some data on the seed bank, but not age-specific germination or survival rates, were collected in 20 studies (fig. 15.3B). Ten studies measured at least some age-specific vital rates for seeds, under either field or lab conditions. Finally, four studies both measured age-specific rates and incorporated them into some form of population modeling. The first category included studies that carried out viability or germination trials with seeds collected from soil cores but not with seeds of known age, in addition to several cases in which few data were available on dormancy but some form of seed bank was still included in a population model. Although the proportion of studies falling into each category was similar for both rare and nonrare species (fig. 15.3B), it is notable that we were unable to locate any studies for rare plants that both measured seed vital rates and incorporated them into a quantitative population model.

Thus, while seed banks have clearly received a fair amount of attention in demographic studies of both rare and nonrare plants, data adequate for carefully assessing the effects of dormancy on population persistence are rarely available, even for common and well-studied species. This is not surprising, considering how difficult and time consuming it can be to collect such data (e.g., Kalisz 1991). Indeed, plant demographers usually have to make various assumptions about seed survival and germination to include seed banks in their analyses at all (e.g., Gross et al. 1998).

GENERALIZATIONS TO COMPENSATE FOR IGNORANCE

Given the lack of empirical data on seed bank demography, what can we do to address dormancy in PVAs of rare and poorly studied species? An alternative to intensive, long-term studies of seed bank dynamics for individual rare species is to rely on general patterns of variation in seed survivorship and germination among habitats and life-history strategies. Such generalities, if robust, could be used to predict the existence or absence of a substantial seed bank and perhaps even something about its governing vital rates. A considerable body of theory has sought to predict optimal seed dormancy rates as functions of adult life span, habitat variation, and other factors (reviewed by Rees 1997). This theoretical work has largely been concerned with understanding the evolutionary trade-offs between the costs of seed dormancy, which result from the negative effects of delayed reproduction on deterministic population growth, and the benefits of increased buffering from environmental variation (e.g., Ellner 1985a,b, 1997). Dormancy can be understood as a mechanism of escape from unfavorable conditions in time, just as dispersal represents escape in space (Levin et al. 1984). A crucial point to reemphasize here, and one that will assume added importance later, is that increased dormancy (i.e., decreased germination rate) has a direct, immediate cost for individuals and for population growth: a nongerminating seed is delaying reproduction and increasing the possibility of death before ever reproducing. As a result, theory predicts that the strength of selection for dormancy in a particular plant population should vary depending on the need for buffering imposed by the habitat and the degree of investment in other life-history traits, such as long adult life span and long-distance dispersal, that are also mechanisms for persisting in variable environments.

Theoretical predictions about optimal dormancy rates are broadly intuitive. Greater dormancy should be favored as environmental variability increases (Ellner 1985a,b); one of the earliest and simplest evolutionary models of dormancy predicts that the optimal germination rate should decline linearly with the probability of complete reproductive failure in any given year (Cohen 1966). Similarly, sibling-sibling competition among germinating seeds or strong inhibition of seedlings by established vegetation can also select for higher dormancy (Ellner 1986). At the same time, good dispersers should rely less on dormancy than species with seeds that are not well adapted for dispersal (Levin et al. 1984). Both dormancy and dispersal ability, however, may also be affected by trade-offs with adult life span and seed size and shape (Venable and Brown 1988). Thus, long-lived adults buffer against the effects of poor years, obviating the need for dormant stages. Large seeds are less likely to be buried at soil depths conducive to long-term dormancy than small seeds. They are also generally better provisioned, and thus able to establish more successfully under a range of environmental conditions. Small-seeded species are therefore more likely to evolve strong dormancy. At the same time, small seeds are also more likely to disperse well, potentially reducing selection for dormancy.

While the basic predictions of theoretical models are relatively straightforward, it is not easy to untangle this set of potential trade-offs and arrive at simple rules for predicting when seed banks should be important. In fact, providing strong empirical support for these predictions has proved extremely difficult. Thompson et al. (1998) found that mean seed longevity in northwest Europe, measured on a qualitative scale, correlated with habitat type; species occurring in presumably less variable environments, such as woodlands, had lower dormancy on average than those from more disturbed habitats, such as pastures. Pake and Venable (1995) showed a weak but significant inverse relationship between variability in reproductive success over a ten-year period and average germination rate for a group of desert annuals, suggesting that species more subject to environmental variability exhibit greater dormancy. Similarly, Rees (1993) found a weak inverse relationship between adult longevity and seed longevity. For seed traits, Bekker et al. (1998) showed that average burial depth correlated well with a composite measure of seed size and shape, and that burial depth was a weak but significant predictor of longevity.

While these correlations generally support theoretical expectations, they explain only small amounts of the variation in seed longevity, and are generally significant only when multiple factors are first accounted for, such as seed weight (Pake and Venable 1995; Rees 1993) or phylogeny (Rees 1993). In some cases, the data required for making a good prediction about dormancy, such as long-term variability in reproductive success, may be no faster or easier to collect than actual vital rates for seeds. Finally, and perhaps most critically, the seed demography data available to test these predictions are both limited and heavily biased toward agricultural weeds. Of the 33 studies that Baskin and Baskin (1998, table 7.5) cite as measuring seed longevity under field conditions, only 11 were ten or more years in duration. Figure 15.4A summarizes results from 14 of these studies, including 9 of the studies that lasted ten years or more. For many of the 179 species included in these studies, seed longevity equaled, and almost certainly exceeded, the length of the study, making it difficult to resolve actual differences in dormancy between species. Worst of all from the perspective of a PVA practitioner, only 25 of the species included in these studies were not weeds (fig. 15.4B).

DOES LACK OF KNOWLEDGE REALLY MATTER?

The preceding review emphasizes the lack of good data on seed banks for most plant species and the lack of powerful generalities that would allow us to predict seed bank importance from easily observed features of the environment or aboveground life history. Given this dearth of information, how should one proceed in constructing PVAs for plant species? A more specific and useful set of questions would include, How important are seed bank dynamics in determining population growth or extinction rates? If seed banks are sometimes important, can we predict the ways in which ignoring them will change the results of a PVA? For which life histories are these omissions most serious? Using a set of

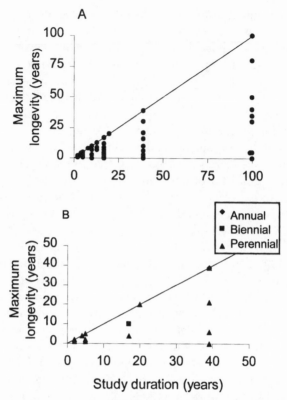

Fig. 15.4 Estimates of seed longevity for 179 species (references compiled in Baskin and Baskin 1998). We report maximum measured longevity as a function of study duration. A, Results for all species. B, Results for the 25 nonweedy species included in this sample.

simulation models, we next explore each of these issues. Although the results do not provide clear-cut answers, they do suggest how to approach such problems for species of concern.

We developed a stochastic matrix model to simulate a range of plant life histories in variable environments. The model is parameterized with the means and standard deviations (SD) of each demographic rate. Except for fecundity, each demographic rate was drawn from a beta distribution (bounded by zero and one) in each year of a simulation. To create reasonable, bounded distributions for seed production (i.e., no negative numbers and no infinitely large values), we again drew values from a beta distribution and rescaled the values to reflect the desired mean, SD, and maximum. Variations in all vital rates were correlated with a single environmental driving variable, E, that fluctuated randomly. In some simulations we also included first-order autocorrelations in environmental variation (i.e., positive correlations in the state of the environment from year to year).

To make these simulations more realistic, as well as sensitive to seed bank effects, we started with small population sizes (ten adult plants), enforced a crude but realistic form of density-dependence by establishing a cap on total adult population size (100 individuals), and allowed simulations to run for 100 years. In different simulations we systematically varied mean seed germination and survivorship rates. When varying the means of these probabilities, we also rescaled SDs to keep variation proportional to means (Doak et al. 1994; Link and Hahn 1996). We also varied the SD of adult survivorships, altering these values from the estimated rate to the maximum possible value (Evans et al. 1996). To summarize simulation results, we use three common measures of population health: probabilities of extinction over set time horizons, times to extinction, and population growth rates.

To parameterize the model, we needed sets of demographic information including data on seed banks and estimates of both the mean and variation in each demographic rate. We found only a handful of studies with estimates of the demographic rates of both adults and seeds under field conditions, particularly for nonweedy taxa. We constructed matrices based on published data for two of these species, each with high seed dormancy but with otherwise contrasting life-history strategies: *Collinsia verna*, a winter annual that occurs in floodplain forests of the eastern United States (Kalisz 1991), and *Calathea ovadensis*, a short-lived perennial herb found in tropical rain forests (Horvitz and Schemske 1995; see tables 15.1 and 15.2). The *Collinsia* data set consisted of demographic rates measured in two different years, while four years of data were available for *Calathea*. In both cases, spatial variabil-

Table 15.1 Mean Matrices for Simulated *Collinsia* and *Calathea* Life Histories

$$
\begin{array}{c}
 & \text{Seed} & \text{Plant} \\
\text{Seed} & \left[\begin{array}{cc} (1 - g_0)s_s & s_{ad}f(1 - g_n) \\ \text{Plant} & g_0 s_s & s_{ad}f g_n \end{array}\right.
\end{array}
$$

Parameter	Symbol	Mean	Standard Deviation
Fecundity	f	19.18	11.77
Survival from germination to reproduction	s_{ad}	0.226	0.0157
Germination rate of newly produced seed	g_n	0.282	0.1407
Germination rate of seeds in the seed bank	g_o	0.07	0.071
Seed survival	s_s	0.115	0.0778

Notes: Collinsia demography is based on table 2 in Kalisz (1991), for transect 3 only, assuming a census immediately following germination. We assumed that the same survivorship rate applied to both germinating and nongerminating seeds.

Table 15.2 Mean Matrix and Demographic Rates for Simulated *Collinsia* and *Calathea ovadensis* Life History

$$\begin{array}{c} \\ \text{Seed} \\ \text{Seedling} \\ \text{Plant} \end{array} \begin{array}{ccc} \text{Seed} & \text{Seedling} & \text{Plant} \\ \left[\begin{array}{ccc} s_s(1-g) & 0 & s_a f \\ s_s g(1-gr_1) & s_j(1-gr_2) & s_a sr \\ s_s gr_1 g & s_j gr_2 & s_a(1-sr) \end{array}\right] \end{array}$$

Rate Description	Symbol	Mean	Standard Deviation	Correlation
Fecundity	f	9.052	4.622	−0.044
Adult survival	s_a	0.831	0.064	0.802
Juvenile survival	s_j	0.096	0.019	−0.997
Growth from juveniles to adults	gr_2	0.809	0.166	−0.475
Shrinking from adult to juvenile	sr	0.005	0.010	0.333
Growth from seed to plant	gr_1	0.034	0.047	−0.983
Germination of seeds	g	0.284	0.237	−0.955
Seed survival	s_s	0.599	0.00	1

Notes: *Calathea* demography is based on Horvitz and Schemske (1995), for plot 2 only. We collapsed all adult size classes into a single category (plants), then decomposed the transition probabilities into the underlying demographic rates based on a post-breeding census (i.e., the plant to seed transition was the product of adult survivorship and fecundity; transitions out of the seed class incorporated both seed survivorship and germination rates). We assumed that the same survivorship rate applied to both germinating and nongerminating seeds.

ity in demographic rates was also measured, but to simplify the analysis, we used data from a single plot for which the mean deterministic matrix yielded a positive growth rate. While a number of excellent demographic analyses have been published on both of these species, the modeling methods and especially the goals of our models are quite different from that of these previous analyses (Kalisz 1991, 1997; Kalisz and McPeek 1992, 1993).

With the data available for most plant species, there are two broad classes of mistakes that one might make about seed demography: misestimation of seed survivorship patterns and misestimation of seed germination patterns. In either case, errors range from identifying the existence of any seed bank to specifics of age-dependent survivorship or growth. While both survivorship and germination rates are likely to be misunderstood together, we addressed these problems one at a time. Furthermore, we did not address the issue of incorrect age-dependence in either seed survivorship or germination (Kalisz 1997). Rather we concentrate on the more egregious problems of incorrectly estimating average germination or survivorship rates and their interplay with inaccurate knowledge of adult survival variation.

Misunderstanding of Seed Survivorship Rates

Underestimation of seed survivorship will obviously lead to pessimistic PVA results, while overestimation will give rosier pictures of population health. For the *Collinsia* model, these results were quite striking, with rapid decreases in the number of simulations suffering extinction, some increases in extinction times, and steady increases in population growth with increasing seed survival (fig. 15.5A). Note that, even when mean growth is positive, the combination of environmental variation and a cap on population growth rate can lead to extinction probabilities that are quite high (Lande 1993). One peculiarity in interpreting time to extinction (T_{ex}) results should be noted: the decline in T_{ex} at higher seed survival rates is a consequence of fewer populations suffering extinction. In this situation most extinctions happen rapidly, leading to lower T_{ex} values (Lande and Orzack 1988; Kalisz and McPeek 1993). The estimated seed survival, 0.115, is low enough that all populations would be expected to become extinct over 100 years. Marginally higher seed survival rates would result in substantially longer persistence times for many populations, while more than doubling this rate would be necessary to afford a substantially higher probability of continued persistence. These results parallel those of Kalisz and McPeek (1992, 1993) and Kalisz (1997), who modeled somewhat different manipulations of seed demography.

At high seed survivorship rates, the *Calathea* models showed a more modest influence of seed survival on population health (fig. 15.5B). At lower values, however, very small errors in survivorship estimation can lead to dramatically different predictions about population safety. Indeed, the estimated seed survival rate of 0.599 is just above the value needed to ensure essentially no chance of extinction, while slightly lower values predict substantial risk. Thus, small errors in seed survivorship estimation would lead to significant misunderstandings about population health and, most practically, the need for active management intervention.

Overall, these results show that the consequences of misestimated seed survival can be quite severe for an understanding of population health. Seemingly paradoxically, the perennial *Calathea* life history, which certainly relies less on seed banks for environmental buffering than does *Collinsia*, shows the potential for poorer predictions in the absence of accurate seed survival data. This sharp influence on *Calathea* persistence probably results from the need for a minimum seed survivorship simply to achieve positive long-run population growth; greater stochastic variation, and hence more important buffering effects at all seed survivals, leads to the more graded response of *Collinsia*.

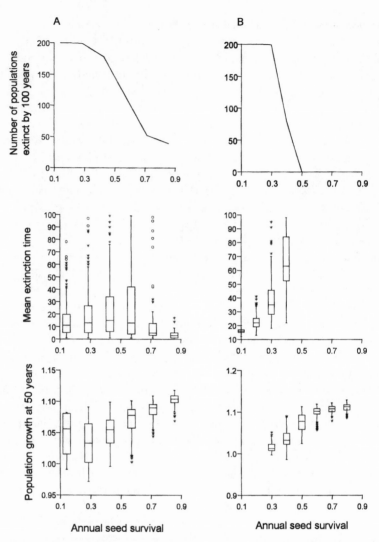

Fig. 15.5 Results of changing annual seed survivorship on simulated plant populations using data (tables 15.1 and 15.2) for *Collinsia verna* (*column A*) and *Calathea ovadensis* (*column B*). Three metrics of population health are shown: number of populations becoming extinct before 100 years (*top panels*), time to extinction for those becoming extinct (*middle panels*), and population growth from year 0 to year 50, $(N_{50}/N_0)^{1/50}$ (*bottom panels*). Results for time to extinction and population growth are shown as *boxplots*. The box is bounded by the 25th and 75th quartiles and is divided at the median, while the *lines* indicate the most extreme values within 1.5 times the interquartile range from the top and bottom of each box (see Sokal and Rohlf 1995 for further information). Results are from 200 simulations for each species for each seed survival level.

The Problem of Estimating Temporal Variation in Demography

While these results seem clear-cut, an added wrinkle should also be considered before moving on to germination rates. As we noted in reviewing the literature on seed dormancy, the major advantage of seed banks is in buffering populations against environmental variation. Thus, correctly analyzing seed bank effects and our ignorance of their dynamics will depend on estimates of variability in other parts of the life cycle, especially the reproductive adults that are most important for deterministic population growth (Caswell 1989; Rees 1994). This dependence forces us to confront two problems. First, we often have no estimates of variance for any life-history stage. Second, when we do have such estimates, they are likely to be biased low after correction for sampling variance (Beissinger and Westphal 1998; Kendell 1998), since they are essentially always based on short series of data. This limitation means that most data sets are unlikely to contain the occasional extremely good or extremely bad years that will drive most variation in vital rates. This problem has, to our knowledge, not been rigorously investigated, but it is closely parallel to the problems of estimating variation in census data (Redfearn and Pimm 1988; McArdle et al. 1990).

Given this likely bias, a sensible precaution is to rerun our sets of simulations with increased levels of temporal variation in other demographic rates, especially adult survivorship (Rees 1994). We manipulated variation in this rate from the observed value up to 90% of the maximum possible for both the *Collinsia* and *Calathea* matrices. For both models, adding variance substantially increased extinction risk (fig. 15.6A and B) and shifted the range of seed survivorship values to which population health was most sensitive. For *Collinsia* this shift means that, if we have underestimated adult variation, there is little concern about error in seed survival estimates, since they will have little influence on population persistence. For *Calathea* exactly the opposite was true; added variation moved the range of high sensitivity into that of estimated seed survival. Additionally, the intuition that increasing environmental variation will lead to more gradual effects of changing seed survival explains the smoother responses of *Calathea* persistence to differences in seed survival with increasing adult variation (fig. 15.6B).

Overall, the results from these simulations suggest that conclusions about the effects of seed survivorship on population health need to be predicated on an understanding of variability in other life stages. However, misestimation of adult variability will not fundamentally change the influence on PVA results of mistakes in estimating seed survival.

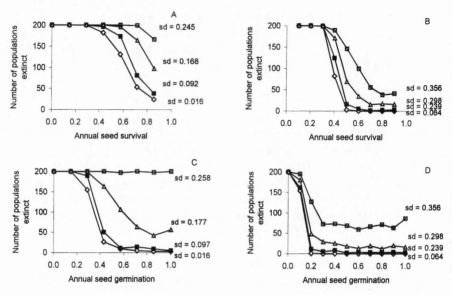

Fig. 15.6 Increasing variation in adult survival alters sensitivity of extinction to seed survival or germination rates. The four lines in each figure are for the estimated and three higher standard deviations of adult survival (sd_{ad}). Results are for A, *Collinsia verna* demography varying across seed survival rates; B, *Calathea ovadensis* demography varying across seed survival rates; C, *Collinsia verna* demography varying across seed germination rates; D, *Calathea ovadensis* demography varying across seed germination rates. See tables 15.1 and 15.2 for empirically estimated values of seed survival and germination.

The Complexity of Misunderstanding Seed Germination

While misunderstanding seed survivorship will influence our estimate of population performance, in some ways the more fundamental rate governing seed banks is germination. As the literature on evolution of dormancy emphasizes, patterns of seed germination should reflect a balance between undelayed reproduction and the safety provided by not simultaneously germinating into a possibly hostile world. Thus, unlike with seed survivorship, our ignorance of germination rates puts us in a potentially odd situation: it is not clear whether overestimating germination rates will yield overly optimistic or overly pessimistic PVA predictions, since some *intermediate* germination rate should be best. Furthermore, this intermediate rate should depend crucially on temporal variation in other parts of the life cycle. Poor estimates of variability in a species' demography will substantially alter estimates of which germination rates are best, or how incorrect estimates of germination may bias our understanding of population health.

To investigate these problems, we performed a series of simulations parallel to those described for seed survivorship, in which we varied both germination rates and the temporal variation in adult survivorship. Using the estimated variation in adult survivorship of *Collinsia* (low SD = 0.016; fig. 15.6), there is no evidence that maintenance of a seed bank is advantageous. Complete, immediate germination of all seeds results in the best population performance, with essentially no probability of extinction over 100 years. A reduction in germination percentage comes at a cost to the probability of extinction, and yields inconsistent gains in the time to extinction for simulations going extinct. However, this simple pattern applies only to simulations using the low estimated variation in adult survival. Increasing variation in adult survival results in poorer population performance for all germination rates but also changes in the optimal germination rate. In particular, for the three highest adult variances examined (SD = 0.258, 0.339, and 0.42), intermediate germination rates were optimal; almost every simulation ended in extinction, but intermediate germination yielded longer times to extinction than did extremely high germination rates (results not shown). These results parallel those of Kalisz and McPeek (1993), who also simulated extreme variation in environments. Thus, we do find evidence for the importance of seed banks, but only if real environmental variation is considerably higher than estimated variation. Most worrisome, understanding whether a mistaken or simplified view of germination biology is optimistic or pessimistic is entirely dependent on knowing the variation affecting adults. Thus, this situation is considerably more complex than that for seed survival, for which higher survival is clearly better.

For *Calathea* simulations, we found no clear advantages for seed banks, even with extremely high adult variation (fig. 15.6). While time to extinction declined with increasing germination rates, these declines were concordant with declines in the number of trials becoming extinct, and there was no intermediate optimum germination rate that minimized extinction probability while maximizing extinction times. For a PVA practitioner, these results are welcome. They suggest that, for all values of adult survival variation, there is a wide range of germination rates over which no perceivable differences in population performance occur. Therefore, mistakes in the estimation of germination rates will not be very important.

Nonetheless, we were puzzled by these results, and tinkered more with the model in an effort to find conditions truly favoring intermediate germination rates. The most obvious target to modify was the estimated correlation structure between demographic rates, which depends on only four years of data and which included rather unlikely, strongly negative correlations. These negative correlations, especially those between adult and juvenile survivorship, helped to buffer the aboveground population against environmental variation, even in the absence of a seed bank. Thus, we ran more simulations with a new correlation structure that resulted in maximum fluctuations in aboveground demography (all correlations equal to 0.9 or −0.9). We also ran a set of simulations in which we added strong autocorrelation in environments ($r = 0.9$) to this modified correlation structure. Results from both these sets of simulations showed that it is nearly impossible to develop a scenario for which it is advantageous to have less than total seed germination.

Our modeling exercise suggests that it is only sometimes possible to find conditions in which population viability is maximized by germination rates that favor a persistent seed bank, and considerable model tuning is often required to achieve these results. Life-history theory shows the clear advantages of seed dormancy, and high dormancy is well documented in many plants. Therefore, it seems odd that we have had such a difficult time illustrating its importance for population health. At least three different factors may account for this apparent contradiction: (1) Estimates of temporal variation in demographic rates, especially adult performance, are typically far too low. (2) Optimal seed dormancy for individual fitness will typically be much higher than that for population persistence. Unless a species consists of many small and extinction-prone subpopulations with strong demic selection, we would expect individual selection to favor higher dormancy than will be best for population performance. This difference will be reinforced by sibling competition (Ellner 1986), but may be altered by dispersal-dormancy trade-offs and metapopulation structure (Cohen and Levin 1987; Kalisz 1997). (3) Germination rates in the field are constrained by enforced dormancy. Because seeds can sense when germination is hopeless, as it often is, germination rates in nature are likely to be much lower than would naïvely seem "optimal" for either population growth or individual fitness. Seeds of many species are able to use quite accurate cues to determine the chances for successful germination and growth, creating correlations between germination and environmental variation that favor lower germination rates (Baskin and Baskin 1998).

While all of these explanations may be important, too few data exist to rigorously assess their ability to explain the mismatch of estimated optimal seed demography and field-estimated demographic rates. Their importance for PVA is simply that estimation of the optimal dormancy rates for population viability is likely to give little guidance on actual seed bank demography.

PVAs FOR REAL PLANTS

While seeds and seed banks are a crucial part of the life cycle and population structure of many plants, few PVAs can call upon high-quality data to understand the demography of seed banks. Thus, we are usually left to make a series of assumptions about how seed demography works and to proceed in the face of ignorance and uncertainty. As many of the chapters in this volume emphasize, most PVAs are conducted with considerable gaps in knowledge, and all must deal with parameter uncertainty (Taylor et al., chap. 12 in this volume; Wade, chap. 11 in this volume). However, we argue that plant seed banks present a particularly troublesome problem that is qualitatively different from those faced in constructing PVAs for most other species. Even without data on some process or complication, a good demographer or good field ecologist can usually make an accurate guess as to whether excluding it from an analysis will be pessimistic or optimistic. Indeed, the majority of biological complications that are frequently excluded from PVAs are known to decrease predictions of population health. Environmental stochasticity, senescence, population ceilings, inbreeding depression, dispersal losses, and mate finding are all left out of many PVAs, with the clear understanding that adding them will only make for more pessimistic results. While leaving out these biological realities obviously can result in less accurate model predictions, there is considerable power in knowing at least the qualitative effects of these omissions (positive or negative), especially when explaining a PVA's results to politicians, judges, or students.

The problem created by seed banks is that they are governed by several parameters that are difficult to understand, and it is not clear how simplified treatments of seed bank processes will influence the qualitative results of PVAs. In particular, underestimation of variation in aboveground performance could make omission of seed dormancy seem optimistic, when in fact it is pessimistic, given an accurate understanding of environmental variability. The conclusions of many PVAs largely revolve around sensitivity calculations of deterministic matrix models (e.g., Crooks et al. 1998; Mills et al. 1999; Mills and Lindberg, chap. 16 in this volume). For species with seed banks, such analyses are likely be mildly to wildly misleading. As we have sought to emphasize throughout this chapter, an understanding of the demography—and hence management—of plants with substantial seed banks is inextricably tied to an understanding of environmental variability. Since variability is just as poorly estimated as seed demography for most species, the plant PVA practitioner is left in an unenviable situation.

So how should one analyze population viability and management for threatened plants? We advocate two general approaches. First, for the majority of plants that require protection and analysis, little or no quantitative demographic data exist. In this situation, beginning to develop a PVA is tantamount to the collection and guesstimation of data for the species. Thus, the question is how to guess about the importance of seed banks and, by extension, how important it is to study them at all. While the work we have reviewed here should caution against broad generalities regarding seed banks, there is a logical classification that may help to prioritize situations in which understanding of seed banks will be critical for population viability (table 15.3). In general, species with shorter and more variable aboveground life spans are most likely to rely on seed banks for population persistence. In addition, the importance of seed banks to population persistence is likely to diminish with increasing population size. Thus, estimation of seed demography is

Table 15.3 The Relative Importance of Understanding Seed Demography in Order to Confidently Conduct a PVA for Different Plant Life Histories

Environmental Variation in Adult Performance	Adult Longevity				
	Annual	Biennial	Short-Lived Forbs and Shrubs	Medium-Lived Trees and Shrubs	Very Long Lived Forbs, Trees, and Shrubs
Extremely high	VI: *Trifolium tricho-calyx**	VI			
High	VI: *Astraga-lus tener* var. *titi**	VI: *Erysi-mum tereti-folium**	VI: *Mimetes hotten-toticus**		
Moderate	VI: *Chori-zanthe pungens* var. *hart-wegiania**	MI	MI: *Oeno-thera del-toides* ssp. *howellii**	NI: *Cupres-sus goveni-ana* ssp. *goveni-ana**	
Low	MI: *Koenigia islandica*	MI	NI: *Epilo-bium lati-folium*	NI: *Cory-phanthus robbin-sorum**	NI: *Wel-witschia mirabilis, Silene acaulis*

Notes: VI = very important to understand seed demography and seed bank dynamics to safely reach management conclusions; MI = moderately important; NI = probably not important. Example species are provided for most of these categories. Asterisks indicate rare or endangered species.

especially crucial in planning reintroduction efforts or the management of critically small populations. In fact, careful monitoring of reintroduction programs may provide some of the best opportunities to collect data on seed dormancy. In these situations, data collection should, if at all possible, quantify seed demography *in nature*, rather than in greenhouse flats or petri dishes. In situ data are considerably more difficult to obtain, but seed behavior in artificial conditions is of very limited value in understanding population viability.

The second approach should be taken in situations when one has some estimates of most demographic rates, but weak information on seed demography or variability of other rates. In this case, we suggest following the route taken in our simulation results: use models that include various assumptions about how seed germination and survivorship operates, in combination with a wide range of assumptions about temporal variation. While one can use a Bayesian framework to incorporate uncertainty (Taylor et al., chap. 12 in this volume; Wade, chap. 11 in this volume), simulating a suite of alternative models that are biologically reasonable may be as or more useful (Burnham and Anderson 1998). It is crucial to take known biases into account (e.g., chronic underestimation of temporal variation) and to explore a wide range of situations. Depending on the life history of the species in question, this type of exploratory modeling is likely to suggest which rates are crucial to estimate, which are not as important, and how robust different management recommendations are likely to be. While we have not emphasized the role of plant PVAs in evaluating management recommendations, it should be clear that analysis of many dangerous and expensive management issues can depend crucially on understanding seed demography (e.g., when to burn or stop the burning of a piece of chaparral or fynbos).

In reviewing PVA concerns for plants, we have emphasized the life-history feature that is most problematic for most species—seed banks. In doing so, we have tried to avoid the easy route of simply reiterating that "more data are needed." Perhaps the most useful lesson from our review and simulations is that too little attention has been given to the interacting problems posed by seed demography and poor data on temporal variability, even though most studies of plants have honestly tried to deal with seed banks. This pattern is part of a larger trend in PVAs. As use of generalized software packages for viability analysis becomes more common, there is a danger that PVA practitioners will stop thinking carefully about the natural history of their species and how different modeling frameworks include or exclude different facets of biology. Especially for plants, different life-history attributes shift the importance of deterministic versus stochastic forces for population viability, as well as the ability to successfully gather the information needed to construct good PVA models. Continuing attention to biological detail is probably as important to the development of useful, credible PVAs as are advances in the general theory of population extinction. While mathematicians are developing new and better ways to conduct PVAs, empirical biologists should constantly critique these tools and use them to flexibly incorporate the idiosyncratic aspects of natural history important to yielding relevant, useful predictions for real species in particular contexts.

LITERATURE CITED

Baskin, C. C., and J. M. Baskin. 1998. *Seeds: ecology, biogeography, and evolution of dormancy and germination.* Academic Press, San Diego, California.

Baskin, J. M., and C. C. Baskin. 1980. The role of seed reserves in the persistence of a local population of *Sedum pulchrellum*: a direct field observation. *Bulletin of the Torrey Botanical Club* 107:429–430.

Beissinger, S. R., and M. I. Westphal. 1998. On the use of demographic models of population viability in endangered species management. *Journal of Wildlife Management* 62:821–841.

Bekker, R. M., J. P. Bakker, U. Grandin, R. Kalamees, P. Milberg, P. Poschold, K. Thompson, and J. H. Wilhelms. 1998. Seed size, shape, and vertical distribution in the soil: indicators of seed longevity. *Functional Ecology* 12:834–842.

Burgman, M. A., and B. B. Lamont. 1992. A stochastic model for the viability of *Banksia cuneata* populations: environmental, demographic, and genetic effects. *Journal of Applied Ecology* 29:719–727.

Burnham, K. P., and D. R. Anderson. 1998. *Model selection and inference: a practical information-theoretic approach.* Springer-Verlag, New York, New York.

Caswell, H. 1989. *Matrix population models.* Sinauer Associates, Sunderland, Massachusetts.

Cohen, D. 1966. Optimizing reproduction in a randomly varying environment. *Journal of Theoretical Biology* 12:119–129.

Cohen, D., and S. A. Levin. 1987. The interaction between dispersal and dormancy strategies in varying and heterogeneous environments. Pages 110–122 in E. Teramoto and M. Yamaguti, editors, *Mathematical topics in population biology, morphogenesis, and neurosciences*, proceedings of an international symposium, Kyoto, November 10–15, 1985. Springer-Verlag, Berlin, Germany.

Crooks, K. R., M. A. Sanjayan, and D. F. Doak. 1998. New insights on cheetah conservation through demographic modeling. *Conservation Biology* 12:889–895.

Doak, D. F., P. Kareiva, and B. Klepetka. 1994. Modeling population viability for the desert tortoise in the western Mojave Desert. *Ecological Applications* 4:446–460.

Ellner, S. 1985a. ESS germination strategies in a randomly varying environment: 1, logistic-type models. *Theoretical Population Biology* 28:50–79.

———. 1985b. ESS germination strategies in a randomly varying environment: 2, reciprocal yield-law models. *Theoretical Population Biology* 28:80–116.

———. 1986. Germination dimorphisms and parent-offspring conflict in seed germination. *Journal of Theoretical Biology* 123:173–185.

————. 1997. You bet your life: life-history strategies in fluctuating environments. Pages 3–24 in H. G. Othmer, F. R. Adler, M. A. Lewis, and J. C. Dallon, editors, *Case studies in mathematical modeling: ecology, physiology, and cell biology.* Prentice Hall, Upper Saddle River, New Jersey.

Epling, C., H. Lewis, and F. M. Ball. 1960. The breeding group and seed storage: a study in population dynamics. *Evolution* 14:238–255.

Evans, M., N. Hastings, and B. Peacock. 1996. *Statistical distributions.* 2d edition. John Wiley and Sons, New York, New York.

Gross, K., J. R. Lockwood III, C. C. Frost, and W. F. Morris. 1998. Modeling controlled burning and trampling reduction for conservation of *Hudsonia montana.* *Conservation Biology* 12:1291–1301.

Hairston, N. G., Jr., S. Ellner, and C. M. Kearns. 1996. Overlapping generations: the storage effect and the maintenance of biotic diversity. Pages 109–145 in O. E. Rhodes Jr., R. K. Chesser, and M. H. Smith, editors, *Population dynamics in ecological space and time.* University of Chicago Press, Chicago, Illinois.

Horvitz, C. C., and D. W. Schemske. 1995. Spatiotemporal variation in demographic transitions of a tropical understory herb: projection matrix analysis. *Ecological Monographs* 65:155–192.

Kalisz, S. 1991. Population dynamics of an age-structured annual: 1, life table analyses of the seed bank and plant cohorts. *Ecology* 72:575–585.

————. 1997. Fragmentation and the role of seed banks in promoting persistence in isolated populations of *Collinsia verna.* Pages 286–312 in M. W. Schwartz, editor, *Conservation in highly fragmented landscapes.* Chapman and Hall, New York, New York.

Kalisz, S., and M. A. McPeek. 1992. Demography of an age-structured annual: resampled projection matrices, elasticity analyses, and seed bank effects. *Ecology* 73:1082–1093.

————. 1993. Extinction dynamics, population growth, and seed banks. *Oecologia* 95:314–320.

Kendall, B. E. 1998. Estimating the magnitude of environmental stochasticity in survivorship data. *Ecological Applications* 8:184–193.

Lande, R. 1993. Risks of population extinction from demographic and environmental stochasticity and random catastrophes. *American Naturalist* 142:911–927.

Lande, R., and S. H. Orzack. 1988. Extinction dynamics of age-structured populations in a fluctuating environment. *Proceedings of the National Academy of Sciences* (USA) 85:7418–7421.

Levin, S. A., D. Cohen, and A. Hastings. 1984. Dispersal strategies in patchy environments. *Theoretical Population Biology* 26:165–191.

Link, W. A., and D. C. Hahn. 1996. Empirical Bayes estimation of proportions with application to cowbird parasitism rates. *Ecology* 77:2528–2537.

McArdle, B. H., K. J. Gaston, and J. H. Lawton. 1990. Variation in the size of animal populations: patterns, problems, and artifacts. *Journal of Animal Ecology* 59:439–454.

McCue, K. A., and T. P. Holtsford. 1998. Seed bank influences on genetic diversity in the rare annual *Clarkia sprinvillensis* (Onagraceae). *American Journal of Botany* 85:30–36.

Mills, L. S., D. F. Doak, and M. J. Wisdom. 1999. The reliability of conservation actions based upon elasticities of matrix models. *Conservation Biology* 13:815–829.

Pake, C., and L. Venable. 1995. Is coexistence of Sonoran Desert annuals mediated by temporal variability in reproductive success? *Ecology* 76:246–261.

Porsild, A. E., C. R. Harrington, and G. A. Mulligan. 1967. *Lupinus arcticus* Wats. grown from seeds of Pleistocene age. *Science* 158:113–114.

Ratsirarson, J., J. A. Silander Jr., and A. F. Richard. 1996. Conservation and management of a threatened Madagascar palm species, *Neodypsis decaryi* Jumelle. *Conservation Biology* 10:40–52.

Redfearn, A., and S. L. Pimm. 1988. Population variability and polyphagy in herbivorous insect communities. *Ecological Monographs* 58:39–55.

Rees, M. 1993. Trade-offs among dispersal strategies in British plants. *Nature* 366:150–152.

———. 1994. Delayed germination of seeds: a look at the effects of adult longevity, the timing of reproduction, and population age/stage structure. *American Naturalist* 144:43–64.

———. 1997. Seed dormancy. Pages 214–238 in M. J. Crawley, editor, *Plant ecology*, 2d edition. Blackwell Scientific, Oxford, United Kingdom.

Rees, M., and M. J. Long. 1993. The analysis and interpretation of seedling recruitment curves. *American Naturalist* 141:233–262.

Roberts, H. A., and J. E. Boddrell. 1985. Seed survival and seasonal pattern of seedling emergence in some Leguminosae. *Annals of Applied Biology* 106:125–132.

Roberts, H. A., and R. J. Chancellor. 1979. Periodicity of seedling emergence and achene survival in some species of *Carduus*, *Cirsium*, and *Onopordum*. *Journal of Applied Ecology* 16:641–647.

Shen-Miller, J., M. B. Mudgett, J. W. Schopf, S. Clarke, and R. Berger. 1995. Exceptional seed longevity and robust growth: ancient sacred lotus from China. *American Journal of Botany* 82:1367–1380.

Silvertown, J., M. Franco, I. Pisanty, and A. Mendoza. 1993. Comparative plant demography: relative importance of life-cycle components to the finite rate of increase in woody and herbaceous perennials. *Journal of Ecology* 81:465–476.

Sokal, R. R., and F. J. Rohlf. 1995. *Biometry*. 3d edition. W. H. Freeman, New York, New York.

Steenbergh, W. F., and C. H. Lowe. 1977. *Ecology of the saguaro: 2, reproduction, germination, and establishment, growth, and survival of the young plant*. National Park Service Scientific Monograph no. 8. U.S. Department of the Interior, National Park Service, Washington, D.C.

———. 1983. *Ecology of the saguaro: 3, growth and demography*. National Park Service Scientific Monograph no. 17. U.S. Department of the Interior, National Park Service, Washington, D.C.

Thompson, K., J. P. Bakker, R. M. Bekker, and J. G. Hodgson. 1998. Ecological correlates of seed persistence in soil in the north-west European flora. *Journal of Ecology* 86:163–169.

Toole, E. H., and E. Brown. 1946. Final results of the Duvel buried seed experiment. *Journal of Agricultural Research* 72:201–210.

U.S. Fish and Wildlife Service. 1998. Endangered and threatened wildlife and plants; final rule listing five plants from Monterey County, CA, as endangered or threatened. *Federal Register* 63:43100–43115.

Venable, D. L., and J. S. Brown. 1988. The selective interactions of dispersal, dormancy, and seed size as adaptations for reducing risk in variable environments. *American Naturalist* 131:360–384.

Wendel, G. W. 1977. Longevity of black cherry, wild grape, and sassafras seed in the forest floor. Forest Service Research Paper NE-375. U.S. Department of Agriculture, Washington, D.C.

16

Sensitivity Analysis to Evaluate the Consequences of Conservation Actions
L. Scott Mills and Mark S. Lindberg

ABSTRACT

Sensitivity analysis can be defined as the set of analytical and simulation-based tools that evaluates how changes in life-history attributes of a demographic model affect population growth or rate of extinction. We describe four approaches to sensitivity analysis: (1) manual perturbation of deterministic and stochastic models, (2) analytical elasticity analysis, (3) life-table response experiment (LTRE), and (4) life-stage simulation analysis (LSA). In a case study using data from snow geese, we address two emerging issues in sensitivity analysis. First, although sensitivity analysis has typically been applied to single populations, it has potential to clarify metapopulation processes by quantifying the importance of among- versus within-population dynamics; our example shows that, relative to other life-history attributes, connectivity among populations has a strong impact on population growth. Second, variation in vital rates has been identified as a critical component of sensitivity analysis, yet the variation usually obtained from field estimates contains sampling variation (variation inherent in estimating vital rates) in addition to the relevant process variation (arising from spatial and temporal variation). Using the snow goose data, we find that conservation inferences based on total variation (sampling + process variation) can differ from those based on process variation alone. Sensitivity analysis is an essential counterpart to population viability analysis, because it goes beyond identifying a problem and helps define the most effective solutions. Advances in both data collection and model evaluation must move forward in tandem to obtain the unique insights possible under sensitivity analysis.

INTRODUCTION

Sensitivity analysis can be defined as the set of analytical and simulation-based tools that facilitates evaluation of how past or future changes in

We thank S. Beissinger and D. McCullough for organizing the PVA conference and editing the book. E. Cooch provided snow goose data and helpful feedback. M. Burgman, J. Citta, K. Crooks, S. Hoekman, M. McCarthy, R. Rockwell, J. Rotella, B. Sandercock, and M. Wisdom provided insightful comments. LSM was supported by NSF grant DEB-9870654.

life-history attributes affect population growth or persistence. It arose from the realization that intuition alone is not enough to predict the effect that changes in individual life-history components will have on the likelihood of a population's reaching a predefined population threshold. In the language of deterministic demographic models, for example, different age- or stage-specific vital rates (i.e., birth, death, immigration, and emigration rates) do not have equal impacts on population growth rate (λ). Although this simple demographic fact has been known a long time (Cole 1954), formal sensitivity analysis has become well developed only recently with the introduction of new analytical and simulation-based techniques.

A frequently cited example of how sensitivity analysis showed intuition to be misleading is the work on declining loggerhead sea turtles (*Caretta caretta*). After years of management focused on the seemingly obvious notion that increasing hatchling survival alone should reverse population decline, sensitivity analysis showed that the most efficient way to reverse the decline of this species is to reduce mortality of the life stages that get killed in shrimp nets (Crouse et al. 1987; Crowder et al. 1994); egg protection programs will also be necessary in areas where egg mortality is high (Grand and Beissinger 1997). Thus, sensitivity analysis on sea turtles informed and even helped change fishery policy (Crowder et al. 1994; Grand and Beissinger 1997). Although population viability analysis (PVA) practitioners are often interested in how best to increase population growth rate for species at risk, sensitivity analysis can be equally valuable to identify ways to decrease population growth rate in introduced or pest species (Shea and Kelly 1998; Citta and Mills 1999) or provide insights into managing maximum sustained yield of a stable population (Caughley 1977).

In addition to its role in directing management, sensitivity analysis can give basic insight into population dynamics and can direct research. For example, a study that documents that "inbreeding affects litter size" or "forest fragmentation affects adult survival" or "acid rain affects hatch probability" can be placed in a more meaningful context with consideration of how the observed changes would be expected to affect population growth.

Sensitivity analysis for conservation decision making has become increasingly sophisticated with recent debate and discussion (Tuljapurkar and Caswell 1997; Akçakaya and Raphael 1998; Ehrlén and van Groenendael 1998; Wisdom et al. 2000), but several issues have yet to be addressed. For readers unfamiliar with sensitivity analysis, we first give a brief overview of the four main sensitivity-analysis approaches that have been used to prioritize conservation management. We then consider two issues—forms of variation in vital rates and the use of field data to evaluate the role of connectivity—that have rarely been incorporated into applications of sensitivity analysis. We explore the effect of these factors on management inferences, using a field data set. These factors may or may not affect rankings of sensitivity analysis for any particular application, but the only way to find out is to incorporate these issues into the analyses.

OVERVIEW OF APPROACHES TO SENSITIVITY ANALYSIS
Manual Perturbation

We term as "manual perturbation" the oldest, and probably most common, approach to sensitivity analysis (a.k.a. "conventional sensitivity" in Cross and Beissinger 2001). With this approach vital rates are manually altered by an amount deemed relevant to the problem at hand. In deterministic applications, net reproductive rate, or annual population growth rate (λ), may be plotted against varying levels of survival or reproductive parameters for different age (or stage) classes. This approach has been used to rank the "importance" of different rates to population growth in a range of species including condors (Mertz 1971), elk (Nelson and Peek 1982), and elephants (Fowler and Smith 1973). The use of stochastic models expands the metrics of sensitivity analysis to include not only how different factors affect population growth, but also how they affect probability of extinction or quasi-extinction (i.e., the probability of reaching some threshold of concern). Although in conventional usage each rate is typically changed by the same proportion or amount (e.g., 10% of the mean values), there is nothing inherent in the method that limits changes to fixed proportional changes. Cross and Beissinger (2001) conducted sensitivity analyses that were scaled relative to the range of values each input variable could assume.

Sensitivity analysis using manual perturbations is not limited to investigating the importance of vital rates alone. Rather, it can include impacts of a range of factors including age structure, density dependence, inbreeding depression, and connectivity (Burgman et al. 1993; Mills et al. 1996; Beissinger and Westphal 1998). For example, sensitivity analysis using manual perturbations has been used in various ways to explore the relative importance of dispersal and other parameters to conclusions of spatially explicit population models (Conroy et al. 1995; Dunning et al. 1995; South 1999; Ruckelshaus et al. 1999).

Also, manual perturbation sensitivity analysis has the unique ability to easily incorporate different age or stage structures. Both deterministic and stochastic analyses can either assume a stable age distribution or input an observed age distribution from the field. In fact, sensitivity analysis can quantify the effect of age structure on population growth or extinction probability (Citta et al. in review).

The strength of manual perturbation sensitivity analysis—that is, the different measures for evaluating "importance" and the myriad of factors that can be considered in unique ways for particular situations—is also its biggest weakness, because there is no standardized metric that can be compared across species or studies. For example, one study may evaluate how 10% changes in demographic rates affect the probability of reaching 20 individuals in 200 years, while another considers how 5% changes affect the probability of total extinction within 50 years. As a result, these approaches can be incredibly valuable for the organism that they target, but the lack of a standardized metric comparable across species or studies is problematic for life-history comparisons and for inferring best management actions by analogy with other species.

Manual perturbation approaches may also be time-consuming, with many simulations of many combinations required. However, McCarthy et al. (1995) have developed an innovative approach to stochastic sensitivity analysis that efficiently reduces the computational effort required to determine quasi-extinction risk for an exhaustive array of model parameters and their interactions. A systematic array of parameter combinations is analyzed via sets of PVA, and logistic regression is used to summarize the relationship between extinction risk and the model parameters. Variables may be ranked according to their importance in predicting extinction or quasi-extinction based on the standardized regression coefficient (Cross and Beissinger 2001). Drechsler (1998) derived a complementary approach to identify parameter combinations that behave in similar ways, effectively reducing the magnitude of the problem to a manageable size for analyses based on extinction or quasi-extinction risk.

Examples of sensitivity analysis based on manual perturbations are as varied as the method itself; some would include Beissinger's analysis (1995) of snail kites (*Rostrhamus sociabilis*), Marmontel et al.'s analysis (1997) of conservation options for the Florida manatee (*Trichechus manatus latirostris*), and Akçakaya and Raphael's study (1998) of management options for northern spotted owls (*Strix occidentalis*). Cross and Beissinger (2001) found that the most sensitive variables in a model of wild dog (*Lycaon pictus*) population dynamics differed when sensitivity was evaluated by manual perturbations of the mean and the range, and when it was evaluated by logistic regression.

Analytical Sensitivity and Elasticity Analysis

Analytical sensitivities and elasticities are calculus-based measures of how infinitesimal changes in individual vital rates will affect population growth. In contrast to manual perturbation, these measures assess sensitivity solely by evaluating changes on λ, so effects on extinction or quasi-extinction cannot be assessed. Analytical sensitivities are calculated from the reproductive value and stable age (or stage) distribution vectors (left and right eigenvectors, respectively) and λ (the dominant eigenvalue) of the population matrix containing stage-specific vital rates (Goodman 1971; Caswell 1989b; de Kroon et al. 2000). Sensitivities measure the absolute change in λ, given an infinitesimal absolute change in a vital rate, while all other vital rates remain constant. Analytical elasticities rescale the sensitivity to account for the magnitude of both λ and the vital rate. Thus, elasticities are "proportional sensitivities" that measure the proportional change in λ given an infinitesimal one-at-a-time proportional change in a vital rate (de Kroon et al. 1986; Caswell 1989b), assuming the population is growing or decreasing at a constant rate and has a stable age distribution (SAD). Elasticities can be added together to predict the joint effect of changes in multiple rates if the changes in vital rates and λ are linearly related, and elasticities of all matrix elements sum to one (de Kroon et al. 1986; Mesterton-Gibbons 1993). Elasticities can be calculated for both matrix elements and "lower-level" components of vital rates that make up matrix elements, such as reproduction and survival components in each element in the top row of a projection matrix (Caswell 1989b; Doak et al. 1994). Elasticities of component vital rates ("lower-level elasticities") do not add to one but can still be ranked. The straightforward interpretations and wide generality of elasticities provide an intuitive metric that assays the relative importance of life-cycle transitions both within and among studies and species (de Kroon et al. 2000).

Because they are operationally defined, easily applied, and comparable across studies, elasticities have been embraced in applied biology. Elasticities are usually calculated from a single population matrix constructed from average, or even "best guess," vital rates. The vital rate in the mean matrix with the highest elasticities is recommended for highest management or research priority (e.g., Crowder et al. 1994; Heppell et al. 1994; Maguire et al. 1995; Olmsted and Alvarez-Buylla 1995). Unlike elasticity analysis assumptions, however, changes due to management actions or natural variation are not infinitesimal, nor do they occur one at a time.

Mills et al. (1999) investigated whether the conservation applications of elasticities are robust to violation of their mathematical definitions and came to three major conclusions. First, if the vital rates in a matrix or a population of interest were different than those of the mean matrix used to calculate elasticities, the rankings of the elasticities could change (see Caswell 1996b). Although changes in rankings may not occur often (Benton and Grant 1996; Dixon et al. 1997; de Kroon et al. 2000), the cases where such shifts do occur may have important conservation consequences because the priorities of research or management may be misdirected (Wisdom et al. 2000). Second, Mills et al. (1999) found that elasticities were good qualitative and quantitative predictors of changes in population growth rate as long as all vital rates changed by the same proportional amount. Unfortunately, vital rates seldom change by equal amounts in nature, and Mills et al. (1999) found that, when different vital rates changed by different amounts, deterministic elasticities were often poor predictors of how population growth would change (see also Caswell 1996a; Ehrlén and Van Groenendael 1998; Wisdom et al. 2000). This last issue—the disconnect between "importance" as assayed by elasticities and "importance" as related to relative variation in a vital rate—is especially disconcerting because vital rates with high elasticity tend to have low levels of variation (Gaillard et al. 1998; Pfister 1998). This implies that basing management inferences solely on elasticities could lead to erroneous predictions about which rates most affect population growth under management.

In short, elasticities are a mathematically elegant metric that can be derived from relatively sparse data in the form of a single matrix of vital rates. However, elasticities alone do not account for how much vital rates have changed in the past, or might change in the future. Also, elasticities are calculated on infinitesimal, one-at-a-time changes, with multiple changes assumed to be additive, and effects of vital rate changes on growth rate assumed to be linear. The calculations rely on asymptotic matrix properties, so a SAD is assumed, although it is possible to calculate elasticities for periodic deterministic vital rates that cycle predictably (Caswell and Trevisan 1994), and from stochastic models (Tuljapurkar 1990; Benton and Grant 1996; Dixon et al. 1997; Grant and Benton 2000). Correlations among vital rates can be incorporated using an extension of elasticity analysis explained by van Tienderen (1995).

Life-Table Response Experiments

The life-table response experiment (LTRE) approach to sensitivity analysis is an extension of analytical sensitivity and elasticity analysis that incorporates vital rate-specific changes or variation. Caswell (1997) explained the approach for a simple, one-way, fixed-effect experimental design as follows. Imagine a "control" set of vital rates and a "treatment" set, perhaps arising from a human-caused perturbation. A population growth rate at SAD is characteristic of the population matrix made up of each set of vital rates. The difference between the two growth rates is a function of how much different vital rates change under the treatment and the effect of changes in each rate on λ. LTRE decomposes the treatment effects on λ into contributions from individual vital rates by taking the product of the vital rate effect (i.e., change in a given rate due to the perturbation) and the analytical sensitivity of λ to changes in that rate (Caswell 1989a, 1997; Brault and Caswell 1993).

LTRE approaches are not limited to fixed "experiments," but rather have been generalized to comparative observations under natural conditions for a variety of matrix models; they can also include known correlation structures among rates (Horvitz et al. 1997). They therefore have become an elegant way of quantifying the fact that "a large effect on a vital rate to which λ is insensitive may contribute much less to variation in λ than a much smaller effect on a vital rate to which λ is more sensitive" (Caswell 1996a, 74).

For example, Levin et al. (1996) and Caswell (1996a) evaluated how different sources of estuarine pollution from sewage, oil, and algae from eutrophication enrichment affected a deposit-feeding polychaete (*Capitella* sp.). The presence of sewage substantially increased *Capitella* population growth from λ = 1.79 without sewage to λ = 4.06 with sewage. With the addition of sewage, fertility increased by almost fourfold, while age at maturity decreased by about one-half. LTRE analysis was used to show that the overall contribution of age at maturity to the substantial increase in λ was three times greater than that of fertility. Survival, which was actually reduced by the addition of sewage, contributed very little to the change in growth rate. Ehrlén and van Groenendael (1998) provided another interesting example of LTRE analysis; for a leguminous herb (*Lathyrus vernus*) with different levels of variance for different rates, vital rate elasticities calculated from the mean matrix were poor indicators of the overall impacts of individual rates on population growth.

LTRE-based approaches are important extensions of analytical sensitivity and elasticity analysis because they can incorporate information on the amount, form, and correlations of changes in different vital rates. However, the approach still incorporates some of the same assumptions of elasticities, including additivity of effects on different growth rates, linearity of the relationship between vital rate and growth rate changes, and asymptotic properties such as SAD. Mills et al. (1999) found that, although LTRE-type approaches led to predictions of population growth that were much better than elasticities alone, nonlinearities between growth rate and vital rate changes did lead to some, usually minor, disparity between actual population growth and that predicted by changing different rates by specified amounts.

Life-Stage Simulation Analysis

Wisdom and Mills (1997) developed a simulation-based approach to sensitivity analysis that might be considered a hybrid of the previous three methods. The approach is called life-stage simulation analysis (LSA; Wisdom et al. 2000) because it uses simulations to evaluate the impact of changes in different vital rates on λ. For purposes of conservation decision making, the user inputs the mean and variation of vital rates observed in the past or those expected in the future under different scenarios of management (see "Discussion" for the need to distinguish past from future changes). Correlations among vital rates are specified from field data if possible, as are the distribution functions for each vital rate (i.e., uniform, lognormal, beta, etc.). A computer program constructs many replicate matrices with each rate in each matrix drawn from the specified distributions. Currently, each matrix is projected asymptotically to a SAD.

Output metrics in LSA include elasticity-based measures (e.g., the frequency of replicates having the same vital rate of highest elasticity, and the differences in elasticity values whenever the rankings of elasticities change across the replicates), as well as other metrics that avoid elasticity entirely (Wisdom and Mills 1997; Wisdom et al. 2000). For example, an LSA approach can calculate the percentage of replicates having positive population growth under different scenarios. Also, by regressing λ on each vital rate as other rates change simultaneously, one can derive the coefficient of determination (r^2), representing the proportion of the variation in the population growth rates over all simulations that is explained by variation in that vital rate. When all main effects and interactions are included, the r^2 values for all rates sum to one.

An intuitive understanding of r^2 for each vital rate can be derived by comparing it to analytical sensitivities and elasticities. When λ is a linear function of the vital rates, the slope of the line equals the analytical sensitivity, and r^2 is a function of both the slope (i.e., analytical sensitivity) and the proportionate variation in that vital rate, adjusted for covariance among vital rate. The same relationships hold for elasticity, if the regression is done on log-transformed data (Brault and Caswell 1993; Horvitz et al. 1997; Wisdom and Mills 1997). Therefore, the simulation-based LSA r^2 corresponds to analytical LTRE approaches, because it accounts for both infinitesimal sensitivity and range in variation of different rates (Caswell 2000; Wisdom et al. 2000). The r^2 metric of LSA is also mathematically similar to the standardized regression coefficient used by McCarthy et al. (1995) in logistic regression of PVA (see "Manual Perturbation" above), because the standardized regression coefficient is equivalent to the square root of r^2 (Snedecor and Cochran 1980, 357).

LSA has been applied to several cases of conservation decision making. For example, Crooks et al. (1998) used LSA to show that adult survival has the greatest influence on cheetah (*Acinonyx jubatus*) population growth, a result that questions whether juvenile mortality alone is limiting growth in this endangered species. In a very different application, Citta and Mills (1999) used LSA to ask which of several potential management scenarios would be more likely to *reduce* population growth of brown-headed cowbirds (*Molothrus ater*), a brood parasite on many passerine birds of conservation concern. They found that egg survival has the greatest importance to population growth of cowbirds under most conditions, implying that egg removal would have the largest effect, and that decreasing adult survival on wintering ranges would have little effect. Because the age structure is skewed toward eggs, however, it is more difficult to initiate a given proportional change in egg survival compared to the same proportional change in adult survival. Collectively, this situation, where egg removal has the largest proportional effect but adult survival is easier to change, points to a more holistic management of land uses to disfavor cowbirds (Citta and Mills 1999).

Thus, LSA has some of the versatility of the stochastic manual perturbation methods, and relies on repeatable metrics (e.g., r^2 and average elasticity measures) that could be compared across species and studies. Like manual perturbation methods and LTRE, it accounts for variation in vital rates. As it has been used to date, however, it shares the disadvantages of both elasticity and LTRE in assuming asymptotic properties. Also, all sensitivity inferences are based on λ; sensitivity measures such as probability of extinction or quasi-extinction cannot be considered using this approach. Although Caswell (2000) argued that LSA cannot be used to make predictions of the future because it incorporates variation in vital rates, we disagree on the grounds that both means and variation can be used to describe the past or to pose scenarios for the future (see "Discussion").

EMERGING ISSUES IN CONSERVATION DECISION MAKING WITH SENSITIVITY ANALYSIS

Next we consider two issues—forms of variation and the role of connectivity—that have rarely been explored but that have strong implications for applying sensitivity analysis to conservation decision making. We provide a case study using field data to demonstrate how these factors can impact conservation conclusions.

All Variation Is Not Created Equal

Three of the four methods (manual perturbation with stochastic models, LTRE, and LSA) are designed to account for how variation in vital rates could affect population growth. However, connections among field data, model input, and conservation applications are often unclear because the total variation in estimates of vital rates includes two forms of variation—process variation and sampling variation—that are distinctly different (White et al., chap. 9 in this volume).

Process variation is caused by spatial and temporal changes in vital rates (Thompson et al. 1998). Spatial variation arises from changes in community composition, habitat quality, and habitat heterogeneity over the landscape, which in turn may be related to environmental conditions such as aspect, slope, precipitation, and successional-stage differences. Temporal variation is the unpredictable change in the environment that impacts a given population through time (Burgman et al. 1993). Weather is often an important driver for temporal variation, but biotic factors such as competition, predation, disease, and human impacts may also be critical. Interactions among factors make it difficult to separate spatial and temporal variation (Burbidge and McKenzie 1989; Doak et al. 1994; Ringsby et al. 1999).

In contrast to process variation, which directly acts on organisms, sampling variation is the uncertainty in parameter estimates that arises from the fact that vital rates are estimates from an incomplete enumeration of individuals. When modeling population dynamics, sampling variation should be removed from total variation so that the focus of a population analysis is on only process variation (Burnham et al. 1987; Link and Nichols 1994; Thompson et al. 1998). Problems due to failure to separate sampling from process variation have recently been noted in the context of population stability (Link and Nichols 1994; Gould and Nichols 1998), monitoring (Caswell et al. 1998; Thompson et al. 1998), and PVA (Beissinger and Westphal 1998; Ludwig 1999). Gould and Nichols (1998) demonstrated that most of the total variation in survival probabilities of three species of birds was the result of sampling variation (see also Hitchcock and Gratto-Trevor 1997). Ludwig (1999) showed that ignoring sampling error, when it is present in the estimates used for PVA analysis, leads to estimates of extinction probability that are positively biased and have confidence intervals that are too small.

The stochastic manual perturbation sensitivity approach can explicitly include both spatial and temporal components of process variation. It captures spatial variation by simulating many different possible populations, and temporal variation using Monte Carlo simulations and calculating probability of extinction (Burgman et al. 1993) or stochastic population growth rate (Tuljapurkar 1997). In application, however, most users do not separate temporal and spatial variation. The other approaches that account for variation in sensitivity analysis, LTRE and LSA, easily accommodate spatial variation, but reliance on asymptotic growth rate at SAD limits the inclusion of temporal variation. LSA could use stochastic projections to model temporal variation, but to date it has only been used to project replicate matrices to SAD. Likewise, numerical differentiation or a stochastic analytical approach can be used to calculate elasticities and LTRE measures under temporal variation (Tuljapurkar 1990; Benton and Grant 1996; Grant and Benton 2000). This approach has seldom been applied to conservation questions, but may have great potential (Dixon et al. 1997). The consequences of failing to separate sampling from process variation in sensitivity analysis has rarely been explored (Gaillard et al. 1998), despite the fact that virtually all sensitivity analyses that include variation have included total variation (i.e., process plus sampling) rather than process variation.

The Value of Connectivity: Among- versus Within-Population Vital Rates

An entire field, metapopulation analysis, is predicated on an overwhelming importance of among-population movement for the persistence of the metapopulation. As Harrison (1994, 177) noted: "It seems necessary to adopt a broader and vaguer view of metapopulations as sets of spatially distributed populations, among which dispersal and turnover are possible but do not necessarily occur . . . A possible way forward is to ask, in each specific case, 'what is the relative importance of among-population processes, versus within-population ones, in the viability and conservation of this species?'"

Sensitivity analysis should be an ideal approach to facilitate an operational definition of metapopulation, because effects of among- versus within-population vital rates on growth rate or extinction can be compared. However, sensitivity analysis has focused primarily on single-population dynamics. In some cases, these single-population analyses may include an asymmetric treatment of movement, where estimates of apparent survival (Lebreton et al. 1992) include the effects of emigration, but estimates of fecundity do not include immigration. Only a few studies have used sensitivity analysis to evaluate the relative importance of among- versus within-population processes for specific populations where relevant field data exists (see Burgman et al. 1993). Most of the examples have used manual perturbation PVA approaches, such as Beier (1993) for cougars (*Puma concolor*), Akçakaya and Baur (1996) for land snails (*Arianta arbustorum*), Lefkovitch and Fahrig (1985) and Fahrig and Paloheimo (1988) for mice (*Peromyscus leucopus*), Pulliam et al. (1992) and Liu et al. (1995) for Bachman's sparrow (*Aimophila aestivalis*), and Lindenmayer and Lacy (1995) for arboreal marsupials (*Trichosurus caninus, Petauroides volans*, and *Gymnobelideus leadbeateri*). Wootton and Bell (1992) calculated deterministic elasticities of within-population vital rates and migration rates between peregrine falcon (*Falco peregrinus anatum*) populations using multiregional matrix models (see Caswell 1989b; Burgman et al. 1993).

ADDING SPECIFIC FORMS OF VARIATION AND CONNECTIVITY TO SENSITIVITY ANALYSIS: A CASE STUDY

The Model System: Lesser Snow Geese

Our objective is to examine how the results of sensitivity analysis might change if types of variation and connectivity were accounted for in an analysis with real data. The challenge, however, is that the data needed for such an analysis typically do not exist for rare or endangered species. Lack of data does not decrease the importance of considering these factors and may guide data collection efforts for the future; it does, however, limit the choice of taxa for our analysis. Therefore, we chose a species, the lesser snow goose (*Anser caerulescens caerulescens*), whose conservation importance comes not from having low and declining numbers, but from having large and increasing numbers. We hope that exploring these issues with this data set will help future data collection and interpretation in other taxa, including threatened ones.

Based on the midwinter index of abundance, the midcontinent population of lesser snow geese (hereafter snow geese) increased by 300% between 1969 and 1994 (Abraham and Jefferies 1997). This dramatic increase is believed to result from changes in agricultural practices on the wintering grounds, where snow geese exploit waste grains, and from increased availability of refuges on wintering and staging areas (Batt 1997). Through excessive feeding pressure, large numbers of snow geese have severely degraded Arctic and sub-Arctic habitats where they stage during migration and breed (Abraham and Jefferies 1997). In addition to direct damage to vegetation resulting from such high snow goose numbers, habitat degradation may also negatively impact other species that breed in these regions.

For these populations of snow geese, as for the brown-headed cowbird example discussed earlier, the question of interest is how best to *decrease* population growth. The final decision of the best course of action will include not only biological expectations, but logistics and social and political feasibility. Sensitivity analysis can inform the biological expectations. Our analysis for snow geese does not consider specific management scenarios.

Rockwell et al. (1997) conducted an elasticity analysis to consider the practical question of what actions would most reduce snow goose population growth. In one of their analyses, they developed a five-stage population projection matrix, parameterized with vital rate estimates based on 12 years of data (1973–1984) from a 30-year study of snow geese breeding at La Pérouse Bay, Manitoba (Cooke et al. 1995). All losses were attributable to mortality; the Rockwell et al. (1997) model contained no connectivity in terms of immigration or emigration. Rockwell et al. (1997) conducted deterministic elasticity analysis for five potential sets of data that accounted for some of the uncertainties in life-history data. In all cases, adult survival contributed more to λ than did any other vital rate. From these results, they considered how much change in growth rate would be achieved under a number of scenarios involving reduced survival or reproductive output.

We analyzed the same 12 years of snow goose data (with some minor exceptions presented below) and used the same basic projection model as Rockwell et al. (1997), except that we included connectivity and specific forms of variation in vital rates. Although we used similar data, our analysis does not explore management options that were considered in Rockwell et al. (1997). The elements in the top row of the matrix are post–birth pulse, age-specific fertilities, incorporating both fecundity (number of eggs laid that survive to fledge) and survival of females that produce those offspring. Age in years is indicated by subscript i, and the projection interval is one year. Thus, fertility in each element of the top row of the matrix is

$$F_i = S_i^* R_i,$$

where S_i is age-specific survival probability and R_i is age-specific fecundity. S_1 is juvenile survival (from fledging to the first anniversary of fledging), and adult survival probability is assumed constant for age classes 2 through 5+. Rockwell et al. (1997) defined the fecundity term (R_i) as

$$R_i = BP_i * (TCL_i/2) * (1 - TNF_i) * P1_i * P2_i * (1 - TBF_i) * P3_i,$$

where BP is breeding propensity or the probability that a female of that age class will breed, TCL is total clutch size reduced by one-half to focus on females only, TNF is total nest failure, $P1$ is egg survival, $P2$ is hatching success, $P3$ is gosling survival from hatching to fledging given that the brood is not totally lost, and *TBF* is total brood failure for the *i*th age class. Thus, fecundity represents the number of birds fledged. Although fecundity comprises a number of different components, some parts do not have reliable parameter estimates, so we do not evaluate importance of individual fecundity components, although it is computationally easy to do.

There are many ways to add connectivity to a matrix model (see Caswell 1989b; Burgman et al. 1993). For demonstration purposes we chose the simple approach of modeling connectivity by multiplying survival probability by the sum of immigration probability and one minus emigration probability $[S_i((1 - E_i) + I_i)]$, where I_i is probability of immigration to the La Pérouse population for age class *i*, and E_i is probability of permanent emigration from the La Pérouse population for age class *i*. Because the field data that we used come from a single population (La Pérouse Bay), emigration can be interpreted as the probability of permanently leaving the La Pérouse population sampled during brood rearing (where snow geese were banded), and immigration represents the proportion of the population that enters La Pérouse from anywhere else. Using these parameters, our projection matrix was

$$
\begin{bmatrix}
[S_1((1 - E_1) + I_1)]R_1 & [S_2((1 - E_2) + I_2)]R_2 & [S_3((1 - E_3) + I_3)]R_3 & [S_4((1 - E_4) + I_4)]R_4 & [S_5((1 - E_5) + I_5)]R_5 \\
S_1((1 - E_1) + I_1) & 0 & 0 & 0 & 0 \\
0 & S_2((1 - E_2) + I_2) & 0 & 0 & 0 \\
0 & 0 & S_3((1 - E_3) + I_3) & 0 & 0 \\
0 & 0 & 0 & S_4((1 - E_4) + I_4) & S_5((1 - E_5) + I_5)
\end{bmatrix}
$$

Methods for parameter estimation and associated variances are reported in Cooke et al. (1995) with the following exceptions. Evan Cooch (Cornell University) provided us with age-specific estimates of means and variances of reproduction and survival parameters when these values were not reported in the original publication, as well as with estimates of emigration probability and variance from capture-recapture models (Burnham 1993). Estimates of immigration probability do not exist, so we made the simplifying assumption of setting the mean and variance for immigration equal to those of emigration probability. This is based on the premise that emigration probabilities for all other populations of snow geese in the midcontinent region are equal to that of La Pérouse Bay, and that the probability of an emigrant immigrating to La Pérouse is equal to the proportional size of that population relative to the overall numbers of snow geese in the midcontinent region. While it is true that setting $I_i = E_i$ would cause these terms to "cancel out" in the matrix, sensitivity analysis using elasticities can be conducted

Table 16.1 Annual Age-Specific Vital Rates Used in Sensitivity Analysis for
Lesser Snow Goose

Scenario	Age Class	R_i	S_i	$E_i = I_i$
Mean	1	0	0.4443	0.3543
	2	0.3305	0.8683	0.02567
	3	0.7761	0.8683	0.02567
	4	0.9822	0.8683	0.02567
	5	1.032	0.8683	0.02567
Low (total variation)	1	0	0.2016	0.0985
	2	0.0846	0.7455	0
	3	0.2946	0.7455	0
	4	0.3167	0.7455	0
	5	0.4979	0.7455	0
High (total variation)	1	0	0.6870	0.6102
	2	0.8480	0.9910	0.1032
	3	1.614	0.9910	0.1032
	4	2.1818	0.9910	0.1032
	5	1.8787	0.9910	0.1032
Low (process variation only)	1	0	0.2289	0.1663
	2	0.1039	0.8282	0
	3	0.3976	0.8282	0
	4	0.4094	0.8282	0
	5	0.6002	0.8282	0
High (process variation only)	1	0	0.6597	0.5424
	2	0.7713	0.9743	0.06566
	3	1.3506	0.9743	0.06566
	4	1.8967	0.9743	0.06566
	5	1.6574	0.9743	0.06566

Notes: The mean rate, as well as the low and high bounds calculated from total variation (process + sampling variation) and from process variation alone, are listed for each annual age class, for fecundity (R), survival (S), emigration (E), and immigration (I).

for I_i and E_i separately, and in the stochastic LSA-based approach immigration and emigration will be drawn as individual random variates for each matrix.

We used the averages of the point estimates for the 12 years of available data as the mean value of survival and emigration rates (table 16.1). Mean R_i (fecundity) values were calculated as the mean of the product of its components.

We estimated total variance (T^2) for the ith age-class mean of each parameter ($\overline{X}_i$) as

$$T_i^2 = \frac{1}{n-1} \sum_{j=1}^{n} (\hat{X}_{ij} - \overline{\hat{X}}_i)^2,$$

where n equals the number of years in the study (12 years) and $\hat{X}_{ij}$ is the estimated parameter value for the ith age class in the jth year of the study. We obtained process variance (in this case temporal variance, τ_i^2) by subtracting average sampling variance from our estimate of total variance (Link and Nichols 1994; White et al., chap. 9 in this volume). Sampling covariance among estimates for each parameter was set at zero because covariance estimates were not available for several parameters. In cases where temporal variance was estimated as a negative value (Gould and Nichols 1998), we assigned a value equal to the smallest positive value that we estimated (0.0004). Based on both total and temporal variance alone (i.e., process variance with sampling variation removed—see also White et al., chap. 9 in this volume), we established a range of vital rates (table 16.1) from which to sample in our sensitivity analysis. For total variance the maximum and minimum were equal to $\bar{X}_i \pm 2.0\,T_i$, and for temporal variance (process alone) the range was equal to $\bar{X}_i \pm 2.0\,\tau_i$. For all parameters we truncated the range of values at a minimum of 0.0, and for probabilities we truncated the maximum value at 1.0.

Sensitivity Analysis for the Snow Goose Model Incorporating Connectivity and Forms of Variation

The effects of connectivity and exact type of variation could readily be evaluated using stochastic manual perturbation approaches, LTRE, or LSA. Our goal here was not to perform a snow goose sensitivity analysis per se (Cooch et al. 2001), but rather to use these data to explore the general issues of how connectivity and type of variation affect sensitivity analysis. For the purposes of demonstration, we use the LSA approach. As described above, the LSA approach uses simulations to calculate traditional elasticities across a range of variation, as well as metrics that bypass elasticities entirely. For the sake of simplicity, in this example we provide elasticities for comparison to previous work and discuss them briefly, but we focus primarily on only one of the LSA sensitivity metrics, r^2, as an index of vital rate "importance." The r^2 metric indicates the proportion of variation in growth rate explained by changes in each vital rate with all other rates changing simultaneously. We let each rate vary independently, although LSA can readily incorporate covariation among vital rates (Hoekman et al. 2000; Wisdom et al. 2000). Moreover, we stress that obtaining information on covariation among rates from the field is highly important (Horvitz et al. 1997; Schmutz et al. 1997), but those data are not yet available.

Currently, the LSA approach can use uniform, beta, poisson, or lognormal distribution for vital rates (Hoekman et al. 2000; Wisdom et al. 2000). Again, to simplify our approach, we elected to use a uniform distribution for all rates. Note that by using the minimum and maximum observed in the field data, we are performing a sensitivity analysis that includes both natural and human-induced variation observed in the past as well as sampling variation, although we shall see what happens when sampling variation is removed. This may be an appropriate starting point to project future changes, assuming that bounds of the past will apply to the future. Alternatively, we could easily relax this assumption of using the past variation to reflect the future and instead simulate specific future management scenarios, change the bounds of variation, or consider different distributions for vital rates (Citta and Mills 1999; Hoekman et al. 2000; Wisdom et al. 2000).

If we perform a deterministic, single population, analytical sensitivity analysis and consider just the mean vital rates and ignore immigration and emigration, our matrix model converges to that of Rockwell et al. (1997) with the same result: adult survival is most important. Based on component elasticities using mean vital rates, values are highest for adult survival (elasticities summed across ages 2 through 5 = 0.832), with much lower elasticity for juvenile survival (0.168) and fecundity (elasticities summed across ages 2 through 5 = 0.168).

How do conclusions change when the analysis accounts for emigration, immigration, and measured variation in vital rates? First we use the LSA approach with total variation, as done in virtually all sensitivity analyses to date that have incorporated variation in vital rates. That is, we followed the usual approach of setting the bounds of vital rates based on total variation including both process and sampling variation observed from field data. Sensitivity analysis using the r^2 metric and the model with connectivity identified juvenile survival as explaining the highest proportion of variation in growth rate, followed by juvenile movement (emigration and immigration), adult fecundity, adult survival, and adult movement (fig. 16.1A). We also performed the sensitivity analysis using total variation and the model without connectivity by using the same model as for calculating elasticities above. Juvenile survival again is ranked highest (r^2 = 0.54), with fecundity and adult survival considerably lower (r^2 = 0.21 and 0.21, respectively). Therefore, we conclude that variation can lead to very different rankings of "importance" than deterministic elasticity analysis—in this case causing juvenile survival to replace adult survival as the highest-ranking rate—because overall change in population growth rate is a function of both the infinitesimal impact of vital rates on growth rate and the amount that a given vital rate can change (see Caswell 1997; Gaillard et al. 1998; Mills et al. 1999; Wisdom et al. 2000). Also, this example demonstrates that connectivity, or movement, is nontrivial in terms of its impact on λ for this population.

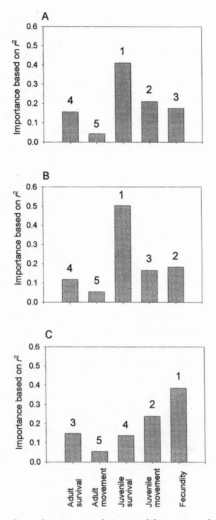

Fig. 16.1 Sensitivity analysis of snow goose data using life-stage simulation analysis (LSA). Relative "importance" of each vital rate to population growth rate is assayed by the proportion of the variance in population growth (r^2) accounted for by changes in each vital rate. *Numbers above bars* give importance rank, with one being the highest. "Movement" refers to connectivity, or immigration and emigration. The deterministic elasticity analysis ranked adult survival highest. However, LSA, including connectivity and forms of variation, led to different rankings: *A*, using total variation (process + sampling); *B*, using process variation only (sampling variation removed); and *C*, using process variation only, under the hypothetical scenario where juvenile survival had the same proportion of process to sampling variation as adult survival had.

How might removing sampling variation from our estimate of total variation alter our conclusions? The range of variation due only to real-world processes is less than total variation (table 16.1), with the decrease varying across vital rates according to the proportion of process variation versus sampling variation. Interestingly, in this case, differences in sensitivity rankings with total variation versus process variation were small (compare fig. 16.1B to 16.1A). Juvenile survival still accounted for the most variation in growth rate, and juvenile movement still had a nontrivial impact on λ. However, these small changes in sensitivity rankings using total versus process variation alone resulted from the fact that the rates with highest total variation tended to have the highest proportion of process to sampling variation. In particular, the highly variable juvenile survival rate was estimated to contain 79% process variation and 21% sampling variation, whereas adult survival had minimal total variation of which only 11% was process and 89% was sampling variation. Therefore, in this particular case, which may be unusual, removing sampling variation did little to change the impact of variation on overall importance.

Given that high total variation in other data sets could be made up of any ratio of process variation to sampling variation, we next consider a different scenario. Suppose that variation in juvenile survival had the same proportion of process to sampling variation as adult survival. In other words, suppose total variation remained unchanged but variation in juvenile survival comprised 11% process variation and 89% sampling variation. With this one hypothetical change, the rankings of importance of different vital rates changed drastically (compare fig. 16.1C to figs. 16.1A and 16.1B). Juvenile survival dropped from highest ranking to fourth, and fecundity went from third highest to first. In this case, conclusions based on sensitivity analysis using total variation (fig. 16.1A) were very different than those based on only process variation (fig. 16.1C). We stress that this last scenario is hypothetical. However, it shows that use of total variation in sensitivity analysis could cloud conclusions, because sampling variation can inflate the "importance" of certain rates. It will not be unusual for this to occur in real population data because some rates will be more difficult to estimate than others (e.g., juvenile survival in passerines).

Other Issues Related to Sensitivity Analysis

Our evaluation of how process variation and connectivity affect sensitivity analysis is not complete without further consideration of the link between the measurement and the modeling of these factors. Process variation can be estimated by subtracting the mean of the sampling variation from total variation if sampling variation is assumed constant (Skalski and Robson 1992; Link and Nichols 1994) or by more complex estimators if sampling variation is not constant (Burnham et al. 1987; Thompson et al. 1998). In either case, components of variance analysis require valid estimates of vital rates. It is difficult to separate sampling from process variation if indices are used instead of valid estimates (Link and Nichols 1994; White et al., chap. 9 in this volume). We think that capture-recapture methods provide some of the most robust approaches for parameter estimation (e.g., Lebreton et al. 1992). Recent advances in capture-recapture approaches have provided new opportunities for estimating movement parameters (e.g., Nichols 1996), which were difficult to estimate in the past. We suggest that more emphasis be placed on estimating vital rates needed to parameterize a matrix model, rather than rushing to conduct a matrix analysis that may provide misleading results. We recognize, however, that appropriately parameterizing a matrix model is a challenging task. Even in a data set as rich as that for the snow goose, some parameters were not rigorously estimated.

It is less clear how demographic stochasticity should be included in sensitivity analysis, especially if the analysis is performed on a different population than the one from which vital rates were estimated. If the population being analyzed is small, and expected to remain small, demographic stochasticity should probably be considered a form of process variation (Lande 1998). If the sampled population from which vital rates are estimated is much smaller than the target population to which modeling will be applied, however, demographic stochasticity in the small population should not be included as process variation in the large population, where demographic stochasticity is unlikely to manifest. Sæther et al. (1998) and Sæther and Engen (chap. 10 in this volume) describe a way to separate demographic from environmental stochasticity.

After process variation is estimated, it can be modeled in a sensitivity analysis. Asymptotic models that account for variation, including LTRE and LSA, properly model only spatial variability, not temporal variability, as process variation. In our snow goose example we used data from one population over 12 years, so that all of the process variation was attributable to temporal variation. By using an approach that depends on asymptotic properties, we must assume that changes in different rates would be constant long enough for the population to reach a SAD. This assumption is probably a stretch, but at least focuses on process variation alone, without being confounded by sampling variation. It remains to be seen how important the violation of this assumption is for the kinds of analyses explored in this work (Citta et al. n.d.).

Finally, our case study does not consider specific management scenarios (Rockwell et al. 1997) nor does it directly address how easy it is, logistically or politically, to change certain rates. This becomes critical when evaluating management alternatives using sensitivity analysis, because we must evaluate not only how changes will affect population growth or persistence, but also how likely those changes are to occur. There is not yet a formal protocol for linking sensitivity analysis with the ability to change different rates, but certainly many investigators have performed sensitivity analyses with an eye toward management possibilities (see for example McKelvey et al. 1993; Lindenmayer and Possingham 1996; Silvertown et al. 1996; Akçakaya and Raphael 1998; Martien et al. 1999). There may be some utility in linking sensitivity analysis with formal decision analysis (Ralls and Starfield 1995). We did not consider management alternatives in our snow goose analysis, but we suggest that the issues we raise—type of variation and connectivity—should be considered in management scenarios evaluated with sensitivity analysis.

DISCUSSION

Sensitivity analysis is critical to understanding population dynamics. In the spirit of this volume of papers, it is an essential extension of PVA because it analyzes what particular actions are most likely to change the population trajectory of the species of interest. Sensitivity analysis is not limited to single-population studies, but can also give useful insights that go to the heart of metapopulation analysis: whether among- versus within-population vital rates are more important to persistence of the metapopulation (see also Wootton and Bell 1992; Hitchcock and Gratto-Trevor 1997). The scale of analysis, or which populations to include and which to properly apply inferences to, remains critically important (Doak et al. 1992; Donovan et al. 1995).

An array of sensitivity analysis approaches is now available, with a range of data requirements and assumptions. Deterministic manual perturbation analysis and analytical elasticities require the least data, but may be more limited for providing insights into future management options because they do not account for variation in vital rates. Stochastic or PVA-based manual perturbation analyses relax nearly all assumptions, but consequently require the estimation of many more parameters. For example, PVA-based approaches do not have to assume asymptotic properties, but if SAD is not assumed, an initial stage distribution must be determined (e.g., we did not have field-based stage distributions even for the well-studied La Pérouse snow goose population!). Of course, parameters that are unknown can and should themselves be considered the targets of sensitivity analysis, but at some point the uncertainty becomes overwhelming. In contrast to PVA-based approaches, LTRE and LSA are, respectively, analytical and simulation-based approaches that use changes in asymptotic population growth rate instead of changes in extinction or quasi-extinction probability. Again, the "best" sensitivity analysis technique will depend on what data are available and what are the most meaningful conservation scenarios.

Caswell (1997, 2000) and Horvitz et al. (1997) make the distinction between prospective and retrospective analysis. Under this artificial dichotomy, elasticities are considered prospective, because they quantify the expected change in growth rate given a specified infinitesimal change in mean vital rates or matrix elements; LTRE and LSA approaches are considered retrospective, because they embrace variation that has occurred in the past. Distinguishing between prospective and retrospective analysis would be useful if it were based on clarifying whether assumptions and inferences are based on potential future changes (i.e., prospective) versus changes that actually have occurred (i.e., retrospective). In strong contrast to the recommendations of Caswell (1997, 2000), however, we do not believe the distinction between retrospective and prospective should hinge purely on whether variation is included in the sensitivity analysis (see also Mills et al. 2001). Any projection into the future is based on both what we know about the past and what we expect in the future, in terms of mean rates, variances, and relationships among parameters. From a conservation decision-making perspective, such projections into the future, inherent in both applied sensitivity analyses and PVA, use information from the past coupled with the changes that are thought to be biologically, politically, and logistically possible under management in the future. These future changes may or may not be very different from changes in the past (Citta and Mills 1999). The LSA technique was developed specifically to avoid the dichotomy of prospective versus retrospective analyses (Wisdom et al. 2000). Instead of implying that mean rates are reliable measures of future dynamics while variances are not, a constructive way forward would be to make explicit whether variation is included in the sensitivity analysis, where the estimates of variation and mean rates are derived, and the rationale for potential future changes in vital rates (Mills et al. 2001).

Obtaining estimates of vital rates, interactions among rates, and variation within rates is not trivial and requires more attention. Indices as opposed to estimates of survival, reproduction, or connectivity are problematic because they may be biased, and because process and sampling variation of indices cannot easily be separated. Recent analytical approaches allow for rigorous estimation of both survival and movement rates that could be used to parameterize matrix models (Lebreton 1992;

Nichols et al. 1992; Brownie et al. 1993). If mark-recapture data are available, Nichols et al. (2000) propose a powerful approach for directly estimating both population growth and sensitivities of vital rates from the data, without assumptions of asymptotic population growth. We believe that this approach has much potential.

Sensitivity analysis is now at a pivotal point. The analytical equations and simulation models are in place to give profound insights to conservation practitioners, but analysis requires data on parameters that are challenging to estimate, especially in a fashion rigorous enough to remove sampling variation. Thus, sensitivity analysis provides an exciting juncture where field biology, biometry, and population analysis converge to provide new insights into populations of concern. A profitable next step would include adaptive management approaches that implement actions or field research based on predictions from sensitivity analysis, monitor the actual impacts, and use that information to refine the models.

LITERATURE CITED

Abraham, K. F., and R. L. Jefferies. 1997. High goose populations: causes, impacts, and implications. Pages 7–72 in B. D. J. Batt, editor, *Arctic ecosystems in peril: report of the Arctic goose habitat working group*. Arctic Goose Joint Venture Special Publication. U.S. Fish and Wildlife Service, Washington, D.C.; Canadian Wildlife Service, Ottawa, Ontario.

Akçakaya, H. R., and B. Baur. 1996. Effects of population subdivision and catastrophes on the persistence of a land snail metapopulation. *Oecologia* 105:475–483.

Akçakaya, H. R., and M. G. Raphael. 1998. Assessing human impact despite uncertainty: viability of the northern spotted owl metapopulation in the northwestern USA. *Biodiversity and Conservation* 7:875–894.

Batt, B. D. J., editor. 1997. *Arctic ecosystems in peril: report of the Arctic goose habitat working group*. Arctic Goose Joint Venture Special Publication. U.S. Fish and Wildlife Service, Washington, D.C.; Canadian Wildlife Service, Ottawa, Ontario.

Beier, P. 1993. Determining minimum habitat areas and habitat corridors for cougars. *Conservation Biology* 7:94–108.

Beissinger, S. R. 1995. Modeling extinction in periodic environments: Everglades water levels and snail kite population viability. *Ecological Applications* 5:618–631.

Beissinger, S. R., and M. I. Westphal. 1998. On the use of demographic models of population viability in endangered species management. *Journal of Wildlife Management* 62:821–841.

Benton, T. G., and A. Grant. 1996. How to keep fit in the real world: elasticity analyses and selection pressures on life histories in a variable environment. *American Naturalist* 147:115–139.

Brault, S., and H. Caswell. 1993. Pod-specific demography of killer whales (*Orcinus orca*). *Ecology* 74:1444–1454.

Brownie, C., J. E. Hines, J. D. Nichols, K. H. Pollock, and J. B. Hestbeck. 1993. Capture-recapture studies for multiple strata including non-Markovian transitions. *Biometrics* 49:1173–1187.

Burbidge, A. A., and N. L. McKenzie. 1989. Patterns in the modern decline of western Australia's vertebrate fauna: causes and conservation implications. *Biological Conservation* 50:143–198.

Burgman, M. A., S. Ferson, and H. R. Akçakaya. 1993. *Risk assessment in conservation biology*. Chapman and Hall, New York, New York.

Burnham, K. P. 1993. A theory for combined analysis of ring recovery and recapture data. Pages 199–213 in J.-D. Lebreton and P. M. North, editors, *Marked individuals in the study of bird population*. Birkhäuser Verlag, Basel, Switzerland.

Burnham, K. P., D. R. Anderson, G. C. White, C. Brownie, and K. H. Pollock. 1987. Design and analysis methods for fish survival experiments based on release-recapture. American Fisheries Society Monograph no. 5. American Fisheries Society, Bethesda, Maryland.

Caswell, H. 1989a. Analysis of life table response experiments: 1, decomposition of effects on population growth rate. *Ecological Modelling* 46:221–237.

———. 1989b. *Matrix population models*. Sinauer Associates, Sunderland, Massachusetts.

———. 1996a. Analysis of life table response experiments: 2, alternative parameterizations for size- and stage-structured models. *Ecological Modelling* 88:73–82.

———. 1996b. Second derivatives of population growth rate: calculation and applications. *Ecology* 77:870–879.

———. 1997. Matrix methods for population analysis. Pages 19–58 in S. Tuljapurkar and H. Caswell, editors, *Structured-population models in marine, terrestrial, and freshwater systems*. Chapman and Hall, New York, New York.

———. 2000. Prospective and retrospective perturbation analyses: their roles in conservation biology. *Ecology* 81:619–627.

Caswell, H., S. Brault, A. J. Read, and T. D. Smith. 1998. Harbor porpoise and fisheries: an uncertainty analysis of incidental mortality. *Ecological Applications* 8:1226–1238.

Caswell, H., and M. C. Trevisan. 1994. The sensitivity analysis of periodic matrix models. *Ecology* 75:1299–1303.

Caughley, G. 1977. *Analysis of vertebrate populations*. John Wiley and Sons, New York, New York.

Citta, J. J., and L. S. Mills. 1999. What do demographic sensitivity analyses tell us about controlling brown-headed cowbirds? *Studies in Avian Biology* 18:121–134.

Citta, J., L. S. Mills, and M. Lindberg. N.d. An experimental field test of sensitivity analysis models. Manuscript.

Cole, L. C. 1954. The population consequences of life history phenomena. *Quarterly Review of Biology* 29:103–137.

Conroy, M. J., Y. Cohen, F. C. James, Y. G. Matsinos, and B. A. Maurer. 1995. Parameter estimation, reliability, and model improvement for spatially explicit models of animal populations. *Ecological Applications* 5:17–19.

Cooch, E. G., R. F. Rockwell, and S. Brault. 2001. Retrospective analysis of demographic response to environmental change: an example in the lesser snow goose. *Ecological Monographs*. In press.

Cooke, F., R. F. Rockwell, and D. B. Lank. 1995. *The snow geese of La Pérouse Bay: natural selection in the wild.* Oxford University Press, New York, New York.

Crooks, K. R., M. A. Sanjayan, and D. F. Doak. 1998. New insights on cheetah conservation through demographic modeling. *Conservation Biology* 12:889–895.

Cross, P. C., and S. R. Beissinger. 2001. Using logistic regression to analyze sensitivity of PVA models: a comparison of methods based on African wild dog models. *Conservation Biology* 15:1335–1346.

Crouse, D. T., L. B. Crowder, and H. Caswell. 1987. A stage-based population model for loggerhead sea turtles and implications for conservation. *Ecology* 68:1412–1423.

Crowder, L. B., D. T. Crouse, S. S. Heppell, and T. H. Martin. 1994. Predicting the impact of turtle excluder devices on loggerhead sea turtle populations. *Ecological Applications* 4:437–445.

de Kroon, H., A. Plaisier, J. van Groenendael, and H. Caswell. 1986. Elasticity: the relative contribution of demographic parameters to population growth rate. *Ecology* 67:1427–1431.

de Kroon, H., J. van Groenendael, and J. Ehrlén. 2000. Elasticities: a review of methods and model limitations. *Ecology* 81:607–618.

Dixon, P., N. Friday, P. Ang, S. Heppell, and M. Kshatriya. 1997. Sensitivity analysis of structured-population models for management and conservation. Pages 471–514 in S. Tuljapurkar and H. Caswell, editors, *Structured-population models in marine, terrestrial, and freshwater systems.* Chapman and Hall, New York, New York.

Doak, D. F., P. M. Kareiva, and B. Klepetka. 1994. Modeling population viability for the desert tortoise in the western Mojave Desert. *Ecological Applications* 4:446–460.

Doak, D. F., P. C. Marino, and P. M. Kareiva. 1992. Spatial scale mediates the influence of habitat fragmentation on dispersal success: implications for conservation. *Theoretical Population Biology* 41:315–336.

Donovan, T. M., R. H. Lamberson, A. Kimber, F. R. Thompson III, and J. Faaborg. 1995. Modeling the effects of habitat fragmentation on source and sink demography of neotropical migrant birds. *Conservation Biology* 9:1396–1407.

Drechsler, M. 1998. Sensitivity analysis of complex models. *Biological Conservation* 86:401–412.

Dunning, J. B., Jr., D. J. Stewart, B. J. Danielson, B. R. Noon, T. L. Root, R. H. Lamberson, and E. E. Stevens. 1995. Spatially explicit population models: current forms and future uses. *Ecological Applications* 5:3–11.

Ehrlén, J., and J. M. van Groenendael. 1998. Direct perturbation analysis for better conservation. *Conservation Biology* 12:470–474.

Fahrig, L., and J. Paloheimo. 1988. Determinants of local population size in patchy habitats. *Theoretical Population Biology* 34:194–213.

Fowler, C. W., and T. Smith. 1973. Characterizing stable populations: an application to the African elephant population. *Journal of Wildlife Management* 37:513–523.

Gaillard, J.-M., M. Festa-Bianchet, and N. G. Yoccoz. 1998. Population dynamics of large herbivores: variable recruitment with constant adult survival. *Trends in Ecology and Evolution* 13:58–63.

Goodman, L. A. 1971. On the sensitivity of the intrinsic growth rate to changes in the age-specific birth and death rates. *Theoretical Population Biology* 2:339–354.

Gould, W. R., and J. D. Nichols. 1998. Estimation of temporal variability of survival in animal populations. *Ecology* 79:2531–2538.

Grand, J., and S. R. Beissinger. 1997. When relocation of loggerhead sea turtle (*Caretta caretta*) nests becomes a useful strategy. *Journal of Herpetology* 31:428–434.

Grant, A., and T. G. Benton. 2000. Elasticity analysis for density dependent populations in stochastic environments. *Ecology* 81:680–693.

Harrison, S. 1994. Metapopulations and conservation. Pages 111–128 in P. J. Edwards, R. M. May, and N. R. Webb, editors, *Large-scale ecology and conservation ecology*. Blackwell Scientific, Oxford, United Kingdom.

Heppell, S. S., J. R. Walters, and L. R. Crowder. 1994. Evaluating management alternatives for red-cockaded woodpeckers: a modeling approach. *Journal of Wildlife Management* 58:479–487.

Hitchcock, C. L., and C. Gratto-Trevor. 1997. Diagnosing a shorebird local population decline with a stage-structured population model. *Ecology* 78:522–534.

Hoekman, S. T., L. S. Mills, D. W. Howerter, J. H. Devries, and I. J. Ball. 2000. Sensitivity analysis of the life cycle of mid-continent mallards. Manuscript.

Horvitz, C. C., D. W. Schemske, and H. Caswell. 1997. The relative "importance" of life-history stages to population growth: prospective and retrospective approaches. Pages 247–272 in S. Tuljapurkar and H. Caswell, editors, *Structured-population models in marine, terrestrial, and freshwater systems*. Chapman and Hall, New York, New York.

Lande, R. 1998. Demographic stochasticity and Allee effect on a scale with isotropic noise. *Oikos* 83:353–358.

Lebreton, J.-D., K. P. Burnham, J. Clobert, and D. R. Anderson. 1992. Modeling survival and testing biological hypotheses using marked animals: a unified approach with case studies. *Ecological Monographs* 62:67–118.

Lefkovitch, L. P., and L. Fahrig. 1985. Spatial characteristics of habitat patches and population survival. *Ecological Modelling* 30:297–308.

Levin, L., H. Caswell, T. Bridges, C. DiBacco, D. Cabrera, and G. Plaia. 1996. Demographic responses of estuarine polychaetes to pollutants: life table response experiments. *Ecological Applications* 6:1295–1313.

Lindenmayer, D. B., and R. C. Lacy. 1995. Metapopulation viability of arboreal marsupials in fragmented old-growth forests: comparison among species. *Ecological Applications* 5:183–199.

Lindenmayer, D. B., and H. P. Possingham. 1996. Ranking conservation and timber management options for Leadbeater's possum in southeastern Australia using population viability analysis. *Conservation Biology* 10:235–251.

Link, W. A., and J. D. Nichols. 1994. On the importance of sampling variance to investigations of temporal variation in animal population size. *Oikos* 69:539–544.

Liu, J., J. B. Dunning Jr., and H. R. Pulliam. 1995. Potential effects of a forest management plan on Bachman's sparrows (*Aimophila aestivalis*): linking a spatially explicit model with GIS. *Conservation Biology* 9:62–75.

Ludwig, D. 1999. Is it meaningful to estimate a probability of extinction? *Ecology* 80:298–310.

Maguire, L. A., G. F. Wilhere, and Q. Dong. 1995. Population viability analysis for red-cockaded woodpeckers in the Georgia Piedmont. *Journal of Wildlife Management* 59:533–542.

Marmontel, M., S. R. Humphrey, and T. J. O'Shea. 1997. Population viability analysis of the Florida manatee (*Trichechus manatus latirostris*), 1976–1991. *Conservation Biology* 11:467–481.

Martien, K. K., B. L. Taylor, E. Slooten, and S. Dawson. 1999. A sensitivity analysis to guide research and management for Hector's dolphin. *Biological Conservation* 90:183–191.

McCarthy, M. A., M. A. Burgman, and S. Ferson. 1995. Sensitivity analysis for models of population viability. *Biological Conservation* 73:93–100.

McKelvey, K., B. R. Noon, and R. H. Lamberson. 1993. Conservation planning for species occupying fragmented landscapes: the case of the northern spotted owl. Pages 424–450 in P. M. Kareiva, J. G. Kingsolver, and R. B. Huey, editors, *Biotic interactions and global change*. Sinauer Associates, Sunderland, Massachusetts.

Mertz, D. B. 1971. The mathematical demography of the California condor population. *American Naturalist* 105:437–453.

Mesterton-Gibbons, M. 1993. Why demographic elasticities sum to one: a postscript to de Kroon et al. *Ecology* 74:2467–2468.

Mills, L. S., D. F. Doak, and M. J. Wisdom. 1999. Reliability of conservation actions based on elasticity analysis of matrix models. *Conservation Biology* 13:815–829.

———. 2001. Elasticity analysis for conservation decision making: Reply to Ehrlén et al. *Conservation Biology* 15:281–283.

Mills, L. S., S. G. Hayes, C. Baldwin, M. J. Wisdom, J. Citta, D. J. Mattson, and K. Murphy. 1996. Factors leading to different viability predictions for a grizzly bear data set. *Conservation Biology* 10:863–873.

Nelson, L. J., and J. M. Peek. 1982. Effect of survival and fecundity on rate of increase of elk. *Journal of Wildlife Management* 46:535–540.

Nichols, J. D. 1996. Sources of variation in migratory movements of animal populations: statistical inference and a selective review of empirical results for birds. Pages 147–197 in O. E. Rhodes Jr., R. K. Chesser, and M. H. Smith, editors, *Population dynamics in ecological space and time*. University of Chicago Press, Chicago, Illinois.

Nichols, J. D., J. E. Hines, J.-D. Lebreton, and R. Pradel. 2000. Estimation of contributions to population growth: a reverse-time capture-recapture approach. *Ecology* 81:3362–3377.

Nichols, J. D., J. R. Sauer, K. H. Pollock, and J. B. Hestbeck. 1992. Estimating transition probabilities for stage-based population projection matrices using capture-recapture data. *Ecology* 73:306–312.

Olmsted, I., and E. R. Alvarez-Buylla. 1995. Sustainable harvesting of tropical trees: demography and matrix models of two palm species in Mexico. *Ecological Applications* 5:484–500.

Pfister, C. A. 1998. Patterns of variance in stage-structured populations: evolutionary predictions and ecological implications. *Proceedings of the National Academy of Sciences* (USA) 95:213–218.

Pulliam, H. R., J. B. Dunning Jr., and J. Liu. 1992. Population dynamics in complex landscapes: a case study. *Ecological Applications* 2:165–177.

Ralls, K., and A. M. Starfield. 1995. Choosing a management strategy: two structured decision-making methods for evaluating the predictions of stochastic simulation models. *Conservation Biology* 9:175–181.

Ringsby, T. H., B.-E. Sæther, R. Altwegg, and E. J. Solberg. 1999. Temporal and spatial variation in survival rates of a house sparrow, *Passer domesticus,* meta-population. *Oikos* 85:419–425.

Rockwell, R. F., E. Cooch, and S. Brault. 1997. Dynamics of the mid-continent population of lesser snow geese: projected impacts of reduction in survival and fertility on population growth rates. Pages 73–100 in B. D. J. Batt, editor, *Arctic ecosystems in peril: report of the Arctic goose habitat working group.* Arctic Goose Joint Venture Special Publication. U.S. Fish and Wildlife Service, Washington, D.C.; Canadian Wildlife Service, Ottawa, Ontario.

Ruckelshaus, M., C. Hartway, and P. Kareiva. 1999. Dispersal and landscape errors in spatially explicit population models: a reply. *Conservation Biology* 13:1223–1224.

Sæther, B.-E., S. Engen, A. Islam, R. McCleery, and C. Perrins. 1998. Environmental stochasticity and extinction risk in a population of a small songbird, the great tit. *American Naturalist* 151:441–450.

Schmutz, J. A., R. F. Rockwell, and M. R. Petersen. 1997. Relative effects of survival and reproduction on the population dynamics of emperor geese. *Journal of Wildlife Management* 61:191–201.

Shea, K., and D. Kelly. 1998. Estimating biocontrol agent impact with matrix models: *Carduus nutans* in New Zealand. *Ecological Applications* 8:824–832.

Silvertown, J., M. Franco, and E. Menges. 1996. Interpretation of elasticity matrices as an aid to the management of plant populations for conservation. *Conservation Biology* 10:591–597.

Skalski, J. R., and Robson, D. S. 1992. *Techniques for wildlife investigations: design and analysis of capture data.* Academic Press, San Diego, California.

Snedecor, G. W., and W. G. Cochran. 1980. *Statistical methods.* 7th edition. Iowa State University Press, Ames, Iowa.

South, A. 1999. Dispersal in spatially explicit population models. *Conservation Biology* 13:1039–1046.

Thompson, W. L., G. C. White, and C. Gowan. 1998. *Monitoring vertebrate populations.* Academic Press, San Diego, California.

Tuljapurkar, S. 1990. *Population dynamics in variable environments.* Springer-Verlag, New York, New York.

———. 1997. Stochastic matrix models. Pages 59–88 in S. Tuljapurkar and H. Caswell, editors, *Structured-population models in marine, terrestrial, and freshwater systems.* Chapman and Hall, New York, New York.

Tuljapurkar, S., and H. Caswell. 1997. *Structured-population models in marine, terrestrial, and freshwater systems.* Chapman and Hall, New York, New York.

van Tienderen, P. H. 1995. Life cycle trade-offs in matrix population models. *Ecology* 76:2482–2489.

Wisdom, M. J., and L. S. Mills. 1997. Sensitivity analysis to guide population recovery: prairie-chickens as an example. *Journal of Wildlife Management* 61:302–312.

Wisdom, M. J., L. S. Mills, and D. F. Doak. 2000. Life-stage simulation analysis: estimating vital-rate effects on population growth for conservation. *Ecology* 81: 628–641.

Wootton, J. T., and D. A. Bell. 1992. A metapopulation model for the peregrine falcon in California: viability and management strategies. *Ecological Applications* 2:307–321.

17

Application of Molecular Genetics to Conservation: New Issues and Examples
Philip W. Hedrick

ABSTRACT

A number of endangered species have declined to very low numbers distributed in a few populations. Particularly in these situations, genetic information is often critical for evaluation of the existing animals and for future management. I give two examples here from endangered fish in the western United States. First, the Gila topminnow remains in only four watersheds of the Gila River drainage. Evaluation of microsatellite variation suggests that there are five management units and that these should be conserved separately. Second, a supplementation program for winter-run chinook salmon, present only in the Sacramento River in California, has been ongoing since the early 1990s. Using data from microsatellite loci, it appears that returns from these hatchery-reared fish are consistent with random expectations for both female and male parents.

Although highly variable loci, such as microsatellite loci, are revolutionizing conservation biology, data from these loci need to be evaluated carefully. First, the magnitude of differentiation measures may be quite small because these loci often have very high within-population heterozygosity. For example, maximum G_{ST} values for populations with no alleles in common at highly variable loci may be small and, at maximum, less than the average within-population homozygosity. As a result, measures that are variation independent are recommended for highly variable loci. Second, bottlenecks or a reduction in population size can generate large genetic distances in a short time for these loci. In this case, the genetic distance may be corrected for low variation in a population, and tests to detect bottlenecks are advised. Third, statistically significant

Thanks to my collaborators on the Gila topminnow research, Karen Parker, W. L. Minckley, and Ruby Sheffer; my collaborators on the winter-run chinook salmon research, Dennis Hedgecock and Vanessa Rashbrook; to Steven Kalinowski for allowing me to use his unpublished research in figure 17.4; and for funding from the National Science Foundation, California Department of Water Resources, U.S. Fish and Wildlife Service, Arizona Heritage Fund, and the Ullman Professorship. I appreciate the comments of the editors, R. Waples, and an anonymous reviewer on the manuscript.

differences may not reflect biologically meaningful differences either because the patterns of adaptive loci may not be correlated with highly variable loci or because the statistical power with these markers may be very high. As an example of this latter effect, the statistical power to detect a one-generation bottleneck of different sizes for different numbers of highly variable loci is discussed.

INTRODUCTION

Molecular genetic techniques have become a fundamental part of research and management of endangered species and can provide insights not found with other techniques in conservation biology (Smith and Wayne 1996). Perhaps the most common application of molecular techniques is to help identify the unit of conservation. Among populations of a species, there has been an effort recently to define evolutionarily significant units (ESUs) so that management decisions about conservation can be made based on objective genetic (and other) criteria (Waples 1995). In providing guidelines for determining ESUs, Moritz (1994) suggests that they should "show significant divergence of allele frequencies at nuclear loci" (see also Moritz 1999). In addition, molecular techniques have been employed in determining management units (MUs; Moritz 1994) or distinct population segments, levels of subdivision important for management and conservation but not necessarily evolutionary distinctiveness. In addition, molecular techniques have been used to evaluate various evolutionary aspects of the units under consideration. These include examining the amount of genetic variation, effective population size, history of bottlenecks, amount of inbreeding, and paternity.

Below I summarize data from two endangered fish species that illustrate the usefulness of molecular techniques and give insights not possible with other approaches. The first is the Gila topminnow (*Poeciliopsis occidentalis occidentalis*), a small live-bearing fish that exists in the United States in four isolated watersheds in Arizona, and the second is the winter-run chinook salmon (*Oncorhynchus tshawytscha*) that has only one extant population in the Sacramento River in California. Finally, I present several factors that suggest that data from highly variable genetic loci should be used carefully in application to conservation issues.

GENETIC CHARACTERIZATION OF GILA TOPMINNOW POPULATIONS

The Gila topminnow is a federally endangered species that primarily exists only in a few isolated areas in four watersheds in Arizona (fig.

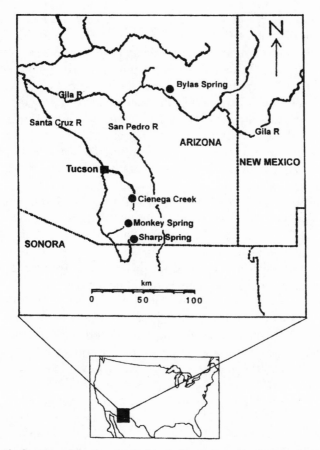

Fig. 17.1 The locations of the sites (*circles*) in the four major watersheds in the United States that contain Gila topminnows.

17.1). It has been the subject of extensive genetic research, first by Vrijenhoek and colleagues (Vrijenhoek et al. 1985; Meffe and Vrijenhoek 1988; Quattro and Vrijenhoek 1989; Quattro et al. 1996) and more recently by me and my colleagues (Sheffer et al. 1997, 1999; Hedrick and Parker 1998; Parker et al. 1999). One of the striking comparisons between these studies is that many of the conclusions from Vrijenhoek's group are not consistent with our work. For example, Vrijenhoek et al. (1985) found more allozyme variation in a sample from Sharp Spring than in a sample from Monkey Spring, and Quattro and Vrijenhoek (1989) found a higher fitness in laboratory experiments for Sharp Spring fish. As a result, they recommended that "the Sharp Spring stock currently offers the best choice for stocking the Gila River system," and the U.S. Fish and Wildlife Service stopped using Monkey Spring fish for reintroductions and began using only Sharp Spring stock. However, Hedrick and Parker (1998) and Parker et al. (1999) found that the sample from Cienega Creek had approximately the same level of genetic variation for major histocompatibility complex (MHC) loci and microsatellite loci as Sharp Spring. Furthermore, Sheffer et al. (1997) found that the fitness of fish from all four watersheds was generally similar and that there was little evidence of inbreeding or outbreeding depression among these populations (Sheffer et al. 1999).

Previously, we surveyed samples from the four watersheds, but here we include data from nine additional populations. Six of these populations are natural, one is a natural recolonization (Santa Cruz), one is the source for many of the early transplants (Boyce-Thompson), and one is a transplant population of unknown origin (Watson Wash; Hedrick et al. 2001; see Minckley 1999 for a discussion of ecological, life-history, and other attributes of the sites).

Figure 17.2 gives an unrooted neighbor-joining tree for these data from five microsatellite loci. The natural populations appear to fall into five groups based on these genetic data. First, the two populations Bylas Spring 1 and Bylas Spring 2, the only samples from the mainstem Gila River, are very similar to each other and different from others. They also have low genetic variation as measured by both expected heterozygosity ($H_E = 0.059$) and observed number of alleles ($n = 1.2$). Second, the Monkey and Cottonwood Springs samples, both in the upper Sonoita Creek watershed, are similar to each other and different from all other natural populations. They have intermediate levels of genetic variation (average of $H_E = 0.207$ and $n = 2.3$). Third, the other samples from the Sonoita Creek drainage (Sonoita Creek, Coalmine Canyon, Red Rock at Cott Tank, and Red Rock Falls) cluster together, with Coalmine and Sonoita Creek being very similar. Except for Red Rock at Cott Tank, which is variable for only one locus, these populations are intermediate in genetic variation. The sample from the recolonized Santa Cruz River population clusters with these nearby Sonoita Creek samples. The Santa Cruz sample is the most variable sample analyzed, with both the highest heterozygosity and number of alleles ($H_E = 0.473$ and $n = 4.4$).

The Cienega Creek and Sharp Spring samples form the fourth and fifth groups. Although they do not appear as different on the phylogenetic tree as the other groups, they have nearly nonoverlapping sets of alleles for two microsatellite loci and do not share any alleles at the MHC locus (not included in the tree). Sharp Spring has higher than average genetic variation ($H_E = 0.283$ and $n = 3.8$) and Cienega Creek has intermediate genetic variation. Based on these data, and other

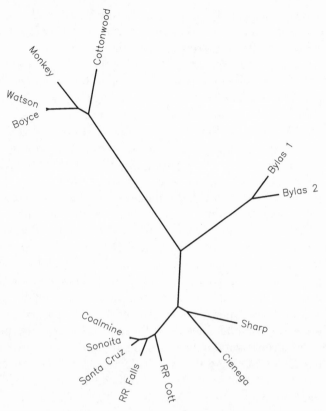

Fig. 17.2 An unrooted phylogenetic tree showing the relationships for the 13 populations of the Gila topminnow based on five microsatellite loci.

factors discussed in Parker et al. (1999) and Hedrick et al. (2001), we recommend that there be five management units based on these five groups for the Gila topminnow.

The samples from Boyce-Thompson and Watson Wash are very similar, suggesting that Boyce-Thompson was the source for the Watson Wash population. Furthermore, the Boyce-Thompson population appears genetically similar to the Monkey Spring population for the microsatellite loci, consistent with its primary origin being from Monkey Spring. Because there is no overlap in alleles from Boyce-Thompson and the Bylas Spring samples for the microsatellite locus *Pooc*-6-10 and the MHC gene, it appears unlikely that there is substantial ancestry from Bylas Spring in the Boyce-Thompson stock, as has been suggested.

Molecular genetics also can help identify previously unrecognized taxonomic differences. In topminnows, the Yaqui topminnow (*P. o.*

sonoriensis) is considered a subspecies, and previous work proceeded under that assumption. For example, Quattro and Vrijenhoek (1989) compared several fitness correlates in the laboratory for samples from Monkey Spring as an example of low allozyme heterozygosity, Yaqui topminnows from Tule Spring as an example of intermediate allozyme heterozygosity, and Sharp Spring as an example of high allozyme heterozygosity. In addition to lack of concordance with the fitness results of Quattro and Vrijenhoek (1989) by Sheffer et al. (1997) within Gila topminnows, we have now examined molecular variation from a sample of Yaqui topminnow from Tule Spring (Hedrick et al. 2001). At seven microsatellite loci and a MHC locus, this sample shares only 2 out of 25 alleles with any of our Gila topminnow samples. For three of the microsatellite loci, the number of repeats is greatly different for these two taxa. For example, at microsatellite locus *Pooc*-OO56, the allele size in the Gila topminnow is between 143 and 153, while Yaqui topminnow is fixed for an allele that is 256 base pairs in length. In addition, Quattro et al. (1996) found that Gila and Yaqui topminnows are substantially different at mitochondrial DNA restriction sites. As a result of this great divergence for three types of molecular markers, it appears that these two subspecies should be considered for ranking as separate species.

Using the seven microsatellites in common to the two taxa, Monkey Spring has $H_E = 0.138$ and $n = 2.0$, Tule Spring has $H_E = 0.443$ and $n = 2.9$, and Sharp Spring has $H_E = 0.202$ and $n = 3.0$. Overall then, the comparison by Quattro and Vrijenhoek (1989) used a sample from what is apparently a different species, and the ranking of genetic variation based on allozyme loci is not consistent with that found for more variable molecular markers.

EFFECTIVE POPULATION SIZE IN RETURNING WINTER-RUN CHINOOK SALMON

Knowledge of the effective population size of natural populations is fundamental to understanding and predicting the extent of genetic variation in endangered species (Waples, chap. 8 in this volume). However, most estimates of effective population size in natural populations are based on demographic information, such as the mean and variance in the number of progeny per parent, or genetic data, such as temporal changes in allele frequency or the extent of linkage disequilibrium (Schwartz et al. 1998).

It is often difficult to acquire either the demographic or genetic information to adequately estimate the effective population size (N_e), but estimates of the number of adults in a population (N) are generally easier to obtain. If the relationship between N_e and N is known, then a general idea of the effective size is possible by estimating N. For example, in their recommendations for evaluating species for different categories of endangerment, Mace and Lande (1991) suggested that generally the ratio N_e/N is approximately 0.2 to 0.5. Nunney (1993) suggested that the ratio over a broad array of mating and demographic situations should be around 0.5 and not less than 0.25 within a generation. In a review of available data, however, Frankham (1995) found that the ratio over a number of species was only around 0.1, although Vucetich et al. (1997) and Waples (chap. 8 in this volume) discuss the possible basis for Frankham's low estimates. Some estimates of the N_e/N ratio are very low because a very large proportion of the adults do not contribute any progeny (Hedgecock et al. 1992).

In organisms with high reproductive potential, such as many plants and invertebrates and a number of fish species, there is an opportunity for the effective population size to be small. Even though the number of offspring is very large, the variance in contribution to the next generation may also be very large. This effect is particularly a concern for endangered fishes for which there may be hatchery programs utilizing a few adults to supplement the natural population with a large number of individuals (e.g., Hedrick et al. 1995; Hedrick et al. 2000a,b).

Winter-run chinook salmon from the Sacramento River, California, drainage are federally listed as endangered. The estimated annual number of spawners dropped from over 100,000 in 1969 to approximately 200 in 1991. One of the major factors causing the initial reduction was the construction of Shasta Dam, which made the traditional spawning grounds for winter-run inaccessible. In addition to the lack of cool-water spawning grounds and adequate spawning gravel, other factors, such as pollution and impediments to upstream and downstream migration, have resulted in further stresses.

Since 1991 the Coleman National Fish Hatchery (CNFH) on Battle Creek, a tributary of the Sacramento River, in a supplementation program for the winter-run chinook salmon has been releasing six- to eight-month-old presmolts with known captured parents. Hedrick et al. (1995, 2000b) evaluated the effects of this program on the effective population size in the first five years of the program, 1991 through 1995. They found that the supplementation did not appear to greatly influence the effective population size of the run, neither significantly reducing nor increasing it. Supplementation was discontinued in 1996 and 1997 because some non-winter-run chinook were mistakenly used as spawners; in addition to winter-run, there are spring, fall, and late-fall runs of chinook in the Sacramento River. In addition, the returning spawners were found primarily in Battle Creek, above the site of the hatchery, and not in the normal spawning area in the mainstem Sacramento River (Hedrick et al. 2000b).

In estimating the effective population size of the released hatchery-raised fish and the overall population, Hedrick et al. (1995, 2000b) assumed that survival and return migration were random. In other words, these estimates were based on family size variance through the juvenile prerelease stage and then random simulation of survival to adulthood. However, Geiger et al. (1997) reported that survival may not be random with respect to family (female-male pairing) in pink salmon (*O. gorbuscha*). The returns of fish in 1997 and 1998 from 1994 and 1995 releases, primarily to Battle Creek, allowed us to determine whether the assumptions of random survival and return were true. This was possible because these returning fish were identified as CNFH winter-run releases and they could be assigned to parents from the hatchery in 1994 and 1995, using 11 variable microsatellite markers. From this information, we are able to estimate directly the effective population size of these released fish, evaluate our earlier predictions, and give a direct measure of the N_e/N ratio. In other words, we now have observed family size variance through adulthood, so do not have to assume random survival through simulation.

Table 17.1 gives the number of progeny released from the different females and males in 1994 (for more details, see Hedrick et al. 2000b).

Table 17.1 The Number of Winter-Run Chinook Salmon Progeny Released and the Number of Returning Spawners from the Different Females and Males in the 1994 Brood Year, and the Ratio of the Proportions of Returns to Releases for a Given Parent

Female	Releases	Returns	Ratio	Male	Releases	Returns	Ratio
3	3,444	10	1.35	B	4,433	9	0.95
4	3,055	5	0.77	C	3,152	9	1.337
5	2,499	7	1.29	D	4,360	16	1.61
6	2,361	6	1.20	E	6,013	8	0.62
7	2,421	3	0.57	F	5,223	15	1.34
8	2,292	2	0.42	G	5,098	6	0.51
9	2,338	5	1.00	H	4,432	10	1.06
11	2,320	7	1.39	I	6,353	16	1.17
12	2,701	3	0.52	J	3,012	3	0.46
13	3,946	8	0.93	K	1,270	1	0.38
14	1,364	2	0.69				
15	3,426	10	1.37				
16	2,855	10	1.64				
17	2,766	7	1.17				
18	3,088	4	0.61				
19	2,470	4	0.75				
Total	43,346	93		Total	43,346	93	

Table 17.2 The Predicted Effective Population Sizes and Observed Population Sizes of Winter-Run Chinook Salmon for 1994 and 1995

Year	Predicted N_e (95% Confidence Interval)	Observed N_e	N_e/N
1994	34.8 (28.1, 41.2)	31.5	1.21
1995	24.5 (16.1, 34.3)	18.0	0.50

Note: Based on 93 returning spawners for 1994 and 23 for 1995.

Of the 29 fish captured in 1994, 26 (16 females and 10 males) produced 43,346 released progeny. Overall, the production was fairly equal among individuals with the proportions of progeny produced for females ranging from 0.032 to 0.092 (mean of 0.062), and for males from 0.029 to 0.147 (mean of 0.1). Also given in table 17.1 are the 93 returning spawners that were assignable to families released in 1994, using the genetic analysis described above (Hedrick et al. 2000c). Every 1994 male and female parent was represented in the returning spawners. Table 17.1 also gives the ratio of the proportions of the returns to released fish for given females and males. For example, for female 3, the ratio is (10/93)/(3,444/43,346) = 1.35. These ratios are in a fairly narrow range from 0.38 to 1.64, suggesting no large differences related to either female or male parent.

In 1995, of the 47 adults captured, 36 (21 females and 15 males) produced progeny and a total of 51,273 progeny was released. The variation in contribution over females was much larger than in 1994, while the male variation was somewhat larger (Hedrick et al. 2000b). Many fewer spawners from the 1995 release returned, and 23 were assignable to families. Both the higher variation in releases and the lower number of returns resulted in higher variation in the ratios of returns to release proportions (Hedrick et al. 2000c).

The predicted effective population sizes are given for the two years in table 17.2, using the simulation approach in Hedrick et al. (1995). The predicted value is larger for 1994, mainly because there were over four times as many returning spawners than in 1995 (see below). For both years, however, the predicted N_e and the observed N_e, based on the family representation of the returning spawners, were fairly close and were within the 95% confidence intervals both years. If we assume that there were 26 parents in 1994 and 36 parents in 1995 (the non-spawning adults were probably not winter-run, and six captive-reared females produced only 1,131 progeny total), then the N_e/N ratios for the two years were 1.21 and 0.50. However, N_e, as we have estimated it using randomly returning individual spawners, is a function of the number of returning spawners (Hedrick et al. 2000c). For example, if a very large number of spawners returned, then N_e would be based on the proportions released from different families and would approach $2N$ as the contributions from different families become more equal.

The overall finding is that it appears, using molecular genetic techniques to identify individual returning spawners to family, that the variance over families is close to that expected at random. As a result, the N_e/N ratio from fish reared in the hatchery and returning as spawners is relatively high and probably would not result in a large reduction in effective population size. This discussion does not include the potential effect of a high contribution from the captive population when there are only a few captive spawners to the overall effective population size (Ryman and Laikre 1991; Hedrick et al. 1995, 2000b).

SOME CAUTIONARY POINTS ABOUT HIGHLY VARIABLE LOCI IN CONSERVATION

Molecular information is often thought to be the type of data that gives definitive answers on particular questions, and as the examples above illustrate, it can give insights generally not possible with other techniques. However, it is important to realize the meaning of these data and not to use them without a proper evolutionary and statistical context. Although at first it may appear that the difference between low variation and highly variable loci is only a matter of degree, a different perspective may sometimes be necessary when these markers are utilized (Hedrick 1999). My emphasis here will be on microsatellite loci, or similar multiple-allelic loci, which presently appear to be the loci of choice for many studies in evolutionary and conservation biology (Jarne and Lagoda 1996). I will not discuss the use of microsatellites in phylogenetic inference, and the various genetic distance measures and the different mutation models that have been suggested for incorporating the number of repeats in microsatellite alleles, a topic already widely discussed elsewhere (Nauta and Weissing 1996; Takezaki and Nei 1996; Goldstein and Pollock 1997).

Measures of Differentiation

Interpretation of the extent of differentiation between populations has been dominated by data from biallelic loci. For example, if populations are variable for two alleles, then we can measure differentiation with an estimate of F_{ST}; the level ranges from $F_{ST} = 0$ (when the populations all share the same allelic frequencies) to $F_{ST} = 1$ (when the populations are fixed for different alleles). When there are multiple alleles, however, the range of G_{ST}, the multiple-allele estimate of F_{ST}, generally does not have a maximum of unity.

To illustrate this, G_{ST} can be defined as

(1a)
$$G_{ST} = \frac{H_T - H_S}{H_T},$$

where H_T is the proportion of heterozygous individuals in the total population and H_S is the average weighted heterozygosity within subpopulations (Nei 1973). With highly variable loci such as microsatellite loci, both H_T and H_S can approach unity, but with Hardy-Weinberg proportions within populations, $H_T > H_S$. As a result, the value of G_{ST} can be small. This can occur if the populations have the same alleles or if they have nonoverlapping sets of alleles. Initially, the latter seems counterintuitive because in the two-allele case, when the populations are monomorphic for different alleles, $F_{ST} = 1$. However, G_{ST} (and other measures determining the proportional amount of variation within subpopulations as compared to the total population) does not specify the identity of the alleles involved. For allozymes or other low-variation loci, this was generally not an issue because the polymorphic populations being compared were usually polymorphic for overlapping sets of alleles.

The magnitude of G_{ST} from expression 1a can also be written as

(1b)
$$G_{ST} = 1 - \frac{H_S}{H_T} < 1 - H_S,$$

where $1 - H_S$ is the homozygosity. From this, it is obvious that the differentiation cannot exceed the level of homozygosity, no matter what evolutionary factor is influencing the amount and pattern of variation. Obviously, if highly polymorphic markers make the level of homozygosity low, the maximum G_{ST} must also be greatly reduced. It is convenient to demonstrate this effect using the estimator G_{ST}, but it appears to be a property of all the common parametric definitions of F_{ST}, all of which are dependent on the within-population variation (Charlesworth 1998).

Jin and Chakraborty (1995) gave a theoretical basis for this observation by deriving the predicted level of G_{ST} over populations, descended from a common ancestral population, when they are completely isolated over time. Asymptotically, this value becomes, for both the infinite allele and stepwise mutation models,

(2a)
$$G_{ST} = \frac{(s - 1)(1 - H_S)}{s - 1 + H_S},$$

where s is the number of populations. At this limit, all populations have nonoverlapping sets of alleles and the genetic distance is maximized.

For example, the standard genetic distance of Nei (1972) is ∞. As in expression 1a, G_{ST} ranges from zero if $H_S = 1$ to unity when $H_S = 0$.

We can also show how differentiation is related to mutation by assuming the equilibrium level of heterozygosity within populations for the infinite-allele model

$$H_S = \frac{4N_e u}{4N_e u + 1},$$

where N_e is the effective population size and u is the mutation rate. Therefore, by substitution, expression 2a becomes

(2b) $$G_{ST} = \frac{s - 1}{s(4N_e u + 1) - 1}.$$

Figure 17.3 plots the relationship of the asymptotic value of G_{ST} and mutation rate for two and ten populations. For high-mutation-rate loci, such as microsatellites, $4N_e u > 1$ and G_{ST} is relatively small. On the other hand, for low-mutation-rate loci, such as allozymes, $4N_e u < 0.1$ and G_{ST} approaches unity. In other words, the size of G_{ST} for isolated populations is strongly influenced by the amount of variation (determined here by mutation) at a locus.

Other approaches can be used to estimate differentiation. For the approach of Weir and Cockerham (1984), the maximal differentiation also appears to be the amount of homozygosity. The rare allele method

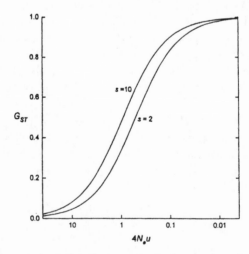

Fig. 17.3 The asymptotic level of G_{ST} between completely isolated populations as a function of the mutation rate for two and ten populations.

of Slatkin (1985) should be useful in identifying the amount of sharing of alleles in different populations. For example, in all the situations discussed in figure 17.2 in which there is complete isolation and no overlap of alleles, the number of migrants per generation (Nm), using the rare allele method, gives values approaching zero. Charlesworth (1998) suggested that absolute measures of differentiation (e.g., Nei 1973) be employed to compare areas of the genome with different levels of diversity, and Nagylaki (1998) suggested different measures of diversity for situations of low and high variation. In addition, using assignment tests with highly variable loci to determine population membership of specific individuals appears to be a powerful approach and may even be used to identify recent migrants or descendants of recent migrants (Davies et al. 1999).

Bottlenecks and Genetic Distance

An advantage of the genetic distance measure of Nei (1972) for the infinite allele model is that it increases linearly with time since the two groups have diverged. If one or both of the two groups have gone through a bottleneck(s) or a substantial reduction in population size, however, the amount of genetic distance may increase very quickly. As a result, the observed amount of genetic distance may not accurately reflect time since divergence, but may be more a function of the reduction in population size.

This effect can be illustrated by assuming that an ancestral population splits into two isolated populations and changes over time by genetic drift. The genetic distance between the two populations at time t (modified from Chakraborty and Nei 1977), assuming no mutation because the number of generations is small, is

(3a)
$$D_t = -\ln \frac{1 - H_0}{[(1 - H_{x,t})(1 - H_{y,t})]^{1/2}},$$

where $H_{x,t}$ and $H_{y,t}$ are the heterozygosities in populations x and y in generation t, and H_0 is heterozygosity in the ancestral population. The heterozygosity in generation t can be written as

$$H_t = \left(1 - \frac{1}{2N_e}\right)^t H_0,$$

assuming, for simplicity, that the effective population size is N_e and is constant in size. If it is assumed that only one population, say population

x, is finite, and the other population is infinite (does not change in allele frequencies), then

(3b)
$$D_t = -\ln\left[\frac{1 - H_0}{1 - H_0\left(1 - \dfrac{1}{2N_e}\right)^t} \right]^{1/2},$$

In this case, the maximum genetic distance is

$$D_t = -\ln(1 - H_0)^{1/2}.$$

Let us assume that the ancestral population had different levels of heterozygosity and that one population lost $(1 - 1/(2N_e))^t$ of its variation. Figure 17.4 illustrates that, when there was high initial heterozygosity, the genetic distance generated can be very large. For example, with a one-generation bottleneck of two individuals (0.75 on the horizontal axis) in one of the two populations, then for H_0 values of 0.3, 0.7 and 0.9, D_t values are 0.051, 0.230, and 0.589, respectively. In other words, for highly variable loci, a bottleneck in one population can generate a large amount of genetic distance. The basis for the high distance value for high heterozygosity is obvious when it is realized that, to have $H_0 = 0.9$, there must be at least the equivalent of ten equally frequent

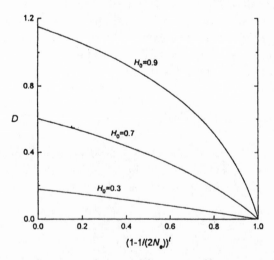

Fig. 17.4 The genetic distance as a function of the amount of heterozygosity lost in one population for three different levels of initial heterozygosity.

alleles. In a bottleneck of two, at most four of these ten alleles can be present, an event that generates large changes in allele frequency.

Corroborating the potential importance of this effect, Paetkau et al. (1997) observed a strong negative correlation of heterozygosity and genetic distance ($r = -0.760$) for the distance measure used above in eight microsatellite loci in North American bear populations. In particular, two island populations on Kodiak Island and Newfoundland had the lowest heterozygosities and generally the largest genetic distances to other groups.

Nei (1987, 240) suggested that the genetic distance can be corrected by assuming that

(4) $$I = J_{XY}/J_X,$$

where J_{XY} is the product of the allelic frequencies in the two populations and J_X is the homozygosity in the population that has not gone through the bottleneck. For example, in Paetkau et al. (1997) the heterozygosity of the brown bears on Kodiak Island is only 0.265, much less than the 0.694 heterozygosity found in the Flathead River sample near the Canada–United States border. The reported standard genetic distance between the samples was a large 1.498. The corrected value, using expression 4, is smaller but still 1.060, supporting the conclusion that there is a large genetic difference between the samples.

Recently, Cornuet and Luikart (1996) developed several statistical tests to identify bottlenecks other than from using lower heterozygosity. Their tests are extensions of Watterson (1978) to determine if an observed distribution of allelic frequencies is more even than neutrality expectation in a sample of a given size with a given number of alleles (see applications in Luikart and Cornuet 1998). Before using the Nei correction, it would be appropriate to determine whether there is statistical evidence for a past bottleneck.

Statistical Significance and Biological Relevance

If there was a statistically significant allozyme difference between groups, it was assumed that this difference was also biologically meaningful because allozyme markers are generally not very variable. In fact, the concordance of genetic distance for allozymes and biological relevance was assumed to be informative enough to suggest guidelines for defining subspecies and species differences based on these genetic distances. In comprehensive reviews, Coyne and Orr (1989, 1997) have suggested that the extent of genetic distance between *Drosophila* species, found by allozymes, is correlated with the degree of postzygotic and prezygotic reproductive isolation as well as biogeography.

Although tests can determine whether genetic differences between groups are statistically significant, short of reproduction isolation it is difficult to provide a threshold for what is a meaningful biological difference between samples within a species and what is not. One approach is to assume that a statistically significant genetic difference occurs only if groups have been separated for some time and that this amount of time will provide adequate opportunity for meaningful biological differences to accumulate. This connection may function adequately for changes that accumulate relatively slowly over time. Some adaptive genetic differences between groups, however, may develop rapidly (e.g., insecticide, herbicide, and antibiotic resistance) and may not be reflected in the degree of neutral genetic differentiation detected by molecular markers.

Let us examine the connection of statistical significance and biological meaningfulness of comparisons of groups (see also Waples 1998; Hedrick 1999). First, there may be no significant statistical and no meaningful biological difference between groups. Second, there may be both, because a significant statistical difference between groups also reflects a meaningful biological difference. In both these cases, statistics based on genetic markers result in an appropriate evaluation of the real biological situation.

However, problems result when statistical significance does not reflect biological meaningfulness, a conflict that can occur in two basic forms. First, there may be no statistical significance when there is actual meaningful biological difference between groups. For allozymes, there was often not enough statistical power to detect real differences between groups because of the low number of polymorphic loci or the low sample size. Using highly variable loci, however, this has become much less of a problem because most organisms are variable for a number of loci, and sample size can often be larger because of less intrusive sampling of individuals.

In some cases, there may be no significant difference based on genetic markers but other, adaptively important, loci may be highly differentiated between populations. For example, in Scots pine (*Pinus sylvestris*) from Finland, molecular markers such as allozymes, RAPDs, and microsatellites all show very little differentiation between northern and southern populations (Karhu et al. 1996). In common experimental environments, however, a number of important adaptive quantitative traits, such as date of bud set, show high levels of genetic differentiation between these populations. In this case, the molecular data appear to be adequately reflecting the high level of gene flow in Scots pine. However, the selective forces between populations are so strong that they overcome the effects of gene flow and result in large adaptive genetic differences between populations. They lead to such low fitness in transplants that it is generally recommended that transplants not be over 100 km different in latitude. In other instances, both selection and gene flow may not be as strong, but could result in similar adaptive differences between populations. In this case, the error is not a typical "false negative" because the result is correct for the neutral nuclear markers. The error results from not assaying the genes involved directly in adaptation, not an easy problem to overcome in most instances.

Second, there may be statistical significance between groups when there is no meaningful biological difference. This conflict threatens to become a major concern in conservation biology as large numbers of highly variable markers become available in many species. For example, data from 5,264 microsatellite loci in humans with average heterozygosity of 0.70 have been published (Dib et al. 1996). With the statistical power provided by so many highly polymorphic loci, very small differences between groups would be statistically significant. This is not a typical "false positive" because the differences detected are real, but they may not reflect a biologically meaningful difference. The high power to detect small differences also means that small artifacts, such as data-collection errors or nonrandom sampling, may be mistaken for biologically important differences (R. Waples, personal communication).

To determine a biologically meaningful difference, we need to define some measure or effect related to the likelihood of the accumulation of significant biological differences, which is not a simple problem. One way to determine the relationship between biological and statistical significance is to evaluate the statistical power to detect a known biological effect. For example, figure 17.5 gives the statistical power to detect a one-generation genetic bottleneck of different sizes compared to the ancestral population for different numbers of loci (S. Kalinowski unpublished data). In this case, each locus had five alleles drawn from a near-neutrality distribution and the sample size was 40. Obviously, as the number of loci increases, the power to detect a bottleneck of a given size is substantially increased. Even with ten loci, the statistical power to detect a bottleneck of size 32 is greater than 0.8. It is unlikely that a one-generation bottleneck of that size would have any biological effects, but there would be high statistical power to detect it with highly polymorphic loci.

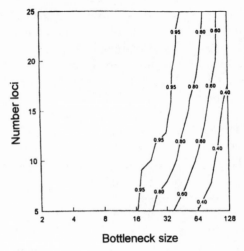

Fig. 17.5 The statistical power to detect a one-generation bottleneck of different sizes for different numbers of polymorphic loci.

CONCLUSIONS

The introduction of highly variable nuclear loci, such as microsatellites, is having a large impact because of their ability to resolve many questions in conservation biology. As illustrated from the examples in the endangered Gila topminnows and winter-run chinook salmon, microsatellite data have given insights that were not possible with earlier molecular or other nonmolecular techniques. However, the interpretation of the significance of variation at these loci within and between groups must be considered carefully. As demonstrated here, the level of differentiation and genetic distance between groups is greatly influenced by the level of heterozygosity for these highly variable loci. In addition, the connection between statistical and biological significance may often be weak. We need to carefully evaluate data from highly variable loci and realize that the information that they give may be quite different from that obtained from less variable markers, such as allozymes.

LITERATURE CITED

Chakraborty, R., and M. Nei. 1977. Bottleneck effects on average heterozygosity and genetic distance with the stepwise mutation model. *Evolution* 31:347–356.

Charlesworth, B. 1998. Measures of divergence between populations and the effect of forces that reduce variability. *Molecular Biology and Evolution* 15:538–543.

Cornuet, J.-M., and G. Luikart. 1996. Description and power analysis of two tests for detecting recent population bottlenecks from allele frequency data. *Genetics* 144:2001–2014.

Coyne, J. A., and H. A. Orr. 1989. Patterns of speciation in *Drosophila*. *Evolution* 43:362–381.

———. 1997. "Patterns of speciation in *Drosophila*" revisited. *Evolution* 51:295–303.

Davies, N., F. X. Villablanca, and G. K. Roderick. 1999. Determining the source of individuals: multilocus genotyping in nonequilibrium population genetics. *Trends in Ecology and Evolution* 14:17–21.

Dib, C., S. Faure, C. Fizames, D. Samson, N. Drouot, A. Vignal, P. Millasseau, S. Marc, J. Hazan, E. Seboun, M. Lathrop, G. Gyapay, J. Morrisette, and J. Weissenbach. 1996. A comprehensive genetic map of the human genome based on 5,264 microsatellites. *Nature* 380:152–154.

Frankham, R. 1995. Effective population size/adult population size ratios in wildlife: a review. *Genetical Research* 66:95–107.

Geiger, H. J., W. M. Smoker, L. A. Zhivotovsky, and A. J. Gharrett. 1997. Variability of family size and marine survival in pink salmon (*Oncorhynchus gorbuscha*) has implications for conservation biology and human use. *Canadian Journal of Fisheries and Aquatic Science* 54:2684–2690.

Goldstein, D. B., and D. D. Pollock. 1997. Launching microsatellites: a review of mutation processes and methods of phylogenetic inference. *Journal of Heredity* 88:335–342.

Hedgecock, D., V. Chow, and R. S. Waples. 1992. Effective population numbers of shellfish broodstocks estimated from temporal variance in allelic frequencies. *Aquaculture* 108:215–232.

Hedrick, P. W. 1999. Perspective: highly variable genetic loci and their interpretation in evolution and conservation. *Evolution* 53:313–318.

Hedrick, P. W., T. E. Dowling, W. L. Minckley, C. A. Tibbets, B. D. DeMaris, and P. C. Marsh. 2000a. Establishing a captive broodstock for bonytail chub (*Gila elegans*). *Journal of Heredity* 91:35–39.

Hedrick, P. W., D. Hedgecock, and S. Hamelberg. 1995. Effective population size in winter-run chinook salmon. *Conservation Biology* 9:615–624.

Hedrick, P. W., D. Hedgecock, S. Hamelberg, and S. J. Croci. 2000b. The impact of supplementation in winter-run chinook salmon on effective population size. *Journal of Heredity* 91:112–116.

Hedrick, P. W., and K. M. Parker. 1998. MHC variation in the endangered Gila topminnow. *Evolution* 52:194–199.

Hedrick, P. W., K. M. Parker, and R. N. Lee. 2001. Using microsatellite and MHC variation to identify species, ESUs, and MUs in the endangered Sonoran topminnow. *Molecular Ecology* 10:1399–1412.

Hedrick, P. W., V. Rashbrook, and D. Hedgecock. 2000c. Effective population size in returning winter-run salmon based on microsatellite analysis of returning spawners. *Canadian Journal of Fisheries and Aquatic Sciences* 57:2368–2373.

Jarne, P., and P. J. L. Lagoda. 1996. Microsatellites: from molecules to populations and back. *Trends in Ecology and Evolution* 11:424–429.

Jin, L., and R. Chakraborty. 1995. Population structure, stepwise mutation, heterozygote deficiency, and their implications in DNA forensics. *Heredity* 74:274–285.

Karhu, A., P. Hurme, M. Karjalainen, P. Karvonen, K. Karkkainen, D. Neale, and O. Savolainen. 1996. Do molecular markers reflect patterns of differentiation in adaptive traits of conifers? *Theoretical and Applied Genetics* 93:215–221.

Luikart, G., and J.-M. Cornuet. 1998. Empirical evaluation of a test for identifying recently bottlenecked populations from allele frequency data. *Conservation Biology* 12:228–237.

Mace, G. M., and R. Lande. 1991. Assessing extinction threats: towards a reevaluation of IUCN threatened species categories. *Conservation Biology* 5:148–157.

Meffe, G. K., and R. C. Vrijenhoek. 1988. Conservation genetics in the management of desert fishes. *Conservation Biology* 2:157–169.

Minckley, W. L. 1999. Ecological review and management recommendation for recovery of the endangered Gila topminnow. *Great Basin Naturalist* 59:230–244.

Moritz, C. 1994. Defining "evolutionarily significant units" for conservation. *Trends in Ecology and Evolution* 9:373–375.

———. 1999. Conservation units and translocation: strategies for conserving evolutionary processes. *Hereditas* 130:217–228.

Nagylaki, T. 1998. Fixation indices in subdivided populations. *Genetics* 148:1325–1332.

Nauta, M. J., and F. J. Weissing. 1996. Constraints on allele size at microsatellite loci: implications for genetic differentiation. *Genetics* 143:1021–1032.

Nei, M. 1972. Genetic distance between populations. *American Naturalist* 106:283–292.

———. 1973. Analysis of gene diversity in subdivided populations. *Proceedings of the National Academy of Sciences* (USA) 70:3321–3323.

———. 1987. *Molecular evolutionary genetics*. Columbia University Press, New York, New York.

Nunney, L. 1993. The influence of mating system and overlapping generations on effective population size. *Evolution* 47:1329–1341.

Paetkau, D., L. P. Waits, P. L. Clarkson, L. Craighead, and C. Strobeck. 1997. An empirical evaluation of genetic distance statistics using microsatellite data from bear (Ursidae) populations. *Genetics* 147:1943–1957.

Parker, K. M., R. J. Sheffer, and P. W. Hedrick. 1999. Molecular variation and evolutionarily significant units in the endangered Gila topminnow. *Conservation Biology* 13:108–116.

Quattro, J. M., P. L. Leberg, M. E. Douglas, and R. C. Vrijenhoek. 1996. Molecular evidence for a unique evolutionary lineage of an endangered Sonoran desert fish (genus *Poeciliopsis*). *Conservation Biology* 10:128–135.

Quattro, J. M., and R. C. Vrijenhoek. 1989. Fitness differences among remnant populations of the endangered Sonoran topminnow. *Science* 245:976–978.

Ryman, N., and L. Laikre. 1991. Effects of supportive breeding on the genetically effective population size. *Conservation Biology* 5:325–329.

Schwartz, M. K., D. A. Tallmon, and G. Luikart. 1998. Review of DNA-based census and effective population size estimates. *Animal Conservation* 1:293–299.

Sheffer, R. J., P. W. Hedrick, W. L. Minckley, and A. L. Velasco. 1997. Fitness in the endangered Gila topminnow. *Conservation Biology* 11:162–171.

Sheffer, R. J., P. W. Hedrick, and A. L. Velasco. 1999. Testing for inbreeding and outbreeding depression in the endangered Gila topminnow. *Animal Conservation* 2:121–129.

Slatkin, M. 1985. Rare alleles as indicators of gene flow. *Evolution* 39:53–65.

Smith, T. B., and R. K. Wayne, editors. 1996. *Molecular genetic approaches in conservation*. Oxford University Press, New York, New York.

Takezaki, N., and M. Nei. 1996. Genetic distances and reconstruction of phylogenetic trees from microsatellite DNA. *Genetics* 144:389–399.

Vrijenhoek, R., M. E. Douglas, and G. K. Meffe. 1985. Conservation genetics of endangered fish population in Arizona. *Science* 229:490–492.

Vucetich, J. A., T. A. Waite, and L. Nunney. 1997. Fluctuating population size and the ratio of effective to census population size. *Evolution* 51:2017–2021.

Waples, R. S. 1995. Evolutionary significant units and the conservation of biological diversity under the Endangered Species Act. *American Fisheries Society Symposium* 17:8–27.

———. 1998. Separating the wheat from the chaff: patterns of genetic differentiation in high gene flow species. *Journal of Heredity* 89:438–450.

Watterson, G. A. 1978. The homozygosity test of neutrality. *Genetics* 88:405–417.

Weir, B. S., and C. C. Cockerham. 1984. Estimating F-statistics for the analysis of population structure. *Evolution* 38:1358–1370.

18

Pedigree Analyses in Wild Populations
Susan M. Haig and Jonathan D. Ballou

ABSTRACT

While rarely used by field biologists, pedigree analyses provide a unique perspective regarding the microstructure of a population. Information on founder contribution, inbreeding, heterozygosity, genome uniqueness, and mean kinship can be gleaned from pedigree analyses, used to better understand current population structure, and can test hypotheses regarding the future potential of the population. Traditionally, pedigree analyses have been used to manage domestic and captive zoo populations. Here we review the use of pedigree analyses in wild populations and present three case studies that illustrate (1) various options for reintroducing Guam rails (*Rallus owstoni*) back to the wild, (2) the benefits and costs of choosing specific golden lion tamarins (*Leontopithecus rosalia*) individuals to be moved among populations, and (3) the diversity of approaches needed to evaluate the complex population structure in red-cockaded woodpeckers (*Picoides borealis*). In each case, the novel application of pedigree analyses, integrated with other population diagnostic tools, provides perspective on their potential utility in future studies of wild populations.

INTRODUCTION

As population viability becomes dangerously low and the interrelatedness among genetic, demographic, and environmental factors is better recognized (Mills and Smouse 1994; Tanaka 1997; Frankham 1998), multiple and often nontraditional means of assessing persistence probabilities are needed. Genetic factors in natural populations are often assessed with molecular tools to estimate genetic diversity within and among populations, genetic distance among populations, the mating system, and perhaps identification of individuals within a population (Haig 1998). These measures provide a helpful starting point in understanding population structure. However, they often fall short in providing the

We thank Susan Daniels and Ken Jones for helpful discussions related to pedigree analyses in wild populations and for providing early drafts of manuscripts.

388

perspective needed to understand the relative importance of specific individuals or in designing recovery strategies for small populations.

Population viability analyses (PVAs), as described throughout this book, provide a broader understanding of various demographic and environmental factors influencing populations. However, PVAs often fall short in evaluating genetic diversity, as most PVA software does not consider genetic factors. Although VORTEX is an exception (Lacy et al. 2000; Allendorf and Ryman, chap. 4 in this volume; Beissinger et al. 2002), it uses simulated random pedigrees, rather than a specific pedigree, to examine loss of genetic diversity over time. Furthermore, most PVAs do not detail individual-specific influences on population persistence.

Pedigree analyses calculate the loss of genetic diversity over time for a defined group of individuals. They provide an important bridge between the descriptors provided by molecular tools and the multifactored assessments carried out in PVAs. Following creation of a simple data set (table 18.1), pedigree analyses describe the current population structure, provide a means for testing hypotheses relating to the genetic importance (or value) of specific individuals in current and future populations, and serve as a tool for monitoring genetic diversity over time.

In this chapter, we call for integration of pedigree analyses, demographic PVA models, and molecular genetic tools. If these methods are integrated properly, field biologists and population managers will have a unique set of tools that allow for iterative and effective evaluations of population structure and management options (fig. 18.1). Thus, we illustrate how pedigree analyses can easily be used by field biologists and provide three unique case studies where integration of pedigree

Table 18.1 Structure of the Standard Input File for Gene Drop Pedigree Analyses

Studbook No.	Sex	Dam	Sire	Alive (A) or Dead (D)
1	F	WILD	WILD	D
2	M	WILD	WILD	D
3	F	WILD	WILD	D
4	F	1	2	D
5	M	1	2	D
6	F	3	2	D
7	F	4	5	A
8	M	4	5	A
9	M	6	5	A
10	F	3	8	A

Notes: WILD = founder. See figure 18.2 for the corresponding pedigree.

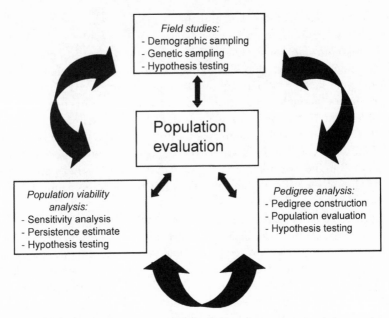

Fig. 18.1 Iterative integration of techniques needed to assess current population structure and potential viability.

analyses with other diagnostic tools results in efficient and more sound recommendations for recovery of small populations.

PEDIGREE MODELS AND METHODS

Pedigree analyses begin by defining each founder allele as unique, so that all subsequent measures of inbreeding, heterozygosity, and so forth are measures of identity by descent from those founders (see table 18.2 for definitions). Thanks to the extensive pedigree modeling that has been carried out for captive populations, the structure of data sets and the software needed to carry out these analyses are simple and straightforward. Perhaps the most direct method of pedigree analysis is to calculate parameters (e.g., inbreeding and kinship coefficients) using path analyses or additive relationship matrices (Ballou 1983). While these methods are amenable to computer programming, they cannot easily provide all parameters that might be desirable. For example, calculation of survival or extinction probabilities for founder alleles is extremely complex and beyond the reach of even the fastest computers for even fairly simple pedigrees (Thompson 1983). These probabilities, however, can be easily estimated through "gene drop" simulations.

The "gene drop" model was initially developed by Jean MacCluer

Table 18.2 Definitions of Genetic Factors Commonly Used in Pedigree Analyses

Factor	Definition Used in Gene Drop	Commonly Used Relationship to Other Factors
Allelic diversity	The number of unique founder alleles present in the living population.	
Founder	An individual who has no ancestors in the pedigree.	
Founder contribution	The proportion of the living population's gene pool that has descended from each founder.	
Founder genome equivalent (FGE)	The number of animals from the founding (source) population that have the same gene diversity as the living population.	$FGE = 0.5/[1 - (GD_t/GD_0)]$ $FGE \approx N_e/t$; N_e is effective population size; t is time.
Gene diversity (GD)	A measure of expected heterozygosity. In pedigree analysis, it usually refers to the proportion of expected heterozygosity retained in the living population (GD_t) relative to the population from which the founders were obtained (i.e., source population; GD_0).	$GD = 1 - 1/(2 * FGE)$ $GD = 1 -$ average MK $GD = GD_t/GD_0$
Genome uniqueness	The proportion of genes in an individual that are not present in others.	
Heterozygosity	The number of heterozygous individuals in the living population. Expected heterozygosity refers to expectations under Hardy-Weinberg equilibrium.	
Inbreeding coefficient	The probability of an individual's receiving two identical copies of unique founder alleles; the decrease in observed heterozygosity relative to the assumed 100% heterozygosity among the founders.	
Mean inbreeding coefficient	The proportional decrease in observed heterozygosity relative to the expected heterozygosity in a population.	
Kinship	The probability that alleles sampled will be identical by descent. It is a measure of relatedness. The kinship of a pair is the same as the inbreeding coefficient of their offspring. Kinship of an individual to itself = 0.5.	

(*continued*)

Table 18.2 (*continued*)

Factor	Definition Used in Gene Drop	Commonly Used Relationship to Other Factors
Mean kinship (MK)	The average coefficient of kinship between a living individual and all nonfounders in the living population. It is a useful descriptor of the importance of an individual to preserving genetic diversity in the population. The lower the MK, the greater the individual's importance.	Average MK = 0.5/ FGE Average MK = 1.0 − GD

Note: See Lacy (1995) for more details.

(MacCluer et al. 1986; Mace 1986) and was subsequently programmed by Georgina Mace (London Zoological Society) and Robert Lacy (GENES; Lacy 1999). Currently, the software is available in a program called SPARKS (Single Population Analysis and Records Keeping System; SPARKS 1996), which is a studbook database management and analysis system. GENES also runs as a stand-alone program and can be downloaded from the Web site www2.netcom.com/~rlacy/home. While SPARKS is primarily geared for use in captive population management, the GENES add-on is amenable to using data from wild populations. For example, data on wild golden lion tamarin (*Leontopithecus rosalia*) births, parentage, intergroup transfers, emigration, and deaths are entered into SPARKS in much the same way data from zoo animals are entered. This provides input data for pedigree analysis using GENES as well as instant calculation of inbreeding coefficients, life tables, age structures, and other default analyses provided by SPARKS (Dietz et al. 2000).

Gene drop pedigree analyses are Monte Carlo simulations in which many (e.g., 10,000) iterations through the model represents sampling an individual's entire genome. Following creation of the standard data set (table 18.1), each founder (an individual who has no ancestors in the pedigree) is assigned two unique alleles at the beginning of a simulation (fig. 18.2). The model then moves alleles through the pedigree, generation by generation, assuming Mendelian inheritance (i.e., each individual receives one allele that is selected randomly from its mother and its father). Following 10,000 iterations, results are summarized as the frequency with which neither, one, or both of two alternate founder alleles survives in the living population. The results generate a distribution of probabilities that provide an estimate of the proportion of a founder's alleles that have survived in the extant living population.

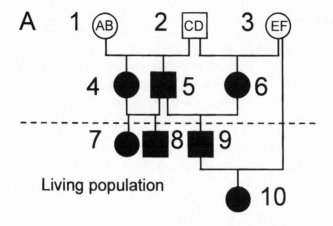

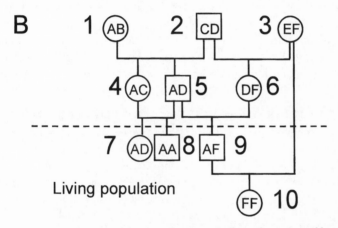

Fig. 18.2 Gene drop pedigree model and analysis for a small population (see table 18.2 for partial input file). A, In this example pedigree, the three founders (individuals with studbook numbers 1–3) have been given six unique alleles (A–F). B, The gene drop model operates by randomly passing founder alleles through the pedigree from parents to offspring with a 50% chance of each allele passing to the offspring. After one iteration of the simulation, individuals 8 and 10 are homozygous, whereas others in the living population (individuals 7 and 9) are heterozygous. Thus, 50% of the heterozygosity is retained (two of four living individuals). In the living population, three unique alleles (A, D, F) have survived, and all three founders are represented, although not equally (as would be optimal): four of eight alleles in the living population are contributed by founder 1, one of eight from founder 2, and three of eight from founder 3. The simulation would continue by returning to the original pedigree (A) and beginning a new iteration of gene drop process. This process would be repeated until 10,000 iterations have been completed to produce probability distributions of the model outputs (see text for details).

Combining gene drop pedigree analysis and additive relationship pedigree analyses, the following parameters can be calculated for the current population or tracked over time: founder contribution, number of unique alleles surviving (allelic diversity), proportional gene diversity, heterozygosity, mean kinship, inbreeding coefficients, and founder genome equivalents (see table 18.2 for definitions).

It is important to keep in mind that these pedigree analyses measure genetic parameters relative to the founding population. For example, measures of genetic diversity (gene diversity, allelic diversity, and heterozygosity) reflect proportional increases or losses relative to the founders, and are not the actual levels that might be found in the population by using molecular markers. This is because the assumptions of the model differ from what is measured via molecular markers. The model begins by assuming 100% heterozygosity among founders (i.e., each founder has two unique alleles), 100% allelic diversity, and the number of founder genome equivalents is equal to the number of founders. Other assumptions of gene drop models include independence among model runs, no linkage, and no selection. These assumptions are violated to the extent that linkage and selection exist in natural populations.

USES OF PEDIGREE ANALYSES IN WILD POPULATIONS

Pedigree analyses are poorly integrated in even the most desperate of population recovery efforts. In a review of 95 translocation and reintroduction programs, Kalmer (1995) found that 47 programs considered some aspect of genetics in designing management efforts. Of those, 36 (38%) used pedigree analyses to design the translocation or reintroduction, but few maintained pedigrees once the initial translocation or reintroduction had taken place.

Using formal pedigree analyses in nonendangered populations is even less common. Perhaps because field biologists associate pedigree analyses with domestic or captive situations, few are aware of the benefits derived from pedigree analyses for wild populations. Because there are so few cases where pedigree analyses have been used to evaluate wild populations, we cannot always illustrate the full range of uses with examples from the wild. Instead, the approaches described below may provide a starting point for field biologists considering compiling pedigree data.

Population Description/Population Monitoring

First and foremost, pedigree analyses describe the microstructure of a population (table 18.1). By quantifying heterozygosity, inbreeding, and the contribution of founders to the living population, we can better understand the status of the population and potential for increasing genetic diversity. Examining the relative importance of single individuals or founder lines to population structure provides an enlightening view not provided by molecular genetics or PVA (Berger and Cunningham 1995; Gompper et al. 1997). Thus, whether the population is threatened or not, understanding that relatively few (or many) individuals are contributing to population structure sheds light on the social structure, demography, and effective size of the population (Laikre et al. 1997). Identifying specific individuals (or founder lines) involved (or not involved) in reproduction further pinpoints sources of diversity. For example, species that live in complex social systems, such as exist for cooperatively breeding birds, often have population structures characterized by a multigeneration, extended-family social system (Stacey and Koenig 1990). Molecular analyses can identify parentage and PVAs can model changes in demography, but pedigree analyses would provide a helpful perspective in identifying which individuals actually contribute genes to subsequent generations and partitioning the effects of their contributions. Carrying out these analyses on an annual basis provides a view of how genetic structure and diversity changes over time, given the birth or death of specific individuals. Comparing pedigree results among species with similar or dissimilar mating systems or population structures will provide an untapped perspective on theoretical and applied issues of genetic diversity. Previously, this measurement of loss of genetic diversity over time could be carried out only by complex models of effective population size or genetic simulation models.

A more classic use of pedigree analyses remains useful in wild populations. Traditionally, pedigree analyses have been used to calculate modes of inheritance of specific traits. Traits can be as simple as eye color or more complex, as in inheritance of resistance to disease (Laikre et al. 1993). Recently, potential success of the Mexican wolf (*Canis lupus baileyi*) and red wolf (*C. rufus*) reintroduction programs were examined by calculating inbreeding coefficients from pedigrees (Kalinowski et al. 1999). Similarly, Bush et al. (1996) traced the origin and effects of hernias in captive and reintroduced golden lion tamarins, using pedigree analyses.

Hypothesis Testing

Quantifying basic pedigree descriptors can lead to unique opportunities for testing hypotheses about population management as well as the evolution of populations over time. For example, specific individuals can be taken out of pedigrees or new individuals (founder lines) added to quantify the impact they have on population structure and genetic diversity over various time periods and in multiple populations. Thus, biologists interested in metapopulation dynamics have the opportunity to understand the impact of greater or lesser gene flow by moving or breeding specific individuals. Those concerned with reintroduction or translocation programs can test hypotheses regarding how many individuals and which ones move into or out of donor and recipient populations, prior to carrying out management activities (Russell et al. 1994; McCullough et al. 1996; Earnhardt 1999).

A novel twist on testing hypotheses using pedigrees was developed to evaluate various training methods used in the reintroduction of golden lion tamarins to Brazil (J. Ballou, B. Beck, and M. Bush, personal communication). Reintroduced animals have been monitored and a pedigree database maintained for the population since the reintroduction program began in 1983. Instead of founder lines being identified by ancestors alone, family lines were defined by the type of training they received prior to reintroduction. By evaluating the subsequent genetic success of various founder-training groups, managers were able to determine the most effective means of preparing animals for reintroductions.

Challenges to Carrying Out Pedigree Analyses on Wild Populations

As with any model, pedigree analyses have drawbacks that need to be considered. First, a pedigree is only as good as its depth (number of generations) and the identification of the individuals in it. Thus, the parentage of all individuals must be understood. Parentage can be identified by long-term monitoring of individuals in the field and/or by molecular analyses (e.g., Hughes 1998). While the process is not simple, relatedness can also be derived using a combination of partial pedigree information and molecular markers (Queller and Goodnight 1989; Haig et al. 1994a, 1995; Geyer and Thompson 1995; Blouin et al. 1996; Taylor et al. 1997; Goodnight and Queller 1999). Further, the program CERVUS infers parentage from codominant molecular data (http://helios.bto.ed.ac.uk/evolgen/cervus/cervus.html; Marshall et al. 1998). When these methods fail, various mathematical means have been developed to derive the relatedness of unknown individuals in a pedigree. For example, maximum avoidance of inbreeding (MAI) schemes are generally used only when complete pedigrees are known. However, Princee (1995) developed guidelines whereby inbreeding avoidance management is focused at the level of social groups rather than specific individuals. This low-intensity genetic management approach provides the opportunity to examine the effects on genetic diversity of varying social-group composition or genetic exchange rates among groups.

A more specific problem arises when a pedigree is known but attributes of specific individuals within the pedigree are unknown. For example, in Przewalski's horse (*Equus przewalskii*), alloyzme data had been collected to examine polymorphisms due to coat color and associated inbreeding coefficients. However, not all animals in the pedigree had been sampled. Thomas (1995) solved this problem by estimating the genotype of nonsampled individuals based on the genotypes of their ancestors, relatives, and descendants using pedigree probability calculations through a process called "peeling" (Cannings and Thompson 1981). In this method, individuals are progressively peeled from the pedigree, and their genotype information is converted into a function related to others remaining in the pedigree. This is repeated until relatedness is estimated for all individuals.

While the best strategy for field biologists is to record parentage of all individuals in their population over time, it is becoming increasingly easier to fill in gaps when individual identification data are missing.

CASE STUDIES

There are relatively few case studies that use detailed pedigree analyses in wild populations, yet those described below illustrate the benefits that can be derived when pedigree information is collected from the beginning of a study.

Designing New Populations: Guam Rails

Guam rails are one of the three native forest birds from Guam that were brought into captivity in 1986 just prior to their extinction in the wild. Devastation of this endemic flightless rail came as a result of introduction of the brown tree snake (*Boiga irregularis*) during World War II (Savidge 1987). In the mid-1980s, 21 birds (founders) were brought into captivity in U.S. zoos, and a pedigree was begun. Once the captive program was established, plans were made to introduce the birds back into the wild as an experimental population on the nearby Mariana island of Rota. Rails had not previously inhabited Rota, so there were no native rails to compete with on the island. Because the snake remained a serious problem on Guam, it was deemed critical to get the birds back in the wild as soon as possible. The first step was to quantify genetic diversity in the wild and captive birds relative to other congenors (Haig and Ballou 1995). Results from allozymes and minisatellite DNA analyses indicated that, while subsequent captive offspring had lost about 6% of their genetic diversity relative to the 21 original founders, their overall genetic diversity was higher than other congenors. Finally, molecular tools were used to identify unknown founders and complete the pedigree (Haig et al. 1994a).

Once the captive pedigree was established, the task at hand was to design a reintroduced population that would represent the genetic diversity of the founders without risking it in the captive population. We used hypothesis testing to evaluate the best approach to the problem. Six hypotheses were tested by creating potential wild populations and running the gene drop model to evaluate the resulting genetic diversity (Haig et al. 1990). Each potential wild population was created by adding offspring to the existing pedigree that represented the reintroduction option to be evaluated. By denoting these offspring as the living population, we were able to assess genetic diversity in the reintroduced population. Options considered included (1) choosing the most fecund breeders to produce chicks for reintroduction, (2) using allozyme data to choose parents that would produce the most genetically diverse chicks, and (3) selecting pedigree management options such as choosing pairs to maximize allelic diversity, founder contribution, or founder genome equivalents. In the end, maximizing founder genome equivalents in the wild population resulted in the highest levels of genetic diversity as measured by heterozygosity, allelic diversity, founder contribution, the number of founders represented in the wild population, and the number of founder genome equivalents. The introduction to Rota was carried out by producing chicks to achieve this goal. This hypothesis-testing approach could be used to evaluate numerous questions regarding the structure or potential of any population—threatened or not. Hence, it illustrates one of the strongest features of using pedigree analyses with wild populations.

Moving Individuals among Populations: Golden Lion Tamarins

The value of using pedigree analyses for hypothesis testing is further exemplified in the challenging job of long-term management of a single, complex pedigree in which there are multiple populations. Simultaneous management of captive and reintroduced golden lion tamarins provides a unique example of this type of decision making.

The golden lion tamarin is a primate from Brazil's Atlantic coastal rain forest that has undergone tremendous population declines since the 1950s (Kleiman et al. 1987). The captive-breeding program began in 1981, and since 1984 the reintroduction program has released approximately 150 animals into available forests in Brazil. Thus, the tamarin pedigree is complex because it is deep, wide, and has to provide for both populations, as the reintroduced population is constantly being derived from the captive population. The decision to move animals from the captive population to the wild reintroduced population must consider how to improve the gene pool of the wild population without diminishing diversity in the captive population. At first glance, the situation appears similar to the Guam rail reintroduction. In this case, however, it is not one new population that will be introduced; instead there is an annual evaluation of which individuals should be released in the wild.

To address this problem, pedigree analyses were used to identify the genetic value of individuals to both captive and reintroduced populations by estimating average mean kinship and gene diversity (Ballou and Lacy 1995; Ballou 1997). The change in gene diversity in the captive or wild populations, as a result of an individual leaving one and joining another, can be measured by comparing its mean kinship in the current captive population with its mean kinship if it were introduced to the wild population. The resulting costs and benefits are plotted so that appropriate management decisions can be made (fig. 18.3). This process

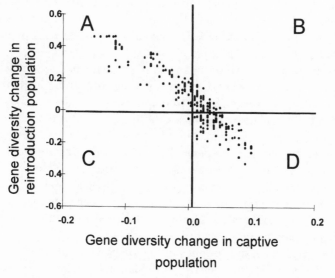

Fig. 18.3 Genetic costs and benefits of introducing individual golden lion tamarins into the wild (from Ballou 1997). Each point is an individual in the captive population. The location of the point shows how gene diversity in the captive and reintroduced populations will change if that individual is moved to the wild. Each quadrant identifies individuals that, if reintroduced, would (A) increase gene diversity of the reintroduction population but reduce diversity of the captive population, (B) improve gene diversity of both populations, (C) decrease gene diversity in both populations, and (D) benefit the captive population but reduce gene diversity in the wild population.

can be done iteratively until a group of candidates is selected or when additional candidates are needed.

The scenario outlined above has quite practical applications for population management, but the approach can also be used to better understand the genetic values of specific individuals in captive or wild populations regardless of their conservation status.

Assessing the Structure and Status of Wild Populations: Red-Cockaded Woodpeckers

Long-term studies of individually marked red-cockaded woodpeckers (*Picoides borealis*) illustrate important insights that can be gained by integrating pedigree analyses and population evaluation tools in recovery of wild populations. Red-cockaded woodpeckers have a unique life history due to a cooperative breeding social system, dependence on late successional trees for nest and roost cavities, limited dispersal, and nonmigratory nature (Walters 1990). Due to habitat fragmentation and loss in the southeastern United States, they were also one of the first species listed under the U.S. Endangered Species Act in 1970 (35 *Federal Register* 16047). The combination of their unusual social system and endangered conservation status has led to detailed analyses of avian population structure.

The cooperative social system in red-cockaded woodpeckers is manifested by up to four offspring, usually male, remaining in their natal territory for up to several years to help raise subsequent broods. This, along with limited dispersal, raises the question of potential levels of inbreeding in the population. Therefore, an obvious question to pursue is whether helpers pass on genes by mating with their mothers. DNA fingerprint analyses of parentage among woodpeckers in North Carolina (Haig et al. 1994b) and South Carolina (Haig et al. 1993a) suggested that pairs were genetically monogamous. Thus, one potential source of inbreeding was eliminated. However, the precarious status of woodpecker populations required further analysis.

Jeffrey Walters and colleagues have closely monitored red-cockaded woodpeckers in the sandhills of North Carolina for over 20 years. Detailed data collected on individually marked birds provided the opportunity for pedigree analyses to evaluate the extent of inbreeding in this population. By examining the relationship among kinship and inbreeding values calculated from the pedigree, natal dispersal distances, and reproductive success, Daniels and Walters (2000) found inbreeding to have a significant effect on population viability. For example, they found close inbreeding significantly reduced reproductive success: survival of inbred fledglings to one year of age was reduced, as were hatching rates for closely related pairs. Furthermore, female fledglings were less likely to disperse from natal territories if closely related males were not breeding on the territory the following year. Thus, dispersal is a trade-off between the risk of inbreeding depression and the challenge of finding a new territory.

The implications of inbreeding in red-cockaded woodpeckers was further investigated in a spatially explicit population dynamic model that simulated the accumulation of inbreeding in populations of 25, 49, and 100 territories (Daniels et al. 2000). They found substantial amounts of inbreeding accumulated in small, closed populations and that moderately high levels of immigration (two or more individuals per year) were required to keep the mean inbreeding level under 0.10.

Understanding the amount and impact of inbreeding in red-cockaded woodpeckers has serious implications for designing recovery strategies. Nowhere was this more obvious than for the tiny population that exists in the Savannah River National Forest in the upper coastal plain/sandhills region of South Carolina. This once robust population had declined to less than ten birds in the late 1980s. Efforts to recover the population would require fairly significant intervention, but could not begin until population structure was understood. The initial pedigree was constructed using records from the field and DNA fingerprinting (Haig et al. 1993a). Pedigree analyses indicated that only 5.4 founder genome equivalents were present in the population even though 15 founders were originally present (Haig et al. 1993b). Kinship coefficients averaged 0.18 and confirmed close inbreeding. Thus, the population was in a more precarious state than census data indicated.

The pedigree analyses contributed to a PVA of the population by identifying the number and sex of breeders and their reproductive success (Haig et al. 1993b). The PVA corroborated the precarious status of this population, suggested by pedigree analyses, and indicated that the population had a mean time to extinction of 41.5 years. Therefore, the best means by which the population could reasonably recover appeared to come from translocating individuals from other populations. The challenge to accomplish this was twofold. First, donor populations are also endangered, so translocations could not result in reduced viability of donor populations. Second, managers at the Savannah River National Forest had limited resources for implementing translocations. Thus, PVA modeling was used to evaluate the costs and benefits of various translocation strategies. The recommendation adopted required annual movement of three females and two males into the population over a ten-year period. These translocations would more than double the mean time to extinction to 84.6 years, would not cause undue harm to donor populations, and were within the realm of what could practically be implemented by U.S. Forest Service personnel. The final step was using molecular markers to identify appropriate donor populations (Haig et al. 1996). Since these analyses, the Savannah River population has slowly begun recovery at little or no cost to donor populations.

SUMMARY AND RECOMMENDATIONS

While still relatively novel, the use of pedigree analyses in wild populations provides a detailed view of population structure, diversity, and potential viability that is otherwise difficult to obtain. Construction of data sets needed for the analyses is straightforward, and user-friendly software has been developed. The resulting analyses take into account the precise structure of the pedigree and provide a unique opportunity to test hypotheses regarding population structure and potential management options. Three case studies illustrated that iterative integration of multiple analytical tools and long-term field data can provide a comprehensive approach to understanding population processes and conservation planning.

The major drawback to using pedigree models is the lack of detailed pedigree data. Often this problem can be mitigated via molecular or mathematical identification of unknown individuals. However, similar to the data-heavy requirements of PVAs, it may be that relatively few species or populations will receive comprehensive pedigree analyses. In particular, it will be difficult to perform pedigree analyses on taxa such as fish, plants, and most invertebrates. In the end, however, biologists conducting long-term monitoring of marked individuals in wild populations will miss a unique opportunity to understand their study species if they are not collecting pedigree data from the onset of their study.

LITERATURE CITED

Ballou, J. D. 1983. Calculating inbreeding coefficients from pedigrees. Pages 509–520 in C. M. Schonewald-Cox, S. M. Chambers, B. MacBryde, and L. Thomas, editors, *Genetics and conservation: a reference for managing wild animal and plant populations.* Benjamin/Cummings, Menlo Park, California.

———. 1997. Genetic and demographic aspects of animal reintroductions. *Supplemento alle Ricerche di Biologia della Selvaggina* 27:75–96.

Ballou, J. D., and R. C. Lacy. 1995. Identifying genetically important individuals for managing genetic diversity in captive populations. Pages 76–111 in J. D. Ballou, M. E. Gilpin, and T. J. Foose, editors, *Population management for survival and recovery.* Columbia University Press, New York, New York.

Beissinger, S. R., J. R. Walters, D. G. Catanzano, K. G. Smith, J. B. Dunning, S. M. Haig, B. R. Noon, and B. M. Smith. 2002. The use of models in avian conservation. *Current Ornithology* 17. In press.

Berger, J., and C. Cunningham. 1995. Multiple bottlenecks, allopatric lineages, and Badlands bison *Bos bison*: consequences of lineage mixing. *Biological Conservation* 71:13–23.

Blouin, M. S., M. Parsons, V. Lacaille, and S. Lotz. 1996. Use of microsatellite loci to classify individuals by relatedness. *Molecular Ecology* 5:393–402.

Bush, M., B. B. Beck, J. M. Dietz, A. J. Baker, E. James, A. Pissinatti, L. G. Phillips, and R. J. Montali. 1996. Radiographic evaluation of diaphragmatic defects in golden lion tamarins (*Leontopithecus rosalia rosalia*): implications for reintroductions. *Journal of Zoo Wildlife Medicine* 27:346–357.

Cannings, C., and E. A. Thompson. 1981. *Genealogical and genetic structure*. Cambridge University Press, Cambridge, United Kingdom.

Daniels, S. J., J. A. Priddy, and J. R. Walters. 2000. Inbreeding in small populations of red-cockaded woodpeckers: analysis using a spatially-explicit simulation model. Pages 129–148 in A. G. Young and G. M. Clarke, editors, *Genetics, demography, and viability of fragmented populations*. Cambridge University Press, London, United Kingdom.

Daniels, S. J., and J. R. Walters. 2000. Inbreeding depression and natal dispersal in red-cockaded woodpeckers. *Condor* 102:482–491.

Dietz, J. M., A. J. Baker, and J. D. Ballou. 2000. Demographic evidence of inbreeding depression in golden lion tamarins. Pages 203–212 in A. G. Young and G. M. Clarke, editors, *Genetics, demography, and viability of fragmented populations*. Cambridge University Press, London, United Kingdom.

Earnhardt, J. M. 1999. Reintroduction programmes: genetic trade-offs for populations. *Animal Conservation* 2:279–286.

Frankham, R. 1998. Inbreeding and extinction: island populations. *Conservation Biology* 12:665–675.

Geyer, C. J., and E. A. Thompson. 1995. A new approach to the joint estimation of relationships from DNA fingerprint data. Pages 245–260 in J. D. Ballou, M. E. Gilpin, and T. J. Foose, editors, *Population management for survival and recovery*. Columbia University Press, New York, New York.

Gompper, M. E., P. B. Stacey, and J. Berger. 1997. Conservation implications of the natural loss of lineages in wild mammals and birds. *Conservation Biology* 11: 857–867.

Goodnight, K. F., and D. C. Queller. 1999. Computer software for performing likelihood tests of pedigree relationships using genetic markers. *Molecular Ecology* 8: 1231–1234.

Haig, S. M. 1998. Molecular contributions to conservation. *Ecology* 79:413–425.

Haig, S. M., and J. D. Ballou. 1995. Genetic diversity in two avian species formerly endemic to Guam. *Auk* 112:445–455.

Haig, S. M., J. D. Ballou, and N. J. Casna. 1994a. Identification of kin structure among Guam rail founders: a comparison of pedigrees and DNA profiles. *Molecular Ecology* 5:109–119.

———. 1995. Genetic identification of kin in Micronesian kingfishers. *Journal of Heredity* 86:423–431.

Haig, S. M., J. D. Ballou, and S. R. Derrickson. 1990. Management options for preserving genetic diversity: reintroduction of the Guam rail to the wild. *Conservation Biology* 4:290–300, 464.

Haig, S. M., J. R. Belthoff, and D. H. Allen. 1993a. Examination of population structure in red-cockaded woodpeckers using DNA profiles. *Evolution* 47:185–194.

———. 1993b. Population viability analysis for a small population of red-cockaded woodpeckers and an evaluation of population enhancement strategies. *Conservation Biology* 7:289–301.

Haig, S. M., R. Bowman, and T. D. Mullins. 1996. Population structure of red-cockaded woodpeckers in south Florida: RAPDs revisited. *Molecular Ecology* 5:725–734.

Haig, S. M., J. R. Walters, and J. H. Plissner. 1994b. Genetic evidence for monogamy in the red-cockaded woodpecker, a cooperative breeder. *Behavioral Ecology and Sociobiology* 23:295–303.

Hughes, C. 1998. Integrating molecular techniques with field methods in studies of social behavior: a revolution results. *Ecology* 79:383–400.

Kalinowski, S. T., P. W. Hedrick, and P. S. Miller. 1999. No inbreeding depression observed in Mexican and red wolf captive breeding programs. *Conservation Biology* 13:1371–1377.

Kalmer, A. 1995. Genetic assessments in animal translocation programmes: reintroductions re-stocking and conservation implications. M.S. thesis, University of Kent, United Kingdom.

Kleiman, D. G., B. B. Beck, J. M. Dietz, L. A. Dietz, J. D. Ballou, and A. F. Coimbra-Filho. 1987. Conservation program for the golden lion tamarin: captive research and management, ecological studies, education strategies, and reintroduction. Pages 959–979 in K. Benirschke, editor, *Primates: the road to self-sustaining populations*. Springer-Verlag, New York, New York.

Lacy, R. C. 1995. Clarification of genetic terms and their use in the management of captive populations. *Zoo Biology* 14:565–578.

———. 1999. GENES: software package for genetic analysis of studbook data. Chicago Zoological Society, Chicago, Illinois.

Lacy, R. C., K. A. Hughes, and P. S. Miller. 2000. VORTEX: a stochastic simulation of the extinction process. Chicago Zoological Society, Chicago, Illinois.

Laikre, L., N. Ryman, and N. G. Lundh. 1997. Estimated inbreeding in a small, wild muskox *Ovibos moschatus* population and its possible effects on population reproduction. *Biological Conservation* 79:197–204.

Laikre, L., N. Ryman, and E. A. Thompson. 1993. Hereditary blindness in a captive wolf population: frequency reduction of a deleterious allele in relation to gene conservation. *Conservation Biology* 7:592–601.

MacCluer, J. W., J. L. Vandeberg, B. Read, and O. A. Ryder. 1986. Pedigree analysis by computer simulation. *Zoo Biology* 5:147–160.

Mace, G. 1986. *Genedrop: computer software for genedrop analyses*. Zoological Society of London, London, United Kingdom.

Marshall, T. C., J. Slate, L. E. B. Kruuk, and J. M. Pemberton. 1998. Statistical confidence for likelihood-based paternity inference in natural populations. *Molecular Ecology* 7:639–655.

McCullough, D. R., J. K. Fischer, and J. D. Ballou. 1996. From bottleneck to metapopulation: recovery of the tule elk in California. Pages 375–403 in D. R. McCullough, editor, *Metapopulations and wildlife conservation*. Island Press, Covelo, California.

Mills, L. S., and P. E. Smouse. 1994. Demographic consequences of inbreeding in remnant populations. *American Naturalist* 144:412–431.

Princee, F. P. G. 1995. Overcoming the constraints of social structure and incomplete pedigree data through low-intensity genetic management. Pages 124–154 in J. D. Ballou, M. E. Gilpin, and T. J. Foose, editors, *Population management for survival and recovery*. Columbia University Press, New York, New York.

Queller, D. C., and K. F. Goodnight. 1989. Estimating relatedness using genetic markers. *Evolution* 43:258–275.

Russell, W. C., E. T. Thorne, R. Oakleaf, and J. D. Ballou. 1994. The genetic basis of black-footed ferret reintroduction. *Conservation Biology* 8:263–266.

Savidge, J. A. 1987. Extinction of an island forest avifauna by an introduced snake. *Ecology* 68:660–668.

SPARKS: Single Population Analysis and Records Keeping System 1.4. 1996. International Species Inventory System (ISIS). Apple Valley, Minnesota.

Stacey, P. B., and W. D. Koenig, editors. 1990. *Cooperative breeding in birds: long-term studies of ecology and behavior*. Cambridge University Press, Cambridge, United Kingdom.

Tanaka, Y. 1997. Extinction of populations due to inbreeding depression with demographic disturbances. *Researches in Population Ecology* 39:57–66.

Taylor, A. C., A. Horsup, C. N. Johnson, P. Sunnucks, and B. Sherwin. 1997. Relatedness structure detected by microsatellite analysis and attempted pedigree reconstruction in an endangered marsupial, the northern hairy-nosed wombat (*Lasiorhinus krefftii*). *Molecular Ecology* 6:9–19.

Thomas, A. 1995. Genotypic inference with the Gibbs sampler. Pages 261–270 in J. D. Ballou, M. E. Gilpin, and T. J. Foose, editors, *Population management for survival and recovery*. Columbia University Press, New York, New York.

Thompson, E. A. 1983. Gene extinction and allelic origins in complex genealogies. *Proceedings of the Royal Society of London*, series B, Biological Sciences, 219:241–251.

Walters, J. R. 1990. Red-cockaded woodpeckers: a primitive cooperative breeder. Pages 69–101 in P. B. Stacey and W. D. Koenig, editors, *Cooperative breeding in birds: long-term studies of ecology and behavior*. Cambridge University Press, New York, New York.

19

Rangewide Risks to Large Populations: The Cape Sable Sparrow as a Case History

Stuart L. Pimm and Oron L. Bass Jr.

ABSTRACT

Very small populations usually go extinct quickly. More surprisingly, larger populations that occur across wide ranges can also become extinct quickly. Understanding the fate of these species is the difficult challenge that this chapter addresses. The species of concern is the Cape Sable sparrow. We explore two methods of calculating the sparrow's risk of extinction. The first employs the idea that one can characterize the natural limits of population-size fluctuations over time on the basis of past experience of the species of concern or some similar species. So armed, one can predict whether the lower limit will encompass such low levels that rapid extinction will be probable. This method failed spectacularly when applied to a situation where it would seem entirely appropriate. The second method identifies the underlying causes of the sparrow's population fluctuations. In particular, we consider the factors that cause its range to shrink and determine its ability to recover from such shocks. By understanding the mechanisms underlying population fluctuations, we conclude that, if poor water management were to continue, it would exterminate the bird.

INTRODUCTION

The reasons for populations numbering a few to a few dozen breeding pairs going extinct quickly are well understood (Pimm 1991). Such populations suffer the problems of finding suitable mates, of many individuals dying before the next breeding season from different causes, loss of genetic variability and its deleterious consequences, and other unavoidable vagaries of birth and death. The importance of these chance factors diminishes quickly as populations become larger. Nonetheless, experience teaches us that much larger populations can also become extinct quickly. Indeed, we know that vertebrate populations numbering in the low thousands of breeding pairs are too rare to enjoy a secure future (Collar et al. 1994; Baillie and Groombridge 1996; Mace 1996). Under-

We thank the National Park Service and the Fish and Wildlife Service for support.

standing the fate of these species is the much more difficult challenge that this chapter addresses.

Large populations may comprise many smaller partially isolated subpopulations constituting a metapopulation. If so, the balance between frequent local extinction and recolonization from surviving populations determines the species' long-term fate (Hanski 1998, chap. 5 in this volume). In such cases, insights from studies of very small populations are of value (Pimm et al. 1993; Pimm and Curnutt 1994). In other cases, an inexorable decline in numbers, perhaps driven by a readily observable reduction in habitat, leads to a clear prediction of a species' demise. Yet other species may be at risk because of the high year-to-year variability in numbers that typify many natural populations (Pimm 1991). In nature, many individuals die from the *same* causes, for example, bad weather. Such natural population fluctuations can prove terminal for a species that is more geographically restricted than in the past.

The case history we present may be typical in requiring answers to all the questions implied by the last paragraph. What is the spatial organization of the population? Are any of its geographically determined subpopulations sufficiently small to warrant concerns over unavoidable vagaries of birth and death? What are the unnatural causes of population decline? How will these causes affect the population in the future? What are the natural causes of population fluctuations, and how can we anticipate the low levels to which they will drive the population in the future?

The species we consider is the Cape Sable sparrow (*Ammodramus maritimus mirabilis*), a drab, olive-brown bird, so obscure and lacking in charisma that it was not discovered until well into this century. First we summarize its natural history and describe the southern Everglades, to which it is restricted (Lockwood et al. 1997; Curnutt et al. 1998; Nott et al. 1998).

Next we explore two methods of calculating the sparrow's risk of extinction. The first employs the idea that one can characterize the natural limits of population-size fluctuations over time from the study of time-series data. So armed, one can predict whether the lower limit will encompass such low levels that rapid extinction will be probable. This is a familiar recipe. It characterizes the papers in Brook et al.'s meta-analysis (2000) of the predictive accuracy of "population viability analysis." One of us has devoted considerable thought to it (e.g., Pimm 1991). This method failed spectacularly even though the Cape Sable sparrow case would seem an entirely appropriate application. The second method identified the causes of the sparrow's population fluctuations, in particular its range contractions and its ability to recover from them. By understanding the mechanisms underlying population fluctuations, we deduce an altogether bleaker picture of the bird's future.

THE CAPE SABLE SPARROW AND THE ECOSYSTEM
ON WHICH IT DEPENDS

The Cape Sable sparrow is considered to be a subspecies of the widespread seaside sparrow, albeit an ecologically and geographically distinct one. It is not a "seaside" sparrow ecologically as it inhabits freshwater rather than saltwater marshes. Although first discovered in 1918 on Cape Sable, vegetation changes after the massive hurricane of September 1935 made the cape unsuitable for it. In addition to its unique habitat, it is geographically isolated in the Everglades region. The nearest surviving subspecies, *A. m. peninsulae*, occurs 300 km to the north.

The U.S. Fish and Wildlife Service included the subspecies in the first list of endangered species on March 11, 1967 (32 *Federal Register* 4001). Its restricted range and the fate of the population on Cape Sable were the primary justifications. The subsequent rapid extinction of the dusky seaside sparrow (*A. m. nigriscens*) in northern Florida lent support to that decision.

Shark River Slough is the primary drainage in the southern Everglades of Florida (fig. 19.1). To the west lies the higher ground of the Big Cypress National Preserve and, to the east, the Atlantic coastal ridge. Expanses of marl prairie lie between the main drainage of Shark River Slough and these two modest ridges. In contrast to the main slough, the prairies are inundated on average only from three to seven months per year. These seasonally flooded wetlands to the east and west of the slough are the ecosystem on which this bird depends.

Currently, nearly all the overland flow in the Shark River Slough drainage originates from the four S-12 gated spillways at the northern boundary of Everglades National Park (fig. 19.1). The east-west distribution of these structures covers about half of the predrainage expanse of Shark River Slough. Historically, most of the overland flow occurred toward the *eastern* edge of Shark River Slough, as suggested by the figure. The S-12 structures, however, are on the *western* edge: it is across these structures that the water now flows.

This artificial hydrology affects the two expanses of marl prairie on either side of the slough in opposite ways. The western marl prairies naturally remained dry for much of the year. They were inundated seasonally by rainfall and natural overflow from the slough. They are now subject to the managed water releases from the S-12 structures.

The southeast corner of the Florida peninsula held the largest expanse of marl prairie. Bounded by the eastern edge of Shark River

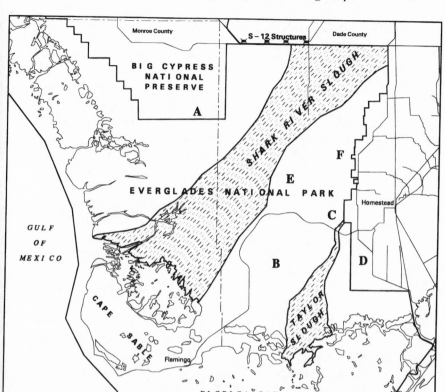

Fig. 19.1 Map of Everglades National Park and adjacent areas. The *bold line* is the boundary of Everglades National Park, and a *thin line* indicates a dike and associated canal. A–F indicate the approximate centers of the subpopulations of the Cape Sable sparrow.

Slough, it spread southeast, encompassing the southern terminus of the Atlantic coastal ridge. It ended at the thin line of mangroves along the northeastern shore of Florida Bay. To the north, the marl prairies once extended in a long arm to central Dade County. This expanse of potential sparrow habitat suffered two major assaults. The more drastic was the conversion of the eastern prairie to residential and agricultural lands.

Much of the remaining prairie, at and around the eastern boundary of Everglades National Park, is overdrained and subject to frequent fires. Fires in the wet season (June to October) are caused by lightning strikes and are generally small and patchy because the ecosystem is already wet. They occur throughout the region. Fires at the end of the dry season (March to late May) are frequently caused by human carelessness and tend to burn large areas along the Everglades eastern boundary but sometimes burn deep into the natural areas.

Curnutt et al. (1998) estimated that nearly half of the original prairie has been destroyed or degraded. As for many species at risk of extinction, the ultimate cause of endangerment is the massive reduction in suitable habitat.

Bass and Kushlan (1982) conducted the first extensive sparrow survey in 1981. We repeated the survey in 1992 and annually thereafter. Across a 1 × 1 km grid of more than 600 sites, we record the number of sparrows seen or heard within a ten-minute interval. We take particular care to visit all locations that might hold sparrows, and we do not observe birds at most of the sites we survey. This suggests that we do not miss many sites that hold birds.

To estimate the actual numbers of sparrows from the number we observed on our survey, we multiply each singing male by 16. This correction is based on the range at which we can detect the sparrow's distinctive song—it encompasses one-eighth of a square kilometer—and on the assumption that one female accompanies each singing male. Work on our intensive study plots confirms this calibration (Curnutt et al. 1998).

Using this calibration, we estimated that the total population of this species was over 6,000 in both 1981 and 1992. The birds are not distributed continuously, but are grouped into six subpopulations of varying sizes (fig. 19.1). Subpopulation A (west of Shark River Slough) was the most numerous in 1981 (~2,700 birds), and B held fewer birds (~2,300). Subpopulation B held more than A in 1992 (~3,000 versus ~2,600). Subpopulation E consistently held ~600 birds. The other three subpopulations held between 100 and 400 birds, although we found no birds in F in 1992.

RISK ANALYSIS 1: A PHENOMENOLOGICAL APPROACH

What is the likelihood that this bird will become extinct? Other things being equal, populations that are highly variable in their numbers from year to year are more likely to go extinct than less variable ones (Pimm et al. 1988; Pimm 1991). The causes of population variability are diverse. They include population factors (birth and death rates), features of the food web in which the species is embedded (e.g., whether it is a trophic specialist or generalist), and fluctuations of the host ecosystem. These factors operate at different scales (Pimm 1991). Estimating the population variance (or, equivalently, the variance in birth and death rates) and dissecting out underlying causes is a critical step in answering the key question about a species' fate. So how do we estimate this variability?

Sæther et al. (2000) have provided statistically rigorous dissections of key population variables, their variances, and their time dependence for various species. For example, they utilized a 20-year record along a 60 km stretch of the dipper's (*Cinclus cinclus*) riverine habitat, a large fraction of the population was color-banded, and the bird is widely distributed, relatively common, and conspicuous. Data-rich, long-term studies to assess population variability directly are a luxury afforded very few conservation biologists, however. For many endangered species, infrequent estimates of population size are often the only information available. For many species, we lack even this information. The urgency of the problem, however, does not allow us to request 20 years of intensive field effort before returning an answer. We might have access to long-term data on surrogate species that are closely related or at least ecologically similar. Using one, or at best a few, estimates of abundance and a surrogate estimate of year-to-year variability, we may be able to predict risk of extinction. This is a familiar tactic (Brook et al. 2000).

As for many other threatened species, there are no long-term data on year-to-year changes in Cape Sable sparrow populations, or indeed on other seaside sparrows. There are, however, substantial long-term records of grassland sparrow numbers in the Breeding Bird Survey (BBS). BBS data are obtained from point counts—a method very similar to the survey method we employ—and grassland sparrows from prairie states are broadly similar in their life-history characteristics.

Curnutt et al. (1996) used BBS data on ten North American grassland sparrows to explore how populations behave simultaneously in space and time. Two well-known relationships guided this exploration. The first is the power law relating variance of population abundance over time to average abundance across a species' geographic range (Maurer 1994). The second relationship examined the increase in a population's variability at a single location over time (Pimm and Redfearn 1988). Curnutt et al. (1996) asked how abundance, variability, and increase in variability change over a species' geographic range and with respect to one another.

For all but one of the species they analyzed, variability increased more slowly than expected with increasing abundance across the species' range. If relative variability were independent of abundance, the slope of the logarithm of standard deviation versus the logarithm of abundance would be one. Most of the species had slopes of ~0.7. This means that, where a species is least common—typically at the edge of its range—its population fluctuations will be relatively more variable. To put this average slope into more accessible terms, a sample of ten observations will span values encompassing approximately ±1.5 standard deviations of the mean for a normally and independently distributed (statistical) population.

First, consider one of the larger subpopulations of Cape Sable sparrows (A or B) and suppose we had counted 200 birds (log 2.3), leading to an estimate of 3,200 individuals (200 birds per km^2 × 16 km^2). The log of the standard deviation of this population would be 0.7 × 2.3 = 1.6, and so the standard deviation would be ~41. A range of ±41 × 1.5 (= 61) would have the population varying between 140 and 260 counted birds or an estimated 2,240 to 4,160 birds. This approximates a twofold span of values over a sample of ten points, that is, over a decade. This hypothetical situation fits comfortably with the experiences of those who count common birds over such intervals.

Now consider a site where the species is much rarer: say a mean count of 10 birds and an estimate of 160 birds. Using the same logic, it would have a standard deviation of 5, and abundances would be between 18 (an estimate of 288 birds) and 2 (an estimate of 32 birds). This is a much greater span of values than in the previous example (i.e., a factor of nine, versus a factor of about two). It is large enough that local extinctions might occur naturally, by chance, at least intermittently over the span of a decade or two. Mean population counts below ten should experience regular periods when the birds would not be counted—and where they might indeed be locally extinct.

We have not missed the significance of the assumptions of normal and independently distributed population sizes in the previous analysis. The population count in one year is likely to be dependent, probably strongly so, on that of the previous year. As a consequence, for most populations, estimates of the variability of population abundances increase with increased length of record (Pimm and Redfearn 1988).

This was also the case for the grassland sparrows. Curnutt et al. (1996) found that, of the seven species with at least ten sampling locations of continuous data over 20 years, six showed significant increases in variability over all time periods. These increases in variability over time would mean that, not only would we expect a sample of 20 years to encompass a wider range of standard deviations than the samples of 10 years exemplified above, but that the standard deviation itself would be larger.

We will not discuss how large the envelope of population fluctuations is with the added complication of increasing variability over time, for this has been done elsewhere (Lande 1993; Ariño and Pimm 1995).

Incorporating these details—or formalizing the mathematics—does not alter the general conclusions about the Cape Sable sparrow. The two largest subpopulations are large enough that, given the normal year-to-year variability seen in other grassland sparrows, we should not expect dangerously low populations within a century (or indeed a much longer interval). In contrast, the smaller subpopulations might well fall below levels where we likely could not count them, and where unavoidable vagaries of birth and death might well doom them to at least local extinction.

Thus, local subpopulations may become extinct, but at least one of the three larger subpopulations (A, B, or E) should be available to naturally restock them. This is an entirely comforting conclusion. It stems from a rough-and-ready estimate of risk, but one certainly appropriate to the amount of information at hand.

This conclusion was rudely shaken in April 1993. The western subpopulation (A in fig. 19.1), which the preceding calculation suggests might vary twofold over a decade, declined to one-seventh of its 1992 abundance in the spring of 1993. It has remained at low levels ever since. Subpopulation D in the southeast corner of the species' range nearly disappeared, and the subpopulations in the northeast (C and F) also declined. Curnutt et al. (1998) provided a detailed analysis to show that these declines were statistically highly improbable, given what we know about year-to-year variation in other sparrow populations.

The result was particularly discouraging to one of us (Pimm), because he had spent much of the previous decade cataloguing and analyzing natural year-to-year variation in population sizes for conservation ends (Pimm 1991). Moreover, he was a founding partner, with John Lawton (Ascot, United Kingdom) of the effort to provide a catalogue of more than 2,000 long-term time series (now available at http://www.sw.ic.ac.uk/cpb/cpb/gpdd.htm). A central objective of this compilation was to provide conservation biologists an accessible set of estimates of natural population variability for population risk assessments.

Worse still was that the assumption of natural variability seemed a particularly sensible one. The Cape Sable sparrow is found almost entirely within Everglades National Park and Big Cypress National Preserve. These adjacent protected areas are very large by the standards of the hemisphere. Only about 20 national parks in Central and South America are as large or larger (Mayer and Pimm 1998). If the method of using natural variability to calculate risk of extinction should apply anywhere, this bird in these national parks would seem to be a good candidate. Why did this approach fail?

RISK ANALYSIS 2: A MECHANISTIC APPROACH

Our surveys showed that the Cape Sable sparrow population on the western side of Shark River Slough declined dramatically since 1992. It has declined similarly since 1981 in the northeast and southeast of its range. Only two subpopulations have remained more or less constant. According to Lockwood et al. (1997), Curnutt et al. (1998), and Nott et al. (1998), there are three reasons. (1) The massive decline in the western subpopulation is the consequence of the inundation of the breeding habitat during the dry season by managed flows over the S-12 structures in 1993, 1994, and 1995. (2) The decline in most of the northeastern subpopulations is due to the very high fire frequencies in these areas over the last decade or more. We erect the plausible hypothesis that the high fire frequency is due, in part, to the high incidence of unplanned human ignitions in the areas adjacent to the park. Moreover, we suggest that unnaturally low levels of water permit high fire frequencies during the breeding season. Water that should have naturally flowed through northeast Shark River Slough to seasonally flood the eastern populations has been diverted to the west through the S-12s. Moreover, the water was prevented from flowing back to the east by a barrier to water flow called the L67-extension. (3) The decline in the lower part of subpopulations C and in D was due to changes in the managed water levels that have locally converted the seasonally flooded prairies that favor the birds to nearly continuously flooded, sawgrass-dominated marshes that the birds avoid.

The next step is to combine the variable area of suitable habitat with a simple demographic model of the sparrow. Such a model needs extensive data on the bird's birth and death rates. This is a time-consuming effort, given the bird's rarity and inaccessibility. (All but one of the subpopulations are reached only by helicopter.) In recent years, we have banded about 100 individuals per year and have found as many nests. The central feature of our model of risk assessment was the availability of suitable breeding habitat. Our studies show this varies considerably from year to year. For this step in risk assessment, we postpone the longer-term changes in vegetation effected by changes in hydrology and fire frequencies.

Lockwood et al. (2001) update an earlier effort (Lockwood et al. 1997) and report on the sparrow's demographic parameters. In brief, the sparrows laid an average of 3.2 eggs per clutch, a number that varied little from year to year or from place to place. About half of the eggs fledged young, and that fraction varied considerably. In particular, it depended on whether the clutch was laid earlier in the year and was almost certainly a first clutch, or whether it was laid later and was likely a second clutch. Rising water levels, which are common later in the year, terminate clutches. There were far fewer second clutches than first clutches, and known third clutches were so few and fledged so few young that they contributed little, if anything, to the population size of the next generation. Maximum likelihood estimates of banded birds showed that 66% of territory-holding males survived from one year to the next. Lockwood et al. (2001) combined the best estimates of these parameters and inferred others, including the survivorship of females and first-year birds. They came to the conclusion that the overall growth rate of the population was, plus or minus a few percent, close to replacement. Those "few percent" are a measure of the rigor of our procedures, for these data are derived from a population that has not changed perceptibly in size over the years during which we collected the data. That is, we estimated parameters consistent with the birds' replacing each other, and they have obliged us by doing so.

For a typical population viability analysis (PVA) we would devote considerable effort to estimating the bird's demographic parameters. We do not. While we applaud rigor and the best possible procedures, we now ask whether tight confidence intervals applied to some parameters make any difference or, worse, obfuscate the critical issues. Perhaps the most important parameters that we need to know, only serendipity will give us. How quickly do birds die when evicted from their homes by fire and flood? And how quickly does the population recover thereafter? These are inherently rare events, for which our detailed estimates are merely a guide, however small the confidence intervals about them.

How quickly do birds die when conditions are bad? Even under the best conditions, 34% of the males are lost from their territories from one year to the next. We have smaller sample sizes for females that suffer the extra stress of producing and carrying eggs. We see only about a quarter of the fledged young the following year, but this must be an underestimate of their survival, for some move to areas away from our extensive network of study sites. Almost certainly, however, young do not survive as well as territory-holding adults. A much greater fraction of birds will likely die under the worst conditions—prolonged, deep flooding of the habitat (which occurred from 1993 to 1995 in the western subpopulation) or extensive fires (such as that which burned most of the eastern subpopulation's habitat in 1989).

We do not have survival estimates during these conditions and think that few studies will ever satisfactorily estimate parameters during rare events—even those that befall common species. We assume conservatively that adult survival (males and females alike) is 66% even in bad years. We assume that 50% of young survive from their hatch year to the next—a number that we feel is almost certainly too high.

How quickly can birds recover when conditions are favorable? Obviously, long-term estimates of parameters give means, not maxima. There are some obvious limits on those maxima, however, a point confirmed by Walters et al. (2000) in their independent review of this study. First, suppose every pair in a subpopulation laid two clutches a year. (We have never seen anything like every pair laying a second clutch even when the conditions remain dry enough, long enough, for them to do so.) Second, suppose that the best fledge rate ever observed in a given year (60% of eggs) applied to both clutches. (We have never seen second clutches fledge the same fraction of eggs as first clutches.) Combined with the optimistic survival rates of the last paragraph, a population could increase at 61% per year. We then assume that these birds could fill up the area available for nesting without any additional mortality during their dispersal. We label this the wildly optimistic scenario.

The best fledge rate ever sustained for a few years in a row at a particular subpopulation was 53%. This was in subpopulation E, where numbers have steadily increased in the last few years. Even there, second clutches were less frequent and less successful than first clutches. Assume that 60% of birds with available habitat lay second clutches and that their success was the same as the first attempt. This leads to a potential growth rate of 34% per year. This is still very optimistic—we label the scenario as such—for second clutches have never been observed to be so frequent or so successful. Reducing the 60% to 50% leads to a maximum growth rate of 24% per year. We label this case plausible.

Certainly we can change other parameters. Reducing the survival of the hatch year birds—a parameter which this and many other studies estimate imprecisely—has the same effect as reducing the number of young that fledge. What matters is the relative rates of increase between years; 1.61 for the wildly optimistic case, 1.34 for the optimistic case, and 1.24 for the plausible case. We now see which of these are consistent with our observations and what the implications are for each subpopulation's risk of extinction.

The Subpopulation West of Shark River Slough (A)

Subpopulation A, west of Shark River Slough, is situated on a low ridge that is particularly vulnerable to flooding. Water depths of more than a few centimeters prevent breeding or terminate it if it has already started (Lockwood et al. 1997). Nott et al. (1998) calculated the extent of available breeding habitat for each of the last 20 years, classifying areas as those that remain dry enough for just one brood to be raised, and those that could produce two (assuming birds were physiologically capable of doing so). We then estimated the number of sparrows produced each year from the breeding and survival parameters scaled by the available habitat under the various scenarios described in the previous section (fig. 19.2). The sparrow numbers start with a guess of 2,000 birds in 1977 and follow deterministically thereafter, with the observed previous 20-year sequence of water levels repeated cyclically into the future. (Beissinger [1995] used a very similar assumption in his risk analysis of the snail kite in the Everglades.) This starting point in 1977 allows the subpopulation to increase to its estimated 2,500 birds by the time of the 1981 survey.

The model caps the subpopulation at 3,500 birds—an estimate of carrying capacity that doesn't strongly enter into the model's results because water levels so rarely allow the birds to breed across the potential range. We determined the cap based on the maximum available habitat and typical maximum observed densities. The estimates of actual habitat conditions show, for example, that in 1977 all 2,000 birds had the chance to raise one brood, but only 11% of them were in places dry enough to raise a second, even had they been physiologically fit enough to do so.

Two of the three scenarios in figure 19.2 (wildly optimistic and optimistic) allow the subpopulation to persist. However, both optimistic scenarios fail to match two features of the rangewide survey of the birds. First, whereas both optimistic scenarios suggest an increase in numbers between 1981 and 1992, the subpopulation in 1992 was estimated to be 7% lower than it was in 1981. Unfortunately, there were no surveys in intermediate years, including some when substantial areas suffered prolonged flooding. Second, these models do not recreate the drop in subpopulation that followed the wet years of 1993 to 1995 inclusive, when the subpopulation estimates fell to fewer than 400 birds. The plausible model predicts fewer birds in 1992 than in 1981, but even it optimistically predicts ~800 birds after these wet years.

An additional validation of the model and its parameters requires the sparrows to persist in the absence of these unnatural events. Were the sparrows predicted to decline, we might suppose the model erred in not allowing the birds to recover quickly enough. A second set of models tested this "what if" alternative. If, during the catastrophic years, the sparrow's habitat had not been flooded early in the season and if 100% of the habitat had been available for one brood, then the subpopulation would have thrived even under the plausible scenario. Indeed, it would have often reached the model's subpopulation ceiling of 3,500 birds (fig. 19.2).

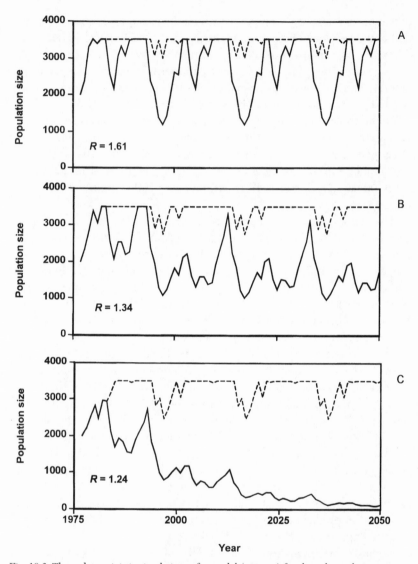

Fig. 19.2 Three deterministic simulations of a model (see text) for the subpopulation west of Shark River Slough. The proportional maximum change from one year to the next (R) varies from 1.61 (wildly optimistic, *A*), through 1.34 (optimistic, *B*), to 1.24 (plausible, *C*). The *solid line* is the known extent of breeding habitat available for first and second clutches over the 20 years prior to 1997. It then repeats the same pattern. This extent is driven by managed water flows. If massive, dry-season releases were prevented, more habitat would be available for second clutches (*dashed lines*). Only the plausible model is consistent with the known subpopulation estimates in 1981, 1992, and subsequent years.

Thus calibrated, we ran our models for more sets of 20 years. The models recycle the exact patterns of habitat availability, whereupon the subpopulation declines toward extinction within 50 years in the plausible scenario (fig. 19.2C). It goes to extinction even in the optimistic one (fig. 19.2B). What if water were not released? The subpopulation dips below its population ceiling periodically, but persists indefinitely even in the plausible scenario.

Our previous analyses indicate that the catastrophic years of 1983, 1984, 1986, 1987, 1993, and 1995 were not *naturally* bad years. They resulted from deliberate, massive dry-season releases of water through the S-12s into Everglades National Park (Nott et al. 1998). The contribution of rainfall to the water levels is small in comparison.

So we conclude that repeating managed water flows with the pattern of the last two decades will drive this endangered species to extinction in the area that once held the largest number of birds. The survey data we have collected since 1997 confirms this speculation. The population has remained under 500 birds, and it is restricted to a few square kilometers of habitat.

The Subpopulations to the North and East of Shark River Slough (C, F)

Managed high-water levels are not an issue in the other sparrow subpopulations; indeed, it is shortage of water that is the problem. Here, frequent fires burn the prairies. Birds are not found in areas that are burned as often as once every two years (Curnutt et al. 1998). We see little point in running risk analyses of these subpopulations. In total, they number a few hundred birds scattered across a wide area that fires burn, in some cases, annually. Thus, the birds are already scarce and the threats to them are self-evident. More important is the question of whether fires that start in this area might spread southward to burn the only area where more than 1,000 birds remain—the southeastern subpopulation.

The Southeastern Subpopulation (B)

Small portions of the area that harbors subpopulation B burn every year, often as consequence of fires from the pinelands to its north. Yet in 1989 nearly half of it burned as a consequence of a massive, dry-season fire. Probably all of subpopulation E burned, perhaps explaining why it is still recovering. Such fires can burn many hundreds of square kilometers in the Everglades. This size dwarfs the sparrow's range: the population in the southeast occupies only about 60 km². The policy of Everglades National Park is not to allow major fires to cross the park roads that divide this population into three parts. Nonetheless, fires of this size are hard to control in practice.

We modeled this area's population using the plausible scenario calibrated above. We "set" small fires (1/40th of the available habitat) every year in 20% of that habitat. Birds within these areas cannot breed successfully that year, but do not suffer any direct mortality. We do not know how many adult birds die in fires, but it surely is more than we have assumed. In the year after fire, we assumed that 50% of the birds can breed in an area, 75% the year after that, and 100% the following year. Curnutt et al. (1998) show that sparrow populations increase for five or more years after fires, so these estimates are also optimistic. Finally, we varied the frequency of severe fires—those that burn 90% of the bird's habitat.

Figure 19.3 provides two sample simulations, with severe fires every 10 years and every 20 years. In the former case the subpopulation quickly goes to extinction; in the latter case it persists. These simulations are typical. With fires on average every 10 years, only 5% of the simulations allowed the subpopulation to increase over a 50-year period. From their original start of 2,000 birds, 50% of the simulations resulted in the subpopulations dropping below 1,000 birds, and 15% fell below 500 birds. Given enough years, all of the model runs encountered a series of fires that drove the subpopulation to extinction.

In sum, the southeastern subpopulation is in danger of extinction

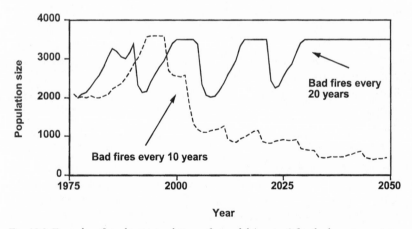

Fig. 19.3 Examples of stochastic simulations of a model (see text) for the largest remaining subpopulation of Cape Sable sparrows. Each year 20% of the habitat is burned, on average, plus there are "bad" fires that burn an average of 90% of the habitat. Such fires every 20 years allow the subpopulation to persist; those every 10 years do not.

from extensive fires that occur as frequently as once in ten years. Given that we have observed such fires in or near this subpopulation at that frequency, we conclude that this subpopulation is also at severe risk of extinction.

CONCLUSIONS

We predict that the Cape Sable sparrow subpopulation west of Shark River Slough will decline to extinction if the pattern of managed flows over the S-12 structures for the last 20 years is repeated. If these unnatural breeding season flows over the S-12s are stopped, this subpopulation should flourish. The subpopulations in the northeast have already declined to near extinction. These declines will continue unless the fire regimes change. On its own, the subpopulation in the southeast runs the risk of extinction because of episodic, large-scale fires. The fate of subpopulation E (now the second-largest subpopulation) is interesting because it may illustrate a population that was burned to oblivion in 1989 and is still recovering.

Our models omit some obvious features. We have not included the effects of prolonged inundation or of frequent fires on the vegetation. These processes alter the vegetation in ways that preclude the birds' use of areas for several years (Curnutt et al. 1998; Nott et al. 1998; Jenkins et al. 2001). Incorporating these impacts would likely lead to even greater concerns about the sparrow's future.

The predictions of our models arise from our knowledge of the bird's breeding biology and of the area's water and fire regimes. They are not "curve-fitting" exercises to the census data. Importantly, the results predict the timing and magnitude of the changes in those data. This confirms that the models are both sensible and sufficiently complete to capture the essential features.

The predicted decline to extinction of the southeastern subpopulation is a prediction of future events and thus not confirmed by our short-term data. Notice a subtle problem: if the currently least-affected population is doomed, why does it hold so many sparrows? Should it not have gone extinct earlier? There are two nonexclusive answers. First, it too is affected by episodic fires—such as the fire in 1989—that might be relatively recent phenomena brought on by management changes, and likely to be outside previous experience. Second, the sparrow has become locally extinct on occasion and then recolonized from other subpopulations. This possibility makes good sense. Years of naturally high water west of Shark River Slough would harm the subpopulation there. Concomitant flooding in the northeastern subpopulations would sup-

422   *Stuart L. Pimm and Oron L. Bass Jr.*

press the frequency of natural fires there, and, consequently, the possibility of the fire's spread to the southeastern subpopulations. In contrast, in dry years, the population west of Shark River Slough would be expected to flourish, even if the eastern subpopulations ran higher than average risks due to fires. Simply, a high-risk year west of the slough would be a low-risk year east of the slough, and vice versa. A complete exploration of these possibilities will require a combined water and fire model.

Nonetheless, we offer the following tentative conclusion: the Cape Sable sparrow will survive only if it has at least three healthy subpopulations. To implement this requirement, the breeding areas west of Shark River Slough must not be flooded in the breeding season, and water levels should be raised in the northeast of Shark River Slough to reduce the incidence of fires.

The general conclusion for conservation is that species even within one of the hemisphere's largest national parks—and possibly its best-funded—are not immune to massive anthropogenic impacts from outside. Everglades National Park is not large enough for calculations of risk based on natural population fluctuations to be sensible. Such calculations demonstrably gave the wrong answer for the Cape Sable sparrow.

Critics may counter that this is a special case. The species occupies a wetland, and perhaps wetlands are uniquely vulnerable to the vagaries of water flows upstream. Perhaps, but we are not convinced. Other large parks have unique problems that cross their boundaries. Fire, and our inclination to suppress small fires and so risk catastrophic ones, is an example that comes to mind for many parks in the western United States. We argue that even for the largest protected areas we must develop mechanistic models of what causes populations to decline. Unless we do so, we will not be able to adequately predict future risks.

LITERATURE CITED

Ariño, A., and S. L. Pimm. 1995. On the nature of population extremes. *Evolutionary Ecology* 9:429–443.

Baillie, J., and B. Groombridge. 1996. *1996 IUCN Red List of threatened animals.* IUCN Species Survival Commission, Gland, Switzerland.

Bass, O. L., Jr., and J. A. Kushlan. 1982. *Status of the Cape Sable sparrow.* Report T-672. South Florida Research Center, Everglades National Park, Homestead, Florida.

Beissinger, S. R. 1995. Modeling extinction in periodic environments: Everglades water levels and snail kite population viability. *Ecological Applications* 5:618–631.

Brook, B. W., J. J. O'Grady, A. P. Chapman, M. A. Burgman, H. R. Akçakaya, and

R. Frankham. 2000. Predictive accuracy of population viability analysis in conservation biology. *Nature* 404:385–387.

Collar, N. J., M. J. Crosby, and A. J. Stattersfield. 1994. *Birds to watch 2.* BirdLife International, Cambridge, United Kingdom.

Curnutt, J. L., A. L. Mayer, T. M. Brooks, L. Manne, O. L. Bass Jr., D. M. Fleming, M. P. Nott, and S. L. Pimm. 1998. Population dynamics of the endangered Cape Sable seaside-sparrow. *Animal Conservation* 1:11–20.

Curnutt, J. L., S. L. Pimm, and B. A. Maurer. 1996. Population variability of sparrows in space and time. *Oikos* 76:131–144.

Hanski, I. 1998. Metapopulation dynamics. *Nature* 396:41–49.

Jenkins, C. N., R. D. Powell, O. L. Bass Jr., and S. L. Pimm. 2001. Demonstrating the destruction of the habitat of the Cape Sable seaside sparrow. Manuscript.

Lande, R. 1993. Risks of population extinction from demographic and environmental stochasticity and random catastrophes. *American Naturalist* 142:911–927.

Lockwood, J. L., K. H. Fenn, J. L. Curnutt, D. Rosenthall, K. L. Balent, and A. L. Mayer. 1997. Life history of the endangered Cape Sable seaside sparrow. *Wilson Bulletin* 109:234–237.

Lockwood, J. L., K. H. Fenn, J. M. Caudill, D. Okines, O. L. Bass Jr., J. R. Duncan, and S. L. Pimm. 2001. The implications of Cape Sable seaside sparrow demography for Everglades restoration. *Animal Conservation* 4:275–281.

Mace, G. M. 1996. Classifying threatened species: means and ends. *Philosophical Transactions of the Royal Society of London*, series B, Biological Sciences, 344: 91–97.

Maurer, B. A. 1994. *Geographical population analysis: tools for the analysis of biodiversity.* Blackwell Scientific, Oxford, United Kingdom.

Mayer, A. L., and S. L. Pimm. 1998. Integrating endangered species protection and ecosystem management: the Cape Sable seaside-sparrow as a case study. Pages 53–68 in G. M. Mace, A. Balmford, and J. R. Ginsberg, editors, *Conservation in a changing world.* Cambridge University Press, Cambridge, United Kingdom.

Nott, M. P., O. L. Bass Jr., D. M. Fleming, S. E. Killeffer, N. Fraley, L. Manne, J. L. Curnutt, T. M. Brooks, R. Powell, and S. L. Pimm. 1998. Water levels, rapid vegetational changes, and the endangered Cape Sable seaside sparrow. *Animal Conservation* 1:21–29.

Pimm, S. L. 1991. *The balance of nature? ecological issues in the conservation of species and communities.* University of Chicago Press, Chicago, Illinois.

Pimm, S. L., and J. L. Curnutt. 1994. The management of endangered birds. Pages 227–244 in C. I. Peng and C. H. Chou, editors, *Biodiversity and terrestrial ecosystems.* Monograph Series, no. 14. Institute of Botany, Academia Sinica, Taipei, Taiwan.

Pimm, S. L., J. M. Diamond, T. R. Reed, G. J. Russell, and J. Verner. 1993. Times to extinction for small populations of large birds. *Proceedings of the National Academy of Sciences* (USA) 90:10871–10875.

Pimm, S. L., H. L. Jones, and J. M. Diamond. 1988. On the risk of extinction. *American Naturalist* 132:757–785.

Pimm, S. L., and A. Redfearn. 1988. The variability of animal populations. *Nature* 334:613–614.

Sæther, B.-E., J. Tufto, S. Engen, K. Jerstad, O. W. Røstad, and J. E. Skåtan. 2000.

Population dynamical consequences of climate change for a small temperate songbird. *Science* 287:854–856.

Walters, J. R., S. R. Beissinger, J. W. Fitzpatrick, R. Greenberg, J. D. Nichols, H. R. Pulliam, and D. W. Winkler. 2000. The AOU Conservation Committee review of the biology, status, and management of the Cape Sable seaside sparrows: final report. *Auk* 117:1093–1115.

20 Population Viability Analysis, Management, and Conservation Planning at Large Scales

Fred B. Samson

ABSTRACT

Conservation planning is a less than refined science. It requires initial, careful design in management for the timely allocation of scarce resources to ensure viability of a diverse set of ecosystems and species at risk. Large-scale planning is a two-step process: (1) assessment that makes use of the best available data and develops strategies to conserve biological diversity by maintaining viable populations, and (2) implementation in the form of habitat preservation, habitat restoration, and active management. This chapter reviews the assessment and implementation phases, and how population viability analysis (PVA) has been used in large-scale conservation planning. It offers suggestions to increase management efficiency and to improve how PVA could be better incorporated into the process.

The goal of large-scale planning is to identify a portfolio of sites and strategies to protect all native species and natural processes representative of the planning area. Achieving this goal requires more than collecting and analyzing unlimited amounts of data; it explicitly requires the application of good business management in using information and assets at appropriate scales. Recommendations include use of the Montreal indicators of ecosystem diversity and species diversity, precautionary conservation, and "options" investment of assets as a management philosophy that reduces costs of assessment to preserve resources for implementation and adaptive management.

INTRODUCTION

Resource managers face a formidable task in ensuring the viability of species within a conservation planning area. Although rarely known, the number of animal and plant species for which viability strategies are required may number in the scores, hundreds, or thousands. Examples of federal conservation planning in the United States include the President's Plan in the Pacific Northwest, management of the temperate rain forest in southeast Alaska, and management of the northern Great Plains in the central United States, among others on public and private lands. A basic tenet in conservation planning is to "save all the pieces," but designing both the coarse- and fine-scale strategies to achieve this is problematic.

Viability is generally considered to be one of the four primary goals of conservation planning and ecosystem management. Typical goals include (1) maintenance of viable populations of all native plants and animals in situ, (2) representativeness of all native ecosystem types across their range of variation, (3) maintenance of ecosystem function and key processes, and (4) the accommodation of human use. These goals are often associated with the concept of ecosystem "sustainability," though a clear definition remains elusive and contentious. A widely accepted indicator to measure the goal is missing. A starting point, therefore, is to consider viability as a measure of the security of individual species, of the health of components of the system that shape and sustain species, and of ecosystem sustainability. It is the combination of these factors that will contribute to species viability through the conservation planning process.

Population viability analysis (PVA) is not a mature science, and neither theorists nor managers have reached consensus on a focus for the planning process. One consistent theme is that scale is important. Beginning with the largest scale has both organizational and ecological advantages. Organizational advantages range from consistency to ensuring the pieces of a conservation strategy fit within goals for sustaining species' distribution and populations. Ecological advantages range from the credible inventory of environmental stressors that influence the distributions and abundances of species to understanding that ecological processes operate at different spatial and temporal scales and that their aggregation from one scale to another is difficult.

This chapter describes the path resource managers take to plan for viability and provides a framework to increase management efficiency that directly serves PVA. An initial model to conduct large-scale assessments to ensure the viability of all species is described and compared to recent examples in bioregional conservation planning. Suggestions are provided to increase management efficiency and to better incorporate PVA into such planning.

THE PROBLEM: VIABILITY FOR SCORES, HUNDREDS, OR THOUSANDS OF SPECIES

There is little agreement on how to develop a good conservation plan, despite the importance of conservation planning. An example is the comparison of conservation planning conducted by the U.S. government—the President's Plans for the Pacific Northwest, southern Appalachians, Sierra Nevada range, and northern Great Plains—and that conducted by the Nature Conservancy, a private conservation organization. Investment in the process of conservation planning by the federal government ranges up to $80 million and may require three to six years for the completion of an individual plan. Quite the opposite, the Nature Conservancy may complete an ecoregional plan within one to two years at a cost in the hundreds of thousands of dollars.

Several factors influence the level of investment and commitment to large-scale conservation planning. Some argue that federal planning is required to respond to larger sets of public interests and must blend social and economic considerations (Grumbine 1990). However, blending legal and policy requirements with economic and political incentives is challenging in developing conservation plans to conserve biological diversity on both public and private lands (Shogren 1998). Fundamental to each large-scale conservation plan is the level of threat to species considered to be at risk, often influenced by ownership patterns in terms of public and private lands. Evaluation of large-scale conservation planning is incomplete without full consideration of the entire process, from assessment through implementation.

The two primary phases in resource management whether on public or private lands are *assessment* and *implementation*. Assessment makes the best use of available data and develops strategies to conserve biodiversity by maintaining viable populations (Poiani et al. 2000). Assessment is a stepwise process requiring an understanding of the levels of biological organization at ecosystem and local scales, the identification and description of scale-dependent ecological processes that shape and sustain species, and the application of population viability concepts to species conservation. The incorporation of conservation recommendations emerging from the assessment is implementation. Implementation may take the form of habitat preservation, restoration of habitat and natural processes, and active management that requires continued human intervention to reverse the threat to species at risk (Foin et al. 1998).

Assessment

Assessment begins with the identification of planning units (fig. 20.1), which are relatively coarse biogeographic divisions that share broadly similar environmental conditions, natural plant and animal communities, and primary ecological processes. The "ecoregion" (Bailey 1996) is the common framework in biodiversity conservation and management at a broad scale (Ricketts et al. 1999). Assessment then typically follows

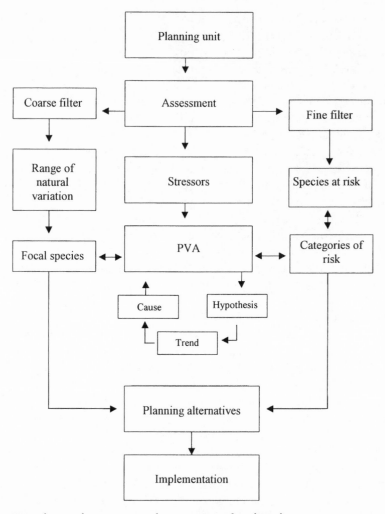

Fig. 20.1 A large-scale conservation planning process, describing the two primary components, assessment and implementation. The assessment process uses the coarse-filter approach to maintain viability for the majority of species. Focal species serve to monitor ecosystem diversity, the components and ecological processes that shape and sustain ecosystems. The fine-filter approach maintains the viability of species and populations not protected by the coarse filter. Categories of risk serve to prioritize species for PVA and to acquire information required for such analyses. Planning alternatives are developed and evaluated using a variety of approaches, including PVA. Implementation applies assessment-based conservation recommendations in the preferred planning alternative.

one of two tracks—a community or broad-scale approach to the conservation of biological diversity known as the "coarse filter," or the detailed assessment of species at risk known as the "fine filter" (Baydack et al. 1999).

The coarse-filter approach (fig. 20.1) suggests that viable populations will be maintained when representative communities are preserved (Hunter et al. 1988) and is based on the distribution of existing communities compared to their pre–intensive settlement patterns (Sharp et al. 1987); the change in disturbance regimes (Risser 1995), and the check on adequate ecological representation of communities (Haufler et al. 1996). The bounded behavior for pre–intensive settlement patterns and disturbance regimes is described by the "range of natural variation" (RNV; Samson 1992). In other words, restoring and maintaining landscape conditions within distributions that organisms are adapted to over evolutionary time is the management approach most likely to produce sustainable ecosystems and their component parts (Haufler 1999).

RNV fails to capture the full meaning and understanding of ecological sustainability without framing the concept in the current and probable future climate change. Current species are adapted to recent climate changes and disturbance regimes and contain a long history in their genetic structure. Building a landscape through the planning process requires considering the current climate period (which may include pre–intensive settlement conditions and influence of native peoples) and extends beyond comparison to pre–intensive settlement (Miller and Woolfenden 1999). The conservation goal is to describe the major successional stages expected within an ecoregion, based on knowledge of disturbance frequencies and rates of vegetation change and development (Parsons et al. 1999).

The coarse-filter approach identifies a set of focal species whose assessment provides substantial information beyond the status of the particular species (fig. 20.1). The aim is to select a small number of focal species whose individual status and trends will collectively allow monitoring of the diversity for an ecosystem. Focal species are required because monitoring and managing for all aspects of biological diversity is impossible. Focal species may include species at risk and other species that represent certain ecological conditions. The ecological literature is rich in examples of organisms that influence and maintain ecosystems (Jones et al. 1994).

The coarse-filter approach to the maintenance of biological diversity contrasts sharply with the fine-filter approach (fig. 20.1) to conserving individual species (Gustafenson 1998). The majority of strategies for individual species are developed either because the species is endangered or because it is a game species (Arnold 1995). It should be recognized that the concepts of species rarity and endangerment (typically due to anthropogenic causes) are not synonymous. Rarity is an expression of the intrinsic pattern of distribution and abundance for a species at a given time, while endangerment refers to factors that may make a species more susceptible to decline or extinction (Morse 1996). Intrinsically rare species are defined as those that have narrow geographic range, high habitat specificity, and small local population size (Primack 1995).

The bridge between the coarse filter and fine filter in terms of species conservation is neither precise nor consistent in application. In principle, those species for which the coarse filter will not result in a high likelihood of maintaining viability must be addressed through the fine filter. The fine-filter approach provides for management of individual species through activities such as the development of conservation strategies that may influence human-related activities, the emulation or restoration of ecological processes, and the conservation of fine-scale ecosystem components (e.g., fens, pools, bogs, caves, etc.).

The criteria to select species at risk (fig. 20.1) for the fine-filter analysis across and within governmental agencies and among conservation organizations are vague and inconsistent. For example, species at risk in the U.S. Geological Survey's *Status and Trends of the Nation's Biological Resources* (Mac et al. 1998) differs from the listing in the U.S. Forest Service/Bureau of Land Management assessment for the interior Columbia River basin (U.S. Forest Service and Bureau of Land Management 2000). Within the U.S. Forest Service, the Pacific Southwest Region applies the Nature Conservancy's global ranking of G1 to G3, the Intermountain Region is considering G1 to G3 and state rankings of S1 to S2, the Northern Region applies G1 to G3 and S1 to S3, while other regions have independent and less well defined criteria. Furthermore, species at risk recognized by conservation organizations—the World Conservation Union (IUCN), the Nature Conservancy, and others—differ in composition.

The selection of species at risk is difficult because no body of knowledge currently exists to unambiguously guide the choice for the fine-filter analysis. The challenge of selecting species at risk is significant, for each selection requires the commitment of resources to establish the environmental threats and to develop and implement conservation measures to ensure viability. The inclusion of species at risk based on administrative versus ecological boundaries inflicts serious costs to conservation (Knopf 1992) and is indicative of the "Noah's ark" problem (Weitzman 1998). The ark metaphor reveals a disproportionate investment of state and federal assets in the United States to conserve a few highly visible and globally secure species at the expense of numerous less well known species with greater needs for protection.

PVA has a limited role in both the coarse-filter and fine-filter approaches in current conservation planning (fig. 20.1). Morris et al. (1999) and Ralls et al. (chap. 25 in this volume) provide a summary of when PVA is appropriate and note that modeling in PVA may lead to misleading results without full recognition of the importance of long-term data. Establishing which species to inventory to gain long-term data for PVA is among the reasons to assign categories of risk to species in need of conservation. Assigning categories of risk can reduce disputes and uncertainties over particular species, which are not easily resolved and may negatively impact species conservation, and establishes priorities for conservation action (Mace and Lande 1991).

Indicators of ecosystem diversity and species diversity are inherent in the Montreal Process (Coulombe 1995), which seeks sustainable management for temperate and boreal forest systems and for grassland ecosystems. Although the Montreal Process was designed for assessing ecological sustainability at national scales, many proposed criteria for ecosystem diversity (e.g., the extent of area by habitat type, age class and successional stage, extent of habitat types protected, and fragmentation) and species diversity (e.g., the number of species at risk and the risk of not maintaining viable populations) are significant to broad-scale conservation planning.

The development of indicators for the conservation of boreal and temperate forests and grasslands is an important and relevant step to the United Nations conventions on biodiversity and climate change (Coulombe 1995). They were developed to evaluate international progress toward sustainability at the national level and to provide the basis for policymakers to facilitate cooperation directed toward the goal of sustainability. Nevertheless, the rigorous evaluation of the qualities of each indicator against ecological theory is lacking in the ecological literature.

The conservation goal for ecoregional planning is to identify a portfolio of sites and a set of strategies to protect all the species and natural processes representative of the ecoregion. Long-term viability will be ensured by protecting multiple viable or recoverable occurrences of species and populations within the ecoregion. Although it may be difficult to identify and measure ecoregion-scale phenomena, indicators developed in the Montreal Process are useful in large-scale conservation planning to better conserve the full expression of biological diversity.

Implementation

Why initiate broad-scale assessment? Concepts of resource management evolve over time, based on scientific knowledge and in response to changing public interests in the conservation of biological diversity and demand for natural products and services. In addition, in the United States, legal requirements impose periodic revisions of land and resource plans. Examples are the National Forest Management Act of 1976, which requires the revision of land and resource plans based on information gained through monitoring, and issues raised through the public involvement required by the National Environmental Policy Act of 1969 (as amended). According to the Montreal Process, indicators will need to be reviewed on a regular basis to reflect new research and to better implement conservation practices required for ecosystem sustainability.

Broad-scale assessment should fully consider coarse- and fine-scale factors that influence species viability. Detecting large-scale ecological patterns can increase the value of recovery planning for species at risk and can help evaluate the need for management in their recovery (Foin et al. 1998). Leopold (1933), in developing the field of wildlife conservation, recognized that some species required wilderness, other species required land management, and still others required active management. The goal of PVA should be better practice and effort in implementation to conserve species, and not neater theory in assessments.

The concluding step in the management course is implementation (fig. 20.1). Hidden traps exist in the decision making. When managers consider a decision, "the mind gives disproportionate weight to the first information it receives. Initial impressions, estimates, or data anchor subsequent thoughts and judgements" (Hammond et al. 1998, 48). In agencies, the sins of commission (doing something) tend to be punished more severely than the sins of omission (doing nothing); this maintains the status quo.

In either business (Slywotzky 2000) or the ecological environment (Kareiva et al. 1999), the "precautionary principle" or "precautionary conservation" is required in the face of high uncertainty and poor information. One alternative is to incorporate precautionary and protective conservation measures early in the planning process. An option to the manager is to frame viability in the planning process as a conservation hypothesis. The hypothesis approach to viability accepts that inventory information is lacking for the vast majority of species.

An example of the efficient initial investment in precautionary conservation is in southeast Alaska (U.S. Forest Service 1997). The recommended system of large reserves and prescriptive management for habitats between reserves is the hypothesis for precautionary conservation. In this case, advantages of the initial investment in precautionary conservation include adequate protection to preclude the listing of species, less litigation on the adequacy of protection, and the focus on stewardship versus project-by-project adjudication.

Mapping of species and ecosystem diversity has become a common strategy in conservation planning (Flather et al. 1997). This often occurs without full consideration of the pattern and process errors that lead to uncertainty in use of geographic information system approaches to conservation-design problems (Rastetter et al. 1992). An overriding issue in implementation is maintaining the consistency required to ensure species and ecosystem conservation practices are in place while permitting the incorporation of new science. Some options to enhance the linkage of consistency and new science include teams to ensure compliance with the assessment, detailed habitat matrices describing patterns in composition and successional stage, maps detailing reserves and linkages between protected areas, or some combination of the above. Protecting ecosystem diversity and species diversity is incomplete without a design for implementing required conservation actions.

HOW HAS PVA BEEN USED IN LARGE-SCALE LANDSCAPE PLANNING?

Species and species population viability are often expressed in conservation planning as the numbers of individuals, rate of successful reproduction, and/or the spatial and temporal distribution of habitats. In discussing viability, it makes little sense to ask whether or not a given population is viable. When the time frame is short and the spatial scale large, the answer is likely yes. For many species, habitat suitable today may become unsuitable in the future (and vice versa), due to land management or natural ecological processes. Time scales must be long enough to fully realize the consequences of management actions and to reflect the life history of the species (Thomas 1990). Exact spatial scales depend on several factors, including the environmental setting, natural subdivisions in the species population, and the species' life-history characteristics.

In conservation planning, statements about population viability that describe the probability of persistence for species and populations in a particular area and over a particular period of time have a reassuring ring. Quantification and objectivity are appealing in conservation planning, where viability in many cases is compared among land-use alternatives. Nevertheless, data needed to make quantitative and objective assessments of viability are rarely available; placing viability in a quantitative mold may be misguided and even misleading when uncertainties in results are not fully acknowledged (Beissinger and Westphal 1998).

Critical in deciding which method of PVA to use is what type of answer is required. A quantitative measure of viability is often necessary when several conservation planning alternatives are to be judged by multiple and conflicting criteria. It may not be necessary to compare alternatives to show which ones exceed a reasonable threshold of acceptability, requiring a less demanding method in PVA. There is no sense investing considerable resources in species and population PVA if uncertainties in input parameters are so great that the range for each conservation planning alternative overlaps.

PVA models that estimate persistence under different scenarios are the exception in recent large conservation planning efforts in the United States. For example, the viability approach in southeast Alaska (U.S. Forest Service 1997) was to use a few "umbrella" species to represent the adequacy of reserves, accompanied by rules that prescribe habitat management between designated reserves. Research for an extended period may provide adequate information to conduct a formal PVA for the seven focal or "umbrella" species.

On the northern Great Plains, adequate long-term census and demographic data permitted species and population PVA for 3 of 119 species at risk: the black-footed ferret (*Mustela nigripes*), black-tailed prairie dog (*Cynomys ludovicianus*), and western prairie fringed orchid (*Platanthera praeclara*; U.S. Forest Service 1999). Conservation principles (Beissinger and Westphal 1998) in the absence of long-term data serve to protect the majority of species at risk on the northern Great Plains. The vast majority of species and populations on public lands in the Pacific Northwest are protected by "expert" opinion (U.S. Forest Service and Bureau of Land Management 1994), the exception being the spotted owl, which is supported by considerable data (*Strix occidentalis*; Lamberson et al. 1994).

Scientific assessments indicate that a step-down planning process is required to achieve ecosystem diversity goals. The linkage of ecosystem diversity and PVA is twofold: identify focal species and require a formal PVA for each one, and retain or manage, to the extent feasible, for the vegetation types and successional-stage abundances at ecoregional scales that fall within the range expected to occur under natural disturbance regimes characteristic of the current climate period. The ecosystem diversity goal in conservation planning is to focus on the mean distribution of stages rather than on the range. Now and in the future, a strong emphasis should be placed on a design to integrate restoration and protective treatments to ensure ecosystem diversity.

In highly fragmented environments, it is highly unlikely that the full range of vegetation types and successional-stage abundances can be maintained. Rather, in discussing viability, management for specific vegetation types and or successional stages may be required to contribute to the persistence of individual species at risk. An example is on the northern Great Plains, where substantial conversion to agriculture and other human uses limits the abundance and distribution of native prairie (U.S. Forest Service 1999). Evidence suggests that management directed toward "low" and "high" seral stages is required to conserve species at risk associated with the prairie relicts (Knopf 1996).

PVA, MANAGEMENT, AND REGIONAL CONSERVATION ASSESSMENTS: WHERE DO WE GO FROM HERE?

Large-scale conservation planning begins with the decision to allocate resources in one of two general directions (fig. 20.2A). One direction is investing assets in assessments at a cost to implementation (fig. 20.2B). Government bureaucracies often carry the responsibility to conduct and implement PVA (Grumbine 1994) and commonly use the hierarchical decision-making process (Daft 1995). Conservation planning based on the hierarchical decision process requires masses of information and extensive analyses to create very detailed and specific solutions to the problem. Managers using this process may fail, not due to the wrong decision, but because they selected the wrong process in decision making—gathering too much information or postponing action too long (Pressey 1998).

Neither strategic planning as practiced now nor adaptive resource management ensures that managers channel their energies into the appropriate issues or projects to conserve species at risk (Weitzman 1998). Thinking through a process from various angles may be more productive than collecting and analyzing unlimited amounts of data, and experimenting with new organizational ideas may be more critical than scientific analysis and discussion (Markides 1999). Never has the idea of process in conservation planning been more important for natural resource managers than it is today—natural resources are limited, environmental uncertainty is high, and promise and product must match to gain scientific and public confidence (Norris 2000).

An alternative approach is to invest in implementation at a cost to assessments (fig. 20.2C). Advantages are the increased level of assets available for habitat preservation, restoration of natural processes, and active management. In business, to conserve assets in response to an uncertain and competitive environment improves the performance at the work site (Amram and Kulatilaka 1999). Another advantage to con-

A. Management approaches

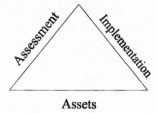

Assets

B. Hierarchical management

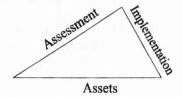

Assets

C. Options management

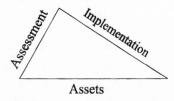

Assets

Fig. 20.2 *A*, Organization for large-scale conservation planning allocates some total amount of fixed resources or assets in one of two general directions (assessment or implementation) and may be hierarchical or options-based. The lengths of the lines on the two sides of the triangle indicate the amount of resources to be invested in either assessment or implementation. *B*, Given fixed assets to invest, hierarchical management makes initial investments in planning and emphasizes assessment (longer line). *C*, The options approach preserves assets to emphasize implementation (longer line). The options approach may provide greater resources and opportunities to apply PVA in large-scale conservation planning. See text for further details of the comparison.

serving initial assets is to allow managers to think more clearly and realistically about the use of available assets to make better decisions (Slywotsky 2000). The most significant advantage is the availability of resources for making an inventory of and gathering demographic information on focal species and species at risk, permitting increased use of "formal" PVAs that provide improved estimates of persistence under different land-management alternatives in conservation planning.

Resources to conserve biodiversity are limited, and their effective use requires one to explore the topic of effectiveness and how effectiveness is measured in organizations (Brunner and Clark 1997). Efficiency is a more limited concept and measures the amount of resources required for a sequence of activities designed to produce some part of a product or service. In business, both effectiveness and efficiency reflect a horizontal rather than a vertical or bureaucratic structure, and organizations obtain scarce and valued resources from other organizations (Williams 1999).

The central question for planning needs to be more "what *options* are gained by moving from point A to point B" and less "*what* is gained by moving from point A to point B" (Lester et al. 1998). In this case, the initial step is to identify the options that exist for investment in information that is required for large-scale conservation planning (fig. 20.2). The influence of competition theory on the community of organizations in an uncertain environment or with limited resources suggests that a larger percentage of the required tasks are better performed by "specialized" organizations, each with a specialized niche or service. Specialist organizations are more competitive than generalists in the area in which their services overlap, and smaller organizations can move faster and are more flexible in response to environmental change (Hannan and Freeman 1997).

The idea of exploring various alternatives to find new and effective processes under the many uncertainties encountered when conducting large-scale conservation planning is compelling. One example of efficient information gathering is on the northern Great Plains in the United States (U.S. Forest Service 1999). Information required to develop the coarse filter came from distinct and specialized sources—a university provided estimates of historic and current vegetation required for the coarse filter, a conservation organization identified areas that are significant for species and communities at risk, and other agencies at no cost provided watershed integrity indices, population-trend information, and specific information on environmental threats to species persistence. The fine-filter conservation approach was developed by a university, conservation organization, and three-federal-agency committee.

Costanza et al. (2000, 150, 153) attempted to move beyond the "environment as a debate" by linking "the way any good business manages its productive assets under uncertainty" to the resolution of environmental problems facing society, particularly sustainable management at appropriate scales. Today's "good" business model is a "wagon wheel" with management at the hub, with science on the rim (Mintzberg and Van der Heyden 1999), and with the clear recognition that up-front asset conservation is essential to success and survival (Williams 1999). Few agencies employ "good" business models, choosing rather to invest in structure and organization (Posner and Rothstein 1994). To move forward, it is essential to link good business models (Eisenhardt and Galunic 2000) to good assessment and implementation in large-scale conservation planning.

CONCLUSIONS

Biological diversity has two primary components—ecosystem diversity and species diversity. PVA in the future may play the key role in conserving ecosystem diversity and species diversity, but change is required in the process of large-scale conservation planning. Like ecology (Peters 1991), conservation planning is a less than refined science. It suffers from a variety of ailments ranging from inconsistent selection of species at risk to inefficient and ineffective organizations, to such a degree that good plans stand out as significant achievements. The premium is often placed on novelty rather than on meticulous scholarship and constructive data.

Significant challenges remain to incorporating PVA approaches into large-scale conservation planning: (1) species at risk must be identified according to widely accepted criteria; (2) the amount and quality of information about natural variability may be insufficient for some ecosystems; (3) indicators in the Montreal Process require refinement and testing, especially at scales appropriate to large-scale planning; (4) most people appear to be interested primarily in the outcome of events, rather than on the process underlying them, which limits the availability of resources and information required to develop and implement PVA models for conservation planning; and (5) most managers and scientists in conservation have not adapted to the changes in today's businesses and continue to use antiquated organizational principles.

Ideally, as Beissinger and Westphal suggested (1998, 835), "funds to implement strategic field studies and validate secondary model predictions should be committed by the time a PVA workshop is held or model developed. If such an approach were implemented, it would result in the kinds of knowledge that would promote species recovery, improve our understanding of critical population processes, and increase the predictive capabilities of future PVA models." Large-scale conservation planning requires a similar if not identical perspective—the initial and careful design in management for the allocation of scarce resources to ensure the viability of ecosystems and species at risk through large-scale conservation planning—to be successful.

PVA, Management, and Conservation Planning 439

LITERATURE CITED

Amran, M., and N. Kulatilaka. 1999. Disciplined decisions. *Harvard Business Review* 77:95–104.

Arnold, G. H. 1995. Incorporating landscape pattern into conservation programs. Pages 309–337 in L. Hanson, L. Fahrig, and G. Merriam, editors, *Mosaic landscapes and ecological processes*. Chapman and Hall, New York, New York.

Bailey, R. G. 1996. *Ecosystem geography*. Springer-Verlag, New York, New York.

Baydack, R. K., H. Campa III, and J. B. Haufler. 1999. *Practical approaches to the conservation of biological diversity*. Island Press, Covelo, California.

Beissinger, S. R., and M. I. Westphal. 1998. On the use of demographic models of population viability in endangered species management. *Journal of Wildlife Management* 62:821–841.

Brunner, R. D., and T. W. Clark. 1997. A practice-based approach to ecosystem management. *Conservation Biology* 11:48–58.

Costanza, R., H. Daley, C. Folke, P. Hawken, C. S. Holling, A. J. McMichael, D. Pimentel, and D. Rapport. 2000. Managing our environmental portfolio. *BioScience* 50:149–155.

Coulombe, M. J. 1995. Sustaining the world's forests: the Santiago Agreement. *Journal of Forestry* 43:18–21.

Daft, R. L. 1995. *Organizational theory and design*. West Publishing, Minneapolis/St. Paul, Minnesota.

Eisenhardt, K. M., and D. C. Galunic. 2000. Coevolving at last: a way to make synergies work. *Harvard Business Review* 49:91–102

Flather, C. H., K. R. Wilson, D. J. Dean, and W. C. McComb. 1997. Identifying gaps in conservation networks: of indicators and uncertainty in geographic-based analyses. *Ecological Applications* 7:531–542.

Foin, T. C., S. Riley, A. L. Pawley, D. A. Ayres, T. M. Carlson, P. J. Hodum, and P. V. Switzer. 1998. Improving recovery planning for threatened and endangered species. *BioScience* 48:177–184.

Grumbine, R. E. 1990. Viable populations, reserve size, and federal lands management: a critique. *Conservation Biology* 4:127–134.

———. 1994. What is ecosystem management? *Conservation Biology* 8:27–38.

Gustafensen, E. J. 1998. Quantifying landscape spatial pattern: what is the state of the art? *Ecosystems* 1:143–156.

Hammond, J. S., R. L. Keeney, and H. Raiffa. 1998. The hidden traps in decision making. *Harvard Business Review* 76:47–55.

Hannan, M. T., and J. Freeman. 1977. The population ecology of organizations. *American Journal of Sociology* 82:929–964.

Haufler, J. B. 1999. Strategies for conserving terrestrial biological diversity. Pages 17–34 in R. K. Baydack, H. Campa III, and J. B. Haufler, editors, *Practical approaches to the conservation of biological diversity*. Island Press, Covelo, California.

Haufler, J. B., C. A. Mehl, and G. J. Roloff. 1996. Using a coarse-filter approach with species assessment for ecosystem management. *Wildlife Society Bulletin* 24:200–208.

Hunter, M. L., G. L. Jacobson Jr., and T. Webb III. 1988. Paleoecology and the coarse-filter approach to maintaining biological diversity. *Conservation Biology* 4:375–384.

Jones, C. G., J. H. Lawton, and M. Shachak. 1994. Organisms as ecosystem engineers. *Oikos* 69:373–386.

Kareiva, P., S. Andelman, D. Doak, B. Elderd, M. Groom, J. Hoekstra, L. Hood, F. James, J. Lamoreux, G. LeBuhn, C. McCulloch, J. Regetz, L. Savage, M. Ruckelshaus, D. Skelly, H. Wilber, K. Zamudio, and NCEAS HCP working group. 1999. Using science in habitat conservation plans. Unpublished report, National Center for Ecological Analysis and Synthesis, University of California, Santa Barbara, California; and American Institute of Biological Sciences, Washington, D.C.

Knopf, F. L. 1992. Faunal mixing, faunal integrity, and the biopolitical template. *Transactions of the North American Wildlife and Natural Resource Conference* 57:330–342.

———. 1996. Prairie legacies: birds. Pages 135–148 in F. B. Samson and F. L. Knopf, editors, *Prairie conservation: protecting North America's most endangered ecosystem.* Island Press, Covelo, California.

Lamberson, R. H., B. R. Noon, C. Voss, and K. S. McKelvey. 1994. Reserve design for territorial species: the effects of patch size and spacing on the viability of the northern spotted owl. *Conservation Biology* 8:185–195.

Leopold, A. 1933. *Game management.* Charles Scribner, New York, New York.

Lester, R. K., M. J. Piore, and K. M. Malek. 1998. Interpretive management: what general managers can learn from design. *Harvard Business Review* 76:86–101.

Mac, M. J., P. A. Opler, C. E. Puckett-Haeker, and P. D. Doran. 1998. *Status and trends of the nation's biological resources.* U.S. Department of the Interior, U.S. Geological Survey, Washington, D.C.

Mace, G. M., and R. Lande. 1991. Assessing extinction threats: toward a reevaluation of IUCN threatened species categories. *Conservation Biology* 5:148–157.

Markides, C. C. 1999. A dynamic view of strategy. *Sloan Management Review* 40: 55–64.

Miller, C. L., and W. B. Woolfenden. 1999. The role of climate change in interpreting historic variability. *Ecological Applications* 9:1207–1216.

Mintzberg, H., and L. Van der Heyden. 1999. Organigraphs: drawing how companies really work. *Harvard Business Review* 77:87–95.

Morris, W., D. Doak, M. Groom, P. Kareiva, J. Fieberg, L. Gerber, P. Murphy, and D. Thompson. 1999. *A practical handbook for population viability analysis.* Nature Conservancy, Arlington, Virginia.

Morse, L. E. 1996. Plant rarity and endangerment in North America. Pages 7–22 in D. A. Falk, C. I. Millar, and M. Olwell, editors, *Restoring diversity: strategies for reintroduction of endangered plants.* Island Press, Covelo, California.

Norris, S. 2000. A year for biodiversity. *BioScience* 50:103–107.

Parsons, D. J., T. W. Swetnam, and N. L. Christensen. 1999. Uses and limitations of historical variability concepts in managing ecosystems. *Ecological Applications* 9:1177–1178.

Peters, R. H. 1991. *A critique for ecology.* Cambridge University Press, Cambridge, United Kingdom.

Poiani, K. A., B. D. Richter, M. G. Anderson, and H. E. Richter. 2000. Biodiversity conservation at multiple scales: functional sites, landscapes, and networks. *BioScience* 50:133–146.

Posner, B. G., and L. R. Rothstein. 1994. Reinventing the government: an interview with change catalyst David Osborne. *Harvard Business Review* 72:133–140.

Pressey, R. L. 1998. Algorithms, politics, and timber: an example of the role of science in a public, political negotiation process over new conservation areas in production forests. Pages 73–87 in R. Wills and R. Hobbs, editors, *Ecology for everyone: communicating ecology to scientists, the public, and the politicians.* Surrey Beatty and Sons, Sydney, Australia.

Primack, R. B. 1995. *Essentials of conservation biology.* Sinauer Associates, Sunderland, Massachusetts.

Rastetter, E. B., A. W. King, B. J. Cosby, G. M. Hornberger, R. V. O'Neill, and J. E. Hobbie. 1992. Aggregating fine-scale ecological knowledge to model coarse-scale attributes of ecosystems. *Ecological Applications* 2:55–70.

Ricketts, T. H., E. Dinerstein, D. M. Olsen, C. J. Loucks, W. Eichbaum, D. Della-Sala, K. Kavanagh, P. Hedao, P. T. Hurley, K. M. Carney, R. Abell, and S. Walters. 1999. *Terrestrial ecoregions of North America: a conservation assessment.* Island Press, Covelo, California.

Risser, P. G. 1995. Biodiversity and ecosystem function. *Conservation Biology* 9: 742–746.

Samson, F. B. 1992. Conserving biological diversity in ecological systems. *Transactions of the North American Wildlife and Natural Resources Conference* 57: 308–320.

Sharp, D. M., G. R. Gunterpergen, C. P. Dunn, L. A. Leitner, and F. Stearns. 1987. Vegetation in a southern Wisconsin agricultural landscape. Pages 137–158 in M. G. Turner, editor, *Landscape heterogeneity and disturbance.* Springer-Verlag, New York, New York.

Shogren, J. F. 1998. *Private property and the Endangered Species Act.* University of Texas Press, Austin, Texas.

Slywotzky, A. J. 2000. The age of the choiceboard. *Harvard Business Review* 78:40–44.

Thomas, C. D. 1990. What do real population dynamics tell us about minimum viable population sizes? *Conservation Biology* 4:324–327.

U.S. Forest Service. 1997. Land and resource management plan: Tongass National Forest. R10-MB-338dd. U.S. Department of Agriculture Forest Service, Alaska Region, Juneau, Alaska.

———. 1999. Land and resource management plan: Dakota Prairie Grassland. U.S. Department of Agriculture Forest Service, Northern Region, Bismark, North Dakota.

U.S. Forest Service and Bureau of Land Management. 1994. Record of decision for amendments for Forest Service and Bureau of Land Management planning documents within the range of the northern spotted owl. U.S. Department of Agriculture Forest Service, Portland, Oregon.

———. 2000. Interior Columbia Basin supplemental draft environmental impact statement. U.S. Department of Agriculture Forest Service, Boise, Idaho.

Weitzman, M. L. 1998. The Noah's ark problem. *Econometrica* 66:1279–1298.

Williams, P. J. 1999. Strategy as options on the future. *Sloan Management Review* 40:73–82, 117–126.

PART **4**

THE FUTURE OF
POPULATION
VIABILITY
ANALYSIS

Our book concludes by examining the future of PVA. PVA models and approaches have changed greatly from their first applications in the early 1980s. Over the next decade we expect PVA to evolve at a similarly rapid rate. However, it is clear from the contributions in this final section that there is not a grand synthesis and meeting of minds about the future directions of PVA. The contributors present a variety of different views about approaches, values, and uses of PVA.

Novel approaches to the construction, use, and interpretation of PVA models are presented by Dan Goodman and by Hugh Possingham and colleagues. Goodman illustrates how to apply Bayesian approaches to develop a risk-assessment paradigm. He shows how various kinds of uncertainty can be built into the process, discusses how to make decisions under such uncertainty, and suggests how to use this approach in recovery planning. Possingham et al. present an alternative method of PVA using a frequentist approach based on decision theory. By way of two examples, this chapter illustrates how PVA can be used to compare and rank management alternatives based on various criteria. Both the risk-assessment paradigm and decision-theory thinking offer insightful ways of conducting PVAs.

Opposing views on the value of increasing model complexity are apparent in contributions from Robert Lacy and Philip Miller, and Donald Ludwig and Carl Walters. Most models ignore projected future changes of human land use, consumption, and population growth. Lacy and Miller argue that a meeting of the minds of sociologists, political scientists, and conservation biologists is needed to develop more complex and realistic PVA models that incorporate social upheaval, war, and land-use change driven by human societies. Ignoring these forces would certainly underestimate the likelihood of extinction. However, Ludwig and Walters argue that we should be cautious about reliance on complex PVA models. Nature is too unpredictable for extinction to be characterized accurately with either simple or complex population models. Increasingly complex PVA tools may be unlikely to produce better insights, but they have a kind of technological appeal to policymakers. Combining

PVA with adaptive management approaches could ameliorate some of these problems. Ludwig and Walters argue, however, that adaptive management has failed because decision makers often are unwilling to conduct the kind of large-scale scientific experiments with ecosystem management that it requires.

The final contribution by Katherine Ralls, Steven Beissinger, and Jean Fitts Cochrane offers a first attempt to develop guidelines for how and when PVA should be used in endangered-species management. They offer a definition for PVA that restricts it to specific types of models, a necessity if guidelines are to be specific and useful. A critically important but often neglected step in the PVA process is defining clear objectives for the modeling exercise. Objectives help define whether a PVA would be helpful or whether some alternative approach would be better, the choice of models to employ, and how to interpret the results. Lastly, Ralls et al. present a valuable set of criteria for judging the quality of a completed PVA.

21

Predictive Bayesian Population Viability Analysis: A Logic for Listing Criteria, Delisting Criteria, and Recovery Plans
Daniel Goodman

ABSTRACT

A tension exists between the desire to find policies that preserve bio-diversity for the very long term and the recognition that our management interventions and predictive capabilities are primarily short term. The latter consideration has motivated arguments that, because of their uncertainty, population viability analyses (PVAs) with long time horizons are not very useful for management decisions; but the former consideration implies that PVAs with long time horizons, provided they are credible, are the performance measure against which management plans must be judged. In this chapter I argue that "uncertainty" does not invalidate a PVA. Quite to the contrary, correct representation of the uncertainty, in a way that can be interpreted literally as a frequency of outcomes, is exactly what is needed to inform management. I outline the approaches to probability theory, statistical inference, and data compilation that will foster that interpretation. The reconciliation between short-term management and long-term goals for endangered species management will be achieved by rethinking what we mean by a "recovery plan." Instead of treating it as a commitment to a fixed set of interventions, the plan should be cast as a commitment to a fixed rule set and a commitment to monitoring, where the rule set consists of "if, then" statements that determine how the selection of interventions will change in response to the results of monitoring. Probabilistic PVAs will play two critical roles in drafting and evaluating such a plan. Short time horizon PVAs, with relatively low uncertainty, will be used to predict the changes in population status in the intervals from one revision of implementation to the next. A long time horizon PVA, distinguished by taking the management system of contingencies and monitoring into account, will be used to evaluate the performance of the system in delivering the desired long-term probability that the population will not go extinct.

INTRODUCTION

Aristotle, in the *Poetics*, ranked poetry, in its universality, as superior to history, because history "relates actual events" whereas poetry and other fiction treat "the kinds of things that might occur." Population viability analysis (PVA) possesses this kind of superiority. The kinship of PVA with fiction raises the natural question, how seriously should we take this? It would be good to know, because PVA is being used in serious applications, such as Endangered Species Act (ESA) decisions.

PVA purports to deliver a prediction. It predicts when a population will go extinct. For a handful of populations this prediction may be taken at face value. This will be for populations that are steadily declining in size, where the decline is owing to readily understood causes (such as overharvest or destruction of habitat), and these causes are expected to persist in their effect on the population. The prediction, then, is essentially an extrapolation—we graph the population size against time, extend the trajectory into the future, and read the predicted time of extinction as the time where the trajectory hits the x-axis. If we are correct in our causal assessment and in our expectation that these causes will operate with constancy for the period in question, this prediction refers to "actual events"—albeit for the future.

Interpretation is not as simple when appreciable uncertainty intrudes on the prediction. The uncertainty may be owing to lack of conviction in the causal assessment, lack of assurance that the causal regime will remain constant for the duration, incomplete information about starting conditions, or the operation of chance in the dominant mechanisms affecting the change in population numbers. Then the prediction deals in "what might occur." The tenuousness of the prediction increases with the span of time until the extinction might occur.

In this chapter I will consider ways of understanding uncertainty in PVA predictions, ways of rationally using uncertain PVA predictions for making decisions, reasons for adopting a long time horizon standard for PVA applications to ESA management, and a proposed framework for making practical short-term ESA decisions with a long-term standard. In the course of the discussion, I will show why a hierarchical Bayes computational procedure is the correct way to calculate extinction risk. This in itself is not new. The argument has been made before (and hasn't persuaded everyone; see Wade, chap. 11 in this volume), but I will express the argument in a stronger form, which may be new.

The strong assertion is that, when a correct empirical prior is used, a Bayes calculation of extinction risk can, in principle, be understood as a literal, objective probability of extinction for a portfolio of managed species. This probability has a concrete frequency interpretation. In this interpretation the uncertainty is part of the risk. This perspective has the potential for removing some of the mystery surrounding extinction-risk analysis in a way that should elevate the role that PVA plays in actual decision making. In particular, it removes the excuse that an analysis can be dismissed just because it is uncertain.

The strong version of the literal correctness of a hierarchical Bayes probability justifies using the results of such a statistical inference in the ways we would use any other probability with a frequency interpretation. Interestingly, then, we can use the inference for predicting the outcome of collecting future data. This allows a "value of information" calculation that can be used for optimization of decisions about the investment in future monitoring and about the merits of deferring an action while more data are collected.

Shortcut schemes for classifying population risk using very limited case-specific data have gained acceptance as a practical approach for making decisions, such as listing decisions, under pressure of time. Adoption of the shortcut methods has coincided in some quarters with a retreat from the belief that PVA in practice has useful predictive meaning. The framework developed in this chapter will be marshaled to argue, to the contrary, that the good performance of properly constructed shortcut schemes of this sort, specifically for listing decisions, can be understood precisely in terms of their function as predictive PVAs within the hierarchical Bayes framework.

But the good performance of shortcut schemes may be restricted to contexts where there is a high premium on speed in arriving at a decision, as in the determination whether a currently unprotected population is at risk of extinction. There is growing political impetus for delisting, now that a large number of populations have been managed under listed status for some period of time. This prompts examination of the decision systems appropriate for delisting and their relationship to management plans. In this application we will find that the shortcut schemes are likely to prove inefficient compared to investment in more detailed PVAs that can resolve components of extinction risk that are attributable to different causes, which offer opportunities for management intervention.

UNCERTAINTY AS FREQUENCY

Our experience of uncertainty is a psychological state, and in this guise it is a measure of security of belief or choice. Subjective Bayesian statistical theory attempts to interpret probability in this way, while still using the probability calculus for quantitative manipulation. Fuzzy set theory adopts a similar interpretation for uncertainty but uses a different machinery, without probability, for manipulating quantities. This is all well and good for a theory of personal decision making, but the subjectivity makes it poorly suited for a scientific treatment of evidence or for public decision making, where the degree of conviction needs to be communicated and shared.

Frequentist statistical theory avoids subjectivity by defining probability as frequency. In this view the statement that a particular outcome has a probability of 0.05 means that this outcome will occur 5% of the time if there are an infinite number of replications. The claim to objectivity is that this frequency is a mathematical fact that is independent of the beliefs of the observer or of the analyst.

FREQUENCY IN FORWARD PROJECTIONS OF A STOCHASTIC PROCESS

Mathematically, we know that some frequency statements can be justified deductively. In the mathematical modeling of stochastic processes we can define a process and then deduce the long-run frequency of outcomes of the process. For example, we might define an idealized representation of population growth as a Brownian-motion process governed by known parameters. Then we can describe the long-run frequency with which replications of that process, initialized at the same starting conditions and governed by the same parameters, will decline below an extinction threshold by a specified time. In principle, this is no more mysterious than deducing the long-run frequency of throwing a seven in repeated rolls of two dice.

Such models have been developed for PVA (Dennis et al. 1991). When plausible parameter values are used, these models show, among other things, that for long mean times to extinction, the variance in time to extinction is rather large. This large variance reflects at least some aspects of the uncertainty, which therefore is large. The large uncertainty, however, does not detract from our ability to say that the probability of extinction by a specified time horizon is a precise fraction.

This bears closer examination. The large variance means that our ability to predict the time to extinction of a single run of the stochastic process is quite weak. Our precise predictions are about the aggregate results from a large sample of repetitions. In application to a PVA this would mean that our ability to predict the actual time to extinction for a particular population is weak, which of course is disappointing. The strong prediction from such a PVA would be about the aggregate results of many trials of the process for this population, which does not seem that informative, since this population will have only one future, and one time to extinction, although we may not know in advance what that time is.

Rather than embark on flights of fancy about multiple futures, an alternative way to understand the strong frequency prediction from such a PVA is to consider a large number of populations, all with identical starting conditions and governed by the same parameters, but located in different places and experiencing independent random perturbations in their population growth. Then a strong frequency prediction that, for example, the probability of extinction within 200 years is 3% has a very concrete interpretation: it means that approximately 3% of the populations will go extinct by that time.

The secure prediction of a 3% failure rate for the aggregate is worth knowing. If 3% seems an undesirably large failure rate, we have the option of choosing a more stringent standard; if the costs of achieving the standard seem too high, we can evaluate the failure rate that will attend a policy of adopting a more affordable standard, and so forth. Thus, we can tame at least one aspect of uncertainty—the process uncertainty.

The next aspect of uncertainty to consider is the uncertainty about starting conditions and parameters. To represent that uncertainty, the replications should encompass a distribution of values for these conditions and parameters and allow their effects to propagate through the forward projection as well. The forward propagation is routinely accomplished by Monte Carlo methods; the sampling of the distribution of parameters and starting conditions can readily be accomplished by Monte Carlo simulation. But determination of the distribution of parameter values that the simulations should sample is a matter of statistical inference.

FREQUENCY IN STATISTICAL INFERENCE

Frequentist statistics maintain the frequency interpretation of probability by quantifying the uncertainty of an inference in terms of the stochastic process whereby the unknown, but fixed, parameter generates data. This provides a measure of the frequency with which the data that were observed will be observed in hypothetical replications of the data-generating process, with the parameter fixed at this particular, but unknown, value. This is a true probability with a concrete interpretation. Since it is a probability of the data, this lends itself to appraising hypotheses concerning the sampling processes that gave rise to the data. But it does not deliver a probability for the value of an imperfectly known parameter, in this case, with this set of data.

Subjective Bayesian statistics will deliver a probability statement about the value of an imperfectly known parameter, conditional on the case-specific data, but this probability does not have a concrete frequency interpretation.

A probability for the parameter with frequency interpretation can be obtained by hierarchical Bayes statistics, if set up properly. This applies Bayes's theorem, with the data that actually were observed, using as a prior the distribution, with frequency interpretation, of values of that parameter in a defined set of cases. This prior distribution is essentially an empirical histogram of parameter values in a concrete collection of cases. For a PVA, the collection of cases defining the empirical prior would be a set of populations for which the values of the relevant demographic and environmental parameters were similar enough that we could not reasonably distinguish their dynamics without case-specific data. The application of Bayes's theorem, then, factors in the data that are specific to the population.

In practice, the prior distributions for demographic PVA parameters will generally be for taxonomically defined sets of populations, with possible regression corrections for factors such as body size. The priors for environmental PVA parameters will generally be for assemblages defined by climate and habitat.

The frequency interpretation of the hierarchical Bayes posterior distribution, for the PVA parameters and starting conditions for a particular population of interest, is that this is the frequency of parameters and starting conditions that we would encounter in the subset of populations that yielded the same data when sampled from a collection of populations whose extinction dynamics would have been judged indistinguishable without population-specific data. A shorter way of saying this is that the posterior distribution for PVA parameters for a population of interest is the distribution of values of PVA parameters in the set of all populations for which we have the same relevant information.

COMBINING FORWARD PROJECTION WITH INFERENCE ON PARAMETERS

If we use hierarchical Bayes procedures to characterize the distribution of PVA parameter values and starting conditions that reflect all the relevant information that we have about a population, and we conduct the PVA by repeatedly sampling this posterior distribution of parameters, and conducting a stochastic projection with each combination of parameter values, this replication, in principle, then captures both the uncertainty about parameters and the uncertainty about the chance dynamical processes. If we use the projection to cumulate a distribution of times to extinction, the frequency interpretation of that distribution will again be based on a collection of populations.

The computed distribution for a given case will be the frequency distribution of times to extinction in a large collection of independent populations for which we had the same information about their relevant parameters and starting conditions. Then, in the long run, managing this collection of populations according to a particular extinction standard, such as 3% probability of extinction in 200 years, will achieve that standard: persistence of 97% of the populations and failure in 3%. Thus, the "sets of cases" envisaged by hierarchical Bayes theory correspond to the management portfolios that are the basis for interpreting the application of probabilistic PVA predictions to population management.

MANAGING WITH UNCERTAINTY

This idealized PVA encompasses a lot of uncertainty, both because of the randomness of the actual process modeled in the forward projection and the imperfect knowledge of the parameter values and starting conditions. Yet we can phrase the PVA results as precise numbers that can be taken literally, because these precise results are statements about the collection of all such populations for which we have the same relevant information, and the law of large numbers applies to the properties— even the stochastic properties—of this collection.

In other words, the policy of managing a large collection of populations to an extinction standard of 3% in 200 years, as gauged by the correctly calculated PVAs for those populations with the management taken into account, should achieve just that: 3% of the populations will be extinct in 200 years. Theoretically, this outcome is not uncertain, thanks to the operation of the law of large numbers for the aggregate of populations. The only uncertainty is about which of the populations will make up that 3%.

Thus we have tamed both aspects of the uncertainty in a PVA prediction—process uncertainty and parameter uncertainty—and turned the prediction into a statement that is useful for management. Because of the acknowledged uncertainty about any particular population, we may not be able to state whether our management will achieve its goal for a given population. But we can create portfolios of populations for which, in the aggregate, we can be assured that implementation of a chosen policy will achieve its goal. This is the way an intelligent investor, insurer, or gambler copes with uncertainty. There is no better model for how an intelligent environmental manager should cope with uncertainty.

CHOICE OF AN EXTINCTION STANDARD

Once we know that PVA calculations can be conducted in such a way that their results can be taken literally, the policy choice of the extinction standard becomes much more significant. If we adopt a policy, for example, of 15% probability of extinction within 100 years, does that really capture our value preferences?

By taking the calculations seriously, we can portray the outcomes in terms of scenarios that illuminate the consequences of the choice. We might date the beginning of civilization to 5,000 years ago, when the Upper and Lower Kingdoms of Egypt were united. Imagine that, at that time, a global policy had been adopted of managing the environment to a standard of 15% probability of extinction within 100 years for all mammalian species. How many mammal species would be left on earth now? The starting number of species would have been about 4,400. Compounding the 15% probability per 100 years over 5,000 years gives a probability of about 0.0003, per species, of surviving to the present. If the extinction dynamics of all the respective species were independent, the probability of no mammals now remaining would be 27%; the probability of more than three species remaining would be about 4%. This doesn't sound good. Our preferred vision of managing the environment for posterity obviously entails very low probabilities of extinction over large time spans.

SCALING TO SHORTER TIME HORIZONS

While our hopes may contemplate a very far future, we prefer the more immediate prospects when drafting a plan for action. Can we translate a long time horizon standard into a short time horizon standard that will accomplish the same result?

The compounding calculation, to go from a shorter to a longer time horizon, presumes that the process has a property—called ergodicity—of consistent and independent behavior from one such time interval to the next. This is to assume that if, for example, the probability of extinction over 100 years is 5%, then the survivors of the first 100 years will themselves have a probability of 5% for going extinct over the next 100 years, and so forth. The equivalent reverse calculation, which involves taking roots of the survival rate rather than raising it to a power, to go from a long time horizon to a shorter time horizon, makes the same assumptions about consistency and independence of the extinction process over the shorter interval.

There are three main kinds of mechanisms that can interfere seriously with consistency and independence of the process of stochastic population dynamics over time intervals on a particular scale. These are population age-distribution effects, population growth trends, and shifts in environmental conditions.

Age-distribution effects tend to wash out on a time scale of a few generations, which, depending on the organism, can be months to centuries. The most common illustration of this effect is a population of a long-lived species that has experienced a recent shift in conditions that has greatly reduced the effective reproduction without much impairing survival rates of adults. While the current generation of adults is in its prime, the rate of decline of the population will be dominated by their mortality rate, which is small; but as this generation approaches senescence, the population will collapse dramatically because the mortality rate then accelerates with age, and reproduction fails to replace these individuals with a similar number in the new generation. Under such circumstances, a PVA result showing a very low probability of extinction over a time span corresponding to the survival time of the first generation will be a misleadingly optimistic prediction of the probability of extinction subsequently.

Population size itself plays a large role in determining the short-term probability of extinction, because the distribution of per capita population growth rates is concentrated around modest values. So, even if a decline rate is persistent, it takes time for a large population to decline to a size at which immediate extinction is likely. Conversely, with a persistent positive rate of increase, it still takes time for a very small population to grow to the level at which it is out of danger of chance imminent extinction. This population-size inertia has an inherent time scale of the length of time that a trend in population growth or decline is liable to persist. PVA results calculated for a time horizon on that scale or shorter will offer a misleading index to the long-term prospects. If the trend is negative, the short time horizon standard will underprotect for the long term. If the trend is positive, the short time horizon standard will overprotect for the long term. Episodes of persistent acceleration in a population trend will magnify this effect.

Finally, treating the random variation in the forward population projection process as a force whose effect on survivors is constant from one interval to the next implies that the probability distribution of this force is the same in every time period. This assumption is contrary to the evidence that we have, on the time scale of human history, about environmental variation. The evidence indicates that environmental variation generally has the property that the measured variance gets larger as the time over which it is measured increases. In time-series parlance, this is called a red power spectrum. The consequence for a PVA is that simple compounding in going from a short time horizon to a long time horizon will underestimate the probability of extinction.

Overall, the pervasiveness of the three nonergodic forces—age structure, trends, and red spectrum environmental variation—will frustrate attempts to translate, by simple inverse compounding, a goal of persistence over millennia into an extinction standard for decades. We conclude from these complications that, if our real ambition for environmental policy is to achieve long-term persistence of biodiversity, we need to express this quantitatively in a long time horizon extinction standard for management, and we need to evaluate management plans with PVAs whose results are calculated for that time horizon.

CAN WE DO IT?

The preceding discussion has been somewhat abstract. Is the recommended program of analysis something that we can actually carry out? The literature on PVA, both theoretical and applied, is voluminous. There have been several review articles. In addition to the materials in this book, some representative particular models can be found in Goodman (1987a,b), Lande and Orzack (1988), Dennis et al. (1991), Gilpin and Hanski (1991), Boyce (1992), Burgman et al. (1993), Mangel and Tier (1993), McKelvey et al. (1993), Pimm et al. (1993), Doak and Mills (1994), NRC (1995), and Caughley and Gunn (1996).

The published arguments for Bayesian PVA rest on the greater clarity and better performance of Bayes decision theory for formulating management criteria (Taylor et al. 1996) and the need for propagating parameter uncertainty through the forward projection (Taylor 1995; Ludwig 1996; Wade, chap. 11 in this volume). The influence of correlations in parameter distributions on the distribution of extinction times is illustrated by Ferson and Burgman (1995). While many statistical approaches are capable of quantifying uncertainty in each of the parameters in a model, only the Bayesian approach lends itself to treating their mutual correlations as a joint distribution.

There is a large literature treating the general problem of uncertainty in environmental decision making (Francis 1997). The merits of the Bayesian approach in the broader context of resource management has been discussed in Thompson (1992), Walters and Ludwig (1994), Ellison (1996), Punt and Hilborn (1997), McAllister and Kirkwood (1998), and Wade (2000).

But there is not a professional consensus in ecology or resource management that the Bayesian approach is necessarily the right one: there is a literature arguing specifically to the contrary (Dennis 1996; Schweder 1998), and there are rich treatments of uncertainty using other approaches (Dennis et al. 1991; Caswell et al. 1998; Ludwig 1999). Indeed, within the discipline of statistics itself the situation appears somewhat confusing. In a magisterial review article on the state of current statistical research, Efron (1998) presented a three-axis graphical representation of the "space" of fundamental statistical theory, with the poles identified as Fisherian, Bayesian, or frequentist. Efron located the current topics of interest in statistical theory as no fewer than 15 distinct locations in that coordinate system, many of them combining elements of two or three of the respective philosophical treatments.

The argument presented in this chapter, that Bayesian analysis could, and should, routinely be conducted so as to deliver results expressed as probabilities with a frequency interpretation, is a minority view, but with distinguished precedents discussed in Goodman (2001a). The mathematical formalisms for applying this approach to PVA are developed in Goodman (2001b). The methods for estimating an empirical prior from comparative data are discussed in Efron (1996) and Goodman (2001a). There are few examples in the literature where an empirical prior distribution, with a strict frequency interpretation, has been used for inference on a population dynamic parameter—for one such example see Liermann and Hilborn (1997).

The compiling of taxonwide comparative data suitable for forming empirical priors for the relevant PVA parameters is a substantial task in its own right, and one that has not received enough attention. Some examples of this enterprise (none, incidentally, undertaken from an overtly Bayesian motivation) can be found in Pauly (1980), Calder (1984), Schmitz and Lavigne (1984), and Myers et al. (1995).

PARAMETERS THAT WILL HOLD FOR THE DURATION

At a minimum, a PVA requires distributions of parameter values for the mean population growth rate and the variance of population growth rate, both as functions of population size (Goodman 1987a,b). Density dependence—the relation of the rates to population size—can be omitted from a PVA model (Dennis et al. 1991), but the model then ceases to be useful for long time horizon applications (Royama 1977). Without density dependence, most combinations of plausible parameter values will result in the population model trending either to short-term extinction or to a biologically meaningless potential for unlimited population growth.

The models predicting short-term extinction without density dependence may be delivering a reasonably correct representation of the prospects for the actual modeled population. A number of ecologically detailed case histories of endangered populations do show persistent downward trends in the population numbers as the dominant source of extinction risk (Caughley 1994; Thomas 1994; Harcourt 1995). But the contrasting configuration of parameter values consistent with probable population growth in the short term confers biologically impossible long-term behavior without density dependence, and the long-term predictions of the PVA then are misleadingly optimistic.

The problem of obtaining correct distributions of parameter values for the long horizon projections is perplexing. The current tradition in PVA is to attempt to estimate the parameters from a time series of censuses for the population in question. Generally, these time series span 10 to 30 years. With such a time series, we can expect to obtain good estimates of the current trend in population growth rate, but environmental driving on a decadal time scale may render a 10- to 30-year sample unrepresentative of the long term (Francis and Hare 1994). A 10- to 30-year sample is surely too small a sample size for reliable estimation of the variance in population growth rate, even if the variance were stationary. And if the census data series begins at a time when the population was already substantially reduced in numbers, the time series will usually present severe obstacles to attempts to estimate carrying capacity and the strength of density dependence from the time series alone (Ginzburg et al. 1990). Model applications that do successfully estimate the strength of density dependence from a time series of censuses generally must rely on extraneous information to establish carrying capacity, using, for example, the assumption common to some fishery and marine-mammal models that at the beginning of the time series the population was at carrying capacity.

The relative weakness of time series of censuses for estimating the variance in growth rate, carrying capacity, and the scaling of density dependence puts much of the burden for inference on the prior. This raises the stakes for the prior being right, rather than just some conventional representation of vagueness or lack of information. Particularly for estimates of carrying capacity, there is a need to look for information not encapsulated in the time series of censuses. Regression relations between effective carrying capacity and the quantity and quality of available habitat hold theoretical promise for developing useful priors for carrying capacity.

The long time horizon that our value system seems to dictate puts further stress on attempts to estimate, from present data, distributions of parameters that will apply for the duration. On this time scale, the forces that are the ultimate causes for extinction may be human population growth, wars, economic instability, and climate change (Lacy and Miller, chap. 23 in this volume). It is not usual for the present generation of PVA to incorporate these sorts of factors as ongoing trends or as contributors to variation in parameter values. Yet we see that logical consistency requires incorporation of these factors. And on reflection it is fairly clear that their influence will generally be to increase the probability of extinction, probably by a considerable margin. This needs to be built into the priors for parameter distributions that will be used for long time horizon PVA projections.

RECONCILING SHORT-TERM MANAGEMENT WITH LONG-TERM GOALS

We can, in principle, conduct long time horizon PVA calculations that are technically correct, in the sense that the predictive distribution correctly represents the distribution of outcomes for the managed portfolio. These PVA predictions are valid for decision purposes, since the predictive distributions correctly reflect the uncertainty. But we will be disappointed in the magnitude of the uncertainty and the concomitant high probabilities of extinction. And we will be frustrated by the mismatch between the scope of the interventions at our disposal in realistic management decisions, compared to the almost cosmic range of causal factors that are beyond our control—and beyond reach of deterministic prediction—but which dominate the long time horizon prospects for extinction. How can we develop a decision system that marries the much more powerful short-term PVA predictions with the reality of our short-term management interventions, while providing reasonable certainty for achieving the long-term conservation goals?

The key lies in recognizing that our short-term management plans are indeed short-term. At the end of some period of perhaps as little as one year, and almost certainly less than 20, we can expect that the status of a protected population will need to be reassessed, the costs and benefits of the management will be reevaluated, and a possibly revised set of management interventions will be put in place for the next period of time. So the management question isn't whether we can select a set of management interventions now that we will commit to for the next interval of several thousand years because our long time horizon PVA shows it will confer an acceptably low probability of extinction for that interval. The management question is whether we can devise a menu of management options, to be selected among at intervals of a few years, where the selection procedure follows some predefined rules guided by monitoring data and short-term PVA predictions about change in population status from one time period to the next, such that this rule system confers an acceptably low probability of extinction for the long term. Here the commitment would be to the predefined rule system for intervention decisions and the updating of the short time horizon PVAs with ongoing data collection.

We know from common experience that, among the list of PVA parameters and variables describing starting conditions, population size is one of the easiest to measure with reasonable accuracy. So much so, that routine estimation of population size is a candidate for regular monitoring. We should also note that population size is a very powerful predictor of population status for short-term PVA, even though its predictive power dissipates over longer time horizons for which other forces come to dominate. Thus, a commitment to monitoring population size could provide sufficient information to fuel a system of periodically revised choices of management action guided by a fixed rule set and short-term PVAs. The coordinating element would be a long time horizon PVA to calculate the long-term probability of extinction with this intervention rule system and monitoring system in place.

We might be pleasantly surprised at the low probability of extinction that a PVA with management contingencies built in will reveal. For example, if the menu of management options includes effective intense interventions that are triggered when the population reaches critically low numbers, such bounding of the population away from extinction is relatively certain. Environmental management systems based on revising the choice of intervention according to updated information have been explored under the label "adaptive management." The most thorough exploration has been with respect to harvest management in commercial fisheries (Walters and Hilborn 1976). The theory is now well developed. Experience with applications has taught us that the system works well only in a highly disciplined institutional setting. This is an important lesson.

Applying that lesson here, to the creation of adaptive recovery plans for delisting, we see that the devalued common use of the term "adaptive management" would not satisfy the performance standards that we should demand for endangered-species management. Unfortunately, loose deployment of the vocabulary of adaptive management has at times been advanced as a rhetorical justification for the absence of a real plan or a real commitment. It is as if the mere announcement that there will be a management experiment, and the acknowledgment that the plan actually will be improvised as events unfold, somehow dignifies the proceedings. Most assuredly that is not what we want here. By "adaptive" we mean that future decisions will be guided by future data, but the rules for making those decisions must be fully specified in advance, for the plan to qualify.

Another institutional consideration is the need for quick decisions when the monitoring shows it is warranted. Speed has not been a common feature of ESA decision making, where crucial procedures, such as listing, drafting recovery plans, and designating critical habitat, can drag on for years.

PRACTICAL SHORTCUTS

The picture that emerges, then, is that the program of analysis proposed in this chapter is at least theoretically feasible, since some solutions have been published for all the pieces of the puzzle. But analyses of this sort are still case-specific research projects—not something that can be mass produced with the current generation of commercial software packages and the current libraries of data compendia.

As a practical matter, the responsible regulatory agencies are sensitive to the perception that data demands for a thoughtful PVA generally exceed the data that are available or that can be quickly collected on a small budget. Similarly, the agencies are reluctant to turn every management decision concerning species at risk into a full scale—and possibly drawn out and contentious—research program in sophisticated and difficult modeling. In particular, the specter of apparently disparate results from alternative models (Mills et al. 1996) is not attractive when litigation is a possibility. The pressure to make many population classification decisions with limited time and budget has motivated development of some shortcut decision systems that supply default values when data are not available, and that use categorical information about populations to supplement quantitative information.

One particularly well-designed system of this kind has been adopted by the National Marine Fisheries Service for administration of their responsibility to limit incidental harvest of marine mammals under the Marine Mammal Protection Act (Wade 1998). This system has the desirable feature of functioning with minimal case-specific data of a sort that usually are available, while offering an incentive for obtaining more and better data. The decision rule uses a lower confidence limit of a key parameter—the estimate of population size—so that the procedure automatically is more conservative with weaker data. Prior to adoption, the details of the system were tuned by studying the performance in simulations using a distribution of parameter values representative of the populations to which it would be applied. The operating characteristics of the system are thus very similar to those of a Bayes decision analysis. A system of this kind could be formalized for a shortcut procedure to classify populations according to extinction risk, bypassing a full PVA. Properly crafted, such a system could be expected to perform well in application for quick risk classification of populations that initially were unclassified, as in ESA listing decisions.

Because a good shortcut system embodies a precautionary principle while operating with very limited population-specific information, its performance will tend, quite properly, to be conservative. This may serve well our actual policy goals with respect to listing decisions that often need to be decided quickly. But, for the same reasons, such shortcut systems can lead to inefficiencies if they are relied upon for delisting decisions and evaluation of recovery plans.

DELISTING AND THE VALUE OF INFORMATION

The three-decade-long history of ESA implementation has provided substantial experience with listing decisions, and the use of science in the decision process has evolved accordingly. There has been little experience with delisting. The logical connections between PVA, delisting criteria, and recovery plans are relatively unexplored. With increasing political pressure for delisting, the scientific basis for such decisions merits closer scrutiny.

Consider, for example, a shortcut classification scheme that relied primarily on population size as the indicator of risk status. For such a system to reflect a precautionary approach, the threshold population level for classification to a status equivalent to the ESA categories of threatened or endangered would have to be set high enough that the classification was protective for a plausibly wide range of the other PVA parameters that affect the risk of extinction. We could thereby be assured that, despite the simplifications of the analyses, we would not frequently misclassify a population that is in fact at considerable risk of extinction. A decision system with this protective feature is designed to keep the frequency of "false negative" determinations to an acceptably low level.

When a decision system is restricted to sparse information, low false negative rates are achieved at the cost of exposure to a high false positive rate. That cost may be acceptable in the crisis atmosphere of a decision whether to list an unprotected population, where it is feared that there is no time to spare for a more detailed analysis or for the gathering of more data. But once the decision to confer protected status has been made, the pressure of time is relieved, so a different cost-benefit structure may come into play in the decision system for delisting.

If we continued to use the example shortcut analysis, based primarily on population size, to define the delisting criterion, this would commit the system to maintain the protected status until the population grew to a conservatively high level. The attainment of such a recovery threshold may be very expensive if the required habitat is not easily made available. Furthermore, the time to achieve that population level may be extremely long. Bearing in mind that the high population threshold of the shortcut analysis system was deliberately set high to provide a margin of safety, we might reasonably ask whether an acceptable degree of safety could be achieved at lower cost in some other way. To address that question we would need a deeper PVA analysis of the source of the risk (Goodman 1987c; Caughley 1994).

THE DIAGNOSTIC VALUE OF A FULL PVA

We may usefully distinguish four components of extinction risk attributable to small population size, small average population growth rate, large variance in population growth rate, and reducible uncertainty in any of the previous three parameters. The roles of population size and population growth as sources of risk and as opportunities for intervention are widely recognized. The importance of variance in population growth is appreciated as a source of risk (Lewontin and Cohen 1969; Goodman 1987a,b), but it has been oddly neglected as a target for management intervention (Goodman 1987c). Feasibility analysis, and the ability to optimize, will depend on the ecological information about the prospects for management to affect the various relevant parameters of a full PVA, and it will also depend on the PVA model's having enough detail to track how the changes in these parameters influence the distribution of time to extinction.

The possible role of reducible uncertainty, as a focus for management intervention, depends on our ability to interpret the distribution of time to extinction from the PVA model as a literal, objective probability of extinction with a concrete frequency interpretation. Recalling that this interpretation applies to a portfolio of managed populations, we understand that a statement, for example, that "this population has 7% probability of extinction within 100 years" means that this population is a member of a management portfolio of populations that presently are indistinguishable in their prospects for persistence, and that 7% of this portfolio of populations really will go extinct in less than 100 years, while the remaining 93% percent will go extinct in more than 100 years. If our policy were to tolerate, for example, a 5% probability of extinction within 100 years, then the risk for this population would be deemed unacceptably high, and the population would be classified as needing protection, as would all the other populations in this portfolio.

Bear in mind that the various populations in this portfolio will go extinct at different times, some before the critical time (100 years), but most after that time. If we could only know which were which, 93% of the populations in the portfolio could safely be delisted. We don't know which populations will go extinct at which time, in part because of irreducible process uncertainty, and in part because of parameter uncertainty that could in fact be reduced with an investment in additional data. The information from additional data would allow us to partition the management portfolio into new subsets that have different respective probabilities of extinction. The probability of extinction within 100 years averaged over all the subsets would still be the same, 7%. But the new subsets that would be distinguishable on the basis of the new information would have different respective probabilities of extinction within 100 years, some higher than 7% and some lower. If some of the new subsets have less than 5% probability of extinction within 100 years, then these subsets can legitimately be reclassified under the hypothetical policy of this example.

In effect, gathering additional data about the population in question allows us to move it to a different management portfolio. Interestingly, the distribution of time to extinction in the original portfolio provides a basis for predicting the probability that new information will justify moving the population to a portfolio of populations that can be delisted. We already saw that, in this example, perfect information (which we acknowledge is unattainable) would confer a 93% probability of delisting. Less complete, but feasible, information obtained from limited additional data gathering will confer a smaller, but still potentially worthwhile, probability of delisting.

For any defined data collection plan, the probability that the planned amount of new data will justify a conclusion of reclassification can be calculated in advance of having the data. This allows a straightforward cost-benefit calculation of whether the cost of the planned data collection is warranted by the expected cost savings from reclassification. Details of such a calculation are illustrated in Goodman (2001b).

DELISTING AND THE NEXT GENERATION OF RECOVERY PLANS

A tension exists between the desire to find policies that preserve biodiversity for the very long term, and the recognition that our management interventions and predictive capabilities are primarily short term. The latter consideration has motivated arguments that long time horizon PVAs, because of their uncertainty, are not very useful for management decisions; but the former consideration implies that long time horizon

PVAs are the performance measure against which management plans must be judged.

Uncertainty does not invalidate a PVA. Quite to the contrary, correct representation of the uncertainty, in a way that can be interpreted literally as a frequency of outcomes, is exactly what is needed to inform management. The reconciliation between short-term management and long-term goals for endangered-species management will come from rethinking what we mean by a recovery plan. Instead of treating this as a commitment to a fixed set of interventions, the plan should be cast as a commitment to a fixed rule set and monitoring, where the rule set consists of "if, then" statements that determine how the selection of interventions will change in response to the results of the monitoring. Probabilistic PVAs will play two critical roles in drafting and weighing the merits of such a plan. Short time horizon PVAs with relatively low uncertainty will be used to predict the changes in population status in the intervals from one revision of implementation to the next. A long time horizon PVA, distinguished by taking the management system of contingencies and monitoring into account, will be used to evaluate the performance of the system in delivering the desired long-term probability that the population will not go extinct.

Shortcut assessment systems that are less demanding of data may substitute satisfactorily for a full PVA in making decisions about classifying extinction risk for a population that is unprotected before the decision. In such a system, the risks attendant upon delay create a trade-off under which the speed advantage of the shortcut system is more important than the cost of the high false positive rate that results from the combination of low precision and a demand for a low false negative rate. The trade-off structure is different for the decision to reduce the risk classification of a population that is already protected. There a full PVA will perform much better than a shortcut system by reducing the false positive rate and by taking into account a richer menu of intervention options for reducing risk.

The incorporation of information updating and revision of intervention in response to the monitoring in this new kind of PVA will redistribute the uncertainty. The specific time until the population reaches a particular size or until a particular intervention will be triggered may still be poorly predicted. But the linkage between population size and the triggering of an intervention will be highly predictable because of the rule set and monitoring. Since population size plays such a large role in the short-term probability of extinction, such a PVA and management system, despite some very large uncertainties about specific timing of events, can provide welcome certainty about the frequency of population sizes and the long-run frequency with which various interventions are triggered. Prediction of intervention frequency gives a valuable index to long-term costs. Prediction of population distribution can focus on the bounding of population size away from extinction. The best plans will be the ones that achieve an acceptably low probability of long-term extinction at the lowest cost.

Shortcut population decision systems that satisfy the precautionary principle may be satisfactory for time-sensitive decisions to classify a currently unprotected population, but the inherent conservatism of such systems, coupled with their very limited measures of population status, will lead to inefficiencies if the same system is used to define delisting or recovery criteria. A detailed PVA, informed by strong, population-specific data, offers a means to identify the specific sources of risk that most threaten the population. This, in turn, will reveal opportunities for finding the least expensive way to reduce the population risk to an acceptable level. Because the cost of continued protection may be high, while the component of the population risk attributed to reducible uncertainty may be large, a value of information analysis from a detailed PVA may well show that a fairly substantial investment in more data and modeling for a delisting decision is optimal on simple economic grounds.

A credible monitoring plan coupled with a credible conditional intervention plan could, on the strength of the plan alone, contribute to substantial risk reduction, provided we really were assured that the monitoring would be implemented and that the proposed decision rules to trigger interventions would be followed. The credibility would be especially important in the interesting situation where the risk reduction attributed to the plan were in itself sufficient to meet a delisting criterion. Smooth functioning of such a system will require some institutional changes to encourage faster response to monitoring and firmer commitment to preparation for contingency plans.

LITERATURE CITED

Boyce, M. S. 1992. Population viability analysis. *Annual Review of Ecology and Systematics* 23:481–506.

Burgman, M. A., S. Ferson, and H. R. Akçakaya. 1993. *Risk assessment in conservation biology.* Chapman and Hall, New York, New York.

Calder, W. A. 1984. *Size, function, and life history.* Harvard University Press, Cambridge, Massachusetts.

Caswell, H., S. Brault, A. J. Read, and T. D. Smith. 1998. Harbor porpoise and fisheries: an uncertainty analysis of incidental mortality. *Ecological Applications* 8:1226–1238.

Caughley, G. 1994. Directions in conservation biology. *Journal of Animal Ecology* 63:215–244.

Caughley, G., and A. Gunn. 1996. *Conservation biology in theory and practice*. Blackwell Scientific, Cambridge, Massachusetts.

Dennis, B. 1996. Should ecologists become Bayesians? *Ecological Applications* 6: 1095–1103.

Dennis, B., P. L. Munholland, and J. M. Scott. 1991. Estimation of growth and extinction parameters for endangered species. *Ecological Monographs* 61:115–143.

Doak, D. F., and L. S. Mills. 1994. A useful role for theory in conservation. *Ecology* 75:615–626.

Efron, B. 1996. Empirical Bayes methods for combining likelihoods. *Journal of the American Statistical Association* 91:538–551, with discussion, 551–563.

———. 1998. R. A. Fisher in the 21st century. *Statistical Science* 13:95–122.

Ellison, A. M. 1996. An introduction to Bayesian inference for ecological research and environmental decision making. *Ecological Applications* 6:1036–1046.

Ferson, S., and M. A. Burgman. 1995. Correlations, dependency bounds, and extinction risks. *Biological Conservation* 73:101–105.

Francis, R. C. 1997. Managing resources with incomplete information: making the best of a bad situation. Pages 513–524 in D. J. Strouder, P. A. Bisson, and R. J. Naiman, editors, *Pacific salmon and their ecosystems*. Chapman and Hall, New York, New York.

Francis, R. C., and S. R. Hare. 1994. Decadal-scale regime shifts in the large marine ecosystems of the north-east Pacific: a case for historical science. *Fisheries Oceanography* 3:279–291.

Gilpin, M. E., and I. Hanski, editors. 1991. *Metapopulation dynamics: empirical and theoretical investigations*. Academic Press, London, United Kingdom.

Ginzburg, L. R., S. Ferson, and H. R. Akçakaya. 1990. Reconstructability of density dependence and the conservative assessment of extinction risks. *Conservation Biology* 4:63–70.

Goodman, D. 1987a. Consideration of stochastic demography in the design and management of biological reserves. *Natural Resource Modeling* 1:205–234.

———. 1987b. The demography of chance extinction. Pages 11–34 in M. E. Soulé, editor, *Viable populations for conservation*. Cambridge University Press, New York, New York.

———. 1987c. How do any species persist? Lessons for conservation biology. *Conservation Biology* 1:59–62.

———. 2001a. Taking the prior seriously: Bayesian analysis without subjective probability. In M. Taper and S. Lele, editors, *The nature of scientific evidence*. University of Chicago Press, Chicago, Illinois.

———. 2001b. Uncertainty, risk, and decision: the PVA example. In J. M. Berkson, L. L. Kline, and D. J. Orth, editors, *Incorporating uncertainty into fishery models*. American Fisheries Society, Bethesda, Maryland.

Harcourt, A. H. 1995. Population viability estimates: theory and practice for a wild gorilla population. *Conservation Biology* 9:134–142.

Lande, R., and S. H. Orzack. 1988. Extinction dynamics of age-structured populations in a fluctuating environment. *Proceedings of the National Academy of Sciences* (USA) 85:7418–7421.

Lewontin, R. D., and D. Cohen. 1969. On population growth in a randomly varying environment. *Proceedings of the National Academy of Sciences* (USA) 62:1056–1060.

Liermann, M., and R. Hilborn. 1997. Depensation in fish stocks: a hierarchic Bayesian meta-analysis. *Canadian Journal of Fisheries and Aquatic Sciences* 54:1976–1984.

Ludwig, D. 1996. Uncertainty and the assessment of extinction probabilities. *Ecological Applications* 6:1067–1076.

———. 1999. Is it meaningful to estimate a probability of extinction? *Ecology* 80:298–310.

Mangel, M., and C. Tier. 1993. A simple direct method for finding persistence times of populations and application to conservation problems. *Proceedings of the National Academy of Sciences* (USA) 90:1083–1086.

McAllister, M. K., and G. P. Kirkwood. 1998. Using Bayesian decision analysis to help achieve a precautionary approach for managing developing fisheries. *Canadian Journal of Fisheries and Aquatic Sciences* 55:2642–2661.

McKelvey, K., B. R. Noon, and R. H. Lamberson. 1993. Conservation planning for species occupying fragmented landscapes: the case of the northern spotted owl. Pages 424–450 in P. M. Kareiva, J. G. Kingsolver, and R. B. Huey, editors, *Biotic interactions and global change*. Sinauer Associates, Sunderland, Massachusetts.

Mills, L. S., S. G. Hayes, C. Baldwin, M. J. Wisdom, J. Citta, D. J. Mattson, and K. Murphy. 1996. Factors leading to different viability predictions for a grizzly bear data set. *Conservation Biology* 10:863–873.

Myers, R. A., N. J. Barrowman, J. A. Hutchings, and A. A. Rosenberg. 1995. Population dynamics of exploited fish stocks at low population levels. *Science* 269:1106–1108.

National Research Council (NRC). 1995. *Science and the Endangered Species Act*. National Academy Press, Washington, D.C.

Pauly, D. 1980. On the interrelationships between natural mortality, growth parameters, and mean environmental temperature in 175 fish stocks. *Journal du Conseil International pour l'Exploration de la Mer* 39:175–193.

Pimm, S. L., J. M. Diamond, T. R. Reed, G. J. Russell, and J. Verner. 1993. Times to extinction for small populations of large birds. *Proceedings of the National Academy of Sciences* (USA) 90:10871–10875.

Punt, A., and R. Hilborn. 1997. Fisheries stock assessment and decision analysis: the Bayesian approach. *Reviews in Fish Biology and Fisheries* 7:35–63.

Royama, T. 1977. Population persistence and density dependence. *Ecological Monographs* 47:1–35.

Schmitz, O. J., and D. M. Lavigne. 1984. Intrinsic rate of increase, body size, and specific metabolic rate in marine mammals. *Oecologia* 62:305–309.

Schweder, T. 1998. Fisherian or Bayesian methods of integrating diverse statistical information? *Fisheries Research* 37:61–75.

Taylor, B. L. 1995. The reliability of using population viability analysis for risk classification of species. *Conservation Biology* 9:551–558.

Taylor, B. L., P. R. Wade, R. A. Stehn, and J. F. Cochrane. 1996. A Bayesian approach to classification criteria for spectacled eiders. *Ecological Applications* 6:1077–1089.

Thomas, C. D. 1994. Extinction, colonization, and metapopulations: environmental tracking by rare species. *Conservation Biology* 8:373–378.

Thompson, G. G. 1992. A Bayesian approach to management advice when stock-recruitment parameters are uncertain. *U.S. Fishery Bulletin* 90:561–573.

Wade, P. R. 1998. Calculating limits to the allowable human-caused mortality of cetaceans and pinnipeds. *Marine Mammal Science* 14:1–37.

———. 2000. Bayesian methods in conservation biology. *Conservation Biology* 14: 1308–1316.

Walters, C. J., and R. Hilborn. 1976. Adaptive control of fishing systems. *Journal of the Fisheries Research Board of Canada* 33:145–159.

Walters, C. J., and D. Ludwig. 1994. Calculation of Bayes posterior probability distributions for key population parameters. *Canadian Journal of Fisheries and Aquatic Sciences* 51:713–722.

22

Decision Theory for Population Viability Analysis

*Hugh P. Possingham, David B. Lindenmayer, and
Geoffrey N. Tuck*

ABSTRACT

We explore two different ways of using population viability analysis
(PVA) and decision theory to help managers choose among specific actions that will enhance the persistence of a threatened species. In the
first case we created a decision tree of management options for the
Leadbeater's possum where different options were sequentially added
to a management scenario and the extinction probability was evaluated
using a traditional PVA. There was no formal mathematical optimization
in this example; it was more a qualitative approach to decision making.
In the second case we used a formal optimization tool, stochastic dynamic programming, to find the optimal state-dependent strategies for
managing fire in two patches of different quality supporting a threatened
species. The best decision in this case depends both on the state of the
habitat in both populations *and* the size of the population. The solution
to the problem was given by prescribing the optimal action for each
state of the system. Compared to the first case study, the stochastic
population model that underlies this analysis was very simple—essentially a Markov chain population model.

INTRODUCTION

Conservation biology is intended to be an applied science. Its objective
is to integrate ecological and socioeconomic knowledge to aid the conservation of the world's biodiversity and ecosystem functions. Unfortunately, this straightforward applied agenda is often dominated by pure

We are grateful to the University of Adelaide theoretical ecology research group (PHLEM)
and the National Center for Ecological Analysis and Synthesis Working Group on Population
Management (a center funded by the NSF, grant DEB-94-21535, the University of California
at Santa Barbara, and the State of California) for valuable comments. In particular we thank
P. Amarasekare, A. Tyre, M. Burgman, M. McCarthy, I. Ball, J. Lawton, P. Kareiva, J.-P. Rodri-
quez, R. Lacy, I. Davies, I. Noble, and K. Shea. This work was partially supported by a Australian Research Council (ARC) grant to Possingham and I. R. Noble, an ARC SPIRT grant to
Possingham and Lindenmayer, and by Environment Australia.

ecology. The field of population viability analysis (PVA) suffers from this same problem—too much theory and not enough problem solving. In this chapter we explore two examples of the way that decision theory can be used with PVA for guiding management action and policy.

PVA traditionally had two general purposes. The original intent was to assess the viability of a population (Shaffer 1981). This assessment of extinction risk was useful and indicated that small populations in protected areas are very vulnerable to chance extinction. Often the goal of PVA was to identify a target population size, or habitat area, large enough for viability. The first uses of PVA were *not* to choose between different management options, but to accurately assess extinction risk (Beissinger, chap. 1 in this volume). A second application of PVA, which also relies on the accurate assessment of extinction probability for a species, was ranking the vulnerability of species (Mace and Lande 1991). Many countries, states, and provinces are required, for legislative and/or conservation priority setting, to maintain and update lists of threatened species. One way of ranking species according to their level of threat would be to do a PVA on every species. While this has the merit of apparent objectivity, PVAs are rarely used for classifying threatened species. This is because there is insufficient data to parameterize PVAs for the vast majority of species. The obsession with accurate estimation of extinction risk reaches its zenith in the United States. In the United States, the Endangered Species Act, the National Forest Management Act, and recovery plans and habitat conservation plans are founded in the ideal that species must be managed for viability. In these legal and planning issues, the accurate assessment of extinction risk, and hence what constitutes a minimum viable population size, is seen as fundamental. The application of PVA for these situations is part of conflict resolution between conservation and development. This approach has some merit for a few charismatic well-studied species like the northern spotted owl (*Strix occidentalis caurina;* Forsman et al. 1996). More recently, several authors have questioned these traditional uses of PVA and the ability of PVA to accurately predict extinction risk for most species (Boyce 1992; Possingham et al. 1993; Taylor 1995; Beissinger and Westphal 1998). There appear to be adequate data to make very accurate estimates of extinction risk for only a very small fraction of threatened species.

An alternative role for PVA is as a support tool to help make management decisions. This role does not invest so much faith in the quantitative estimation of extinction risk, but rather focuses on the relative ability of different management decisions to deliver acceptable conservation

strategies. The key is to use PVA to choose between management options for a fixed budget rather than find the level of action that will
ensure viability (Possingham et al. 1993; Noon and McKelvey 1996).

To illustrate these contrasting approaches, consider how we might
choose among revegetation strategies for a small, threatened bird that
occupies a patchy habitat. The traditional approach is to determine how
much habitat needs to be constructed to ensure viability (defined as an
acceptably low level of extinction probability). An alternative approach
is to assume that a certain amount of revegetation is possible, and then
decide whether that effort should be put in to making corridors to link
existing patches, making new patches, or expanding existing habitat
patches.

There are many examples of situations that might require a choice
between management options. How much of a population should be
brought into captivity, and conversely how much should be reintroduced
(Maguire et al. 1987)? How many animals should be translocated between populations or to new sites to improve viability (Lubow 1996)?
What is the optimal fire regime for a threatened species (Possingham
and Tuck 1998)? Should we create new habitat for a threatened metapopulation or relocate individuals to empty patches (Possingham 1996)?
Should we increase the area of conserved habitat by acquiring cheaper
low-quality habitat or more expensive high-quality habitat (Haight et al.
n.d.)? How can land be assigned to different land uses to conserve a
species?

Decision theory is a way of thinking that can be applied to all management problems. It can be broken down into the following steps (Shea
et al. 1998): (1) Define the objective. In the context of PVA, this is
usually minimizing the probability of population extinction. (2) List a
set of management options that are defined in terms of one or more
control variables (e.g., how much habitat to put where, or when to start
a prescribed burn). (3) Define variables that describe the state of the
system (e.g., population size or habitat state). (4) Describe the dynamics
of the state variables given any choice of management option (e.g., the
population and/or habitat model). (5) Define constraints that bound the
state of the system and the control variable (e.g., a budget, or a maximum amount of area that can be conserved).

In this chapter we present two case studies that illustrate, in very
different ways, the application of decision theory to threatened species
management. The first example uses PVA to choose among different
forest management strategies that will give a threatened arboreal marsupial the best chance of persisting in a single forest block. While the PVA
is detailed and quantitative, the decision theory approach used in this

case study is more qualitative. The second case study uses a quantitative decision theory tool to explore the best fire management strategy for a species that occupies two adjacent habitat patches. In contrast to the first example, the population model is simplified and the decision theory analysis is more complex and explicit.

CASE STUDY 1: FOREST MANAGEMENT FOR LEADBEATER'S POSSUM

The conservation of the endangered Australian marsupial Leadbeater's possum (*Gymnobelideus leadbeateri*) is one of the most contentious forestry issues in Australia (Lindenmayer 1996). To assist in planning for its recovery, a spatially explicit population simulation model was used to estimate the probability of extinction of Leadbeater's possum in forest blocks subjected to different forest management options. Our spatially explicit population model was a modified version of the PVA model ALEX (Possingham and Davies 1995). The results of the model were used to develop a decision tree that encapsulated the consequences of different combinations of decisions.

The controversy over the conservation of Leadbeater's possum is due, in part, to the impact of clearfell logging on a rotation time of 80 to 120 years over 75% of its known distribution (Smith and Lindenmayer 1992), an area of 60 km by 50 km in the central highlands of the state of Victoria in southeastern Australia. The requirement of Leadbeater's possum for nest sites in trees that are over 150 years old, coupled with fires that burned more than 60% of the forest within the range of the species early this century (Noble 1977), places the species at risk. The current management strategy is to not cut areas containing large dead or senescent trees such as old-growth forest, linear strips along streams, steep and rocky areas, and areas burned early this century (MacFarlane and Seebeck 1991). The remainder of the forest matrix is logged using intensive clearfell methods on a rotation time that precludes Leadbeater's possum, while areas that burn in the future may be salvage logged.

Our task was to determine the viability of the species subject to current management options, and to consider the relative benefits of other management options. Options were compared by estimating short-term and steady state extinction probabilities within 150-year periods. Because estimates of extinction probabilities can be very sensitive to changes in parameters (Burgman et al. 1993), we examined how changes in uncertain parameters affected the relative benefits of options. We have limited data on movement of Leadbeater's possum between patches, and are uncertain about the frequency and extent of future fires, so we focused our sensitivity analyses on these processes.

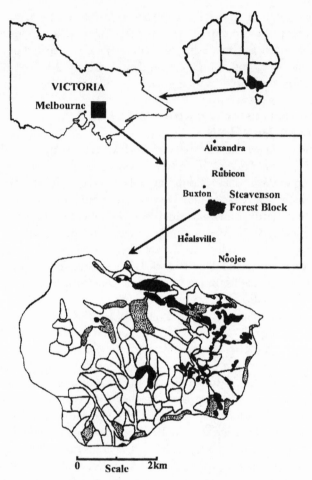

Fig. 22.1 Patches of different habitat used to model the persistence of *Gymnobelideus lead-beateri* in Steavenson Forest Block. The location and size of patches were derived from a geographic information system. *Solid black polygons* are areas of old-growth forest, *lightly stippled patches* are large riparian corridors or steep and rocky areas excluded from logging, and *open polygons* are 50-ha areas that have not been recently logged and are potential reserve sites. Mortality-free movement is allowed between patches that touch, are closer than 200 m, or are connected by a riparian exclusion (*lines*).

Because repeated wildfire may completely destroy suitable habitat in any single forest block, we searched for management options that would enhance species persistence on a block-by-block basis.

The focus of this study was the Steavenson Forest Block (fig. 22.1). Looking at species viability on a block-by-block basis is consistent with

Victorian initiatives to retain species throughout their normal range (Flora and Fauna Guarantee Act, Government of Victoria 1988). Our objective was to find an acceptably low local-extinction probability within realistic constraints for the forest industry.

The following forest management options were explored for the Steavenson Forest Block by changing patch structure and habitat dynamics (Lindenmayer and Possingham 1996): (1) maintain the current strategy (see above) where old-growth is not permanently reserved and salvage logging would be allowed; (2) reserve existing area of old-growth permanently (currently it is reserved unless it burns); (3) do not allow salvage logging in the future (when forest with merchantable timber burned in the past, the majority of the remaining trees, whether dead or damaged, were removed by salvage logging; without salvage logging, some dead and damaged trees may survive, leading to more complex habitat dynamics); (4) create additional reserves of varying size and number; and (5) increase the time interval between clearfell harvesting (if the interval between harvesting or rotation time is increased to exceed 150 years, suitable habitat will develop).

The result of our PVA simulations was the decision tree summarized in figure 22.2. Because the current state of the system (population size and forest structure) influences the extinction probability, we consider two time frames—the next 150 years and a typical 150-year period some time in the future when the forest has reached a quasi-equilibrium state. At the top of the decision tree is the probability of species extinction over a typical 150-year period in the future (with the extinction probability over the next 150 years in parentheses) if we had continued with the previous management strategy. Each step in the decision tree represents a change in management actions—and a progressive decrease in extinction probability for both the typical 150-year period (and the next 150-year period). There are several ways such a decision tree can be constructed. Figure 22.2 is just one example that suits this particular situation where the most palatable actions appear at the top of the tree and alternatives are added sequentially.

When the previous strategy is applied to the Steavenson Forest Block, there is a reasonable chance of persistence of Leadbeater's possum over the next 150 years. However, the extinction probability is projected to increase because no areas of suitable habitat are being set aside to become old-growth, and fires are likely to reduce the suitability of existing old-growth. While the extinction probability over the next 150 years is only 33%, the long-term extinction probability for a typical 150-year period in the future is 100%.

For the Steavenson Forest Block the permanent preservation of

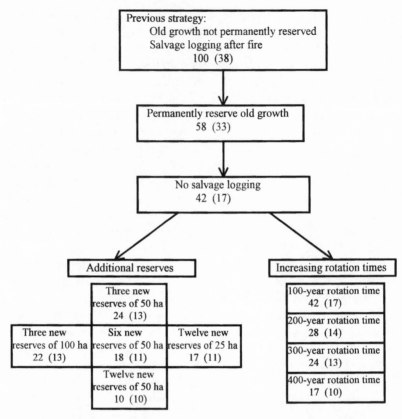

Fig. 22.2 Flow chart showing the impact of different management options on the viability of *Gymnobelideus leadbeateri* in Steavenson Forest Block. The *first number* in each option is the steady state extinction probability in 150 years, that is, the extinction probability for the population over a 150-year period 500 years in the future when the system has approached a quasi-stationary state (simulations were run for 750 years). The *number in parentheses* is the extinction probability in the next 150 years. Moving down the decision tree adds different conservation strategies to the previous strategies.

existing old-growth forest from logging and salvage logging increases Leadbeater's possum viability substantially and is economically efficient. If existing old-growth patches are permanently reserved, the probability of extinction in the first 150 years is 33% and the probability of extinction in subsequent 150-year periods is 58%. If old-growth patches that burn are included in the logging cycle, the species is unlikely to persist (fig. 22.2). In the short term, the only option that increases Leadbeater's possum viability significantly is not to salvage log old-growth forest. Because the remainder of the forest is only 60 years old, it would take at least another 100 years to realize the benefit of other strategies. Given that no salvage logging and permanently protecting old-growth is expedient, we locked these in to the top of the decision tree.

Next we wanted to explore the best ways of further reducing the extinction probability—two popular suggestions are to increase the rotation time and to make more reserves. Extending the rotation time to 200 years or more throughout the forest block improves viability significantly (dropping long-term extinction probabilities from 42% to 28%). However, this option requires an almost complete cessation of logging for the next 150 years! This is unlikely to be acceptable to the timber industry. Setting aside just six 50-ha reserves (about 5% of the total forest block) improves viability even more than increasing the rotation time and reduces logging at most by 5%. These reserves should be placed so they are uncorrelated with respect to fires, a risk-spreading strategy, but are sufficiently close to allow recolonization by Leadbeater's possum. One solution to these apparently conflicting demands is to place reserves on either side of a potential fire barrier, like a rain-forest gully. These conclusions apply to our baseline parameter set—reasonable movement and one-in-100-year fires that burn, on average, 50% of the patches (Lindenmayer and Possingham 1995).

Given that permanent reserves appear to be more efficient than increasing the rotation time, it is interesting to look at the issue of what size of reserve is most effective and to explore the robustness of our conclusions. We set the economic constraint that only a certain number of hectares can be set aside and explored various options, trading off size and number of reserves. We found that one or two very large reserves were not useful, because they might be burned by the same large fire. We found that many very small reserves, each of which is demographically unstable, relies heavily on frequent recolonizations between reserves. Ultimately we recommended an intermediate number of reserves of 50 ha to 100 ha. Such a size, in the absence of fire, has reasonable viability but also enables replication and consequent risk spreading (Lindenmayer and Possingham 1996).

We assessed the sensitivity of our conclusions by changing the parameters for processes about which we were uncertain. The advantages of not salvage logging and increasing the rotation time significantly diminished if movement was restricted, because these management options relied on animals recolonizing patches. If the frequency or extent of fires was increased, the relative ranking of the best management options was not significantly altered. Regarding reserve size and number, we found that a lower recolonization ability favored bigger reserves, and vice versa, but not significantly. For example, even with no potential for recolonization, one big reserve was never the best option. Changing the fire frequency had surprisingly little impact on the optimal reserve size, so while the absolute extinction probability varied, the best decision for a fixed total reservation area remained surprisingly invariant.

In summary, we found that the previous management strategies were inadequate because the long-term extinction probability would have been 100%. The most efficient and effective ways to make long-term population persistence likely were (1) minimizing the extent and frequency of wildfires; (2) reserving several connected 50 ha to 100 ha areas per 5,000–10,000 ha forest block, which are unlikely to burn in the same fire; and (3) no future postfire salvage logging in old-growth forest. Increasing the rotation time was not a useful strategy in this case because it would stop the local industry for 100 years. The recommended actions had a relatively low impact on the industry compared to the increases in Leadbeater's possum population viability they were predicted to achieve. These recommendations are in the process of being implemented.

This case study shows that, by associating each suite of management options with an extinction risk, conservation biologists can provide decision makers with information to weigh against other forest values. Our approach in this example was to informally recognize the social and economic constraints and to provide the best solution within those constraints.

CASE STUDY 2: PRESCRIBED BURNING FOR A THREATENED SPECIES

A central question in conservation is, how should habitat disturbances like fire be managed? Every year park and forest managers must decide whether to promote or suppress fire. In this case study, we use stochastic dynamic programming (SDP) to solve the specific problem of fire management for a threatened, fire-sensitive plant or animal population that exists in two adjacent patches.

Managing habitat disturbance to benefit biodiversity employs the intermediate disturbance hypothesis (Connell 1978). This hypothesis states that species diversity at any one location is maximized when the disturbance frequency is intermediate. This generalization provides little specific guidance for a manager faced with a particular ecosystem, species of concern, or kind of disturbance. More importantly, this hypothesis, like all other ecological theory, is not couched within a decision-making framework. Managers must make decisions within the constraints of time and money. When there are trade-offs between actions, general ecological theory may offer little practical guidance.

Habitat disturbances, like fire, have both direct and indirect effects on threatened species. They often cause a short-term reduction in the size of populations (direct effect), and they usually change the quality of the habitat for a longer period of time, sometimes decades (indirect effect). Population increase for some plant species occurs soon after fire because of the release of space and nutrients that are in short supply in undisturbed habitat (Gill 1996). Many animal species favor recently burned habitat (Friend 1993).

Possingham and Tuck (1998) provided a framework for answering the specific question, how long after the last fire should we wait before imposing the next fire in a single patch of habitat? The objective was to minimize the probability of extinction of the population. They used SDP to determine the best decision about when to burn a single patch (the decision variable), given an estimate of the current size of the population. They concluded that (1) habitat should be burned well after the peak in habitat quality; (2) a patch with a small population should be burned later, if at all; and (3) wildfire frequency has relatively little impact on the optimal fire management strategy.

These rules are applicable to the management of reserves for which a single disturbance regime must be imposed simultaneously across the entire reserve. Often, however, reserves are large enough that different parts can be disturbed at different times. Below we use SDP to determine the optimal fire management strategy when a reserve is managed as two patches, each with different disturbance regimes. The PVA model we use is a very simple population model that is consistent with the crude kinds of population data available for many species.

There are two steps to solving a management problem using SDP. First, we model the stochastic state-space dynamics of the system, in this case the population dynamics of the organism in each patch. Here we use a relatively simple stochastic population model where each population state is not the actual size of the population, but a rough measure of its abundance. Second, we use decision theory to minimize the probability of extinction; SDP equations are formulated, and costs and benefits are defined for this management objective.

The Population Model

Assume that the threatened species of concern exists in two patches and let its population size in each of these patches be denoted x_1 and x_2. Given that abundance data on threatened species are often poor, we assume that the abundance of the species in each patch is in one of seven classes. Class 0 represents local extinction, and class 6 is the maximum abundance of a patch (fig. 22.3). These abundance classes are not in-

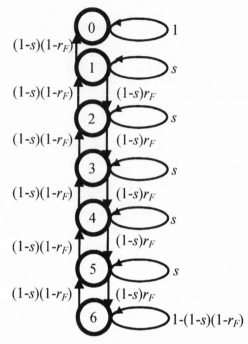

Fig. 22.3 Population transition probabilities for a single patch. The parameter s is the probability that the population does not change. The parameter r_F is the probability that the population moves up one state, if it changes, and the time since the last fire was F years (fig. 22.4).

tended to represent a linear measure of real abundance, but some scaled measure of abundance such that the population grows to the next class with about equal probability (e.g., a natural log scale). For many species we are more likely to have this sort of qualitative data than to have exact counts. We describe first a model for the dynamics of a population in a single patch, and then the way in which populations in the two patches interact.

Let s be the state-independent probability of the population staying in its current abundance state, regardless of what else happens. By making s bigger we model a more stable population. Let r_F be the probability the population grows and moves up one state, assuming it is *not* staying in its current state, if it has been F years since the last fire. When there is a fire, the population falls by one state (after any other transitions for that year). The transition probabilities for a particular patch are illustrated in figure 22.3.

The suitability of the habitat changes with the time since the last fire

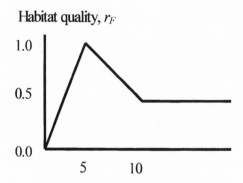

Fig. 22.4 The relationship between the time (years) since the last fire in a patch, F, and the patch's habitat quality, r_F.

by changing the chance of the population growing as a function of the time since the last fire, r_F. We assume that r_F takes the form shown in figure 22.4 with very low habitat quality immediately after a fire, followed by a rapid rise to a peak in habitat quality five years after a fire ($r_5 = 1.0$), then a decline to the habitat quality of mature habitat ($r_{10} = 0.5$). This habitat-quality dynamic is typical for a species that favors early or midsuccessional habitat. Using these definitions of s and r_F and the state-transition diagram in figure 22.3, it is possible to define the probability of a population in a single patch moving from any state to any other state.

Now we need to define how the populations in the two patches interact. To model the dynamics of a population in two patches simultaneously, we need to make two sets of assumptions. First, we need to model the degree of environmental correlation between the two patches; if one patch experiences a bad year (and the population declines), does the other population also experience a bad year? Second, we need to model the movement of individuals between the two patches. Here we assume that the patches are environmentally correlated, and that a good year for one patch is a good year for the other. If both patches are in the same successional state (i.e., the same time since the last fire), then when it is a population-growth year for one patch, it is also a population-growth year for the other. If the better patch has a year of population decline, the worse patch also has a year of population decline, but there is a chance that the patch in the better successional state will have a population-growth year while the patch in the worse successional state has a population-decline year. Assuming, without loss of generality, that patch 1 is better habitat than patch 2 and ignoring boundary conditions, the abundance transition probabilities can be written

$$P((x_1 + 1, x_2 + 1)/(x_1, x_2)) = (1 - s)r_{F2}$$

(the probability of both patches increasing their abundance state is the probability of a change of state occurring, $1 - s$, times the probability the worse patch increases its population size, r_{F2});

$$P((x_1 + 1, x_2 - 1)/(x_1, x_2)) = (1 - s)(r_{F1} - r_{F2})$$

(the probability the better patch increases its abundance state but the worse patch declines in abundance);

$$P((x_1, x_2)/(x_1, x_2)) = s$$

(the probability neither patch changes its abundance state); and

$$P((x_1 - 1, x_2 - 1)/(x_1, x_2)) = (1 - s)(1 - r_{F1})$$

(the probability both patches decrease in their abundance state), where $P((x_1, x_2)/(y_1, y_2))$ is the probability that the population state moves from (y_1, y_2) to (x_1, x_2), and F_1 and F_2 are the time since the last fire in patches 1 and 2, respectively.

We consider two extreme forms of population connection between the two habitat patches to illustrate the range of possible behaviors. We assume that either (1) individuals are unable to move between patches (e.g., plants with poor seed dispersal) or (2) the populations are connected and some individuals are able to select the better patch in which to live (e.g., many species of birds). For the latter case we assume that, if patch 1 is the better patch ($r_{F1} > r_{F2}$, at a particular time), then the majority of animals are in that patch, so $x_1 = \max(x_1, x_2)$ and $x_2 = \min(x_1, x_2)$. We also assume that movement occurs before the population transitions each year.

Now we have a full description of the stochastic two-patch population dynamics, and we will determine the state-dependent strategy that minimizes extinction probability. Traditionally, we would carry out Monte Carlo simulations and try different management strategies until we found the best one. We will use an optimization tool to find the exact optimal strategy numerically, without simulation. Because the decision is whether to burn a patch, for every possible time since fire for that patch and the other patch, while taking into account the population sizes in each patch, finding the precise optimal strategy by simulation would take an extremely long time with current computational power.

The Decision Tool: Stochastic Dynamic Programming

The objective of an SDP is to determine the state-dependent optimal decision for controlling a stochastic process (Intriligator 1971). PVA models are often stochastic so SDP is the logical choice of optimization tool. To use the method, we need to define payoff values for achieving a certain state for the system, describe the management options mathematically, and define the dynamic programming equation. Our payoff in this example will be one if the population persists and zero otherwise. The optimal management strategy is found by back-stepping through time and choosing the optimal decision for each year, assuming that later decisions are made optimally.

For our problem there are four possible decisions in the strategy set $a \in A = \{0, 1, 2, 3\}$: burn neither patch ($a = 0$), burn patch 1 ($a = 1$), burn patch 2 ($a = 2$), and burn both patches ($a = 3$).

To determine the best decision for every possible state of the system, we work backwards and start from the penultimate decision before some terminal time. Let the terminal time at which our success is assessed be $t = T$. If the population is extinct at time T, we gain no points; if it is extant we gain one point, so

$$J_T(x_1, x_2, F_1, F_2) = 0 \quad \text{if } x_1 = x_2 = 0$$
$$= 1 \quad \text{otherwise,}$$

where $J_t(x_1, x_2, F_1, F_2)$ is the value of being in state (x_1, x_2, F_1, F_2) at time t. To find the best decision for the penultimate time $(T - 1)$, we express the value of being in state (x_1, x_2, F_1, F_2) as a function of the value of being in each state at the terminal time $(J_T(x_1, x_2, F_1, F_2))$, weighted by the probability of moving to each of these terminal states. These probabilities come from our population model, and depend on the current state of the system and the decision that is chosen from the strategy set. This generates the optimal strategy if we are only interested in one year ahead. To determine the best long-term strategy, the back-stepping method is repeated until an equilibrium strategy is found. This is the best state-dependent long-term strategy. Here we report the optimal decision, given the size of the population in each patch and the time since the last fire in each patch, when we are 50 time steps from the terminal time. After this time we have reached a stable optimal long-term fire management strategy that minimizes the chance of extinction over a 50-year time frame.

SDP Results

We do not explore all possible scenarios here but focus on how the optimal strategy changes with or without movement of organisms be-

Time since last fire in patch 2, F_2

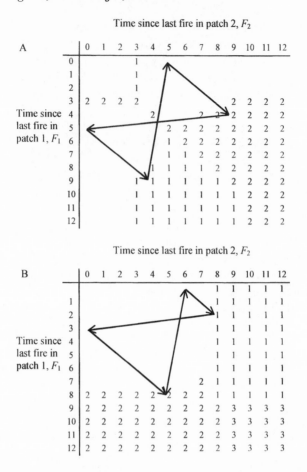

Time since last fire in patch 2, F_2

tween patches. Our point of departure is a baseline scenario in which there is a 50% chance of a patch changing state ($s = 0.5$), and fire reduces the population size in a patch by one state.

When there is movement between patches by a habitat-selecting organism, the optimal decision for each possible combination of times since last fire (assuming the population sizes $x_1 = x_2 =$ state 2) is shown in figure 22.5A. The results are that (1) the strategy is almost entirely independent of the state of the population; (2) the best strategy is never to burn both patches; and (3) the patch that should be burned always has lower habitat quality compared to the other patch. This means that, given our population movement rules, the burned patch will always have the lower population size.

To further interpret figure 22.5A, consider starting with patch 1 three

Time since last fire in patch 2, F_2

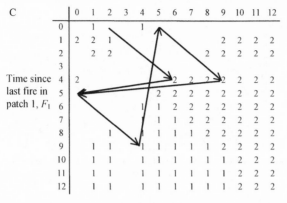

Fig. 22.5 The optimal decision for each combination of times (years) since last fire, F_1 and F_2. Optimal strategy *1* indicates the best option is to burn patch 1, *2* indicates the best strategy is to burn patch 2, and *3* indicates the best strategy is to burn both patches. A *blank* indicates neither patch should be burned. The *arrows* indicate a trajectory of times since last fire for each patch, assuming no wildfire and that the optimal strategy is implemented. *A,* The baseline scenario with movement of organisms between patches. Burning both patches is never optimal. The optimal strategy set is almost completely independent of the population sizes in each patch. *B,* A scenario with no movement of organisms and zero-quality old-growth habitat. *C,* A scenario with movement of organisms and zero-quality old-growth habitat.

years after its last fire and patch 2 eight years after its last fire, ($F_1 = 3$, $F_2 = 8$). The best decision in this state is to do nothing, so we move to state ($F_1 = 4$, $F_2 = 9$), at which point patch 2 is burned, taking us to state ($F_1 = 5$, $F_2 = 0$). Continuing this way, we find that the system traces out a permanent cycle that takes the state of the system deterministically between the states ($F_1 = 4$, $F_2 = 9$), ($F_1 = 5$, $F_2 = 0$), ($F_1 = 9$, $F_2 = 4$) and ($F_1 = 0$, $F_2 = 5$). Remarkably, this cycle is approached regardless of the initial state of the system! This optimal long-term burning strategy is graphed from the perspective of the time since the last fire in each patch in figure 22.6.

If there is no movement of organisms between patches, then the optimal strategy is to never burn either patch! The diagram equivalent to figure 22.5A is filled with blank spaces. The qualitative and quantitative differences are enormous. A detailed explanation of why it might be optimal never to burn is presented in Possingham and Tuck (1998). Briefly, when the patches are disconnected, each population is independent, so the best strategy is independent of the state of each patch. For

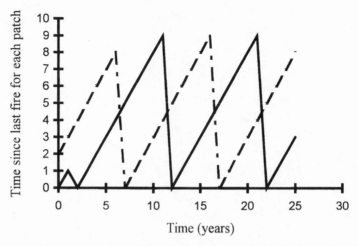

Fig. 22.6 Population trajectories in relation to the time since the last fire for each patch (*solid line* is patch 1; *dashed line* is patch 2) beginning with patch 1 burned recently and patch 2 burned two years ago. This trajectory is derived from starting at state ($F_1 = 0$, $F_2 = 2$) in figure 22.5A and increasing the time since last fire for each patch until the optimal decision is to burn. The trajectory assumes no wildfires.

this set of parameters, the decrease in abundance caused by initiating a fire is likely to outweigh the benefits of eventually achieving a high habitat quality five years after fire.

From this example we can draw some important conclusions. (1) The optimal strategy changes when there is movement of organisms between patches. (2) With movement between patches, the best strategy is to unsynchronize the two patches as fast as possible so that the time since the last fire is perfectly out of phase. The effect is that one patch is good when the other patch is bad. (3) The optimal strategy forces the system into a stable deterministic cycle that is almost completely independent of the size of the population. This result contrasts with Possingham and Tuck's finding (1998) that the decision to burn a single patch depends on the size of the population in that patch. This last conclusion occurred in the movement case, because patches were burned only when they had low population sizes. The patch in the better state had a higher population size and was unburned.

Let us now assume that habitat quality declines to zero when fire does not occur for ten or more years, ($r_{10} = 0.0$ in fig. 22.4). Regardless of whether the organism can move between patches, the patches have to be burned, because unburned patches have zero habitat quality. Any population left in an unburned patch is doomed to rapid local extinction.

When there is no movement of the organism between patches, the optimal strategy for an initial state of ($F_1 = 2$, $F_2 = 2$) is shown in figure 22.5B. Each patch is burned after eight years regardless of the state of the other patch. The probability of the population persisting for 50 years was less than 20%, regardless of the successional state of either patch. (The terminal reward values of the SDP are the extinction probabilities.)

The optimal strategy when there is movement between patches is shown in figure 22.5C. The set of states through which the system cycles is very similar to the previous case: ($F_1 = 4$, $F_2 = 9$), ($F_1 = 5$, $F_2 = 0$), ($F_1 = 9$, $F_2 = 4$) and ($F_1 = 0$, $F_2 = 5$). The probability of persistence over 50 years now exceeds 40%, regardless of which state we start from, a large improvement over the no movement case.

In contrast to the previous example, the no movement and optimal movement strategies are quite similar. Both show that patches should be burned eight or nine years after a fire. The big difference is that, with movement, the strategy forces an asynchronous cycling of patch states.

We examined a number of examples and found that, when there is movement between patches, the extinction probability with optimal decision making is a lot lower compared to the case without movement. Furthermore, the optimal strategy is quite robust to changes in the state of the population and variation in the habitat quality function (fig. 22.4). The insights developed here, and in Possingham and Tuck (1998), are just a first step toward a general theory of disturbance management for single-species conservation.

CONCLUSION

Can SDP be used to find the best conservation strategy for any problem? In theory the answer is yes, although in practice it is no. Any process that can be represented as a Markov chain (and almost any population model can be) can have an objective maximized using SDP (Mangel and Tier 1993). Realistically, however, describing complex population models using Markov chains can be difficult. More importantly, the SDP approach works only where the state space of the entire system being managed is low. The size of the state space in case study 2 was $7 \times 7 \times 13 \times 13$. If we had attempted to apply the same tool to the Leadbeater's possum example with ten patches, which each have a population abundance and experience a time since fire or logging, the number of possible states of the system would be a minimum of 10^{20}! While formulating such a problem is possible, solving it with existing constraints on computer speed and storage capacity would be impossible. There are mathematical programming algorithms for deterministic population models to find the best management strategy where the number of possible states is very large; however, the relevance of deterministic models to the conservation of small populations is tenuous (Hof and Raphael 1997; Hof and Bevers 1998).

In summary, there are a range of approaches for applying decision theory to conservation problems, and each approach has an optimal context. This chapter presents just two of many possible applications of decision theory to endangered-species management. It represents an attempt to illustrate for managers the merits of decision theory and expose the dearth of applied theory for conservation biology. We believe more models are needed to solve problems; we need more of a new kind of ecologist—the theoretical applied ecologist.

LITERATURE CITED

Beissinger, S. R., and M. I. Westphal. 1998. On the use of demographic models of population viability in endangered species management. *Journal of Wildlife Management* 62:821–841.

Boyce, M. S. 1992. Population viability analysis. *Annual Review of Ecology and Systematics* 23:481–506.

Burgman, M. A., S. Ferson, and H. R. Akçakaya. 1993. *Risk assessment in conservation biology*. Chapman and Hall, London, United Kingdom.

Connell, J. H. 1978. Diversity in tropical rainforests and coral reefs. *Science* 199: 1302–1310.

Forsman, E. D., S. DeStefano, M. G. Raphael, and R. J. Gutiérrez. 1996. Demography of the northern spotted owl. *Studies in Avian Biology* 17:1–122.

Friend, G. R. 1993. Impact of fire on small vertebrates in mallee woodlands and heathlands of temperate Australia: a review. *Biological Conservation* 65:99–114.

Gill, A. M. 1996. How fires affect biodiversity. Pages 47–55 in *Biodiversity and fire: the effects and effectiveness of fire management*. Department of Environment, Sports and Territories, Canberra, Australia.

Haight, R. G., K. Ralls, P. A. Kelly, and H. P. Possingham. N.d. Designing habitat protection strategies for a core population of the San Joaquin kit fox. Manuscript.

Hof, J., and M. Bevers. 1998. *Spatial optimization for managed ecosystems*. Columbia University Press, New York, New York.

Hof, J., and M. G. Raphael. 1997. Optimization of habitat placement: a case study of the northern spotted owl in the Olympic Peninsula. *Ecological Applications* 7: 1160–1169.

Intriligator, M. D. 1971. *Mathematical optimization and economic theory*. Prentice Hall, Englewood Cliffs, New Jersey.

Lindenmayer, D. B. 1996. *Wildlife and woodchips: Leadbeater's possum as a test case of ecologically sustainable forestry*. New South Wales University Press, Sydney, Australia.

Lindenmayer, D. B., and H. P. Possingham. 1995. Modelling the impacts of wildfire on the metapopulation behaviour of the Australian arboreal marsupial, Leadbeater's possum, *Gymnobelideus leadbeateri*. *Forest Ecology and Management* 74: 197–222.

————. 1996. Ranking conservation and timber management options for Leadbeater's possum in southeastern Australia using population viability analysis. *Conservation Biology* 10:235–251.

Lubow, B. C. 1996. Optimal translocation strategies for enhancing stochastic metapopulation viability. *Ecological Applications* 6:1268–1280.

Mace, G. M., and R. Lande. 1991. Assessing extinction threats: toward a reevaluation of IUCN threatened species categories. *Conservation Biology* 5:148–157.

MacFarlane, M. A., and J. H. Seebeck. 1991. Draft management strategies for the conservation of Leadbeater's possum, *Gymnobelideus leadbeateri*, in Victoria. Department of Conservation and Environment, Melbourne, Australia.

Maguire, L. A., U. S. Seal, and P. F. Brussard. 1987. Managing a critically endangered species: the Sumatran rhino as a case study. Pages 141–158 in M. E. Soulé, editor, *Viable populations for conservation*. Cambridge University Press, Cambridge, United Kingdom.

Mangel, M., and C. Tier. 1993. A simple direct method for finding persistence times of populations and application to conservation problems. *Proceedings of the National Academy of Sciences* (USA) 90:1083–1086.

Noble, W. S. 1977. *Ordeal by fire: the week a state burned up*. Hawthorn Press, Melbourne, Australia.

Noon, B. R., and K. S. McKelvey. 1996. Management of the spotted owl: a case history in conservation biology. *Annual Review of Ecology and Systematics* 27: 135–162.

Possingham, H. P. 1996. Decision theory and biodiversity management: how to manage a metapopulation. Pages 391–398 in R. Floyd, A. Sheppard, and P. De Barro, editors, *The proceedings of the Nicholson Centenary Conference 1995*. CSIRO Publishing, Canberra, Australia.

Possingham, H. P., and I. Davies. 1995. ALEX: a population viability analysis model for spatially structured populations. *Biological Conservation* 73:143–150.

Possingham, H. P., D. B. Lindenmayer, and T. W. Norton. 1993. A framework for the improved management of threatened species based on population viability analysis. *Pacific Conservation Biology* 1:39–45.

Possingham, H. P., and G. Tuck. 1998. Fire management strategies that minimize the probability of population extinction for mid-successional species. Pages 157–167 in D. Fletcher, L. Kavaliers, and B. Manly, editors, *Statistics in ecology and environmental monitoring 2*. University of Otago Press, Dunedin, New Zealand.

Shaffer, M. L. 1981. Minimum population sizes for species conservation. *BioScience* 31:131–134.

Shea, K., P. Amarasekare, M. Mangel, J. Moore, W. W. Murdoch, E. Noonburg, A. Parma, M. A. Pascual, H. P. Possingham, C. Wilcox, and D. Yu. 1998. Management of populations in conservation, harvesting, and control. *Trends in Ecology and Evolution* 13:371–374.

Smith, A. P., and D. B. Lindenmayer. 1992. Forest succession, timber production, and conservation of Leadbeater's possum (*Gymnobelideus leadbeateri* Marsupalia: Petauridae). *Forest Ecology and Management* 49:311–332.

Taylor, B. L. 1995. The reliability of using population viability analysis for risk classification of species. *Conservation Biology* 9:551–558.

23

Incorporating Human Populations and Activities into Population Viability Analysis

Robert C. Lacy and Philip S. Miller

ABSTRACT

Population viability analysis (PVA) uses population biology models to assess probability of extinction, threats, and effects of possible actions. Yet the conservation of biodiversity is mostly a matter of addressing processes that revolve around human populations. Unfortunately, experts who model human systems (e.g., human demography, local economics and resource use, industrial activities, social systems, and political systems) rarely interact with biologists who use PVA to model wildlife populations. We propose that it is possible and necessary to integrate analysis of human systems with PVA. The linkages between human demographic, economic, and social systems and wildlife population biology must be identified; these often include direct harvest of the species, reduction of habitat amount or quality, and habitat fragmentation. Detailed quantitative models of the relevant human systems, when available, can generate inputs into a PVA model of the wildlife population. Integrating understanding of human and natural systems will require a broader array of expertise, models, data, and perspectives than has been applied to most PVAs to date.

POPULATION VIABILITY ANALYSIS AS A HOLISTIC ASSESSMENT OF THREATS

Population viability analysis (PVA) was developed to model and understand the multiple, interacting threats to population persistence. PVA models can be used to examine harvest, habitat degradation, habitat loss and fragmentation, impacts of exotic species, increased environmental variation, catastrophic impacts, demographic uncertainty, disrupted breeding systems, and the genetic problems of random drift and in-

The ideas presented in this chapter are the result of ongoing discussions with a biodiversity network of Jenna Borovansky, Onnie Byers, Susie Ellis, Ronda Fisher, George Francis, Gayl Ness, Phil Nyhus, Emmanuel Raufflet, Ulysses Seal, Harrie Vredenburg, Frances Westley, John Williams, and other colleagues. Specific work outlined here was supported by a grant to Frances Westley from the Social Sciences and Humanities Research Council of Canada.

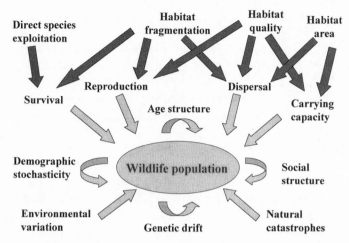

Fig. 23.1 Primary factors impacting the viability of wildlife populations that need to be considered in development and application of PVA models. Factors connected directly to the wildlife population by *gray arrows* are natural processes that affect the dynamics of all wildlife populations. The four human processes listed across the top of the diagram impact or modulate the natural processes of wildlife population dynamics.

breeding depression (Shaffer 1981; Gilpin and Soulé 1986; Soulé 1987; Boyce 1992; Lacy 1993/1994). Figure 23.1 shows a schematic diagram of some of the processes that must be understood and incorporated into PVA models if we are to make meaningful assessments of population viability within realistic representations of the natural world.

Because of the multiplicity and the complexity of the threats to viability, the predictive capabilities of PVA can be improved by recruiting expertise from many fields to address these threats. These disciplines include wildlife management, population ecology, community ecology, landscape analysis, geography and geographic information systems (GIS), genetics, and others outside the realm of the natural sciences such as statistics, modeling, decision analysis, and conflict resolution. Despite arguments about the importance of one discipline or another, PVA is fundamentally an analysis of multiple interacting factors. Therefore, PVA is necessarily synthetic and holistic, rather than reductionistic. PVA models are complex and diverse, as they must be, because the essence of population viability cannot be captured with a few elegant equations or broadly applicable theories.

NEED FOR SOCIAL SCIENCE EXPERTISE

Although PVAs have been complex and multifaceted since the first models of Shaffer (1981), most PVA models still omit the primary driving force in population viability: humans. Whereas the processes in the lower part of figure 23.1 are natural, the forces listed across the top of the figure are not intrinsic to natural systems. Rather, the dominant threats to the viability of natural populations are directly or indirectly caused by human activities.

Population viability models can be sophisticated representations of biological systems. However, most PVA models make simplifying assumptions about human roles in the system. We usually ask, What happens to the probability of population persistence (or some other measure of viability) *if* humans do not change in number, distribution, or activity patterns over time? Yet the assumption of stasis in human systems is naïve and unrealistic. Changing human numbers, activities, and impacts are why we have a field of conservation biology and why we go to such lengths to develop PVA models for projecting dynamics of wildlife populations. In most PVAs, the dominant human impacts shown in figure 23.1 are treated as though they were the original or ultimate causes of species vulnerability. Yet those processes are just the intermediaries, or links, between a diversity of activities of the human population and the population biology of the species of concern in the PVA. For example, habitat destruction is not a spontaneous process. PVAs that consider habitat destruction only as an abstract process disconnected from the economic and social forces driving it are not likely to contribute much to the amelioration of the impacts.

As conservation biologists trained in areas such as wildlife biology, ecology, genetics, or evolutionary biology, we should appreciate that there are experts in other fields who have data and tools that we need. We need to find the people who share our concern for the natural environment and ask for their help and expertise. Social scientists model processes like the effect of a change in a political system on the rate of agricultural development, timber harvest, or mineral exploration, and the effect of access to global markets on changes in land use as people shift from subsistence production for local use to commodities production for external markets. Other social scientists are concerned with issues such as the interactions between human demographic trends, economics, the likelihood of civil war, and mass relocation of peoples. Demographic and economic forecasters can project how changing national economies will change human birth and death rates, in turn affecting numbers and age structure of people, and trends toward urbanization versus movements for land rights and redistribution. All of these things matter; in fact they are at the core of why we have conservation problems and why addressing those problems is important.

IMPACTS OF HUMAN ACTIVITIES ON POPULATION VIABILITY

In recent years, social scientists concerned with how people fit into conservation assessments and actions have engaged some population biologists in a series of discussions about how PVA concepts can be expanded to incorporate the ultimate threats to viability of natural populations—human populations and their activities. This Strategic Network on the Social and Scientific Challenges to Biodiversity Conservation, funded by a grant from the Social Science and Humanities Research Council of Canada, explored how natural scientists and social scientists can combine expertise to develop more useful applications of PVA. In this chapter, we describe what we have learned about integrating human numbers, distribution, and activities into PVAs on wildlife populations by collaboration in this network.

It is useful to first examine the population biology processes that are central to most PVAs (lower part of fig. 23.1), and then to identify those human systems that impact wildlife population dynamics, and the mechanisms by which these impacts occur. The biological processes fall loosely into several categories. There are processes that are intrinsic to population dynamics. These include the stochastic problems of sampling in small populations (e.g., demographic stochasticity and genetic drift) and processes resulting from population-specific characteristics (e.g., age structure, breeding system, and inbreeding depression; Lacy 2000). In addition, there are largely unavoidable natural processes that are driven by forces external to the population—environmental stochasticity and natural catastrophes—that can strongly impact population dynamics. Perhaps most importantly, there are the fundamental, deterministic processes driving changes in population size: reproduction, survival, carrying capacity of the habitat, and dispersal (emigration and immigration). These deterministic components of population growth are the only factors considered by most classical models of population growth (Leslie 1945; Deevey 1947; Birch 1948).

If human populations were not expanding and increasingly dominating environments around the earth with their activities, then the dynamics of most wildlife populations could be modeled adequately by entering basic life-history parameters and habitat descriptors into a PVA model that incorporated all of the above biological factors. A deterministic or perhaps stochastic application of a life-table projection would be sufficient to assess population vulnerability, and conservation options could be evaluated through sensitivity analysis that included relevant demographic rate variables. To extend our use of population biology models for conservation purposes, however, we need to know whether population biology processes are modulated by, and continuing to change because of, human activities. Thus, we need to specify the primary linkages between human activities and population biology processes.

Human activities negatively impact demographic rates through four primary modes: direct species exploitation, reductions in habitat quality, reductions in habitat area, and habitat fragmentation. Direct exploitation can take the form of harvesting for local use or external markets, incidental killing, sport collecting, or even retaliation out of spite or protest. Exploitation, if documented and quantified, can be easily incorporated into PVA models as increases in age- and sex-specific mortality.

Habitat quality can be reduced by human impacts on food, predators, competitors, disease, water availability, and basic geophysical resources, and on cover, nest sites, or other microhabitats required for the completion of life history. These effects may be caused by introduced exotic plants and animals, the use of natural resources by humans, discharge of wastes, alteration of the habitat matrix through landscape development or agricultural conversion, and regional or global changes (e.g., the use of chlorofluorocarbons and the emission of greenhouse gases). Habitat area is reduced by conversion or loss of unprotected habitat under changing socioeconomic conditions. Reduced habitat amount and quality can be combined in projections of carrying capacities in PVA models, and can alter dispersal among habitat patches. Habitat fragmentation can occur either through the direct subdivision of a formerly large, contiguous habitat into multiple smaller units, or by the erection of barriers to movement (e.g., roads or canals) between existing habitat patches in an already fragmented landscape. The detrimental effects of habitat fragmentation might be incorporated into a PVA as increased age- or sex-specific dispersal mortality (which would be imposed in addition to the normal mortality of resident organisms) or as decreased rates of migration among patches.

The stochastic processes intrinsic to population dynamics—demographic stochasticity, environmental variation in demographic rates, genetic drift, and inbreeding—are less likely to be the most important entry points through which human activities modulate population viability. Although the nature of these intrinsic processes is unchanged by human activities, their expression may be increased. How stochastic processes interact to threaten small populations has been the focus of much attention in conservation biology over the past two decades (Franklin 1980; Shaffer 1981; Gilpin and Soulé 1986; Soulé 1987; Simberloff 1988; Lacy 1997), but not without controversy (Caughley 1994; Hedrick et al. 1996). Human activities most directly modulate stochastic population processes and natural variation in the environment by causing reductions in population sizes. Any PVA model that includes demographic stochasticity and inbreeding effects as functions of population size will capture some of these impacts. Although some conservation biologists have thought that these components are the primary emphasis of PVA (e.g., Burgman and Lamont 1992; Caughley 1994; Saltz 1996), the impact of these indirect effects in PVA models is usually quite minor compared to the direct effects of changing fecundity and mortality (e.g., Seal and Lacy 1989; Ellis et al. 1992; Fisher et al. 1999).

Human activities can modulate stochastic processes with or without changing population size. For example, human-modified landscapes may experience increased fluctuations in the biotic and physical environment because of reduced species diversity or removal of ecosystem functions, which normally buffer environmental perturbations. The resulting greater seasonal, interannual, and spatial variation in demographic rates may further threaten population viability. Catastrophes, such as disease epidemics or severe weather patterns, may increase because of human activities, such as the transport of organisms, reduced diversity, and climate change. Biologists tend to avoid or ignore these arenas, but a change in the local economic base, a shift in cultural and ethical values, or a transmigration of people can have far greater impacts on the viability of wildlife populations than does inbreeding depression, changing numbers of predators, or the intrinsic maximum rate of increase of a population. Moreover, most of the biological rates and processes that PVA models include are greatly altered by human systems.

Most links between human activities and the dynamics of wildlife populations are fairly obvious, even if they have been inadequately studied and ignored in many PVAs. However, some links between human activities and population processes are more subtle. For example, there are currently no more than a few Sumatran rhinoceroses (*Dicerorhinus sumatrensis*) within each of a number of isolated protected areas. Field surveys in Malaysia in 1995 found tracks of only 1 juvenile among 35 sets of tracks, and only 1 of 21 adult females captured in the prior decade was pregnant (AsRSG 1996). If the population were breeding as expected for a rhinoceros species, about 30% of adult females should be pregnant at any time, and about 15% of the animals should be under two years of age. It is possible that a scarcity of mates is causing a near cessation of breeding over much of the fragmented range. Thus, one impact of habitat destruction outside of parks and poaching within parks may be a considerable reduction in reproduction. The disruption of breeding systems as an indirect effect of human activities can be incorporated into many PVA models via Allee effects (e.g., Groom 1998;

Groom and Pascual 1998), but rarely do PVAs incorporate strong density dependence (Mills et al. 1996; Lacy 2000).

INCLUDING HUMAN PROCESSES IN PVA

The diagram of population biology in figure 23.1 is driven by primary demographic processes and impacted by human threats; it describes what needs to be considered in any comprehensive PVA. Although we recognize that human actions are the reason that biodiversity is under siege, and therefore that PVAs are needed, usually the assessment of human activities in PVAs is quite crude. For example, at a recent workshop on the conservation of Humboldt penguins (*Spheniscus humboldti*), it was recognized that a major threat to the population was the number of penguins killed each year in fishing nets (Araya et al. 1999). The best we were able to do to quantify this threat in Chile was to count the number of penguin carcasses washing up on beaches in the past few years. Yet the fishing industry along the Chilean and Peruvian coasts is changing. No attempt was made to project the impacts of change, for example, by estimating the effects of a conversion from local fishing boats using small nets to major international factory fleets working farther offshore.

In PVAs on the Florida panther (*Puma concolor coryi*; Seal and Lacy 1989; Seal et al. 1992), habitat loss was crudely estimated to lie between no loss and a loss of half the panther habitat remaining on private lands. The future carrying capacity of the habitat for panthers in south Florida may be critical to the debates about whether genetic supplementation is needed and whether some panthers should be translocated to northern Florida or southern Georgia (Maehr et al., chap. 14 in this volume). Extensive data are available for projecting changing land use by humans (e.g., Pearlstine et al. 1995), and the preferences of panthers for various landscape features have been documented (Maehr and Cox 1995). Yet PVAs to date only minimally attempt to project the impacts of human population growth, development, and rapidly changing land-use patterns. Maehr et al. (chap. 14 in this volume) describe briefly how data on changing land use could be used to inform PVAs and lead to more holistic conservation plans for the Florida panther.

It is perhaps due to the narrowness of expertise among those who develop and use PVA that detailed and sophisticated analyses of factors such as variation in age-specific fecundity rates under different population densities are included, whereas questions such as how many hectares of habitat will remain or how many animals will be killed by poachers in the next decade are not considered. We often treat these fundamental threats in models by simply assuming that rates will remain constant. The specificity of models, the data quality demanded, and the sophistication of analysis typically all decline we as move from the purely biological factors (bottom of fig. 23.1) to the processes involving interactions with humans (top of fig. 23.1). Thus, the rigor of consideration of various factors in PVAs may be inversely related to the importance of the factors to wildlife population viability.

Nonbiologist colleagues have argued that we can do a better job of including consideration of changing human activities in PVAs, but often those voices are excluded or ignored when PVAs are applied. For example, as population biologists, we may not know how to project the likely future effects of poaching, but it might be possible to establish the social and economic determinants of rates of harvest of wild species by humans for subsistence use, local markets, and export (e.g., Thorbjarnarson and Velasco 1999). We need to broaden our conceptual model of PVA to include analyses of human systems.

The links between human and natural systems can be explicitly modeled within large multicompartment models (Costanza et al. 1993). For example, the Forest Land Oriented Resource Envisioning System (FLORES) project of the Center for International Forestry Research (CIFOR) in Indonesia is examining interactions between economics, agriculture and industry, road building, social processes, and forest biodiversity in Indonesia (Vanclay 1998). However, we believe that it will usually be more feasible and productive to develop separate models for the human dimensions and the wildlife population biology and ecology. Each of the models would then take information derived from the others as input and, in return, would contribute its output to other systems as their input. Projections from one system model might project the trends in a critical resource or process as a time series, and those trends would then define limiting factors influencing processes in the next model along a chain of causality.

Human systems that are relevant to species conservation include human demographics, economic systems, social systems, systems of governance and politics, activities of corporations involved in the major industrial activities in the area, agricultural systems, and ethical and moral systems (table 23.1). Although we cannot immediately incorporate all of these disciplines into a PVA on a specific species, we must begin to address those factors that have the most critical effects.

The impacts on wildlife viability of many of the systems listed in table 23.1 can be dissected and analyzed. First, we need to outline the plausible linkages between systems and to relate them to the primary entry points to the wildlife population biology at the top of figure 23.1. We then need to find and recruit the expertise needed to analyze each level

Table 23.1 Some Human Systems That Can Impact Viability of Wildlife
Populations

System	Process or Activity
Human demographics	Population growth rate (fertility, mortality)
	Age structure
	Distribution and movement
Economics	Subsistence practices, hunting
	Local markets
	Nonlocal markets and commodity prices
Industry	Extractive industries (timber, mining, fisheries)
	Agriculture
Geography, sociology, cultural anthropology	Urbanization
	Transportation and access
	Religion and ethical beliefs
	Cultural practices
Political system	Governance
	Landownership
	War

of the system. Initial discussions among experts from the diverse fields that study these processes will be difficult. We speak different technical languages, have different mental models, ask questions differently, and get our data from very different sources. It takes time and continuing effort to recognize what each perspective has to offer, and how to develop collaborations. Finally, we need to synthesize the understanding from multiple models into a holistic picture of the conservation issues, threats, and options.

The chain of connections between human processes in table 23.1 and the viability of a wildlife population in figure 23.1 can be broken down into four components: (1) the numbers of humans, (2) the distribution of humans, (3) the activities of those humans, and (4) the impacts of those activities on the fundamental inputs into the population biology of the species (i.e., direct exploitation and habitat degradation, reduction, and fragmentation). If we can marshal expertise to address these components—certainly no small task—then we can produce better predictive models of the viability of natural populations in a human-dominated world.

The first step in this process is to understand the changing size of human populations. Conservation biologists rightly bemoan the fact that we often have few data with which to assess wildlife demography. Yet for *Homo sapiens* there are massive databases on demographic trends. Detailed demographic data and projections for almost all nations are available from sources such as the U.S. Census Bureau (McDevitt 1998; available at http://www.census.gov/ipc/www/wp98.html) and the

United Nations Population Information Network (http://www.undp.
org/popin/). Sometimes specific conservation issues and species of concern will also require estimates of local deviations from national patterns.
Local demographic data and trends may be available from published
sources, the Internet, or governmental agencies. If not, estimating local
trends may require survey of the relevant population and application of
demographic models by experts in human populations (e.g., Ness 1997;
Stover and Kirmeyer 1997).

Knowledge of the spatial distributions of humans is necessary. For
example, a PVA was completed on the Indian rhinoceros in Jaldapara
National Park, India, to provide local park managers and national admin-
istrators with guidance on conservation strategies (Walker and Molur
1994). Near the conclusion of the PVA workshop, it was revealed by a
team of human demographers who had surveyed local villages that the
number of households surrounding the park would likely double over
the coming decade, largely as a result of people fleeing poverty in nearby
Bangladesh. Unlike the existing community of small-scale farmers, these
immigrants would be mostly landless and, therefore, heavily dependent
on the park as a source of firewood, grazing lands, and other resources.
A PVA for the rhinoceros that ignored this pending influx of people
would clearly have overlooked important threats. The wildlife biologists
lacked the knowledge necessary to make a meaningful PVA, although
we would have remained confident in the value of our forecasts had the
human demographers not revealed the inadequacy of our PVA.

Predicting changes in the distribution of people should consider geography, resource access, legal restrictions, macroeconomic forces, industry, government policy, landownership and tenure systems, and
more. Fortunately, conservationists do not have to develop such fields,
but can make use of the expertise in other disciplines. Existing models
of human dispersal are more sophisticated and based on more data than
are the models of wildlife dispersal that biologists are using in PVAs.

After knowing changes in numbers and distributions of humans, we
need to understand the activities of the people on the landscape. Trends
in agricultural practices, local economy and technology, nonlocal markets (which may spur resource exploitation far beyond what would be
useful locally or, alternatively, lure families away from harvest of local
resources), and belief systems can all change how humans exploit wildlife and affect habitat. An overview of the impacts of humans on the
environment is provided in Turner et al. (1990), while discussions of
some of the principles determining how humans utilize natural resources are available in Hanna et al. (1996), Buck (1998), and Ostrom et
al. (1999). Examples of analyses of the impacts of governmental policies,

property rights, economic policies, and value systems on biodiversity, ecosystems, and harvested species are provided in Perrings et al. (1995a,b).

Broad considerations of human systems are useful to identify the kinds of human processes that might be important, but details of human impacts in a specific habitat of concern have to be obtained from local knowledge or focused study. For example, Coomes (1995) described how national and international markets have driven extraction of forest products from the Amazon rain forest, while Coomes and Barham (1997) described the microeconomic determinants of livelihood and resource use by peoples living within the Amazonian rain forest. Clayton et al. (1997) developed a detailed spatial model of harvesting babirusa (*Babyrousa babyrussa*) in Sulawesi, Indonesia, that accounted for changing road conditions and hunters' opportunity costs, in an attempt to evaluate the viability of this endemic and endangered wild suid subjected to intense hunting pressures. Such expertise and models from disciplines outside of conservation biology can provide an understanding of linkages between human systems and their effects on wildlife populations.

EXAMPLES OF INTEGRATING HUMAN ACTIVITIES INTO PVA

Together with colleagues from other disciplines, we have experimented with methods for incorporating human dynamics into PVAs on wildlife species. We briefly describe some of these attempts.

Grizzly Bears in the Central Rockies Ecosystem of Canada

Recently, the Eastern Slopes Grizzly Bear Project (ESGBP) invited more than 60 colleagues to participate in a population and habitat viability assessment (PHVA) workshop, with facilitation by the World Conservation Union/ Species Survival Commission's (IUCN/SSC) Conservation Breeding Specialist Group (Herrero et al. 2000). The focus of the workshop was the conservation of the population of grizzly bears (*Ursus arctos horribilis*) in the central Rockies ecosystem of Canada (the area in and around Banff National Park). At the workshop, PVA projections were made first on the assumption that bear mortality would persist at about the rate observed in recent years. This assumption—that data from recent field studies provide the best estimates for future demographic rates—is the underlying basis for most PVAs. Yet it is very unlikely that mortality of grizzly bears will remain constant. Most mortality of bears in the central Canadian Rockies is due to humans, primarily legal hunting, illegal hunting, "self-defense" killing by hunters who see

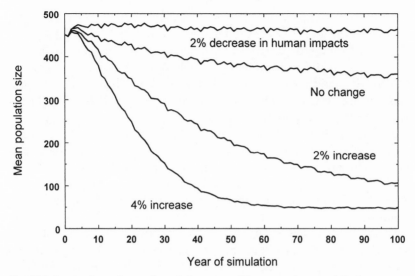

Fig. 23.2 Projected population of grizzly bears over 100 years in the central Rockies eco-system of Canada, with varying impacts of human activities on bear mortality as the local human population increases by 4% per year for the next 10 years. In the four cases, it was assumed that bear mortality would increase 4% per year, increase 2% per year, increase not at all, or decrease 2% per year.

a bear and decide to kill it, and animal-control killings by park authori-ties of bears that have become problems because of frequent encounters with humans.

At the workshop, it was projected that the human population of the Calgary area adjacent to the central Canadian Rockies would grow 4% per year. Most of this growth will be due to immigration from elsewhere in Canada. The expected social and demographic profile of the immi-grants was predominantly young people who are drawn to the area be-cause of its natural attractions. Thus, the utilization of bear habitat by people for hunting, hiking, skiing, and other activities was projected to grow at a rate at least as fast as the human population.

The projection of changes in the human population was used to mod-ify the bear PVA by assuming, first, that the mortality rate of bears due to adverse encounters with humans would increase linearly with the number of humans in the area (fig. 23.2, bottom line). This modified PVA projected that the bear population would crash rapidly due to un-sustainable killing resulting from an increased frequency of encounters between bears and people. Further analyses assumed that the rate of increase of human activities on bear mortality could be held to one-half the rate of growth in numbers of humans. The trend looked better, but mortality was still predicted at unsustainable levels. As the number of humans in the area increases, human-caused mortality must be kept as it is today, or even decreased, to maintain a viable population of grizzly bears in the region. This preliminary analysis of likely impacts of changing numbers and activities of humans is useful, as it points out that a major thrust of management must be to reduce the lethality to bears of encounters with people.

Furthermore, the number of bears killed is influenced by the percentage of each bear's home range that is within 500 m of roads and trails, the number of logging roads in the area providing access to bear habitat, how people respond when they encounter bears (i.e., shooting the bears with guns versus deterring them with pepper spray), and other factors that can be quantified and modeled. An expanded PVA can be provided by the addition of complex spatial models of human-induced alterations of grizzly bear habitat. Extensive geographic information system (GIS) data exist for the central Rockies ecosystem, such as standard vegetation classifications and spatial distributions of human land use and mean road density. Data on bear demographics and habitat use have been collected over the past 25 years (ESGBP 1998), so a significant opportunity exists for a productive synthesis of these two sets of information. However, to date these data have not been synthesized in the context of a quantitative population risk assessment.

During the course of the PHVA workshop, a team of participants led by Rick Mace, Dave Mattson, and Mark Boyce developed an algorithm for using GIS and animal telemetry data to predict grizzly bear mortality risk as a function of selected map variables. Logistic regression techniques similar to those described by Mace et al. (1999) were used to derive an expression for the relative probability that an animal would die in a specific area defined by map features, namely elevation, political jurisdiction, and mean road density. Preliminary mortality risk analyses indicate that the highest probability of grizzly bear mortality in British Columbia and Alberta occurs at lower elevations with high mean road density. While perhaps intuitive, these results nevertheless demonstrate the power of combining GIS and demographic data to derive functional relationships between the spatial characteristics of a given habitat, use of the habitat by humans, and population dynamics of wildlife.

Perhaps the greatest potential for this technique lies in its direct application to wildlife population risk projections. Given some data on how the landscape will change over time—primarily due to human use—one can incorporate future changes into a logistic model and GIS maps. If we estimate that road density will increase by 20% in the next decade, this process can identify the distribution of roads, the types of habitat they will influence, and, consequently, how the risk of grizzly bear mortality will change over time and space. This function can then be entered into a population viability model to more realistically simulate metapopulation dynamics in a landscape increasingly modified by humans.

Mountain Gorillas in Eastern Africa

The mountain gorilla (*Gorilla gorilla beringei*), totaling about 650 individuals, is restricted in distribution to two essentially isolated locations in eastern Africa: the Bwindi Impenetrable National Park in extreme southwestern Uganda, and the Virunga Volcanoes region straddling the borders between Rwanda, the Democratic Republic of the Congo (DRC), and Uganda. Large-scale civil unrest in the area that began in 1994 produced huge numbers of refugees that encroached into the periphery of Rwanda's Parc National des Volcans and the DRC's Parc National des Virungas (Plumptre et al. 1997). This movement of refugees resulted in measurable increases in gorilla mortality and loss of forest habitat, both directly through removal of biomass for fuelwood and indirectly due to the close proximity of the refugee camps to prime gorilla habitat. It also posed a significant risk of transmission of diseases into the gorilla populations. At a 1997 PHVA workshop convened by the Uganda Wildlife Authority (Werikhe et al. 1998), participants worked together to model the effects of future episodes of war and the associated displacement of refugees on the viability of the mountain gorilla population inhabiting the Virunga Volcanoes region. This effort required describing the impacts of war in terms of ecological effects (i.e., habitat quality and area, and direct exploitation), which could then be translated into gorilla population demographic impacts (i.e., mortality, fecundity, and carrying capacity). Deliberations among workshop participants, including representatives from local wildlife management agencies, resulted in estimates of major civil unrest in the area with a frequency of 30 years and duration of 10 years. The effects of war for gorillas were defined in terms of decreased female reproductive success, and increased juvenile and adult mortality during war years. In addition to the direct impact on the gorillas, the model assumed a cumulative reduction in available habitat as a result of the mass displacement of refugees during and after these events. The models also included a series of catastrophic disease events that were assumed to increase in frequency as human population density increased during and following periods of civil unrest. Estimates of disease risk were derived after lengthy deliberations among a group of wildlife veterinarians dealing specifically with mountain gorilla health issues and their application to population management.

The modeling efforts of this workshop demonstrated the significant demographic impacts that periodic war and disease can have on mountain gorilla populations—even when those events are episodic and, in some instances, infrequent (fig. 23.3). Despite our recognition of the relative simplicity of these results, they nevertheless served to improve our understanding of the connections between wildlife populations and the human populations with which they interact. Moreover, our ability to communicate these connections to those charged with managing the wildlife resource was likewise improved. This analysis helped to stimulate renewed efforts on the part of national agencies and international nongovernmental organizations to assess the actual impacts of the recent conflicts on the local gorilla populations and surrounding habitat.

Tree Kangaroos in Papua New Guinea

All six species of tree kangaroo (*Dendrolagus* spp.) in Papua New Guinea (PNG) are endemic to the island of New Guinea (Flannery et al. 1996). Population declines and habitat loss have accelerated throughout the country in the past few decades. If PNG's tree kangaroos are to survive into the 21st century, an aggressive national conservation plan must be created and implemented. Because 96% of the land in the country, and the wildlife on it, is privately owned, the success of a recent viability assessment workshop (Bonaccorso et al. 1999) was critically dependent on participation by local landowners. The information most crucial to an expanded PVA existed only in verbal form in the local community.

Researchers currently studying PNG tree kangaroos identify hunting by local villagers for the food and pet trade as perhaps the primary threat to the future viability of the species. The Conservation Breeding Specialist Group put together a team of human demographers and social scientists to conduct detailed interviews with nearly a dozen landowners from across the country; some were conducted in the villages before the workshop, and others during the meeting. Interview data were used to generate estimates of the number and size of villages in or near tree kangaroo habitat, the number of households per village, the number of hunters per household, and the number of animals removed annually per hunter. The team then attempted to estimate the annual total rate of animal extraction for a given tree kangaroo population in a given area. This rate of removal could then be added to the estimated baseline mortality to assess the direct impacts of human hunting on tree kangaroo population viability.

The models demonstrated that current hunting rates can be a factor in the viability of tree kangaroo populations in PNG, particularly when

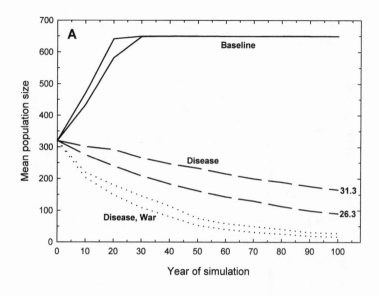

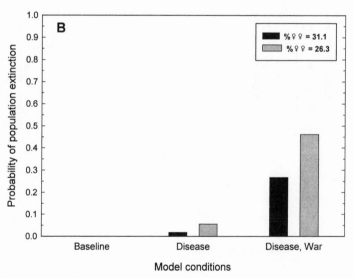

Fig. 23.3 *A*, Projections for a simulated mountain gorilla population in the Virunga Volcanoes region under baseline demographic conditions (*solid lines*), with the inclusion of catastrophic disease events (*dashed lines*), and with the addition of disease and periodic war
among human populations inhabiting the region (*dotted lines*). Each set of models consists
of a pair of projections showing alternate measures of female breeding success (31.3% or
26.3% of adults breeding in a given year) as part of a larger demographic sensitivity analysis. *B*, Probabilities of extinction of the Virunga mountain gorilla population predicted from
each of the scenarios in A.

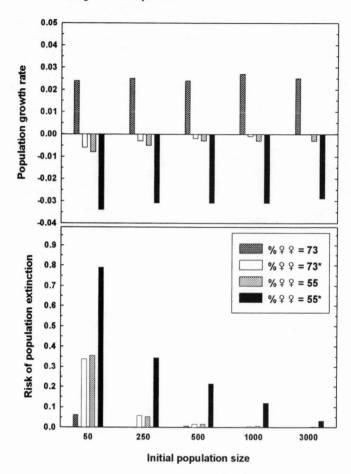

Fig. 23.4 Stochastic population growth rate (*top*) and probability of extinction (*bottom*) for simulated Papua New Guinea tree kangaroo (*Dendrolagus* spp.) populations under alternate levels of female breeding success (73% or 55% adults breeding in a given year) and additional mortality of adult females and juveniles of both sexes resulting from hunting by local human populations (models designated by * in the key).

those populations are quite small (fig. 23.4). Furthermore, workshop participants were able to observe the serious demographic consequences of preferential hunting of adult females over males in these polygynous species, an unexpected result to those engaged in hunting. Despite our interest in developing increasingly insightful PVA models, it is important to recognize that neither sophisticated demographic models nor comprehensive field data may be necessary to generate insights vital to the conservation decision-making process.

Additional focus was directed toward the country's most critically endangered species, the tenkile, or Scott's tree kangaroo (*Dendrolagus scottae*). This species has been reduced to 50 to 100 individuals in a single locale within the Torricelli Mountains of northwestern PNG. The recent precipitous population decline and imminent risk of extinction revealed by the data collection and population modeling were instrumental in the immediate formation of a multidisciplinary "rapid response team" at the workshop. This team is composed of representatives from zoo and wildlife management organizations in PNG and Australia who traveled to the Torricellis to update the conservation status of this rare taxon. As a result of the PVA workshop and subsequent work of this team, a two-year moratorium on hunting the species has been signed by representatives of all 13 local villages.

CONCLUSIONS

The preceding examples show how PVA can provide insights into the dynamics of interactions between human systems and wildlife viability and, therefore, can become more useful for conservation. Although there are examples in the literature and within this volume of PVAs that incorporate information from models of human systems, we have just begun to explore how best to integrate models of human systems with habitats, ecosystems, and wildlife populations. We are convinced that such integration is essential to conservation solutions that will succeed in an increasingly human-dominated world. Rather than being the exception, we hope that this kind of thinking will become routine. PVAs need to treat the dynamics of human populations and human activities as seriously as we treat the dynamics of wildlife population biology.

Even with access to a wide array of expertise, we are not likely to be able to develop holistic models that provide complete understanding of all relevant systems. We cannot do that even for population biology, yet we recognize that there is value to specifying, analyzing, and utilizing the knowledge we do have. To the extent we can expand our understanding of the human as well as natural forces affecting viability, we will be able to more effectively conserve biodiversity.

LITERATURE CITED

Araya, B., D. Garland, G. Espinoza, A. Sanhuesa, R. Lacy, A. Teare, and S. Ellis, editors. 1999. *Taller análisis de la viabilidad del hábitat y de la población del pingüino Humboldt (Spheniscus humboldti): Borrador del informe.* Conservation Breeding Specialist Group, IUCN/SSC, Apple Valley, Minnesota.

Asian Rhino Specialist Group (AsRSG). 1996. Report of a meeting of the IUCN SSC Asian Rhino Specialist Group. Sandakan, Malaysia, November 29–December 1, 1995.

Birch, L. C. 1948. The intrinsic rate of natural increase of an insect population. *Journal of Animal Ecology* 17:15–26.

Bonaccorso, F., P. Clark, P. S. Miller, and O. Byers. 1999. *Conservation assessment and management plan for the tree kangaroos of Papua New Guinea and population and habitat viability assessment for Matschie's tree kangaroo.* Conservation Breeding Specialist Group, IUCN/SSC, Apple Valley, Minnesota.

Boyce, M. S. 1992. Population viability analysis. *Annual Review of Ecology and Systematics* 23:481–506.

Buck, S. J. 1998. *The global commons: an introduction.* Island Press, Covelo, California.

Burgman, M. A., and B. B. Lamont. 1992. A stochastic model for the viability of *Banksia cuneata* populations: environmental, demographic, and genetic effects. *Journal of Applied Ecology* 29:719–727.

Caughley, G. 1994. Directions in conservation biology. *Journal of Animal Ecology* 63:215–244.

Clayton, L., M. Keeling, and E. J. Milner-Gulland. 1997. Bringing home the bacon: a spatial model of wild pig hunting in Sulawesi, Indonesia. *Ecological Applications* 7:642–652.

Coomes, O. T. 1995. A century of rain forest use in western Amazonia: lessons for extraction-based conservation of tropical forest resources. *Forest and Conservation History* 39:108–120.

Coomes, O. T., and B. L. Barham. 1997. Rain forest extraction and conservation in Amazonia. *Geographical Journal* 163:180–188.

Costanza, R., L. Wainger, C. Folke, and K.-G. Mäler. 1993. Modeling complex ecological economic systems. *BioScience* 43:545–555.

Deevey, E. S., Jr. 1947. Life tables for natural populations of animals. *Quarterly Review of Biology* 22:283–314.

Eastern Slopes Grizzly Bear Project (ESGBP). 1998. Grizzly bear population and habitat status in Kakanaskis country, Alberta. Report to the Department of Environmental Protection, Natural Resources Service, Alberta. Prepared by ESGBP, University of Calgary, Calgary, Alberta.

Ellis, S., K. Hughes, C. Kuehler, R. C. Lacy, and U. S. Seal. 1992. *'Alala, akohekohe, and palila population and habitat viability assessment reports.* Captive Breeding Specialist Group, IUCN/SSC, Apple Valley, Minnesota.

Fisher, A., E. Rominger, P. Miller, and O. Byers. 1999. *Population and habitat viability assessment workshop for the desert bighorn sheep of New Mexico (Ovis canadensis).* Conservation Breeding Specialist Group, IUCN/SSC, Apple Valley, Minnesota.

Flannery, T. F., R. Martin, and A. Szalay. 1996. *Tree kangaroos: a curious natural history.* Reed Books, Melbourne, Australia.

Franklin, I. R. 1980. Evolutionary change in small populations. Pages 135–149 in M. E. Soulé and B. A. Wilcox, editors, *Conservation biology: an evolutionary-ecological perspective.* Sinauer Associates, Sunderland, Massachusetts.

Gilpin, M. E., and M. E. Soulé. 1986. Minimum viable populations: processes of species extinction. Pages 19–34 in M. E. Soulé, editor, *Conservation biology: the science of scarcity and diversity.* Sinauer Associates, Sunderland, Massachusetts.

Groom, M. J. 1998. Allee effects limit population viability of an annual plant. *American Naturalist* 151:487-496.

Groom, M. J., and M. A. Pascual. 1998. The analysis of population persistence: an outlook on the practice of viability analysis. Pages 4–27 in P. L. Fiedler and P. M. Kareiva, editors, *Conservation biology for the coming decade.* 2d edition. Chapman and Hall, New York, New York.

Hanna, S. S., C. Folke, and K.-G. Mäler, editors. 1996. *Rights to nature: ecological, economic, cultural, and political principles of institutions for the environment.* Island Press, Covelo, California.

Hedrick, P. W., R. C. Lacy, F. W. Allendorf, and M. E. Soulé. 1996. Directions in conservation biology: comments on Caughley. *Conservation Biology* 10:1312–1320.

Herrero, S., P. S. Miller, and U. S. Seal. 2000. *Population and habitat viability assessment (PHVA) workshop for the grizzly bear of the central Rockies ecosystem (Ursus arctos horribilis).* Conservation Breeding Specialist Group, IUCN/SSC, Apple Valley, Minnesota.

Lacy, R. C. 1993/1994. What is population (and habitat) viability analysis? *Primate Conservation* 14/15:27–33.

———. 1997. Importance of genetic variation to the viability of mammalian populations. *Journal of Mammalogy* 78:320–335.

———. 2000. Considering threats to the viability of small populations. *Ecological Bulletins* 48:39–51.

Leslie, P. H. 1945. On the use of matrices in certain population mathematics. *Biometrika* 33:183–212.

Mace, R. D., J. S. Waller, T. L. Manley, K. Ake, and W. T. Wittinger. 1999. Landscape evaluation of grizzly bear habitat in western Montana. *Conservation Biology* 13:367–377.

Maehr, D. S., and J. A. Cox. 1995. Landscape features and panthers in Florida. *Conservation Biology* 9:1008–1019.

McDevitt, T. M. 1998. *World population profile: 1998.* U.S. Bureau of the Census, Report WP/98. U.S. Government Printing Office, Washington, D.C.

Mills, L. S., C. Baldwin, M. J. Wisdom, J. Citta, D. J. Mattson, and K. Murphy. 1996. Factors leading to different viability predictions for a grizzly bear data set. *Conservation Biology* 10:863–873.

Ness, G. D. 1997. *Population and strategies for national sustainable development.* Earthscan Publications, London, United Kingdom.

Ostrom, E., J. Burger, C. B. Field, R. B. Norgaard, and D. Polikansky. 1999. Revisiting the commons: local lessons, global challenges. *Science* 284:278–282.

Pearlstine, L. G., L. A. Brandt, W. M. Kitchens, and F. J. Mazzotti. 1995. Impacts of citrus development on habitats of southwest Florida. *Conservation Biology* 9: 1020–1032.

Perrings, C. A., K.-G. Mäler, C. Folke, C. S. Hollings, and B.-O. Jansson, editors. 1995a. *Biodiversity conservation: problems and policies.* Kluwer Academic Publishers, Dordrecht, Netherlands.

———. 1995b. *Biodiversity loss: economic and ecological issues.* Cambridge University Press, Cambridge, United Kingdom.

Plumptre, A. J., J.-B. Bizumuremyi, F. Uwimana, and J.-D. Ndaruhebeye. 1997. The effects of the Rwandan civil war on poaching of ungulates in the Parc National des Volcans. *Oryx* 31:265–273.

Saltz, D. 1996. Minimizing extinction probability due to demographic stochasticity in a reintroduced herd of Persian fallow deer *Dama dama mesopotamica. Biological Conservation* 75:27–33.

Seal, U. S., and R. C. Lacy. 1989. Florida panther population viability analysis. Report to the U.S. Fish and Wildlife Service. Captive Breeding Specialist Group, IUCN/SSC, Apple Valley, Minnesota.

Seal, U. S., et al. 1992. Genetic management strategies and population viability of the Florida panther (*Felis concolor coryi*). Report to the U.S. Fish and Wildlife Service. Captive Breeding Specialist Group, IUCN/SSC, Apple Valley, Minnesota.

Shaffer, M. L. 1981. Minimum population sizes for species conservation. *BioScience* 31:131–134.

Simberloff, D. 1988. The contribution of population and community biology to conservation science. *Annual Review of Ecology and Systematics* 19:473–511.

Soulé, M. E., editor, 1987. *Viable populations for conservation.* Cambridge University Press, Cambridge, United Kingdom.

Stover, J., and S. Kirmeyer. 1997. DemProj: a computer program for making population projections: version 4 user's manual. Policy Project, Washington, D.C.

Thorbjarnarson, J., and A. Velasco. 1999. Economic incentives for management of Venezuelan caimans. *Conservation Biology* 13:397–406.

Turner, B. L., II, W. C. Clark, R. W. Kates, J. F. Richards, J. T. Mathews, and W. B. Meyer, editors. 1990. *The earth as transformed by human action: global and regional changes in the biosphere over the past 300 years.* Cambridge University Press, Cambridge, United Kingdom.

Vanclay, J. K. 1998. FLORES: for exploring land use options in forested landscapes. *Agroforestry Forum* 9:47–52.

Walker, S., and S. Molur. 1994. Population and habitat viability analysis (PHVA) workshop for Indian/Nepali rhinoceros. Zoo Outreach Organisation/Conservation Breeding Specialist Group, Coimbatore, India.

Werikhe, S., L. Macfie, N. Rosen, and P. S. Miller. 1998. Can the mountain gorilla survive? Population and habitat viability assessment workshop for *Gorilla gorilla beringei.* Conservation Breeding Specialist Group, IUCN/SSC, Apple Valley, Minnesota.

24 Fitting Population Viability Analysis into Adaptive Management

Donald Ludwig and Carl J. Walters

ABSTRACT

Adaptive management and population viability analysis are commonly advocated as tools for preservation of biodiversity, but each technique has severe technical limitations. Population viability analysis suffers from a requirement for detailed and accurate knowledge in the face of numerous complications due to biological complexity, the complexity of human responses and motivations, and the inherent complexity of the problem of inference. Adaptive management requires evaluation of the future effects of policies over a high-dimensional state space. Moreover, the large risks associated with experimental policies may often preclude their adoption.

The limitations of these methods imply that they cannot offer a simple, technical solution to the difficult problem of conservation of biodiversity. However, they are important tools that help us organize our thinking and recognize the limitations of our present knowledge and attitudes. We should reexamine the attitudes that have led us to overstress technical and economic approaches to important decisions.

INTRODUCTION

Population viability analysis (PVA; Shaffer 1981; Gilpin and Soulé 1986; Soulé 1987; Simberloff 1988; Boyce 1992; Burgman et al. 1993) and adaptive management (AM; Holling 1978; Walters 1986, 1997; Halbert 1993; Gunderson et al. 1995; Taylor et al. 1997) are often advocated as tools to help preserve biodiversity. Each method has attracted enthusiastic proponents who are uninitiated into the technical difficulties in their implementation. Can these two ideas be combined to help in the difficult task of preserving the world's species and ecosystems? The appropriateness and success of the combination will depend on the ways that the two concepts are defined and implemented.

We believe that AM can introduce a much-needed idea of uncertainty into PVA. At a minimum, it may help us to understand why a policy of "no surprises" in endangered-species legislation is potentially extremely destructive, and why legislation that mandates unequivocal scientific advice to guide policy decisions is unworkable. However, we

511

don't think that experimental approaches to management of endangered species—even if feasible—will be acceptable on either political or economic grounds. Some of the basic attitudes that underlie AM are more appropriate than the scientistic and economistic approach that is currently so popular. However, a technocratic application of AM and PVA is likely to fail, just as it fails wherever ecological and environmental issues are paramount.

HISTORY OF ADAPTIVE MANAGEMENT

The idea of adaptive management emerged in 1974 when the team of Buzz Holling, Carl Walters, and Ray Hilborn visited the International Institute for Applied Systems Analysis (IIASA) headed by Howard Raiffa. Raiffa was a great proponent of statistical decision theory for business problems (Raiffa and Schlaifer 1961; Raiffa 1968). He helped stimulate developments in statistics that resulted in Bayesian decision theory (Chernoff and Moses 1959; Berger 1985; Lindley 1985; Morgan and Dowlatabadi 1996). The starting point of the Bayesian theory is recognition and characterization of our uncertainty. This is then combined with projections of the likely effects of proposed policies on the system itself and on our knowledge of the system. It requires us to compute the value of acquiring new information and to use it in assessing the merits of various policies.

Holling was (and still is) an entomologist. He realized that problems of managing the spruce budworm were not merely problems of pest control or forest management, but that the management in fact involved the whole social system of Atlantic Canada. For Holling then and now, problems of resource management are problems of human perceptions and human society. Changes in the "managed system" cannot be separated from changes in the human society: each is managing the other. Similar, but perhaps not such encompassing, ideas are often termed "holism."

Holling and his collaborators naturally thought of experimentation as the most reliable means of reaching an understanding of resource systems. When this idea is combined with Raiffa's idea of assessing plans on the basis of what we can learn from implementing them, it follows that management should rely on experimental design and monitoring in dealing with resource systems. To attempt to include all of the essential components, Holling devised workshops where experts and practitioners from very diverse fields could plan the experiments. Computers, though quite rudimentary by present standards, were an important tool for communication among the diverse disciplines. The resulting models were never intended to have much predictive value. Their main merit was their suggestiveness: what were the minimum ingredients to obtain the gross qualitative behavior? Would this behavior be displayed in more realistic elaborations? What are the critical areas of uncertainty that we should seek to clarify by means of well-designed experiments?

What sets this approach apart from more conventional experimental management is the idea of experimenting on the system as a whole, rather than attempting to analyze the individual parts. We will never be able to understand the individual parts well enough to synthesize the overall system behavior. Moreover, whole ecosystem experiments may be the only feasible ones. Although one can point to many successes of reductionist science, past experience may not be a reliable guide to future problems. The choice between the reductionist and holistic approaches is primarily a matter of faith.

Experience in implementing AM is surveyed in Halbert (1993), Taylor et al. (1997), and Walters (1997). It is fair to say that the ratio of action to talk about AM is extremely low (Boyce, chap. 3 in this volume). There are many possible factors that impede the implementation of AM. It is difficult to communicate such a broad and novel vision beyond the circle of those involved in its conception. Perhaps more important, experiments are time-consuming and often extremely expensive. The expense must be measured not only in direct expenditures, but in foregone opportunities. The costs are not only financial direct costs, but social and political costs. If we shut down a fishery to see if the stock will increase in abundance, hundreds or thousands of people may lose employment. It is very difficult to balance a certain present loss against a merely possible or plausible future gain or loss, based mainly on the hunches of a few experts.

In spite of the lack of actual implementations of AM, it has become a very popular goal of management agencies. Perhaps it is used as a defensive measure by bureaucrats anxious to create an image of change without actually changing very much. Perhaps it is appealing because the alternatives seem so bleak. Boyce (1992, 1997), Lindenmayer et al. (1993), and Beissinger and Westphal (1998) advocated AM in conjunction with PVA. However, they supplied few details on how it might be implemented. Thus, their recommendations stand more as a gesture of support than as a solid basis for further work.

TECHNIQUES FOR IMPLEMENTING ADAPTIVE MANAGEMENT

To understand some of the technical difficulties in the way of an implementation of AM, it may be helpful to consider a simple case (Ludwig and Walters 1982). This study, though 17 years old, may be one of the few for which the theory has been worked out.

Knowledge as a State Variable

In light of Raiffa's insight that we must consider our own knowledge as part of the system to be investigated, we must include that knowledge as part of the collection of state variables. Ludwig and Walters (1982) considered a fishery model with a single variable (stock size) to represent the biological system. We postulated a Ricker stock-recruitment relationship that has two unknown parameters. Their expected values (or maximum likelihood estimates) must be added to the list of state variables. The tentative total is three, including the stock size. But expected values are not sufficient to describe our state of knowledge; we must include variables to describe our uncertainty about these variables. At a minimum, this includes a parameter covariance matrix. This matrix is symmetric and of order two. Hence there are three additional variables that must be included. The total is now six state variables. If the number of ecological state variables is n and the number of parameters is p, the total number of state variables is $n + p + p\,(p + 1)/2$. For example, if $n = 5$ and $p = 10$, then the total number of state variables is 70. The number of state variables is approximately proportional to p^2 if p is large.

The Curse of Dimensionality

Up to this point the problem seems large, but well within the capability of today's computers. Computers can easily handle systems of hundreds of linear equations; nonlinear equations may pose more difficulties. To devise effective strategies, we must consider the effects of present actions on future states. We must evaluate the consequences over the state space for each action being considered. Since we can't predict the future, we must weight each possible outcome by the probability of its occurrence. This involves evaluation of the consequences over a grid in the state space. A fair resolution might be 10 points in each direction. This would not be enough for high accuracy, but at least might capture the qualitative behavior. If the dimension of the state space is 6, then we need approximately 10^6 evaluations per action. If the possible actions can be described by a single variable, with 10 grid points, then the total number of evaluations might be 10^7. For a more realistic problem with 70 state variables (see above), the grid in state space would have 10^{70} points, and the grid in action space might have 100 points. The total number of evaluations is now at 10^{72}. As a comparison, the number of electrons in the universe is on the order of 10^{30}.

Richard Bellman, the great popularizer of dynamic programming, termed this growth in computational burden the "curse of dimensionality." The situation is actually a bit worse than this for PVA, since each "evaluation" might involve a calculation of an extinction probability.

Each calculation of an extinction probability might be quite computer intensive, and Monte Carlo simulation methods are commonly employed. In our fisheries example, we had the benefit of an efficient root-finding method for the evaluation. We also took a number of numerical shortcuts to reduce the number of calculations to the capabilities of a mainframe computer in 1980. There are many ways to reduce the computational burden far below the estimates we have just given. There are methods for representing the state of our present information in terms of a few million points, even for problems of high dimensionality. The fact remains, however, that efficient implementation of AM will require great insight to reduce the problem to its essentials and great ingenuity to make the calculations feasible.

Social, Political, and Economic Feasibility

Aside from the computational difficulties in finding an effective AM plan, under what circumstances is experimental management appropriate? An experiment always involves some sort of risk. In the case of fisheries, if harvesting is drastically curtailed, there is the risk of social and economic disruption with dire political consequences. If harvests are drastically increased, there is the risk of stock collapse. That risk always exists, but the manager is less likely to be blamed if management policy (however misguided) has not changed much recently. There are analogous risks for management of threatened or endangered species or ecosystems. In view of the large risks, there must be some prospect of substantial gains if experiments are undertaken.

Theory (or a bit of reflection) shows that timid experimental actions are never justified. In ecological experiments, it is commonly appreciated that, if the treatment differs little from the control, it is unlikely that the results will be capable of showing any effect of the treatment, and consequently the whole effort is wasted. But we have seen that drastic treatments involve large risks. They may nevertheless be justified if a relatively short experiment has a good chance of producing a substantial amount of information. By a "substantial amount of information," we mean that the new information makes it clear that a change in current policy is desirable, and that the new policy makes a substantial improvement in the performance of the system.

In view of this long list of requirements, it is not surprising that experimental management is seldom applied. This does not imply that it is never justified, but the various costs and benefits must be carefully appraised. An additional complication is that, when economic criteria are employed, there is a tendency to apply standard economic discount factors to future benefits. The propriety of doing this has been hotly debated; for an introduction see Heal (1997). It is fair to say that economic criteria will seldom favor experimental management or indeed any sort of management for the long term.

FUTURE PROSPECTS FOR PVA AND AM

In light of the many technical and practical difficulties in implementing AM, it is not likely that the full-blown method will be applied very frequently. Likewise, sufficient data to implement a full-blown PVA are unlikely to be available (Beissinger and Westphal 1998; Ludwig 1999; Ralls et al., chap. 25 in this volume). That does not imply that the underlying ideas and methods have no use. The difficulties in implementing what might appear to be the most straightforward approaches force us to reappraise our requirements and expectations.

PVA Is a Forecasting Problem

Difficulties in implementing PVA suggest that we look at other comparable situations and learn from experience. Are there other cases where clear scientific conclusions are difficult or impossible to reach with available data and methods, but that have important practical consequences? A good parallel is the case of weather prediction (Beissinger and Westphal 1998). Weather prediction is of great practical importance, as events of the past few years have convincingly demonstrated. The theory underlying weather prediction is fairly well worked out. It involves fluid dynamics and thermodynamics, as well as substantial solar inputs and topographical complications. Weather cannot be predicted more than a few days in advance because the longer you want to predict, the more detailed and accurate your worldwide data must be. The underlying dynamics are chaotic, so that the requirements for accuracy and detail increase extremely sharply as the desired prediction interval lengthens.

Another case of prediction that is not very accurate is economic forecasting. In spite of the obvious importance of economic forecasts and the enormous volumes of data that are available, the record for economic forecasts is poor. We suspect that part of the reason is that the underlying theories are too simple: they fail to take into account psychological and social motivations for human behavior. Of course, theories that do attempt to take these complexities into account appear to be unworkable—at least that is what the economists tell us (for a discussion, see Scitovsky 1976).

In spite of their inadequacies as precise predictors, weather and economic forecasting are extremely useful—in part because we recognize that they are not very accurate and probably cannot be made more accurate. The prudent conclusion is that we should be cautious in making decisions whose outcomes are sensitive to the weather or to general economic conditions. It is not difficult to extrapolate to the case of threatened or endangered species and ecosystems. We cannot predict very well what the outcome of various policies will be, and therefore we should use great caution in undertaking policies that may be harmful. Steps in that direction are suggested by Goodman (chap. 21 in this volume) and Ralls et al. (chap. 25 in this volume). Goodman advocates a Bayesian approach to calculations of risks. Such calculations appear in Ludwig (1996). The effect of the Bayesian approach is to increase the estimated risks by orders of magnitude over the "best estimate" obtained from least-squares fitting. Caution should be exercised in such cases, since the prior distribution that is assumed for parameters is very influential. This leaves great latitude for disagreements about such prior assumptions. In any case, the more difficulty we have in predicting, the greater should be the caution that we exercise. This is a truly important conclusion from work on PVA.

One might argue that public policy will not be responsive to scientific input unless it is delivered with an air of authority, even at the risk of making firm assertions that later turn out to have been ill founded. This objection has lost a great deal of its force in light of the public debate on issues of global climate change. Prediction of climate is quite a different science than weather prediction because there is no attempt to predict specific events at specific times. Instead, considerations of gross energy balance or gross carbon balance in the earth's atmosphere dominate. Hence, scientists can state with a fair amount of certainty that increases in human consumption of fossil fuels have altered the earth's climate, and these effects will amplify in the future. There is still plenty of room for scientific disagreement about climate predictions and the role of humans in influencing the climate. It is clear that these disagreements will continue, but they do not prevent implementation of political responses to such predictions. Within the last few decades, starting with the debate on acid rain, governments have at least paid lip service to the necessity for care about how we use the atmosphere. The opinions of atmospheric scientists carry greater weight now because they were hesitant to make premature judgments. Conservation ecologists can achieve similar credibility if their claims are properly qualified in the light of numerous uncertainties.

AM Can Make a Difference

If one takes an appropriately modest view of the contributions that PVA can make, what becomes of AM? For the reasons given above, we doubt that full assessment and experimentation will be carried out. But that is not the only contribution that AM can make. The idea of exploring various alternatives and trying to find policies that are robust under the many uncertainties is still as compelling as ever. Guided by Holling's insight, perhaps we can recognize that our problems of management are not merely external ones, but deeply involve our attitudes toward each other and toward the rest of the world.

Having recognized the limitations of high-tech and computationally intensive approaches to management, the question arises, Why did such ideas ever seem plausible? The idea of making important decisions by some sort of mathematical calculation goes back to the Enlightenment of the 17th and 18th centuries (Daston 1988). The tradition continues in modern decision theory, and it is firmly entrenched in economic theory. There is nothing wrong and much to be gained from such techniques— the harm comes from surrendering our judgment to mechanical calculations. Philosophers call the attempt to extend scientific methods and reasoning to all facets of human culture "scientism." An attempt to reduce all values and judgments to economics is called "economism." A similar attitude was called "the arrogance of humanism" by Ehrenfeld (1981). These attitudes are characteristic of our times. In retrospect, it is clear that they have done enormous harm, and they are capable of harm on an even larger scale if they are not modified.

Perhaps it is time to reassess many of our entrenched ideas. Why was it a goal for so many years to find "maximum sustained yields"? Most of us now realize that sustained yields are seldom achieved, and when they are achieved, they are achieved only by restraint. Why should we attempt to find a "minimum viable population"? Why should we try to find the smallest possible protected areas? Why should we try to guarantee "no surprises" when making plans for extended periods? What scientific evidence is there that there are thresholds of habitat destruction below which no harm is done? Such concepts are convenient political fictions that have no place in scientific investigations or prescriptions.

LITERATURE CITED

Beissinger, S. R., and M. I. Westphal. 1998. On the use of demographic models of population viability in endangered species management. *Journal of Wildlife Management* 62:821–841.

Berger, J. O. 1985. *Statistical decision theory and Bayesian analysis*. Springer-Verlag, New York, New York.

Boyce, M. 1992. Population viability analysis. *Annual Review of Ecology and Systematics* 23:481–506.

Boyce, M. 1997. Population viability analysis: adaptive management for threatened and endangered species. Pages 226–236 in M. S. Boyce and A. Haney, editors, *Ecosystem management: applications for sustainable forest and wildlife resources.* Yale University Press, New Haven, Connecticut.

Burgman, M. A., S. Ferson, and H. R. Akçakaya. 1993. *Risk assessment in conservation biology.* Chapman and Hall, London, United Kingdom.

Chernoff, H., and L. E. Moses. 1959. *Elementary decision theory.* John Wiley and Sons, New York, New York. Reprint, Dover Publications, New York, New York, 1986.

Daston, L. 1988. *Classical probability in the enlightenment.* Princeton University Press, Princeton, New Jersey.

Ehrenfeld, D. 1981. *The arrogance of humanism.* Oxford University Press, New York, New York.

Gilpin, M. E., and M. E. Soulé. 1986. Minimum viable populations: processes of species extinction. Pages 19–34 in M. E. Soulé, editor, *Conservation biology: the science of scarcity and diversity.* Sinauer Associates, Sunderland, Massachusetts.

Gunderson, L. H., C. S. Holling, and S. S. Light. 1995. *Barriers and bridges to the renewal of ecosystems and institutions.* Columbia University Press, New York, New York.

Halbert, C. L. 1993. How adaptive is adaptive management? *Reviews in Fish Biology and Fisheries* 1:261–283.

Heal, G. 1997. Discounting and climate change. *Climatic Change* 37:335–343.

Holling, C. S. 1978. *Adaptive environmental assessment and management.* John Wiley and Sons, New York, New York.

Lindenmayer, D. B., T. W. Clark, R. C. Lacy, and V. C. Thomas. 1993. Population viability analysis as a tool in wildlife conservation policy: with reference to Australia. *Environmental Management* 17:745–758.

Lindley, D. V. 1985. *Making decisions.* 2d edition. John Wiley and Sons, New York, New York.

Ludwig, D. 1996. Uncertainty and the assessment of extinction probabilities. *Ecological Applications* 6:1067–1076.

———. 1999. Is it meaningful to estimate a probability of extinction? *Ecology* 80: 298–310.

Ludwig, D., and C. J. Walters. 1982. Optimal harvesting with imprecise parameter estimates. *Ecological Modelling* 14:273–292.

Morgan, M. G., and H. Dowlatabadi. 1996. Learning from integrated assessment of climate change. *Climatic Change* 34:337–368.

Raiffa, H. 1968. *Decision analysis: introductory lectures on choices under uncertainty.* Addison Wesley, Reading, Massachusetts.

Raiffa, H., and R. Schlaifer. 1961. *Applied statistical decision theory.* Division of Research, Graduate School of Business Administration, Harvard University, Boston, Massachusetts.

Scitovsky, T. 1976. *The joyless economy: an inquiry into human satisfaction and consumer dissatisfaction.* Oxford University Press, New York, New York.

Shaffer, M. L. 1981. Minimum population sizes for species conservation. *BioScience* 31:131–134.

Simberloff, D. 1988. The contribution of population and community biology to conservation science. *Annual Review of Ecology and Systematics* 19:473–511.

Soulé, M. E., editor. 1987. *Viable populations for conservation.* Cambridge University Press, Cambridge, United Kingdom.

Taylor, B., L. Kremsater, and R. Ellis. 1997. *Adaptive management of forests in British Columbia.* Ministry of Forests, Forest Practices Branch, Victoria, British Columbia.

Walters, C. J. 1986. *Adaptive management of renewable resources.* Macmillan, New York, New York.

———. 1997. Challenges in adaptive management of riparian and coastal ecosystems. *Conservation Ecology* 1 (1). http://www.consecol.org/vol1/iss2/art1.

25

Guidelines for Using Population Viability Analysis in Endangered-Species Management

Katherine Ralls, Steven R. Beissinger, and Jean Fitts Cochrane

ABSTRACT

Biologists use a variety of population viability analysis (PVA) models to interpret past population trends, evaluate likely threats to a population, and project future population trends. PVA is widely used to gain a better understanding of the population biology of an endangered species, and to estimate the relative risks and conservation values of alternative management options. However, there are no consensus guidelines on why, when, or how PVA should be used in endangered-species management. Moreover, the population models used in PVA are not the only useful approach to recovery planning. We describe PVA and alternative approaches to gain understanding, estimate risk, and make decisions. We stress that it is essential to define objectives before choosing a specific method for addressing a management problem. We then provide guidelines to help biologists (1) decide whether a PVA would be helpful for a given species, (2) choose and develop an appropriate model, (3) conduct the analysis, and (4) interpret the results. We also develop criteria for judging the quality of a completed PVA. Although many PVAs are conducted during brief workshops and never revisited, we argue that PVA is more useful as a long-term, iterative process coupled with an adaptive management approach to species recovery.

INTRODUCTION

Not all scientists have agreed on a definition of population viability analysis (PVA), a consensus approach for conducting these analyses, or standards for judging the quality of a completed PVA. This situation is quite normal and familiar to scientists. It is healthy and productive for scientists in any rapidly developing field to have a variety of opinions, use a variety of approaches, and criticize each other's work. However, this aspect of science is confusing to many people involved with endangered-

We appreciate helpful suggestions by M. Boyce, R. Lande, D. McCullough, M. Reed, and A. Starfield. R. Akçakaya, S. Andelman, M. Burgman, H. Possingham, and P. Sjögren-Gulve kindly contributed unpublished manuscripts.

species management, including decision makers, biologists, politicians, and judges. They need guidance on when and how to use PVA, and how to judge the quality of a completed PVA (e.g., Warshall 1994). The greater the degree of consensus on advice that conservation biologists give, the more influence we will have on policy and management. After more than a decade of experience with PVA, we believe that it is time for scientists to provide consensus guidelines on the use of PVA for endangered-species management. These guidelines will have to be periodically updated to incorporate advances in our understanding of the extinction process and improved capabilities for modeling populations.

This chapter is a first step toward the development of such guidelines. We suggest a working definition of PVA, discuss the use of PVA and alternative approaches with respect to common management objectives for recovering endangered species, and offer guidelines for evaluating a completed PVA. Our aim is to synthesize and promote consensus rather than provide original insights. Thus, we draw heavily on recent reviews and position papers on PVA to guide our analysis (Boyce 1992; Ralls and Taylor 1997; Beissinger and Westphal 1998; Groom and Pascual 1998; Reed et al. 1998, in press; Morris et al. 1999; Akçakaya and Sjögren-Gulve 2000; Burgman and Possingham 2000).

WHAT IS A PVA?

In the preface and first chapter of this book, the editors discuss the lack of standard definition of what makes up a PVA. Before we can make recommendations on whether and how to use PVA, or can develop criteria for judging the quality of a PVA, we need a working definition of PVA. This book has considered PVA in the broad sense. It includes many different kinds of analyses that can provide insight into the viability of a population, ranging from models that estimate rate of population growth or extinction (e.g., Maehr et al., chap. 14 in this volume; Mills and Lindberg, chap. 16 in this volume; Possingham et al., chap. 22 in this volume) to qualitative analyses that depend primarily upon expert opinion, discussed by Samson (chap. 20 in this volume) and later in this chapter. PVA has even been used to describe workshops that have attempted to examine all threats affecting a population and collate data. A broad definition of PVA would create a mix of apples and oranges that would contribute little to solving the pressing issues of when and how to use PVA wisely. Properly defining a problem is the first step toward its resolution.

The term "population viability analysis" was first used to describe quantitative modeling exercises that estimated the risk of extinction within a specified time period (Gilpin and Soulé 1986). The output of a probabilistic estimate of the risk of extinction from a stochastic model differentiated PVA from earlier deterministic models that had been used to guide the recovery of endangered species (Beissinger, chap. 1 in this volume). We believe it useful to limit the term to similar modeling efforts.

Therefore, we define PVA as an analysis that uses data in an analytical or simulation model to calculate the risk of extinction or a closely related measure of population viability, such as the proportion of simulated populations that end above some size after some specified period of time. Reed et al. (in press) have proposed a similar definition for PVA. Restricting the term PVA by directly connecting it to a model that estimates the risk of extinction is necessary to differentiate PVA from the many other kinds of analyses that do not quantitatively or directly estimate such a rate.

By restricting our definition of PVA, we have excluded different kinds of analysis that can provide important insights on the risk or vulnerability of a population, which might complement a PVA model. Our definition of PVA excludes qualitative assessments based on verbal models (e.g., Ruggiero et al. 1994) or expert opinion (e.g., FEMAT 1993), genetic models that attempt to estimate effective population size (Waples, chap. 8 in this volume), and demographic models focused on questions other than extinction risk, such as analyses of population growth rate (e.g., Caswell 2001; Mills and Lindberg, chap. 16 in this volume). These approaches may be useful and important analyses to apply to evaluate conservation problems. For example, population genetic models can estimate the potential for loss of genetic diversity (Allendorf and Ryman, chap. 4 in this volume; Haig and Ballou, chap. 18 in this volume; Waples, chap. 8 in this volume) but do not make direct demographic projections of the rate of extinction. Such models can, however, be used to develop genetic goals for the acceptable loss of heterogeneity (Allendorf and Ryman, chap. 4 in this volume) that can be used in conjunction with PVA models that estimate extinction rate. Furthermore, most of these techniques are fundamentally different from PVA in their approach and data requirements. Adopting a broad definition of PVA that included these other kinds of analyses would require us to develop a very different and massive set of guidelines that were specifically tailored to each type of analysis. Finally, a useful assessment of viability for a species or for hundreds of species as part of a regional plan would likely use many kinds of analyses, including PVA, to evaluate management alternatives.

WHEN SHOULD WE MODEL POPULATION VIABILITY?

When should PVA be used to make decisions for managing endangered species? Because PVA involves modeling, we can gain some insight into this question by considering the circumstances under which building a formal model is useful. Two conditions must be met before predictions from quantitative models are likely to be helpful. First, an objective should be stated that could be met by constructing a model. For instance, we might want to know what pattern of habitat acquisition would be most helpful to conserve a species. In other words, we should know what questions we want the model to help answer. Second, there should be sufficient data to support a model designed to meet that objective. There are many endangered-species management problems for which one or both of these conditions are not met (table 25.1). Urgent management needs and research priorities can often be identified without constructing a quantitative model (see "Alternatives to Conducting a PVA").

If modeling seems like a feasible approach to the stated problem, we must decide if data are of sufficient quantity and quality to use or build a model. If sufficient data are lacking to make decisions without modeling, we may lack sufficient data to build a model. Models, of course, can be and are often built with incomplete or missing data. In fact, management decisions are usually made with incomplete data, and models can be used to see how much difference obtaining the missing data

Table 25.1 Situations When It Might Be Advisable to Choose an Alternative to Conducting a PVA as the Recommended Conservation Action

Situation	Recommended Alternatives
The distribution, abundance, or natural history of the population or species is very poorly known.	Undertake field studies to gather basic information about the species.
The reasons for the decline of the population are very poorly known.	Undertake field studies to determine the reasons.
The population is known to be small and declining, and the reasons for the decline are well known.	Undertake management actions to stem the decline in situ and consider if ex situ conservation measures are appropriate.
There is no specific question or objective in mind that PVA might be able to answer.	Consider alternative ways to make decisions, such as expert panels, rules of thumb, decision analysis, development of principles, or other methods discussed in this chapter.
Very few data required by PVA models are available, but it would be helpful to obtain, evaluate, and synthesize the available data; promote interactions among interested parties; and develop a plan of action.	Consider other ways to accomplish these goals, such as a workshop, developing guidelines, using an alternative method of risk assessment, or developing simple models that address specific questions.

might make (Starfield et al. 1995; Starfield 1997). Nevertheless, we often possess so few data about an endangered species that modeling, particularly with complicated PVA models, may be premature or even misleading (Noss et al. 1997; Ralls and Taylor 1997; Ruckelshaus et al. 1997; Beissinger and Westphal 1998; Groom and Pascual 1998). Green and Hirons (1991) found that data suitable for population modeling were available for only 2% of threatened bird species. If there are few basic data on the distribution, abundance, demography, dispersal, habitat use, natural history, and/or reasons for the decline of the species, perhaps a top priority should be to gather such information (table 25.1).

The key to deciding whether there are enough good data available to support the development or use of a model is to clearly define the objectives of the model. General objectives, such as estimating the risk of extinction over the next 100 years, are most useful when trying to classify or manage a group of species according to the same standard. For example, healthy marine-mammal populations in the United States are managed to have a 95% probability of remaining above maximum net productivity level for 20 years (Taylor et al. 2000). Specific objectives tend to be more useful when models are employed to choose between recovery options for a single species. In these cases, the model and its outputs need to be tailored to the specific task.

DIFFERENT QUESTIONS SUGGEST DIFFERENT APPROACHES TO MODELING RISK

Scientists build models to help answer questions that are motivated by specific objectives. The first PVA models that were developed had objectives and attempted to answer questions regarding the true or *absolute risk* of extinction: "What are the minimum conditions for the long-term persistence and adaptation of a species or population in a given place?" (Soulé 1987, 1). Hence, among the first definitions of PVA was "analyses that estimate minimum viable populations" (Gilpin and Soulé 1986). Soon, however, biologists began to expand the use of PVA models to help answer other questions related to endangered-species management (Beissinger, chap. 1 in this volume). These questions, such as "Which of these management plans would be most beneficial to this species?" often concerned comparisons of the *relative risk* of extinction rather than estimating the absolute risk of extinction. Frequently such questions can be answered without estimating minimum viable population size, which cannot be done with accuracy (Ludwig 1996, 1999). Furthermore, some scientists objected to the concept of estimating the "minimum viable population size" because population size is only one of the factors influencing the probability of extinction (Gilpin and Soulé 1986).

Different types of models are required to estimate the absolute and relative risk of extinction. A model that attempts to estimate absolute risk, or the minimum viable population size necessary for long-term persistence and the maintenance of a population's ability to adapt genetically to changing environment, should strive to develop an accurate representation of reality. The ideal model would be comprehensive and include all factors influencing the probability of extinction (fig. 25.1). Such a model would incorporate a wealth of data that are not available

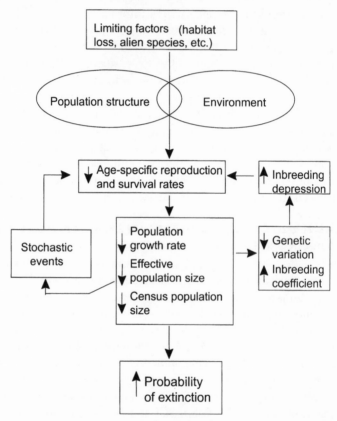

Fig. 25.1 Simplified representation of an ideal PVA model. Limiting factors, including habitat loss and degradation, invasion of exotic species, overharvesting, and so forth, interact with population structure and environmental variation to affect demography, population growth rates, and the probability of extinction. Population structure includes the age structure, sex ratio, behavioral interactions, distribution, physiological status, and intrinsic birth and death rates. Environment includes habitat as well as extrinsic factors that vary over time, such as weather, competition, predators, and food abundance. *Arrows within boxes* indicate an increase or decrease. Adapted from Soulé and Mills (1998).

for most, if any, endangered species. Genetic concerns and catastrophes must be included because of the role that these processes can play in population viability when a long time period is considered (Mangel and Tier 1994; Allendorf and Ryman, chap. 4 in this volume). A model that is intended to be an accurate representation of reality should be judged by its accuracy, and its ability to predict the future should be extensively validated. The construction of models that can accurately estimate absolute extinction risk and the collection of the data necessary for these models are, for most species, still beyond our capabilities (Ludwig 1996, 1999). The ideal model is only approximated to some degree by various case studies. The improvement of such models is an important ongoing research effort.

Fortunately, many important endangered-species management questions can be answered without accurate estimates of the absolute risk of extinction. Models designed to help answer such questions need only compare differences among management options (i.e., estimate relative risks of extinction), so that management options can be ranked according to their probable benefits to a population. This type of model is best viewed as a tool for generating hypotheses, conducting thought experiments, and providing support for management decisions rather than an accurate portrait of reality (Starfield 1997; Beissinger and Westphal 1998). Such a model may be less comprehensive than a model intended to estimate absolute risk of extinction. For example, modeling inbreeding depression may be unnecessary if any management plan that would minimize the probability of extinction would also minimize the severity of inbreeding depression. Furthermore, depending upon the specific question being asked, the time period modeled may be much shorter than that required for a population to adapt genetically to major environmental changes. Given that such a model has an appropriate conservation objective, it should be judged by the extent to which it meets that objective, not by how well it represents reality (Starfield 1997). For example, did a model built to give insight into a problem do so? Did users of a model built to help evaluate management options make decisions that were more likely to improve the prospects for the species of concern than they would have made without using it?

In practice, models range along a continuum from those that attempt to create comprehensive representations of nature to those that are extremely simplified versions of reality. Before choosing or designing a PVA model and determining where to be along this continuum, one must decide (1) what is the objective of the analysis, (2) whether a model is needed to achieve that objective, and (3) whether the objective of the model is best met by estimating relative or absolute risk of extinction.

POTENTIAL OBJECTIVES AND USES OF PVA

PVA can be used to help meet many common objectives of endangered-species management, such as synthesizing information, ranking risks and benefits of management alternatives, and assessing impacts of habitat loss under the Endangered Species Act (table 25.2). By meeting one or more of these objectives, PVA can identify and prioritize conservation actions and facilitate decision making.

The first group of objectives is concerned with organizing information, gaining understanding, and developing a plan of action. These objectives often can be achieved for less cost without using PVA or other analytical methods (see the section "Alternatives to Conducting a PVA"). PVA can be expensive, especially if it involves developing a custom model, which is often the best way to address a particular problem, or if convening a workshop is necessary.

The second group of objectives is concerned with setting priorities and ranking management alternatives. Estimates of the relative risk of extinction are adequate and often helpful for meeting these objectives. We agree with the emerging consensus that PVA models are most appropriate for helping to meet this group of objectives, because the models are best used to estimate relative extinction risks (Ralls and Taylor 1997; Beissinger and Westphal 1998; Groom and Pascual 1998; Reed et al. 1998).

The third group of objectives in table 25.2 is related to actual or

Table 25.2 Common Approaches and Objectives of Endangered-Species Management

Approach	Objective
Organize information.	Synthesize data.
	Gain understanding of species and/or system.
	Determine research needs.
	Develop action or recovery plan.
Estimate relative risk.	Set priorities among species/groups/areas.
	Rank risks and benefits of management alternatives.
	Evaluate sensitivity of population dynamics to set research priorities.
Estimate absolute risk.	Determine habitat, reserve, or population size required to prevent extinction.
	Develop criteria for listing, reclassifying, delisting, or recovery.
	Assess impacts under the Endangered Species Act.
	Classify species by degree of vulnerability.
	Estimate likelihood of extinction given current conditions.
Manage adaptively.	Continuously improve both model and management by experiments and monitoring.

Note: PVA and various alternative methods can be used to help meet these objectives.

implied standards of the absolute extinction risk. As noted above, currently we rarely have much confidence in estimates of the absolute risk of extinction made with PVA or any other method. Nevertheless, there are important management questions related to predicting the true risk of extinction that must be addressed (Shaffer et al., chap. 7 in this volume). Thus, we must sometimes do our best either to estimate absolute risk or to evaluate the likelihood of falling above or below some absolute-risk threshold. Our best effort, however, may not be to construct a rigorous, quantitative model. The key in these situations is choosing an approach that minimizes the risk of incorrectly judging the level of risk to the species.

The method used to assess risk and estimate the likelihood of extinction should match the type and accuracy of available data. For example, if the most accurate data are on the species' current geographic range and its use of various habitats, risk assessment based on trends in the amount of suitable habitat would likely be more reliable than estimates of extinction resulting from a demographic PVA model. Likewise, analysis of population trends would be more suitable for estimating extinction risk for a species for which there were good data on population size over time but limited demographic data. Since different approaches for assessing risk may give different results, it may be helpful to try several approaches or different model types, and to evaluate how consistent model predictions were with the available data (Hilborn and Mangel 1997).

WHAT TYPE OF PVA MODEL(S) SHOULD WE USE?

Creating a model often involves a necessary trade-off between generality, precision, and realism (Levins 1968). Analytical models (e.g., the Levins metapopulation model; Hanski, chap. 5 in this volume) that were created to develop theory and heuristic (conceptual) models will often be most general, but they are difficult to apply to real management situations. Statistical models (e.g., incidence function models) may yield precise model outputs for some situations, but their generality may extend only to a specific set of data and they may have limited applicability due to assumptions of the models (Hanski, chap. 5 in this volume). The most realistic models are often system-simulation models (e.g., stochastic single-population models or spatially explicit models) that can be developed for specific species, scenarios, or landscapes (Beissinger and Westphal 1998), and allow the incorporation of particular mechanisms of interest (e.g., changes in demographic rates or land use). However, these models often contain so many parameters that must be estimated for each subcomponent or system of interest that they have less generality. Furthermore, these models often yield less precise results because they are stochastic, so that model outputs are in the form of probability distributions rather than a single result, and because we can rarely estimate all the model parameters from data specific to the system of interest. Therefore, the choice of a PVA model often will be limited by the objectives of the analysis, the data that are available, and the assumptions that are realistic to make.

Incidence function models are among the simplest models that yield an estimate of the risk of extinction (Hanski 1999). They require data on whether the species is present in a given patch of habitat for many patches in the landscape of interest (Hanski, chap. 5 in this volume; Harrison and Ray, chap. 6 in this volume). Estimates of colonization and extinction parameters can be obtained from a time series of such data collected over a number of years, and population persistence can be estimated by determining the minimum amount of occupied habitat that will support the species (Hanski et al. 1996). Demographic rates are not explicitly incorporated in the model, but are implicitly incorporated through the processes of extinction and colonization. Occupancy data can be collected more quickly and at larger scales than the data required to estimate demographic rates. Incidence function models may be appropriate for species, such as butterflies (Hanski et al. 1996), amphibians (Sjögren-Gulve and Ray 1996), and plants (Harrison and Ray, chap. 6 in this volume), in which extinction and colonization occur frequently and for which demographic rates may be difficult to measure (e.g., Doak et al., chap. 15 in this volume). However, incidence function models typically assume that the amount and distribution of habitat is not changing rapidly (Hanski, chap. 5 in this volume), which is not likely to be true for many endangered species. Furthermore, these models have limited ability to evaluate specific demographic or management mechanisms that might account for colonization and extinction in patches (e.g., edge effects, Allee effects, or specific causes of reproductive failure) because the models are not mechanistic (Beissinger et al. 2002).

Most other types of PVA models require estimates of demographic rates and may require other kinds of data as well. The main types of demographic models used in conservation are deterministic single-population models, stochastic single-population models, metapopulation models, and spatially explicit models that often are individual-based (Beissinger and Westphal 1998). Data requirements for each type of model increase greatly as models become more complex and more realistic (table 25.3). Deterministic single-population models require the least amount of data but do not yield probabilistic estimates of extinc-

Table 25.3 Minimum Data Required for the Dominant Types of Demographic Models Used in PVA

Data Type	Data Needs	DSP	SSP	Meta	Space
Demographic	Age or stage structure	x	x	x	x
	Age of first breeding	x	x	x	x
	Mean fecundity for each age or stage	x	x	P	P
	Mean survival for each age or stage	x	x	P	P
	Variance in fecundity		x	x	x
	Variance in survival		x	x	x
	Carrying capacity and density dependence		x	P	P
	Variance in carrying capacity		x	x	x
	Frequency and magnitude of catastrophes		x	x	x
	Covariance in demographic rates		x	x	x
	Spatial covariance in rates			P	P
Landscape	Patch types			x	x
	Distance between patches			x	x
	Area of patches			x	x
	Location of patches				x
	Transitions among patch types				x
	Matrix types				x
Dispersal	Number dispersing			P	P
	Age class and timing of dispersal			x	x
	Density dependent or independent dispersal			x	x
	Dispersal-related mortality			x	x
	Number immigrating			P	P
	Movement rules				x

Notes: DSP = deterministic single population, SSP = stochastic single population, Meta = metapopulation, and Space = spatially explicit. An x indicates data are estimated for the population as a whole, and a P indicates data are estimated on a per patch basis.
Source: Adapted from Beissinger and Westphal 1998.

tion, so we do not consider them PVA models. They can, nevertheless, be quite useful for understanding population trends and how different demographic factors affect them (Mills and Lindberg, chap. 16 in this volume). Deterministic population models can yield important insights into the demographic factors that affect the outcomes of more complex PVA models.

Choosing among PVA model types depends upon the quality and quantity of data that are available, the assumptions of the models, and the objectives of the analysis. Stochastic single-population models require less data than other types (table 25.3), although their data requirements are still onerous, but their utility is limited if spatial processes are important. Patch metapopulation models can be a useful way to incorporate spatial processes, but have less ability to include mechanistic processes than spatially explicit models and require nearly as much information (table 25.3).

Legislative directives (Shaffer et al., chap. 7 in this volume) or the objectives of the modeling exercise may require an estimate of the absolute risk of extinction. In this case, we want to build as realistic a PVA model as possible, and it may be important to incorporate uncertainty directly in the model. Uncertainty can be explicitly incorporated in demographic parameters using Bayesian or non-Bayesian statistical methods (Ludwig 1996; Goodman, chap. 21 in this volume; Sæther and Engen, chap. 10 in this volume; Wade, chap. 11 in this volume). Uncertainty about model structure and model parameters can be explored using sensitivity analysis (Mills and Lindberg, chap. 16 in this volume), but sensitivity analysis alone is often inadequate. Confidence intervals should be provided for important model outputs such as the probability of extinction. Furthermore, PVAs that attempt to estimate absolute risk should be interpreted very conservatively. We should take a precautionary approach and add a safety factor to help compensate for our uncertainty about modeling results. Safety factors should be included in the list of modeling assumptions to enable their consideration during scientific review. PVAs that attempt to estimate absolute extinction risk should receive particular scrutiny during the peer-review process (Meffe et al. 1998).

Most PVAs use a single model. "The norm in PVA appears to be to put together a reasonable representation of one's best understanding of a population's dynamics, with only a handful of authors justifying their choices" (Groom and Pascual 1998, 16). This is unfortunate, because consideration of multiple models lends rigor and credibility to the results of a PVA. There may be several possible models with different management implications (Taylor 1995; Pascual et al. 1997). For example, Crone and Gehring (1998) examined data on a threatened plant that occurred at six sites on an island. They found that models made very different predictions with and without the incorporation of spatial heterogeneity into the model structure (i.e., using individual growth rates observed at each site versus using a mean growth rate across sites for each site). The model without spatial heterogeneity predicted that the plant would be extinct within 100 years, while the model with spatial heterogeneity predicted that the population would stabilize and persist at two of the six sites. Many years of monitoring would be required to determine which model more accurately represented the population dynamics of the plant, because insufficient data were available to test the predictive capabilities of the models. If data on the state of the population were available, they could be compared to quantitative predictions of the models and Akaike's information criterion (AIC) could be used to test which model is best supported by the data (Hilborn and Mangel 1997; Burnham and Anderson 1998).

BEYOND MODEL BUILDING: TESTING MODELS AND MANAGING ADAPTIVELY WITH PVA

When making important management decisions, it is unwise to have confidence in the predictions of a PVA model or any other system-simulation model that has not been tested to determine its accuracy and ability to make predictions (Oreskes et al. 1994; Bart 1995). Testing a PVA model to corroborate how well it performs is a particularly difficult challenge (Belovsky et al., chap. 13 in this volume), and useful tests vary by model type. We can test model predictions or outputs, and verify model assumptions and input parameter estimates by comparing them with newly or independently collected data. Potential tests vary by model type.

The primary prediction from a PVA model is the probability of extinction, which is almost impossible to test in the field because we are usually studying one or only a few populations (Beissinger and Westphal 1998; Belovsky et al., chap. 13 in this volume). Comparing a real population trajectory to the average trajectory from model output (e.g., Brook et al. 1997, 2000) is a weak way to test a PVA model, because it is the value used for the variance or stochasticity in the model that results in differences among replicate model runs and greatly influences the probability of extinction. Also, because error bars around predicted trajectories of stochastic models are often very large, we are unlikely to reject a PVA model using this test unless the observed trends differ radically from the predicted trends.

Secondary predictions from PVA models may be easier to test (i.e., patch occupancy rates, the frequency of colonization and extinction events, or movement patterns of individuals). For example, McCarthy et al. (2000) predicted the occurrence of treecreepers in forest fragments and the number of extinction and colonization events to compare to output from a metapopulation model. Nevertheless, it can be difficult to find meaningful tests because (1) the impacts of demographic stochasticity can cause local extinctions and affect the outcomes in particular fragments, (2) a short time period is used in testing model predictions (several years) compared to the time periods used in model projections (typically 50 to 100 years) and the rarity of catastrophic events, and (3) there are few events in the field to compare to model outcomes (e.g., extinction and colonization). Another approach is to compare the fit of different model structures or parameters (Hilborn and Mangel 1997;

Burnham and Anderson 1998), but sometimes this exercise can result in equally good fit among models with radically different structures (Pascual et al. 1997).

Despite the difficulties in testing PVA models, it is important to find portions of the model that can be tested and to verify estimates used to parameterize models, especially ones based on educated guesses. Field recovery efforts and adaptive management could be directly linked to PVA through the process of model testing (Beissinger and Westphal 1998). These activities are rarely undertaken once a PVA model has been developed. However, model results should be used with caution until model testing is under way.

Although many PVAs are conducted during brief workshops, or by a small group of experts, and never revisited, PVA is more useful as a long-term, iterative process coupled with an adaptive management approach to species recovery (Boyce 1993; Maehr et al., chap. 14 in this volume). As Groom and Pascual (1998, 21) argue, "Judicious use of PVA can be a significant aid in discrimination among management options and guide refinements to our management strategies and data gathering protocols." PVA models and their projections are best viewed as testable hypotheses—a perspective compatible with an adaptive management approach. PVA models that are easily modified or replaced with improved versions are best suited to adaptive management applications.

ALTERNATIVES TO CONDUCTING A PVA

PVA is not the only method for meeting the objectives listed in table 25.2. Like PVA, other approaches have strengths and weaknesses. The best method to use in a particular situation will depend on the management objective, the quality and quantity of available data, legal mandates, and other factors.

We often need to synthesize data about a species, gain a better understanding of the threats it faces, and develop a plan of action. These objectives can often be met without a PVA model. Individuals knowledgeable about a particular species frequently agree about pressing research and management needs and can produce an action plan (e.g., U.S. Fish and Wildlife Service recovery plans and World Conservation Union [IUCN] action plans). However, if knowledgeable individuals disagree or if the information at hand is overwhelming or confusing, a structured process for organizing information and diagnosing probable threats may be helpful. A simple, step-by-step process to identify the factors driving a species toward extinction is outlined and illustrated with case studies by Caughley and Gunn (1996). A related approach called "species-centered environmental analysis" involves preparation of a diagram summarizing information about the environmental factors that would be expected to affect population size in a species (James et al. 1997). Analysis of demographic factors can also help identify the causes that limit growth of a population. For example, one could compare the demographic traits of a population with those of a related well-known species (Peterson and Silvy 1996) or explore the sensitivity of deterministic matrix population models (Wisdom and Mills 1997; Mills and Lindberg, chap. 16 in this volume).

Another frequent task is to group species or areas according to degree of threat in order to set conservation priorities. Various rule-based or point-scoring systems have been developed to accomplish this task. The IUCN criteria for classifying species as critically endangered, endangered, or vulnerable are probably the best-known rule-based system (IUCN 1994). Keith (1998) suggested modifications of the IUCN criteria for use with vascular plants, and Allendorf et al. (1997) proposed criteria for salmon. Other rule-based systems have been devised by the Nature Conservancy (Master 1991) and the U.S. Forest Service (Cleaves 1994). Scoring systems have been used to rank species according to degree of threat in New Zealand, Australia, and the United States (Millsap et al. 1990; Todd and Burgman 1998; Carter et al. 2000). Wikramanayake et al. (1998) described a point-scoring method for identifying and ranking areas with the greatest potential for conserving large mammals that are widely distributed, such as tigers. They scored 159 tiger conservation units based on measures of habitat integrity, poaching problems, and population status, and then ranked them into three categories that reflected decreasing likelihood of long-term population persistence. To ensure a representative sample of various ecosystems, they used a hierarchical method grouping tiger habitat types within bioregions and conservation units within habitat types.

Rule-based and point-scoring systems, although they can be useful in some situations, have problems. One problem with rule-based systems is that species within the same category may in fact have quite different risks of extinction; a species judged endangered by one criterion does not necessarily have the same risk of extinction as a species judged endangered by another criterion (Beissinger et al. 2000). Furthermore, point systems often require that disparate quantities, such as population-size points and taxonomic-distinctiveness points, must be assigned scores on a similar scale and summed. Given and Norton (1993) used multivariate statistical techniques with point scores to group species in multidimensional space and to eliminate some of the problems associated with point-scoring systems.

Another problem with current rule-based and point-scoring systems is that they rarely incorporate the inevitable uncertainty in data. For example, categories may be assigned that have specific breakpoints, such as 500 individuals. We may know that a species has a population size of about 500, but we are unlikely to know whether there are 499 or 501 individuals, and one or two individuals will not have a major impact on risk of extinction. Todd and Burgman (1998) suggested the use of fuzzy sets to incorporate both probabilistic uncertainty and a variation in expert opinion into such systems to facilitate consideration of a range of plausible outcomes and their reliability. Taylor et al. (chap. 12 in this volume) and Goodman (chap. 21 in this volume) have used Bayesian statistical techniques with PVA models to incorporate uncertainty into criteria for ranking species vulnerability.

PVA occasionally has been used to estimate the amount of habitat critical to a species' survival (Lamberson et al. 1994; Bart 1995). However, managers in both the United States and Australia are charged with maintaining "viable" populations of species of many taxa (e.g., plants, invertebrates, and vertebrates). Given the large numbers of species and the sparse data available, it would be impossible to conduct a PVA on every species affected by a bioregional management plan (Samson, chap. 20 in this volume), let alone the more than 1,700 species listed as threatened or endangered by the U.S. Fish and Wildlife Service (1999). The U.S. Forest Service has experimented with expert opinion organized in different frameworks and combined with modeling to varying extents (FEMAT 1993; Marcot 1994; Ruggiero et al. 1994; Samson, chap. 20 in this volume).

One alternative is to develop rules of thumb. A "species equity formula" was developed in Australia to define a target habitat area for mammals and birds (H. Possingham and S. Andelman, personal communication), and another protocol has been developed for plants (Burgman et al. 2001). Other rules of thumb have been developed using genetic criteria (Allendorf and Ryman, chap. 4 in this volume; Lande, chap. 2 in this volume; Waples, chap. 8 in this volume) and for specific groups, such as primates (Dobson and Lyles 1989). Like rule-based systems, demographic rules of thumb can be evaluated and improved by the use of PVA modeling and by testing their assumptions.

Given great uncertainty in estimates of most vital rates, analysis of population trends may be a more robust method for estimating risk (Goodman, chap. 21 in this volume). Taylor et al. (1996) used this approach in a decision-analysis framework based on population trend data.

Like demographic data, population census and trend estimates can be highly variable and uncertain. The specific management objectives, including acceptable standards of risk, will help determine which approach is most suitable for given species (Goodman, chap. 21 in this volume).

When data or time constraints preclude development of a PVA, expert opinion and structured decision-analysis frameworks may be useful for estimating risks (Harris and Kochel 1981; Thibodeau 1983; Maguire 1986; Marcot 1994; Ralls and Starfield 1995). These methods are ultimately subjective, however, and can contain as much or more uncertainty as other indicators of risk. Thus, decisions based on expert opinion should receive the same scrutiny and review as those based on PVA.

A second alternative to PVA is the use of "principle-centered recovery" (Beissinger et al. 1997) to develop species-specific guidelines for the evaluation of proposed management actions (fig. 25.2). Principles are "fundamental truths that have universal application" (Covey 1989). Useful principles are based on generally accepted knowledge. For example, small populations are usually at greater risk of extinction than are large ones. Specific objectives or management actions that need to be undertaken can then be developed for each principle. An example of two principles that were important in solving conflicts over Mariana crows (*Corvus kubaryi*) is shown in figure 25.2 (National Research Council 1997). The power of this approach is that, once all participants have agreed to the guidelines for conducting recovery, it becomes much easier to develop consensus on the required actions. This approach has proven particularly useful with very highly endangered species (e.g., the Puerto Rican parrot, *Amazona vittata*) for which data are sparse. Often in such situations, resource agencies, nongovernmental conservation organizations, and other interested parties have differing views on how best to recover the species. Principle-centered recovery also lends itself to adaptive management, in which management actions are taken, new data are generated, and new or revised strategies or tasks are proposed and evaluated against the guidelines (fig. 25.2).

CRITERIA FOR EVALUATING A PVA

We have discussed when and how PVAs can be used effectively to help make management decisions. However, rarely will we have all of the data or understanding of a system to create the perfect PVA model. How can we evaluate a completed PVA?

In table 25.4 we suggest a list of questions for evaluating a PVA. The answers to many of these questions will be yes for a high-quality PVA

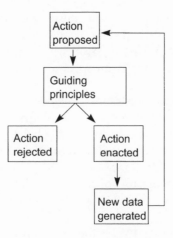

Principle 1: The core wild population(s) is (are) critical to recovering a species, and its (their) security must be the highest priority. The most secure populations will serve as the source of individuals for long-term recovery, and management actions should not appreciably increase the risks to core populations.

Principle 2: Multiple viable wild populations should be maintained to reduce the risk of extinction from catastrophe, disease, and other stochastic factors. A single population, no matter how large, is not immune to extinction from chance events. Each additional population decreases the likelihood of extinction by catastrophe.

Fig. 25.2 Principle-centered recovery (Beissinger et al. 1997) illustrated by two principles used to solve a crisis over the management of Mariana crows (National Research Council 1997). These simple statements were needed because recovery efforts for crows had become paralyzed by conflicts among agencies over the desirability of translocating nestlings from Rota to Guam. About 500 birds were thought to exist on Rota, while less than 20 survived on Guam. The Guam birds had barely reproduced during the past decade, due to nest predation by the introduced brown tree snake (*Boiga irregularis*). Although Rota apparently is snake-free, crow populations have declined by 50% over the past decade due to destruction of its forested habitat. Once interested parties agreed upon the two principles, management objectives became clear. The Mariana crow population on Rota is critical to the recovery of the species, and its security must be the highest priority (principle 1). Every effort must be made to ensure that Rota remains snake-free and that adequate habitat is protected for the crow. It would be unwise to translocate large numbers of nestlings to Guam (even if they could survive the depredations of the snake) without better knowledge of crow demography and population trends on Rota. Finally, effort should be made to save the Guam population because, should it go extinct, another population would have to be established outside the historic range of the crow (principle 2).

Table 25.4 Yes/No Questions for Evaluating a PVA: The Answers to Many of These Questions Will Be Yes for a High-Quality PVA and No for a Low-Quality PVA

Objectives
Are the objectives of the modeling exercise clearly stated?
Is it clear whether the model is intended to produce a relative or an absolute measure of risk?
Are the objectives met?

Model structure
Are multiple model structures evaluated, including different processes or interactions and various forms of density dependence?
Did modelers attempt to validate or demonstrate why the model is acceptable?
Is the complexity of the model appropriate for the quality and quantity of data?
Is the structure of the model appropriate for the questions being asked?
Is the model described clearly and in enough detail that someone else could replicate it?
Are the assumptions of the model clearly stated and discussed?

Data and parameter estimation
Are the sources of the data adequately described?
Are the reasons for choosing particular data sets explained and justified?
Are the limitations of the data discussed?
Are vital rates estimated using robust methods?
Is sampling variance removed from estimates of variance for vital rates to yield an estimate of the process variance of interest?
Is the number of years of study sufficient to experience the full range of environmental conditions to develop accurate estimates of variance in vital rates?

Analysis of model outcomes
Are uncertainties in model choice and parameter estimates clearly explained and explicitly analyzed whenever possible?
If the model is intended to produce an estimate of absolute risk, is uncertainty explicitly incorporated?
If the model is intended to produce an estimate of relative risk, is there an adequate sensitivity analysis?
Is more than one time horizon considered?
Is more than one measure of population viability (performance criterion) considered?
Are the choices of time horizons and performance criteria justified?
Do the short-term performance criteria enable managers to evaluate the probable condition of the populations remaining at the end of the time period?
Are the entire distributions of the performance criteria presented?

Handling unknown factors
Is the possibility of density compensation or depensation at small population sizes adequately treated?
If genetic effects are not modeled, are the implications of this omission justified and/or discussed?
If catastrophes are not modeled, are the implications of this omission justified and/or discussed?
If catastrophes are modeled, are the reasons for considering particular scenarios explained and justified?

Interpretation
Do the authors understand and explain why they got the results they did?
If the uncertainty of some or all parameter estimates is not incorporated into the model, do the authors recognize the limitations of estimating extinction probabilities from a single PVA based on point estimates of these parameters?
Are the results of the modeling discussed in relation to other factors that were not included in the model (e.g., behavior, spatial distribution of populations, cost of management strategies)?
Are the results of the modeling used in a heuristic manner and/or treated as hypotheses to be tested in the field rather than as facts?
Are the limitations of the conclusions that can be drawn from the modeling exercise clearly discussed?

Peer review and model testing
Was the PVA peer-reviewed?
Was the peer review adequate?
Were any of the model outputs tested or model inputs verified using new or independently gathered data?

and no for a low-quality PVA. Below we discuss the major characteristics of PVA that should be considered in evaluating its quality.

Objectives

The objectives of a good PVA will be clearly stated. Whether any analysis was adequate cannot be determined without a clear statement of its objectives. Specific objectives will be more helpful than general ones, such as "estimate the risk of extinction during the next 100 years." Knowing whether the model was intended to estimate relative or absolute risk of extinction is important, because models that are adequate for estimating relative risk may not be adequate for estimating absolute risk.

Model Structure

Modelers should attempt to demonstrate why their model is appropriate for the questions being asked. The complexity of the model should be appropriate for the amount of data that are available, such as whether to (1) use age-, stage-, or individual-based models and (2) incorporate stochasticity, density dependence, and spatial structure in the model. Beissinger and Westphal (1998) and Groom and Pascual (1998) discuss these issues in more detail. As a general rule of thumb, we suggest that, if there are few data, the model should be simple. Modelers should not have to guess about the estimates for a large number of parameters in a complicated model. Assumptions of a PVA model should be clearly stated, so that others can judge whether they are appropriate. A good approach to determining the effects of model structure on management recommendations is to use and evaluate several models (Pascual et al. 1997; Starfield 1997; Groom and Pascual 1998).

One aspect of model structure that can have important affects on the outcome of PVA models is how the model incorporates density dependence and carrying capacity (Ginzburg et al. 1990; Mills et al. 1996). Opposing recommendations for recovering wild dogs (*Lycaon pictus*) resulted from PVA models that differed primarily in whether density dependence was incorporated (Burrows et al. 1994; Ginsberg et al. 1995). Ignoring density effects and assigning a carrying capacity to act as a reflecting boundary may seem realistic for species with densities much lower than can be supported by the available habitat. However, this structure may not allow simulated populations to recover rapidly from low numbers or to remain near carrying capacity because vital rates are drawn randomly. Detecting Allee effects from field data is extremely difficult and may not be feasible on theoretical grounds (Lande, chap. 2

in this volume). Generally, little is known about the nature or form of density dependence for most endangered species (Brook et al. 1997). All of these concerns suggest that several alternative forms of density dependence should be incorporated into PVA models and tested, if possible, against an independent set of data on the state of the population using AIC and related approaches (Hilborn and Mangel 1997; Burnham and Anderson 1998).

Finally, science should be replicable. Thus, models should be described well, so that any knowledgeable reader could replicate them. Ideally, other scientists should be able to obtain a copy of the software program of the model.

Data and Parameter Estimation

Data on most endangered species are limited, and PVAs have to be based on whatever information is available. However, a good PVA should describe the sources of the data used to parameterize the model, the reasons for choosing particular data sets rather than others, and the limitations of the data. If more than one set of data is available, modelers should employ sensitivity analysis to understand the implications of using different data estimates for model parameters (Mills and Lindberg, chap. 16 in this volume).

Perhaps no single part of a PVA model needs more justification than the choice of means and variances for the vital rates (Beissinger and Westphal 1998). Recent advances in the application of statistical techniques to the estimation and decomposition of vital rates provide new methods of deriving these rates for PVA models. Sophisticated analytical techniques are now available that can estimate survival from mark-recapture information (Lebreton et al. 1992; White et al., chap. 9 in this volume). A similar method was recently applied to develop better estimates of the rates of dormancy and survival in a rare orchid (Shefferson et al. 2001). These approaches can separate the variance of vital rates into the portions due to process variance, which is of interest to modelers because it represents both environmental and demographic variation, and sampling variance due to random error generated by the sampling protocol (Gould and Nichols 1998; White et al., chap. 9 in this volume). Process variance itself can be divided into demographic and environmental components (Sæther and Engen, chap. 10 in this volume). In addition, the ability to develop useful estimates of the variance in vital rates will depend upon the number of years that studies have been conducted. Variance in population size does not begin to stabilize, if at all, until 8 to 20 years of data have been collected (Pimm 1991).

This suggests that accurate estimates of the variance in vital rates may require at least several generations of study, which could exceed 10 to 20 years in long-lived organisms (Beissinger and Westphal 1998).

Analysis of Model Outcomes

There are several ways to present PVA model results that greatly affect the quality of a PVA. Modelers must make decisions on what output measures of viability to present, how they should be displayed, and what time frames are used in the analysis. Typical outputs of PVA models include the proportion of simulated populations that went extinct, average time to extinction, and ending population size. All estimators of viability depend on the period of time analyzed.

To illustrate the uncertainties in PVA projections, it is useful to present a full range of model results rather than point estimates (Beissinger and Westphal 1998; Reed et al. 1998). Point estimates of extinction rate and mean time to extinction at 50 years, for example, are much less informative than presenting the complete distribution of ending population sizes and extinction times. This is often done in the form of a cumulative probability function for the measure of interest, and is known as the quasi-extinction function (Ginzburg et al. 1982). Quasi-extinction functions can be compared among management options or time frames, or as quasi-extinction contour graphs that incorporate the time component (Groom and Pascual 1998). Another approach is to estimate a population prediction interval based on the earliest projected time to extinction (Sæther and Engen, chap. 10 in this volume).

The time period chosen to evaluate viability should be justified. Shorter time periods will minimize error propagation, but have limited utility to help us understand processes that require longer time frames (e.g., catastrophes) or species that have long generation times. Also, as the time period of the population projection is lengthened, the probability of extinction will increase (Lande, chap. 2 in this volume). Therefore, the choice of time periods for projections will depend upon the life history of the organism, and we recommend evaluating multiple time horizons ranging from short (e.g., 10 and 20 years) to long (e.g., 50, 100, or 200 years).

Uncertainty in model outcomes will occur regardless of the measures and time frames used in the analysis (Goodman, chap. 21 in this volume), so good PVAs will analyze and present this uncertainty. Analysis of uncertainty in model performance should include a sensitivity analysis to understand the influence of individual parameters on model outcome (Mills and Lindberg, chap. 16 in this volume), but can also include approaches to model building based on Bayesian statistics (Wade, chap. 11 in this volume; Goodman, chap. 21 in this volume). Attention should also be paid to interactions between parameters and the influence of functions or relationships built into the model, as well as basic model structures (e.g., density dependence).

Handling Unknown Factors

Unfortunately, we often know very little about some of the important factors affecting risk of extinction, such as the way in which population dynamics change when population size becomes very small; the probable severity of inbreeding depression and other genetic risks; and the nature, magnitude, and frequency of future catastrophic events (Allendorf and Ryman, chap. 4 in this volume; Hedrick, chap. 17 in this volume; Lande, chap. 2 in this volume). Very small populations may have low reproductive rates for various reasons: perhaps males cannot locate all the estrous females or perhaps some minimum density is necessary to stimulate reproduction (i.e., the Allee effect or depensation; Lande, chap. 2 in this volume). When population densities are low, however, the remaining individuals may have greater access to resources, leading to larger litter sizes, or improved juvenile survival (i.e., compensation). Inbreeding depression increases extinction risk, yet its magnitude varies from species to species and even from population to population within a species.

Currently, decisions on how to handle these unknown parameters are made by each modeler. Often, these parameters are simply ignored and excluded from the model. We could consider adopting a series of default values that could be used for these parameters in models that require them for species with different life histories. This might help reduce the large variability in the way that these parameters are treated. Default values have been used in some management schemes. For example, management of marine mammals in the United States is legally based on models that use a set of default maximum population growth rates for pinnipeds and cetaceans (Taylor et al. 2000). If better data are available for a particular marine mammal species, these data are used instead of the default values.

Another adjustment for handling unknown parameters is to project future populations for a shorter time (Beissinger and Westphal 1998). Projecting over a shorter time minimizes the propagation of errors with each time step. However, acceptable standards of risk for management decisions must be adjusted accordingly. For example, a 0.05% probability of extinction within 50 years may give the same degree of protection as a 5% probability of extinction within 500 years, but we may be more confident in model projections for the next 50 years than in projections for 500 years into the future.

No matter how these unknowns are treated in the model, a good PVA should discuss their possible effects on model outcomes and their interpretation.

Interpretation

Interpretation of a PVA should be tied to its objectives. PVA results are useful only if they help improve management decisions. As with any modeling exercise, PVA results should be interpreted as the logical outcome of our beliefs about a system (Starfield 1997). Model projections are not the "truth"; they are hypotheses about what could happen in the future, given our knowledge and assumptions about a species. Often the process of building and testing a model provides useful, perhaps qualitative, insights about the biology and management of a species. However, our crystal ball is always cloudy. Even our best models reflect past experience and assume that things will be the same in the future. We cannot be sure about the extent to which the future will resemble the past: new diseases may be introduced, habitats may change due to changes in climate, or species on which the species of concern depends may be lost. We can never quantify the future, which remains a combination of the known and the unknowable (Rosenhead 1989). Thus, modeling results should be only one of the factors considered when making a management decision. Such factors will vary from case to case but might include our degree of confidence in the model and its assumptions, behavioral details that could not be included in the model, and the cost and feasibility of the management options under consideration.

Peer Review and Model Testing

Once a PVA model has been completed, additional scrutiny is required to evaluate the quality of the effort, to corroborate the model's ability to make predictions, and to verify that the model inputs are correct. Evaluation through peer review lends credibility to a PVA. If a PVA is published in a scientific journal, it has undergone a normal level of peer review. Some PVAs will be used to influence major conservation decisions, however, and these should receive more extensive peer review. If additional peer review is enacted, the process should be described, so that readers can evaluate its adequacy. While helpful, the peer-review process is not perfect, and the fact that a PVA has been peer-reviewed does not guarantee that it is adequate. Finally, attempts should be made to test model predictions or verify that the inputs are reasonable, by conducting additional field studies or analyzing independently gathered field data that were not used in model building.

CONCLUSIONS

PVA has an influential role in assisting decision making to conserve endangered species and their habitats. Unfortunately, PVA has a credibility gap because sometimes past applications used inadequate data or were based on inappropriate models. The future for this tool depends on how well PVA models are constructed and how well scientists can present the uncertainty in model results to policymakers.

We have drawn on the growing literature in this area and our own experiences to offer a definition of PVA and a set of guidelines that should be useful to evaluate the quality of a completed PVA. We view these recommendations as a first step toward building consensus among conservation biologists on what constitutes a PVA and what is required to conduct a good PVA. We hope that future efforts can build from this foundation to develop policy benchmarks for the use and implementation of results from PVA models.

LITERATURE CITED

Akçakaya, H. R., and P. Sjögren-Gulve. 2000. Population viability analyses in conservation planning: an overview. *Ecological Bulletin* 48:9–21.

Allendorf, F. W., D. Bayles, D. L. Bottom, K. P. Currens, C. A. Frissell, D. Hankin, J. A. Lichatowich, W. Nehlsen, P. C. Trotter, and T. H. Williams. 1997. Prioritizing Pacific salmon stocks for conservation. *Conservation Biology* 11:140–152.

Bart, J. 1995. Acceptance criteria for using individual-based models to make management decisions. *Ecological Applications* 5:411–420.

Beissinger, S. R., S. R. Derrickson, and N. F. R. Snyder. 1997. Principle-centered management of endangered species: an alternative to PVA. Abstracts, 1997 meeting of the Society for Conservation Biology, Victoria, British Columbia.

Beissinger, S. R., J. M. Reed, J. M. Wunderle Jr., S. K. Robinson, and D. M. Finch. 2000. Report of the AOU Conservation Committee on the Partners in Flight species prioritization plan. *Auk* 117:549–561.

Beissinger, S. R., J. R. Walters, D. G. Catanzaro, K. G. Smith, J. B. Dunning Jr., S. M. Haig, B. R. Noon, and B. M. Stith. 2002. Modeling approaches in avian conservation and the role of field biologists. *Current Ornithology* 17. In press.

Beissinger, S. R., and M. I. Westphal. 1998. On the use of demographic models of population viability in endangered species management. *Journal of Wildlife Management* 62:821–841.

Boyce, M. S. 1992. Population viability analysis. *Annual Review of Ecology and Systematics* 23:481–506.

———. 1993. Population viability analysis: adaptive management for threatened and endangered species. *Transactions of the North American Wildlife and Natural Resources Conference* 58:520–527.

Brook, B. W., L. Lim, R. Harden, and R. Frankham. 1997. Does population viability analysis software predict the behaviour of real populations? A retrospective study on the Lord Howe Island woodhen *Tricholimnas sylvestris* (Sclater). *Biological Conservation* 82:119–128.

Brook, B. W., J. J. O'Grady, A. P. Chapman, M. A. Burgman, H. R. Akçakaya, and R. Frankham. 2000. Predictive accuracy of population viability analysis in conservation biology. *Nature* 404:385–387.

Burgman, M. A., and H. P. Possingham. 2000. Population viability analysis for conservation: the good, the bad, and the undescribed. Pages 97–112 in A. G. Young and G. M. Clark, editors, *Genetics, demography, and viability of fragmented populations*. Cambridge University Press, London, United Kingdom.

Burgman, M. A., H. P. Possingham, A. J. J. Lynch, D. A. Keith, M. A. McCarthy, S. D. Hopper, W. L. Drury, J. A. Passioura, R. J. Devries. 2001. Decision support for setting the size of plant conservation target areas. *Conservation Biology* 15: 603–616.

Burnham, K. P., and D. R. Anderson. 1998. *Model selection and inference: a practical information-theoretic approach*. Springer-Verlag, New York, New York.

Burrows, R., H. Hofer, and M. L. East. 1994. Demography, extinction, and intervention in a small population: the case of the Serengeti wild dogs. *Proceedings of the Royal Society of London,* series B, 256:281–292.

Carter, M. F., W. C. Hunter, D. N. Pashley, and K. V. Rosenberg. 2000. Setting conservation priorities for land birds in the United States: the Partners in Flight approach. *Auk* 117:541–548.

Caswell, H. 2001. *Matrix population models*. 2d edition. Sinauer Associates, Sunderland, Massachusetts.

Caughley, G., and A. Gunn. 1996. *Conservation biology in theory and practice*. Blackwell Scientific, Cambridge, Massachusetts.

Cleaves, D. A. 1994. *Assessing uncertainty in expert judgments about natural resources*. Southern Forest Experiment Station General Technical Report SO-110. U.S. Department of Agriculture Forest Service, New Orleans, Louisiana.

Covey, S. R. 1989. *The seven habits of highly effective people*. Simon and Schuster, New York, New York.

Crone, E. E., and J. L. Gehring. 1998. Population viability of *Rorippa columbiae:* multiple models and spatial trend data. *Conservation Biology* 12:1054–1065.

Dobson, A. P., and A. M. Lyles. 1989. The population dynamics and conservation of primate populations. *Conservation Biology* 3:362–380.

Forest Ecosystem Management Assessment Team (FEMAT). 1993. *Forest ecosystem management: An ecological, economic, and social assessment*. U.S. Department of Agriculture Forest Service, Washington, D.C.

Gilpin, M. E., and M. E. Soulé. 1986. Minimum viable populations: processes of species extinction. Pages 19–34 in M. E. Soulé, editor, *Conservation biology: the science of scarcity and diversity*. Sinauer Associates, Sunderland, Massachusetts.

Ginsberg, J. R., K. A. Alexander, S. Creel, P. W. Kat, J. W. McNutt, and M. G. L. Mills. 1995. Handling and survivorship of African wild dog (*Lycaon pictus*) in five ecosystems. *Conservation Biology* 9:665–674.

Ginzburg, L. R., S. Ferson, and H. R. Akçakaya. 1990. Reconstructability of density dependence and the conservative assessment of extinction risks. *Conservation Biology* 4:63–70.

Ginzburg, L. R., L. B. Slobodkin, K. Johnson, and A. G. Bindman. 1982. Quasi-extinction probabilities as a measure of impact on population growth. *Risk Analysis* 2:171–181.

Given, D. R., and D. A. Norton. 1993. A multivariate approach to assessing threat and for priority setting in threatened species conservation. *Biological Conservation* 64: 57–66.

Gould, W. R., and J. D. Nichols. 1998. Estimation of temporal variability of survival in animal populations. *Ecology* 79:2531–2538.

Green, R. E., and G. J. M. Hirons. 1991. The relevance of population studies to the conservation of threatened birds. Pages 594–633 in C. M. Perrins, J.-D. Lebreton, and G. J. M. Hirons, editors, *Bird population studies: relevance to conservation and management.* Oxford University Press, New York, New York.

Groom, M. J., and M. A. Pascual. 1998. The analysis of population persistence: an outlook on the practice of viability analysis. Pages 4–27 in P. L. Fiedler and P. M. Kareiva, editors, *Conservation biology for the coming decade.* 2d edition. Chapman and Hall, New York, New York.

Hanski, I. 1999. *Metapopulation ecology.* Oxford University Press, New York, New York.

Hanski, I., A. Moilanen, T. Pakkala, and M. Kuussaari. 1996. The quantitative incidence function model and persistence of an endangered butterfly metapopulation. *Conservation Biology* 10:578–590.

Harris, L. D., and I. H. Kochel. 1981. A decision-making framework for population management. Pages 221–239 in C. W. Fowler and T. D. Smith, editors, *Dynamics of large mammal populations.* John Wiley and Sons, New York, New York.

Hilborn, R., and M. Mangel. 1997. *The ecological detective: confronting models with data.* Princeton University Press, Princeton, New Jersey.

International Union for Conservation of Nature (IUCN). 1994. *IUCN Red List categories.* IUCN, Gland, Switzerland.

James, F. C., C. A. Hess, and D. Kufrin. 1997. Species-centered environmental analysis: indirect effects of fire history on red-cockaded woodpeckers. *Ecological Applications* 7:118–129.

Keith, D. A. 1998. An evaluation and modification of World Conservation Union Red List criteria for classification of extinction risk in vascular plants. *Conservation Biology* 12:1076–1090.

Lamberson, R. H., B. R. Noon, C. Voss, and K. S. McKelvey. 1994. Reserve design for territorial species: the effects of patch size and spacing on the viability of the northern spotted owl. *Conservation Biology* 8:185–195.

Lebreton, J.-D., K. P. Burnham, J. Clobert, and D. R. Anderson. 1992. Modeling survival and testing biological hypotheses using marked animals: a unified approach with case studies. *Ecological Monographs* 62:67–118.

Levins, R. 1968. The strategy of model building in population biology. *American Scientist* 54:421–431.

Ludwig, D. 1996. Uncertainty and the assessment of extinction probabilities. *Ecological Applications* 6:1067–1076.

————. 1999. Is it meaningful to estimate a probability of extinction? *Ecology* 80: 298–310.

Maguire, L. A. 1986. Using decision analysis to manage endangered species. *Journal of Environmental Management* 22:345–360.

Mangel, M., and C. Tier. 1994. Four facts every conservation biologist should know about persistence. *Ecology* 75:607–614.

Marcot, B. G. 1994. Analyzing and monitoring population viability. Pages 401–413 in R. J. Kendall and T. E. Lacher, editors, *Wildlife toxicology and population modeling*. Lewis Publishers, Boca Raton, Florida.

Master, L. L. 1991. Assessing threats and setting priorities in conservation. *Conservation Biology* 5:559–563.

McCarthy, M. A., D. B. Lindenmayer, and H. P. Possingham. 2000. Testing spatial PVA models of Australian treecreepers (Aves: Climacteridae) in fragmented forest. *Ecological Applications* 10:1722–1731.

Meffe, G. K., P. D. Boersma, D. M. Murphy, B. R. Noon, H. R. Pulliam, and D. M. Waller. 1998. Independent scientific review in natural resource management. *Conservation Biology* 12:268–270.

Mills, L. S., S. G. Hayes, C. Baldwin, M. J. Wisdom, J. Citta, D. J. Mattson, and K. Murphy. 1996. Factors leading to different viability predictions for a grizzly bear data set. *Conservation Biology* 10:863–873.

Millsap, B. A., J. A. Gore, D. E. Runde, and S. I. Cerulean. 1990. Setting priorities for conservation of fish and wildlife species in Florida. *Wildlife Monographs* 111: 1–57.

Morris, W., D. Doak, M. Groom, P. Kareiva, J. Fieberg, L. Gerber, P. Murphy, and D. Thompson. 1999. *A practical handbook for population viability analysis*. Nature Conservancy, Arlington, Virginia.

National Research Council. 1997. *The scientific bases for preservation of the Mariana crow*. National Academy Press, Washington, D.C.

Noss, R. F., M. A. O'Connell, and D. D. Murphy. 1997. *The science of conservation planning: habitat conservation under the Endangered Species Act*. Island Press, Covelo, California.

Oreskes, N., K. Shrader-Frechette, and K. Belitz. 1994. Verification, validation, and the confirmation of numerical models in the earth sciences. *Science* 263:641–646.

Pascual, M. A., P. Kareiva, and R. Hilborn. 1997. The influence of model structure on conclusions about the viability and harvesting of Serengeti wildebeest. *Conservation Biology* 11:966–976.

Peterson, M. J., and N. J. Silvy. 1996. Reproductive stages limiting to productivity of the endangered Attwater's prairie chicken. *Conservation Biology* 10:1264–1276.

Pimm, S. L. 1991. *The balance of nature?* University of Chicago Press, Chicago, Illinois.

Ralls, K., and A. M. Starfield. 1995. Choosing a management strategy: two structured decision-making methods for evaluating the predictions of stochastic simulation models. *Conservation Biology* 9:175–181.

Ralls, K., and B. L. Taylor. 1997. How viable is population viability analysis? Pages 228–235 in S. T. A. Pickett, R. S. Ostfeld, M. Shachak, and G. E. Likens, editors, *The ecological basis of conservation*. Chapman and Hall, New York, New York.

Reed, J. M., L. S. Mills, P. Miller, K. S. McKelvey, E. S. Menges, R. Frye, J. B. Dunning Jr., S. R. Beissinger, and M.-C. Anstett. In press. Use and emerging issues in population viability analysis. *Conservation Biology*.

Reed, J. M., D. D. Murphy, and P. F. Brussard. 1998. Efficacy of population viability analysis. *Wildlife Society Bulletin* 26:244–251.

Rosenhead, J. 1989. Robustness analysis: keeping your options open. Pages 192–218 in J. Rosenhead, editor, *Rational analysis for a problematic world*. John Wiley and Sons, Chichester, United Kingdom.

Ruckelshaus, M., C. Harway, and P. Kareiva. 1997. Assessing the data requirements of spatially explicit dispersal models. *Conservation Biology* 11:1298–1306.

Ruggiero, L. F., G. D. Hayward, and J. R. Squires. 1994. Viability analysis in biological evaluations: concepts of population viability analysis, biological population, and ecological scale. *Conservation Biology* 8:364–372.

Shefferson, R. P., B. K. Sandercock, J. Propper, and S. R. Beissinger. 2001. Estimating dormancy and survival of a rare herbaceous perennial using mark-recapture models. *Ecology* 82:145–156.

Sjögren-Gulve, P., and C. Ray. 1996. Using logistic regression to model metapopulation dynamics: large-scale forestry extirpates the pool frog. Pages 111–137 in D. R. McCullough, editor, *Metapopulations and wildlife conservation*. Island Press, Covelo, California.

Soulé, M. E., editor. 1987. *Viable populations for conservation*. Cambridge University Press, Cambridge, United Kingdom.

Soulé, M. E., and L. S. Mills. 1998. No need to isolate genetics. *Science* 282:1658–1659.

Starfield, A. M. 1997. A pragmatic approach to modeling for wildlife management. *Journal of Wildlife Management* 61:261–270.

Starfield, A. M., J. D. Roth, and K. Ralls. 1995. "Mobbing" in Hawaiian monk seals (*Monachus schauinslani*): the value of simulation modeling in the absence of apparently crucial data. *Conservation Biology* 9:166–174.

Taylor, B. L. 1995. The reliability of using population viability analysis for risk classification of species. *Conservation Biology* 9:551–558.

Taylor, B. L., P. R. Wade, D. P. DeMaster, and J. Barlow. 2000. Incorporating uncertainty into management models for marine mammals. *Conservation Biology* 14:1243–1252.

Taylor, B. L., P. R. Wade, R. A. Stehn, and J. F. Cochrane. 1996. A Bayesian approach to classification criteria for spectacled eiders. *Ecological Applications* 6:1077–1089.

Thibodeau, F. R. 1983. Endangered species: deciding which species to save. *Environmental Management* 7:101–107.

Todd, C. R., and M. A. Burgman. 1998. Assessment of threat and conservation priorities under realistic levels of uncertainty and reliability. *Conservation Biology* 12:966–974.

U.S. Fish and Wildlife Service. 1999. *Endangered Species Bulletin* 24:32.

Warshall, P. 1994. The biopolitics of the Mt. Graham red squirrel (*Tamiasciurus hudsonicus grahamensis*). *Conservation Biology* 8:977–988.

Wikramanayake, E. D., E. Dinerstein, J. G. Robinson, U. Kranth, A. Rabinowitz, D. Olson, T. Matthew, P. Hedao, M. Conner, G. Hemley, and D. Bolze. 1998. An ecology-based method for defining priorities for large mammal conservation: the tiger as case study. *Conservation Biology* 12:865–878.

Wisdom, M. J., and L. S. Mills. 1997. Sensitivity analysis to guide population recovery: prairie-chickens as an example. *Journal of Wildlife Management* 61:302–312.

ABOUT THE CONTRIBUTORS

H. Resit Akçakaya is a senior scientist at Applied Biomathematics. His current research interests include metapopulation dynamics, population viability analysis, and the effects of landscape on species persistence. He has over 50 publications in conservation biology and theoretical ecology, and is a coauthor of 2 textbooks, *Risk Assessment in Conservation Biology* and *Applied Population Ecology*. One of the principal architects of the RAMAS library of software, he designed software used in threatened-species classification, risk assessment, and developing models for PVA.

Fred W. Allendorf is professor of biology at the University of Montana. His primary research interests are conservation and evolutionary genetics. He has published over 150 articles on the genetics and conservation of fish, amphibians, mammals, and plants. He is a past president of the American Genetic Association, serves on the Montana board of the Nature Conservancy, and has served on the editorial boards of *Conservation Biology, Evolution, Molecular Biology and Evolution, Molecular Ecology,* and *Journal of Heredity.*

Jonathan D. Ballou is population manager at the Smithsonian Institution's National Zoological Park in Washington, D.C., and adjunct associate professor of zoology at the University of Maryland. His research interests focus on genetic problems in small populations, particularly the effects of inbreeding and outbreeding in captive populations. He has developed management strategies for maintaining genetic variation in captive and wild populations, and serves as population genetics adviser to numerous endangered-species recovery efforts worldwide, including golden lion tamarins, Florida panthers, black-footed ferrets, California condors, and Przewalski's horse. Ballou is an active conservation educator and presents courses in conservation biology to biologists and agency personnel throughout the world.

Oron "Sonny" L. Bass Jr. is supervisory wildlife biologist at the South Florida Natural Resources Center in Everglades National Park. He is responsible for inventory and monitoring of the park's wildlife. In his 25-year career at Everglades, he has studied invertebrates, amphibians, reptiles, birds, and mammals, with his primary work on determining the status of federally listed threatened and endangered species occurring in the National Park Service areas of south Florida. He has served as a member of the U.S. Fish and Wildlife Service Florida panther and Cape Sable seaside sparrow recovery teams. He is presently involved in the evaluation of select wildlife population responses to the federally funded Everglades restoration initiative.

Steven R. Beissinger is professor of conservation biology and chair of the Department of Environmental Science, Policy, and Management at the University of California, Berkeley, and a research associate of the University of California Museum of Vertebrate Zoology, Smithsonian Institution, and USDA International Institute of Tropical Forestry. Beissinger teaches courses in conservation biology, demography and genetics of small populations, and behavioral and population ecology. His research integrates behavioral ecology, population dynamics, and modeling of endangered or exploited species, especially birds, in the United States and New World tropics. Beissinger is a fellow of the American Ornithologists' Union, and serves on the Marbled Murrelet Recovery Team and the U.S. National Committee of Diversitas. He is senior editor of *New World Parrots in Crisis: Solutions from Conservation Biology* (Smithsonian Press, 1992) and on the editorial boards of *Conservation Biology* and *Current Ornithology*.

Gary E. Belovsky formerly was professor of ecology and head of the Conservation Biology Program in the Department of Fisheries and Wildlife at Utah State University. Currently he is director of the University of Notre Dame Environmental Research Center and professor in the Department of Biological Sciences at the University of Notre Dame. His early research focused on optimal foraging by herbivores of a wide range of body sizes (bison to grasshoppers in North America and livestock, kangaroos, and rabbits in Australia). He also examined the foraging strategies of hunter-gatherer peoples. He has extended the foraging research to examine population dynamics of grasshoppers, including competition between grasshoppers, avian and spider predation on grasshoppers, and the role of grasshopper herbivory in nutrient cycling. He is also working on the harvesting and population dynamics of brine shrimp in the Great Salt Lake, and using brine shrimp as an experimental test of minimum viable population theory. He is an editor for *Conservation Biology* and *Evolutionary Ecology Research*.

Mark S. Boyce is professor of biological sciences at the University of Alberta and holds the Alberta Conservation Association Chair in Fisheries and Wildlife. Educated at Yale and Oxford Universities, he previously held academic positions at the University of Wisconsin and the University of Wyoming. His research interests include population and habitat ecology of wildlife, especially building models with management application. He has studied models for the northern spotted owl, wolves and ungulates in Yellowstone National Park, and whales in the southern ocean. He has published extensively on population viability analysis and recently served a term as editor in chief of the *Journal of Wildlife Management*.

Jean Fitts Cochrane has worked with the U.S. Fish and Wildlife Service since 1984, primarily as an endangered-species biologist. She completed her Ph.D. dissertation on population viability and cumulative effects analysis for gray wolves, through the University of Minnesota Conservation Biology Program. She has worked on population models for various species, including woodland caribou, spectacled eiders, short-tailed albatrosses, and African elephants.

Daniel F. Doak is professor of ecology and conservation biology at University of California, Santa Cruz. His research includes the application of demographic models to a variety of basic and applied problems for plant and animal populations. His published work on population viability analyses includes treatments of grizzly bears, sea otters, desert tortoises, and cheetahs. As a field biologist, however, he concentrates on smaller and less fierce taxa, including work on the demography of long-lived arctic plants and investigation of factors that limit populations of rare and endangered plants in central California. His interests have increasingly focused on influence of environmental variability on population processes and on community properties.

Steinar Engen is professor in statistics at the Norwegian University of Science and Technology with a D.Phil. degree from the Department of Biomathematics at the University of Oxford. He has published 75 articles, mainly in statistical modeling and statistical analysis of community structure and stochastic population modeling. In 1978 he published a book entitled *Stochastic Abundance Models*. He has worked with a large number of collaborators from various fields of biology such as population dynamics, population genetics, and evolution.

Alan B. Franklin is a research associate at the Colorado Cooperative Fish and Wildlife Research Unit at Colorado State University, adjunct professor at Humboldt State University, and adjunct assistant professor at University of Minnesota. His primary research interests are in animal population biology, especially the mechanisms that affect population dynamics.

Michael Gilpin, who has authored 6 books and over 100 papers in the areas of island biogeography, population dynamics, community structure, and evolution, has for the last decade devoted himself primarily to conservation biology. He has done population viability analysis and reserve design for a number of species, for example, the desert tortoise, the Stephen's kangaroo rat, the bull trout, the Concho River water snake, and the black-tailed prairie dog. He is currently most interested in developing software tools for conservation biology decision making.

Daniel Goodman is professor of ecology at Montana State University. His primary research interests are environmental modeling and environmental statistics; his primary professional interest is the application of modeling and statistics to actual environmental decision making.

Susan M. Haig is a wildlife ecologist for the U.S. Geological Survey's Forest and Rangeland Ecosystem Science Center in Corvallis, Oregon, and professor of wildlife ecology at Oregon State University. Her interests include understanding population structure and the role dispersal plays within a species and across a landscape. This includes molecular and pedigree analysis of small populations, often in Micronesia, as well as investigating the role dispersal plays in waterbird species across landscapes such as the Great Basin. Haig has led international recovery efforts for the piping plover since 1985 and recently coordinated definition of recovery goals for 50 species in the U.S. National Shorebird Conservation Plan.

Ilkka Hanski is professor of ecology at the University of Helsinki and director of the Metapopulation Research Group in the same university. His primary research interest is metapopulation biology, though he has also contributed to general population ecology and to community ecology. He has published nearly 200 articles and has written and edited several books. He has served on many international scientific committees and on the editorial boards of the *American Naturalist, Theoretical Population Biology, Oikos, Oecologia,* and *Trends in Ecology and Systematics.*

Susan Harrison is professor of environmental science and policy at the University of California, Davis. Her research focuses on the spatial and landscape aspects of population dynamics, community structure, and conservation. She is currently studying the ecology and management of grasslands and chaparral on serpentine soils in California, in particular the persistence of rare plants and the effects of grazing on diversity. She also directs the Natural Reserve System at University of California, Davis.

Philip W. Hedrick is Ullman Professor of Biology at Arizona State University. His primary research interests are conservation and evolutionary genetics, and he has published 4 books and over 150 articles on these topics. He is a past president of the American Society of Naturalists, is president of the American Genetics Association, and served on the editorial boards of *Conservation Biology, Evolution, Journal of Theoretical Biology, Animal Conservation, Conservation Genetics,* and *Journal of Heredity.*

Thomas S. Hoctor is a Ph.D. candidate in the Department of Wildlife Ecology and Conservation at the University of Florida. He received a B.A. in history and science from Harvard University in 1989 and a master's in forest resources and conservation from the University of Florida in 1992. His primary research interests include the landscape implications for designing functional reserve systems, spatially explicit landscape and population models, and regional conservation planning. He is also co–principal investigator on the Florida Greenways Project and the Southeastern EPA Landscape Ecological Analysis Project.

Erik S. Jules is professor of plant ecology at Humboldt State University. His research has focused on the effects of landscape pattern on population and community process. This work includes analyses of habitat fragmentation on the population structure of long-lived understory herbs, as well as plant-pollinator interactions. He is currently conducting a study of the landscape-level spread of an introduced tree disease, focusing on the dynamics of human facilitation of the invasion.

Robert C. Lacy is a population biologist in the Department of Conservation Biology of the Chicago Zoological Society. He is actively involved in basic research, development of applied tools for conservation and population management, and collaborative efforts to resolve conservation problems. His research includes studies on inbreeding depression and hybridization. He has developed techniques of pedigree analysis, and he wrote the software used to guide genetic analysis and management of breeding programs for species in zoos around the world. He developed the VORTEX program for population viability analysis, and has worked with wildlife agencies in numerous countries, assisting with population analysis and management for wild populations of threatened species.

E. Darrell Land is the Panther Section leader for the Florida Fish and Wildlife Conservation Commission. He has worked with this endangered cat since 1985 and has written or coauthored numerous articles on Florida panther ecology and prey species.

Russell Lande is professor of biology at the University of California, San Diego. He develops and applies theories of genetics, evolution, ecology, and conservation biology. Current interests include genetic mechanisms of speciation, stochastic population dynamics, and partitioning species diversity. He received the Sewall Wright Award from the Society of American Naturalists, and is past president of the Society for the Study of Evolution, member of the American Academy of Arts and Sciences, and a MacArthur fellow.

Chad Larson formerly was a technician at Utah State University, where he began working, as an undergraduate, on brine shrimp in the Great Salt Lake, including studies on demographics, population dynamics and extinction, and overwinter survival of brine shrimp cysts, and compiling a database of brine shrimp literature. Currently he is a technician at the University of Notre Dame.

Ingrid K. Latchis was programs coordinator at Defenders of Wildlife in Washington, D.C. Her primary interest areas are spatial ecology and conservation of the Canada lynx, as well as population genetics and modeling. She has a master's degree in conservation biology from the University of Maryland, College Park, and a B.A. from the University of Colorado.

Mark S. Lindberg is assistant professor of wildlife biology at the University of Montana. His primary research interests are quantitative population ecology and studies of marked bird populations. He is particularly interested in the role of immigration and emigration in population dynamics and studies of waterfowl ecology. He is widely published in diverse professional journals, including ecological, statistical, and ornithological journals.

David B. Lindenmayer is a jointly appointed senior research fellow at the Centre for Resource and Environmental Studies and reader/associate professor in the Department of Geography at the Australian National University. He has published widely in the fields of conservation biology and landscape ecology, including habitat fragmentation effects, wildlife corridor biology, testing PVA models, and the biology and ecology and conservation of Australian mammals and birds. He is coauthor of more than 220 scientific papers, including 4 books, 2 of which were winners of the Whitely Award for Conservation Biology (1997 and 1999). He was cowinner (with Hugh Possingham) of the Eureka Prize for Environmental Research in 1999.

Donald Ludwig is professor emeritus of mathematics and zoology at the University of British Columbia. He worked for some years on problems of wave propagation and diffraction, but during the past 25 years has worked on problems in ecology and conservation, often in collaboration with Carl Walters, Ray Hilborn, and C. S. Holling. In recent years, his interests have focused on the broader connections between science, economics, and public policy.

David S. Maehr is assistant professor of conservation biology at the University of Kentucky, Department of Forestry. His research interests include the recovery and restoration of endangered, threatened, and extirpated large mammals, including the Florida panther, black bear, and wapiti. He is the author of over 80 technical papers and is the author of 2 books, *Florida's Birds* and *The Florida Panther: Life and Death of a Vanishing Carnivore*.

Dale R. McCullough is professor of wildlife biology in the Ecosystem Sciences Division of the Department of Environmental Science, Policy, and Management at the University of California, Berkeley, where he holds the A. Starker Leopold Endowed Chair. He also holds an appointment as research conservationist in the Museum of Vertebrate Zoology at Berkeley. He has long been interested in the ecology, behavior, and conservation of wildlife, particularly large mammals. His interests integrate the traditional approach of wildlife management with modern conservation biology. Recently his research has focused on conservation genetics, paleobiology, and global change. He is the author, coauthor, or editor of 7 previous books. Three of these books received book-of-the-year awards from The Wildlife Society.

Chad Mellison is a master's candidate in the Department of Fisheries and Wildlife at Utah State University. He is studying the effects of corixid predation on brine shrimp and environmental effects that determine the distribution of corixids in the Great Salt Lake.

Philip S. Miller is a program officer with the Conservation Breeding Specialist Group of the IUCN's Species Survival Commission. Academically trained as a population geneticist and conservation biologist, he has developed expertise in the application of simulation modeling techniques to population and habitat viability assessments across a diverse taxonomic range of wildlife and within an equally diverse human cultural background. Current research programs include development of quantitative and process-based tools to include human population data and stakeholder domains in biodiversity conservation programs, and the study of captive-population pedigree data for more rigorous genetic management of endangered species.

L. Scott Mills is associate professor in the Wildlife Biology Program at the University of Montana. His research interests include the causes and consequences of forest fragmentation for vertebrate populations, and the interface between population ecology and conservation genetics. He is currently applying population modeling and genetic tools to his field studies of snowshoe hares, Canada lynx, and small mammals in fragmented landscapes.

Stuart L. Pimm is professor of conservation biology at the Center for Environmental Research and Conservation and Columbia University. His principal interests are the loss of biological diversity and the science needed to prevent further losses. His 3 books include *The Balance of Nature? Ecological Issues in the Conservation of Species and Communities*. He serves on several editorial boards, including the board of reviewing editors for *Science*. He also serves on the advisory boards to Conservation International's Center for Applied Biodiversity Sciences, and the Union of Concerned Scientists.

Hugh P. Possingham is professor of mathematics and zoology and director of the Centre for Conservation Biology at the University of Queensland. His primary research interest is population modeling. He has published over 100 refereed articles that cover areas of pure behavioral, population, and community ecology, and more recently theoretical applied ecology. In 1999 he was awarded the POL Eureka Prize for Environmental Research (with David Lindenmayer) for their research on population viability analysis, and in 2000 he was awarded the inaugural Fenner medal by the Australian Academy of Science. Currently he is chair of two federal advisory commitees: the Australian Biological Resources Study and the National Biological Diversity Committee. He serves on the editorial boards of the *American Naturalist, Animal Conservation, Conservation Biology, Oryx,* and *Ecological Management and Restoration*.

Katherine Ralls is a senior research biologist for the Smithsonian Institution's National Zoological Park in Washington, D.C. Her research interests cover a broad range of topics related to improving the scientific basis for endangered-species management, including approaches to making robust decisions based on limited and uncertain data. She has worked on inbreeding depression in captive populations and several threatened and endangered species, including San Joaquin kit foxes, California sea otters, California condors, and Hawaiian monk seals. She serves on numerous scientific advisory committees for government agencies and has received several awards in recognition of her research contributions.

Uma Ramakrishnan is a graduate student at University of California, San Diego. She is interested in conservation genetics and behavioral conservation, and the role of mating systems in conservation decisions.

Chris Ray is a postdoctoral research fellow at the Biological Resources

Research Center of the University of Nevada. Her work includes theoretical and empirical studies of demography and genetics in structured populations. She has developed models to predict population viability for several sensitive species, as well as a general model of metapopulation dynamics that has been applied to a wide variety of systems. She founded the Bristlecone Institute for Ecological Research, which promotes long-term empirical studies on the interplay between demography and genetics in fragmented populations.

Nils Ryman is professor of population genetics at the Division of Population Genetics, Stockholm University. His research has focused primarily on the genetic structure of natural populations, the genetic effects of human actions on such populations, and related conservation genetics issues. He was one of the pioneers in addressing the problems of inbreeding and genetic drift in hatchery populations of fish and the genetic effects of harvesting and supplementing wildlife populations. Current research interests include the estimation of effective population size, the genetics of populations with overlapping generations, and the interpretation of temporal gene frequency shifts.

Bernt-Erik Sæther is professor in population ecology at the Norwegian University of Science and Technology. He has worked on the adaptive significance of variation in life-history characteristics, using both an experimental and a comparative approach, population biology of ungulates, and stochastic effects on avian population dynamics. In recent years, he has focused on conservation ecology, often working in the interface between theory and empiricism. He has been involved in science policy matters nationally and internationally, both as leader and member of several steering and coordinating boards of research programs.

Fred B. Samson is a wildlife ecologist for the U.S. Forest Service. He received a B.A. and an M.S. from Indiana University and a Ph.D. from Utah State University. His primary conservation activities are large-scale planning and applied biodiversity conservation. He has published over 90 articles and is coeditor of *Prairie Conservation: Preserving North America's Most Endangered Ecosystem*, *Ecology and Conservation of Great Plains Vertebrates*, and *Ecosystem Management: Selected Readings*. He has been involved in and recognized for practical conservation planning for more than two decades.

Mark Shaffer is the senior vice president of programs at Defenders of Wildlife. His areas of professional interest include biodiversity conservation, environmental stewardship, and integrating ecology, economics, and public policy. He has also worked for the U.S. Fish and Wildlife Service, the Wilderness Society, and the Nature Conservancy and has written numerous publications on population viability analysis. He received a Ph.D. from Duke University and a B.S. from Indiana University of Pennsylvania.

Tanya M. Shenk is a researcher for the Colorado Division of Wildlife and faculty affiliate at Colorado State University in the Department of Fishery and Wildlife Biology. Her primary interests are population demographics and parameter estimation. She currently works on spatial and temporal variation in the demography and movement patterns of Preble's meadow jumping mouse, a threatened species under the ESA. She serves as a science adviser to the U.S. Fish and Wildlife Service Recovery Team for the mouse. She also leads the postrelease monitoring program for the Colorado lynx reintroduction project, focusing on survival, reproduction, habitat use, and movement patterns of the reintroduced lynx.

William J. Snape III is vice president for law and litigation at Defenders of Wildlife, where he has been a staff attorney since 1990. He is author and editor of *Biodiversity and the Law* and has written many articles on domestic and international natural resources conservation issues. He was formerly adjunct professor at the University of Baltimore School of Law and now teaches at American University's Washington College of Law.

Michael E. Soulé is professor (emeritus) in environmental studies, University of California, Santa Cruz, and science director of the Wildlands Project. He helped found the first university in Malawi and has taught in Samoa, at the University of California, San Diego and Santa Cruz, and at the University of Michigan. He was a founder of the Society for Conservation Biology and the Wildlands Project and has been the president of both. He has written and edited a number of books on biology, conservation biology, and the social context of contemporary conservation and has published on various subjects, including population and evolutionary biology, population genetics, island biogeography, environmental studies, biodiversity policy, and ethics. He continues to do research on the genetic basis of fitness and viability in natural populations, on the impacts of keystone species, and on the social causes of the destruction of nature worldwide and to write about biology, ethics, education, and conservation.

Barbara L. Taylor is a biologist at the Southwest Fisheries Science Center. Her primary research interests are in conservation, particularly of marine species, conservation genetics, and decision analysis. She has been especially interested in developing analytical methods that allow an easy transition between data and management decisions. She chairs the group responsible for listing decisions and documentation for cetaceans for the IUCN Red List, is a scientific adviser to the Marine Mammal Commission, and is a delegate to the International Whaling Commission.

Diane Thomson is a doctoral student at the University of California, Santa Cruz. Her research has two main foci: the demography of rare plants, and the community-level effects of introduced species. The former work includes detailed studies of rare plants in central California, while the latter involves a study of the behavioral and population mechanisms by which introduced honey bees influence native pollinators and plants.

Geoffrey N. Tuck is a resource modeler at the Commonwealth Scientific and Industrial Research Organisation (CSIRO) Division of Marine Research located in Hobart, Tasmania. His main areas of research are fisheries stock assessment and conservation biology. He has written extensively covering various areas of stock assessment, including catch-rate analysis and simulation modeling, and more recently tag-recapture analysis and the potential of marine protected areas. He has actively participated in population assessments for orange roughy, southern bluefin tuna, Indian Ocean tunas and billfish, Patagonian toothfish, and threatened southern ocean albatross populations. Currently, he is a scientific representative of the Sub-Antarctic Fisheries Assessment Group (SAFAG) and the Commission for the Conservation of Antarctic Marine Living Resources (CCAMLR).

Peter A. Van Zandt is a university Ph.D. fellow at the University of Southwestern Louisiana, studying the effects of salinity on a native Louisiana iris's growth, reproduction, resource allocation, seed germination, and population genetic structure. He completed his master's degree at Utah State University, working on population variability and persistence in brine shrimp.

Paul R. Wade is a wildlife biologist at the National Marine Mammal Laboratory, Alaska Fisheries Science Center, in Seattle, Washington. His research interests focus on the population dynamics and ecology of marine mammals, the conservation biology of marine vertebrates, and the use of modeling and statistics in conservation and management. His initial research was on the abundance and population dynamics of spotted and spinner dolphins in the eastern tropical Pacific, and their conservation status. More recently, he has worked on the population dynamics of gray, bowhead, and northern right whales, plane and small-boat surveys for small cetaceans such as harbor porpoise and Hector's dolphins, and research on establishing acceptable levels of marine mammal bycatch in fisheries. He is a member of the U.S. delegation to the Scientific Committee of the International Whaling Commission, the Cetacean Specialist Group of the IUCN Species Survival Commission, and the spectacled eider and Steller's eider recovery teams.

Carl J. Walters is professor of fisheries and zoology at the University of British Columbia. His research interests are in adaptive policy design for renewable resource management, fisheries stock assessment, and marine ecosystem modeling and sustainable management. He has published 145 papers and 2 books on these subjects. He is a fellow of the Royal Society of Canada.

Robin S. Waples is senior scientist at the Northwest Fisheries Science Center in Seattle. His research interests include patterns of gene flow and genetic differentiation in marine fish, genetic admixture analysis, estimation of effective population size using indirect genetic methods, analysis of spatial and temporal population genetic structure in salmon, and genetic interactions of hatchery and wild salmon. Since 1990 he has been involved with developing the scientific basis for listing determinations, recovery planning, and appropriate uses of artificial propagation of Pacific salmon under the U.S. Endangered Species Act. He is on the editorial board of *Conservation Biology* and *Conservation Genetics*.

Laura Hood Watchman is the director of habitat conservation for Defenders of Wildlife in Washington, D.C. At Defenders, she leads advocacy and research on habitat conservation plans under the Endangered Species Act. She is also engaged in promoting science-based large-scale conservation planning in multiple contexts, including federal lands, states, and growing cities. She has a master's degree in zoology from the University of Washington, and a B.S. from Princeton University.

Gary C. White is professor of fishery and wildlife biology at Colorado State University. His primary research interests are quantitative ecology, estimation of population parameters including development of Program MARK, and compensatory mortality in mule deer populations. He has published over 150 articles and coauthored 2 books on analysis of radio-tracking data and monitoring vertebrate populations. He is a member of the Mexican spotted owl recovery team, past associate editor of the *Journal of Wildlife Management*, and the 51st recipient of The Wildlife Society's Aldo Leopold Award and Medal for Distinguished Service to Wildlife Conservation in 2000.

INDEX

absolute probabilities of extinction, 70, 77, 239–41, 525–27, 528–29, 532
abundance
    capture-recapture estimation, 174, 175, 183
    mark-recapture estimation, 221
    MNA (minimum number known alive), 174, 175
Acinonyx jubatus (cheetah), 300
adaptive evolution
    genetic variance and, 29–30
    population size needed for, 51
adaptive management, 46–47, 460–61, 511–18, 534
    contributions from, 517–18
    history, 512–13
    implementing, techniques for, 513–16
adder (*Vipera berus*), 60
age-distribution effects, 455
Agelaius phoeniceus (blackbirds), 63
Aimophila aestivalis (Bachman's sparrow), 349
Akaike's information criterion (AIC), 176, 532
Akçakaya, H. R., 100, 341, 349
ALEX, 45, 46, 473
Allee effect, 18, 22, 91, 183, 495, 540
Allendorf, F. W., 65, 535
allozymes, genetic distance measured in, 378
Amazona vittata (Puerto Rican parrot), 537
American bison (*Bison bison*), 30
Ammodramus maritimus mirabilis (Cape Sable sparrow), 406–22
Anderson, D. R., 189
Anser caerulescens caerulescens (lesser snow goose), 349–58
Arabis petraea, 58
Arianta arbustorum (land snail), 349
Artemia franciscana (brine shrimp), 257, 260–61, 265–69
Asian elephant (*Elephas maximus*), 34
Aster furcatus, 62
Astragalus clevelandii, 113–19

Atlantic Large Whale Take Reduction Plan, 126
Atwood, J. L., 100

babirusa (*Babyrousa babyrussa*), 500
Bachman's sparrow (*Aimophila aestivalis*), 349
Ballou, J. D., 8, 57, 58, 69
band-recovery models, 178–79, 183
Barham, B. L., 500
Barker, R. J., 179
Barrett, S. C. H., 110, 112
Barton, N. H., 154, 155
Baskin, C. C., 321
Baskin, J. M., 321
Bass, O. L., Jr., 410
Baur, B., 349
Bayesian population viability analysis, 213–36, 242, 245, 248, 532, 536
    binomial estimation, 232–34
    frequentist statistics compared, 215–17, 229–31
    hierarchical, 452–53
    mark-recapture analysis, 234–36
    methods, 219–24
    MLE (maximum likelihood point estimates) PVA compared, 223–24
    parameter estimation, 220–24
    posterior probability distribution, 217–18, 224–27, 234
    predictive, 447–66
    prior distribution, 217–19, 234
bears
    brown bear (*Ursus arctos*), 65, 67–68, 71, 77–78, 194–95
    grizzly bear (*Ursus arctos horribilis*), 6, 125, 128, 192, 500–503
    heterozygosity and genetic distance in North American populations, 381
Beier, P., 349
Beissinger, S. R., 70, 77, 236, 240, 340, 341, 438, 513

563